VOLUME FIVE HUNDRED AND FORTY FOUR

METHODS IN ENZYMOLOGY

Regulated Cell Death Part A: Apoptotic Mechanisms

METHODS IN ENZYMOLOGY

Editors-in-Chief

JOHN N. ABELSON and MELVIN I. SIMON
Division of Biology
California Institute of Technology
Pasadena, California

ANNA MARIE PYLE
Departments of Molecular, Cellular and Developmental Biology and Department of Chemistry Investigator
Howard Hughes Medical Institute
Yale University

Founding Editors

SIDNEY P. COLOWICK and NATHAN O. KAPLAN

VOLUME FIVE HUNDRED AND FORTY FOUR

METHODS IN ENZYMOLOGY

Regulated Cell Death Part A: Apoptotic Mechanisms

Edited by

AVI ASHKENAZI
Cancer Immunology, Genentech, Inc., San Francisco, CA, USA

JUNYING YUAN
Department of Cell Biology
Harvard Medical School
Boston, MA, USA

JAMES A. WELLS
Departments of Pharmaceutical Chemistry and Cellular & Molecular Pharmacology, University of California – San Francisco, CA, USA

AMSTERDAM • BOSTON • HEIDELBERG • LONDON
NEW YORK • OXFORD • PARIS • SAN DIEGO
SAN FRANCISCO • SINGAPORE • SYDNEY • TOKYO
Academic Press is an imprint of Elsevier

Academic Press is an imprint of Elsevier
525 B Street, Suite 1800, San Diego, CA 92101-4495, USA
225 Wyman Street, Waltham, MA 02451, USA
The Boulevard, Langford Lane, Kidlington, Oxford, OX5 1GB, UK
32 Jamestown Road, London NW1 7BY, UK

First edition 2014

ISBN: 978-0-12-417158-9
ISSN: 0076-6879

For information on all Academic Press publications
visit our website at store.elsevier.com

CONTENTS

CONTRIBUTORS

Veronica G. Anania
Department of Protein Chemistry, Genentech Inc., South San Francisco, California, USA

David Andrews
Department of Biological Sciences, Sunnybrook Research Institute, Toronto, Canada

Avi Ashkenazi
Cancer Immunology, Genentech, Inc., San Francisco, California, USA

Gregory H. Bird
Department of Pediatric Oncology and the Linde Program in Cancer Chemical Biology, Dana-Farber Cancer Institute, and Department of Pediatrics, Children's Hospital Boston, Harvard Medical School, Boston, Massachusetts, USA

Craig R. Braun
Department of Pediatric Oncology and the Linde Program in Cancer Chemical Biology, Dana-Farber Cancer Institute; Department of Pediatrics, Children's Hospital Boston, and Department of Cell Biology, Harvard Medical School, Boston, Massachusetts, USA

Xiaoke Chi
Department of Biological Sciences, Sunnybrook Research Institute, Toronto, and Department of Chemical Biology, McMaster University, Hamilton, Canada

Charles S. Craik
Department of Pharmaceutical Chemistry, and Graduate Program in Chemistry and Chemical Biology, University of California, San Francisco, California, USA

Kevin Dagbay
Department of Chemistry, University of Massachusetts, Amherst, Massachusetts, USA

Peter M. Eimon
Department of Electrical Engineering and Computer Science, Massachusetts Institute of Technology (MIT), Cambridge, Massachusetts, USA

Scott J. Eron
Department of Chemistry, University of Massachusetts, Amherst, Massachusetts, USA

Jeanne A. Hardy
Department of Chemistry, University of Massachusetts, Amherst, Massachusetts, USA

Mark G. Hinds
School of Chemistry, and Bio21 Molecular Science and Biotechnology Institute, The University of Melbourne, Parkville, Victoria, Australia

Bradley T. Hyman
MassGeneral Institute for Neurodegenerative Disease, Department of Neurology, Alzheimer's Disease Research Laboratory, Massachusetts General Hospital, Harvard Medical School, Charlestown, Massachusetts, USA

Olivier Julien
Department of Pharmaceutical Chemistry, University of California, San Francisco, California, USA

Young-Wook Jun
Graduate Program in Chemistry and Chemical Biology, and Department of Otolaryngology, University of California, San Francisco, California, USA

Justin Kale
Department of Biochemistry and Biomedical Sciences, McMaster University, Hamilton, and Department of Biological Sciences, Sunnybrook Research Institute, Toronto, Canada

Akiko Koto
Department of Genetics, Graduate School of Pharmaceutical Sciences, The University of Tokyo, Tokyo, Japan

Erina Kuranaga
Laboratory for Histogenetic Dynamics, RIKEN CDB, Kobe, Japan

Marc Kvansakul
La Trobe Institute for Medical Science, La Trobe University, Bundoora, Victoria, Australia

Brian Leber
Department of Biochemistry and Biomedical Sciences, and Department of Medicine, McMaster University, Hamilton, Canada

Susan Lee
Department of Pediatric Oncology and the Linde Program in Cancer Chemical Biology, Dana-Farber Cancer Institute, and Department of Pediatrics, Children's Hospital Boston, Harvard Medical School, Boston, Massachusetts, USA

Jennie R. Lill
Department of Protein Chemistry, Genentech Inc., South San Francisco, California, USA

Di Lin*
Department of Chemistry, University of Massachusetts, Amherst, Massachusetts, USA

Min Lu
Cancer Immunology, Genentech, Inc., San Francisco, California, USA

Peter D. Mace
Biochemistry Department, University of Otago, Dunedin, New Zealand

Masayuki Miura
Department of Genetics, Graduate School of Pharmaceutical Sciences, The University of Tokyo, and CREST, JST, Tokyo, Japan

Charles W. Morgan
Department of Pharmaceutical Chemistry, and Graduate Group in Chemistry and Chemical Biology, University of California, San Francisco, California, USA

*Present address: College of Pharmacy, Purdue University, West Lafayette, IN, USA.

Shigekazu Nagata
Department of Medical Chemistry, Graduate School of Medicine, Kyoto University, Kyoto, Japan

Pradeep Nair
Cancer Immunology, Genentech, Inc., San Francisco, California, USA

Yu-ichiro Nakajima
Stowers Institute for Medical Research, Kansas, Missouri, USA

Samantha B. Nicholls
MassGeneral Institute for Neurodegenerative Disease, Department of Neurology, Alzheimer's Disease Research Laboratory, Massachusetts General Hospital, Harvard Medical School, Charlestown, Massachusetts, USA

Sean Petersen
Cancer Immunology, Genentech, Inc., San Francisco, California, USA

Victoria C. Pham
Department of Protein Chemistry, Genentech Inc., South San Francisco, California, USA

Qui T. Phung
Department of Protein Chemistry, Genentech Inc., South San Francisco, California, USA

Stefan J. Riedl
Program in Cell Death and Survival Networks, NCI Designated Cancer Center, Sanford-Burnham Medical Research Institute, La Jolla, California, USA

Stéphane G. Rolland
LMU Biocenter, Department Biology II, Ludwig-Maximilians-University, Munich, Germany

Guy S. Salvesen
Program in Cell Death and Survival Networks, NCI Designated Cancer Center, Sanford-Burnham Medical Research Institute, La Jolla, California, USA

Julia E. Seaman
Department of Pharmaceutical Chemistry, University of California, San Francisco, California, USA

Banyuhay P. Serrano
Department of Chemistry, University of Massachusetts, Amherst, Massachusetts, USA

Nirao M. Shah
Department of Anatomy, University of California, San Francisco, California, USA

Jean Philippe Stephan
Protein Chemistry Department/Discovery Oncology Department, Genentech Inc., South San Francisco, California, USA

Jun Suzuki
Department of Medical Chemistry, Graduate School of Medicine, Kyoto University, Kyoto, Japan

Cheryl Tajon
Department of Pharmaceutical Chemistry, and Graduate Program in Chemistry and Chemical Biology, University of California, San Francisco, California, USA

Kiwamu Takemoto
PRESTO, JST, Tokyo, and Department of Physiology, Graduate School of Medicine, Yokohama City University, Yokohama, Japan

Elizabeth K. Unger
Department of Anatomy, and Program in Biomedical Sciences, University of California, San Francisco, California, USA

Sravanti Vaidya*
Department of Chemistry, University of Massachusetts, Amherst, Massachusetts, USA

Elih M. Velázquez-Delgado†
Department of Chemistry, University of Massachusetts, Amherst, Massachusetts, USA

Loren D. Walensky
Department of Pediatric Oncology and the Linde Program in Cancer Chemical Biology, Dana-Farber Cancer Institute, and Department of Pediatrics, Children's Hospital Boston, Harvard Medical School, Boston, Massachusetts, USA

James A. Wells
Department of Pharmaceutical Chemistry, and Department of Cellular and Molecular Pharmacology, University of California, San Francisco, California, USA

Arun P. Wiita
Department of Pharmaceutical Chemistry, and Department of Laboratory Medicine, University of California, San Francisco, California, USA

Yoshifumi Yamaguchi
Department of Genetics, Graduate School of Pharmaceutical Sciences, The University of Tokyo, and PRESTO, JST, Tokyo, Japan

Yunlong Zhao
Department of Chemistry, University of Massachusetts, Amherst, Massachusetts, USA

*Present address: Biotechnology Department, MS Ramaiah Institute of Technology, Bangalore, India.
†Present address: Department of Structural Biology, St. Jude Children's Research Hospital, Memphis, TN, USA.

PREFACE

Cell turnover is a fundamental feature of metazoan biology. Severe damage to cellular integrity usually causes passive, nonregulated cell death. In contrast, more confined disruption can lead to more deliberate cell elimination, through specific mechanisms of Regulated Cell Death. In these two volumes of *Methods in Enzymology*, we aim to highlight the current molecular understanding of the major processes of Regulated Cell Death and to illustrate basic and advanced methodologies to study them. Volume A focuses on the most extensively studied mode of cell death—apoptosis. Volume B covers several nonapoptotic mechanisms, including necroptotic and autophagic cell death. In Volume A, Chapters 1–4 cover various aspects of the cell-intrinsic apoptosis pathway, including the Bcl-2 protein family and mitochondria. Chapters 5 and 6 discuss death receptors and the extrinsic pathway. Chapters 8–14 cover caspases—the apoptotic protease machine. Chapter 15 highlights how apoptotic cells are recognized and cleared by other cells, while Chapter 16 features the zebrafish as a versatile genetic model organism for studying apoptosis. We hope these chapters will be conceptually informative and practically useful for readers interested in the current understanding and key open questions in each area as well as in experimental strategies and techniques to interrogate the many facets of apoptotic cell death including signaling, execution, and regulation.

Avi Ashkenazi
Junying Yuan
James A. Wells

CHAPTER ONE

Examining the Molecular Mechanism of Bcl-2 Family Proteins at Membranes by Fluorescence Spectroscopy

Justin Kale[*,†], **Xiaoke Chi**[†,‡], **Brian Leber**[*,§], **David Andrews**[†,1]

[*]Department of Biochemistry and Biomedical Sciences, McMaster University, Hamilton, Canada
[†]Department of Biological Sciences, Sunnybrook Research Institute, Toronto, Canada
[‡]Department of Chemical Biology, McMaster University, Hamilton, Canada
[§]Department of Medicine, McMaster University, Hamilton, Canada
[1]Corresponding author: e-mail address: david.andrews@sri.utoronto.ca

Contents

Abstract

The Bcl-2 family proteins control apoptosis by regulation of outer mitochondrial membrane permeabilization. Studying the Bcl-2 family is particularly difficult because the functional interactions that regulate apoptosis occur at or within intracellular membranes. Compared to other biophysical methods, fluorescence spectroscopy is well suited to study membrane-bound proteins as experiments can be performed with intact membranes and at protein concentrations similar to those found in cells. For these reasons, fluorescence spectroscopy has been particularly useful in studying the

Methods in Enzymology, Volume 544
ISSN 0076-6879
http://dx.doi.org/10.1016/B978-0-12-417158-9.00001-7

regulation of membrane permeabilization by Bcl-2 family proteins. Here, we discuss four fluorescence-based assays used to study protein dynamics at membranes, with a focus on how these techniques can be used to study the Bcl-2 family proteins.

1. INTRODUCTION

The Bcl-2 family of proteins regulates permeabilization of the outer mitochondrial membrane (OMM). In most cell types, once the OMM is permeabilized, the cell is committed to undergoing programmed cell death (Budd, Tenneti, Lishnak, & Lipton, 2000). The sequence of events leading to permeabilization of the OMM begins with prodeath signals triggering posttranslational modifications of activator BH3-only proteins, such as the cleavage of Bid to cBid (comprised of a p7 and p15 fragment, the latter also referred to as tBid), that target them to the OMM where they bind to and activate the pore-forming proteins Bax and Bak. Activation of Bax and Bak results in their oligomerization within the OMM followed by permeabilization of the OMM and release of intermembrane space proteins such as cytochrome *c* and SMAC that act in downstream apoptotic pathways, culminating in cellular apoptosis (Shamas-Din, Kale, Leber, & Andrews, 2013). The antiapoptotic proteins, such as Bcl-2 and Bcl-X_L, inhibit apoptosis by binding to and sequestering both BH3-only activators and Bax/Bak (Bogner, Leber, & Andrews, 2010).

Significant research has been focused on determining the structure of Bcl-2 family proteins. X-ray crystallography and nuclear magnetic resonance (NMR) spectroscopy have revealed that the Bcl-2 family proteins share a highly conserved core structure (Petros, Olejniczak, & Fesik, 2004). These studies have provided insight into how the Bcl-2 family proteins bind to each other and suggest how they may interact with membranes. However, the current relatively static structures for the Bcl-2 family are for proteins without the lipid bilayer required for functional interactions of several of the Bcl-2 family proteins (Leber, Lin, & Andrews, 2007). Determining the structures of proteins within a membrane mimetic environment using X-ray crystallography and NMR spectroscopy is particularly difficult. These techniques require a large amount of protein in a sample environment that mimics but differs significantly from that of the cell and typically includes detergents that can alter the functions of the Bcl-2 family proteins (Hsu & Youle, 1997). As an example, unlike native Bax, detergent-treated

Bax can cause permeabilization of the OMM when added to isolated mitochondria (Antonsson, Montessuit, Lauper, Eskes, & Martinou, 2000), can form oligomers that can be cross-linked in the absence of membranes (Zhang et al., 2010) and has undergone a conformational change that is a prerequisite for Bax activation (Yethon, Epand, Leber, Epand, & Andrews, 2003). Additionally, the zwitterionic detergent CHAPS can prevent the authentic interaction of Bax and tBid (Lovell et al., 2008), further reinforcing the need to study the Bcl-2 family proteins in the absence of detergents.

Fluorescence-based techniques are well suited to study protein dynamics at membranes under physiological conditions in the absence of detergents (Kale, Liu, Leber, & Andrews, 2012). Fluorescence spectroscopy allows observation of protein:protein- and protein:membrane-binding dynamics in real time, while gathering information about the kinetics and affinities of these interactions that cannot be measured using typical structural techniques due to complications from the membrane (Perez-Lara, Egea-Jimenez, Ausili, Corbalan-Garcia, & Gomez-Fernandez, 2012; Satsoura et al., 2012). Additionally, by using an environment-sensitive probe, it is possible to determine the environment of specific residues as they undergo conformational changes within membranes (Malhotra, Sathappa, Landin, Johnson, & Alder, 2013; Shamas-Din, Bindner, et al., 2013). Initial fluorescence-based studies of the Bcl-2 family proteins have used a simple *in vitro* system to study the dynamic interactions that occur at, on, and within membranes.

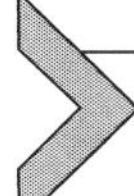

2. AN *IN VITRO* FLUORESCENCE-BASED LIPOSOME SYSTEM

The functional interactions of the Bcl-2 family proteins occur in membranes. Interaction of cBid with the membrane causes the p7 and p15 fragments of cBid to dissociate, whereupon the p15 fragment (tBid) undergoes a conformational change, that does not occur in solution and permits binding between tBid and Bax within the membrane (Shamas-Din, Bindner, et al., 2013; Shamas-Din, Kale, et al., 2013). Binding between Bax and cBid or Bax and Bcl-X_L requires membranes for an interaction to occur as an interaction is not detected in solution (Billen, Kokoski, Lovell, Leber, & Andrews, 2008; Lovell et al., 2008). Therefore, to study the function of these proteins, a biochemical system is required that includes a phospholipid bilayer that separates two distinct aqueous compartments mimicking that of the cytoplasm and the interior of cellular organelles. We and others (Bleicken et al., 2010; Landeta et al., 2011; Ren et al., 2010; Shamas-Din, Bindner, et al., 2013; Shamas-Din, Kale, et al., 2013) have used different variations of liposome or proteoliposome-based systems to study the core mechanism of Bcl-2 family protein regulation of membrane permeabilization. All of these systems lack the detergents typically required for biochemical and structural studies of membrane proteins. For our studies, fluorescently labeled purified full-length recombinant proteins and artificial membranes in the form of liposomes with a composition mimicking that of the OMM are used. This system is free of any other complicating factors such as unknown binding partners that may be present at the OMM or within the cytoplasm.

To use fluorescence to study proteins at membranes, it is essential to make judicious choices of fluorophore, type of measurement, and instrument. Fluorescence measurements require excitation of the fluorophore by illuminating the sample with a specific wavelength of light and then recording the emission from the fluorophore. Upon excitation, after some period of time, termed the fluorescence lifetime (typically 1–10 ns), the fluorophore returns to the ground electronic state via emission of a photon at a lower energy, and thus longer wavelength, than the illuminating (excitation) light. Because the emitted fluorescence is of much lower intensity than the excitation light, the system must be free of molecules that absorb the emitted light, and fluorescence contaminants that may interfere with the emission signal from the fluorophore. Molecules such as quenchers that provoke nonradiative decay of the fluorophore must also be avoided as they change the fluorescence properties of the dyes. If these conditions are met, changes in both fluorescence lifetime and emission intensity can provide specific information about the underlying biochemical properties of the protein the fluorophore is attached to (Lakowicz, 2006).

2.1. Expression and purification of Bcl-2 family proteins

2.1.1 Expression of Bax, Bcl-X$_L$, and Bid

1. *Escherichia coli* are transformed (BL21-AI, New England Biolabs for cBid and Bax; DH5α, New England Biolabs, for Bcl-X$_L$) with the full-length Bax, Bcl-X$_L$, or Bid expression plasmid, plated on LB-ampicillin agar, and then incubated overnight at 37 °C. Bax and Bcl-X$_L$ are expressed with a carboxy-terminal intein-chitin binding domain (IMPACT expression system, New England Biolabs) and Bid is expressed with an amino terminal 6 × histidine tag.

2. The next day, a single colony is picked and used to inoculate 100 mL of LB-ampicillin and grown overnight at 30 °C with shaking. Then 1–3 L of LB-ampicillin is inoculated with the overnight culture (20 mL for each liter of LB-ampicillin) and grown at 37 °C with shaking until the bacterial growth is in log phase (OD_{600} is typically between 0.6 and 0.8) at which point protein expression is induced with either arabinose (0.2%, w/v, BL21-AI cells) or IPTG (1 m*M*, DH5α cells). Bacteria are then incubated for 3–5 h at 30 °C with shaking, harvested using centrifugation, and stored at −20 °C. We find that, for both Bax and Bcl-X_L, a longer expression time (5 h) yields more recombinant protein.

2.1.2 Purification of Bax and Bcl-X_L

1. The bacterial pellet is resuspended in either Bax or Bcl-X_L lysis buffer (10 mL for each 2.5 g of bacterial pellet) and lysed via French press. Lysed cells are centrifuged at 20,000 × *g* and the cell lysate is incubated with 1.5 mL of chitin bead slurry (New England Biolabs) for 2 h at 4 °C while rotating.
2. The cell lysate and resin slurry is then loaded into an Econo-Pac chromatography column (BioRad, Cat. #: 732-1010EDU) and the lysate passed through (three to four times) before the resin is washed with 50 mL of either Bax or Bcl-X_L wash buffer. The column is equilibrated with either Bax or Bcl-X_L cleavage buffer (10 mL) and capped with ~1 mL of cleavage buffer remaining on top of the chitin beads followed by incubation for 24–36 h at 4 °C. The cleavage buffer contains hydroxylamine, which causes cleavage of the intein-chitin binding domain allowing full-length Bax and Bcl-X_L to be eluted. Bax or Bcl-X_L is then eluted with cleavage buffer (4 × 1 mL fractions). The majority of the protein is typically within fractions 1 and 2.
3. **a.** Bax: A 0.2-mL bed volume DEAE–Sepharose column is prepared and equilibrated with 2.5 mL of Bax-cleavage buffer (without hydroxylamine). Bax elution fraction 1 and 2 are pooled and passed through the column three times which removes additional contaminants that bind to the column.

   b. Bcl-X_L: A 0.3-mL bed volume high-performance phenyl Sepharose column is equilibrated with 3 mL of Bcl-X_L wash buffer. Bcl-X_L elution fractions 1–4 are pooled and applied to the column where Bcl-X_L binds and the column is washed with 5 mL of Bcl-X_L wash buffer (no PMSF in the buffer is needed). Bcl-X_L is eluted (3 × 1 mL fractions) with Bcl-X_L wash buffer that does not contain NaCl or PMSF.

4. Both Bax and Bcl-X_L are dialyzed against 3 × 1 L of dialysis buffer (4 °C with stirring). After dialysis, the protein can be aliquoted and stored at −80 °C or can be labeled with fluorescent dyes (see Section 2.2).

2.1.3 Purification of cBid

1. The bacterial pellet is resuspended in Bid-lysis buffer (10 mL for each 2.5 g of bacterial pellet) and lysed via French Press. Lysed cells are centrifuged at 20,000 × *g*, and the cell lysate is incubated with 0.8 mL Ni-NTA agarose slurry (Qiagen) for 1.5 h at 4 °C while rotating.
2. The cell lysate and resin slurry is then added to a Poly-Prep column (BioRad, Cat. #: #731-1550EDU) and the lysate passed through three times. The column is washed with 50 mL of Bid-wash buffer, and Bid is eluted with 10 mL of Bid-elution buffer (collecting 5 × 1 mL fractions). The first two fractions typically contain the highest concentration of Bid and are pooled together. At this point, Bid can be cleaved to cBid (see below, step 3), or if labeling Bid with a fluorescent dye, the pooled fractions are first dialyzed 3 × 1 L against dialysis buffer at 4 °C with stirring and then labeled (see Section 2.2), followed by Bid cleavage and a final dialysis step.
3. To produce cBid, the pooled Bid elutions are adjusted to contain 40 m*M* HEPES, 1 m*M* EDTA, 10 m*M* DTT and incubated with 500 U of recombinant human caspase-8 (Enzo Life Sciences, Cat. #: BML-SE172-5000), and incubated for 48 h with rotating in the dark at room temperature. The cBid sample is next dialyzed against 3 × 1 L of dialysis buffer (4 °C with stirring) and then aliquoted and stored at −80 °C.

2.1.4 Buffer recipes

2.1.4.1 Bax

Bax-lysis buffer: 10 m*M* HEPES pH 7.0, 100 m*M* NaCl, 0.2% (w/v) CHAPS, 1 m*M* PMSF, DNase, RNase

Bax-wash buffer: 10 m*M* HEPES pH 7.0, 500 m*M* NaCl, 0.5% (w/v) CHAPS

Bax-cleavage buffer: 10 m*M* HEPES pH 7.0, 200 m*M* NaCl, 0.1% (w/v) CHAPS, 100 m*M* Hydroxylamine. It is important to check the pH of the cleavage buffer after adding hydroxylamine as it decreases the pH, which will lead to insufficient yield of protein.

2.1.4.2 Bcl-X_L

Bcl-X_L-lysis buffer: 20 m*M* Tris pH 8.0, 500 m*M* NaCl, 1% (w/v) CHAPS, 1 m*M* PMSF, DNase, RNase

Bcl-X_L-wash buffer: 20 m*M* Tris pH 8.0, 200 m*M* NaCl, 0.2% (w/v) CHAPS, 20% (v/v) glycerol, 1 m*M* PMSF

Bcl-X_L-cleavage buffer: 20 m*M* Tris pH 8.0, 200 m*M* NaCl, 0.2% (w/v) CHAPS, 20% (v/v) glycerol, 1 m*M* PMSF, 100 m*M* Hydroxylamine. It is important to check the pH of the cleavage buffer after adding hydroxylamine as it decreases the pH, which will lead to insufficient yield of protein.

2.1.4.3 Bid

Bid-lysis buffer: 10 m*M* HEPES pH 7.0, 100 m*M* NaCl, 10 m*M* imidazole, 1 m*M* PMSF, DNase, RNase

Bid-wash buffer: 10 m*M* HEPES pH 7.0, 300 m*M* NaCl, 10 m*M* imidazole, 1% (w/v) CHAPS

Bid-elution buffer: 10 m*M* HEPES pH 7.0, 100 m*M* NaCl, 200 m*M* imidazole, 0.1% (w/v) CHAPS, 10% (v/v) glycerol

2.1.4.4 Dialysis

Dialysis buffer (for Bax, cBid, and Bcl-X_L): 10 m*M* HEPES pH 7.0, 100 m*M* NaCl, 0.1 m*M* EDTA, 10% (v/v) glycerol

Extensive dialysis is needed to remove CHAPS which can alter the function and binding interactions of Bcl-2 family proteins. We typically dialyze our purified protein samples for a minimum of 4 h against 1 L of buffer, followed by dialysis overnight (~12–16 h) against 1 L of buffer and a final dialysis against 1 L of buffer for a minimum of 4 h. The use of spin-concentrator columns should be avoided, in our experience, as they severely attenuate the function of the Bcl-2 family proteins.

2.2. Site-specific protein labeling

The fluorescence-based techniques we use to study the Bcl-2 family require purified recombinant proteins labeled with a fluorophore at a specific location. There are two main options for labeling proteins, thiol or primary amine labeling. Cysteine residues are less abundant than lysines in most protein sequences, thus we most frequently create single-cysteine mutants to label the protein as this approach minimizes the number of mutations required. There is a full spectrum of fluorescent probes available for purchase, which have different spectral properties that can be ordered with attached thiol reactive moieties such as iodoacetamide or maleimide derivatives. Dyes must be chosen that are compatible not only with your protein of choice but also with the system and equipment available.

In the methods reported below, the proteins were labeled with the low-molecular weight fluorescent probes DAC (*N*-(7-dimethylamino-4-methylcoumarin-3-yl) maleimide; Anaspec, Cat. #: 81403) and NBD (*N*,*N*′-dimethyl-*N*-(Iodoacetyl)-*N*′-(7-nitrobenz-2-oxa-1,3-diazol-4-yl) ethylenediamine; Molecular Probes, Cat. #: D-2004). The small size of these dyes is a distinct advantage as they rarely perturb protein function; however, measurements of NBD fluorescence require a sensitive instrument as the quantum yield (ratio of photons emitted to photons absorbed) is low. Moreover, excitation of DAC requires an ultraviolet light source and both the excitation and emission of this dye overlap endogenous fluorophores in cells typically limiting its use to liposome-based systems. Many brighter (higher quantum yields and extinction coefficients) fluorescent dyes have molecular weights above 1 kDa, and in our experience these larger dyes frequently change the function of the protein they are attached to.

Initially, it is best to follow the labeling protocol included by the manufacturer when labeling your protein of interest; however, it is often necessary to deviate from these conditions to get labeling that is both specific and efficient.

1. For Bid, Bax, and Bcl-X_L, the protein-labeling reaction is performed in a HEPES-based buffer at pH 7.0–7.5 (10 m*M* HEPES, 200 m*M* NaCl, 0.4%, w/v CHAPS). This pH range allows the cysteines to be most-reactive while decreasing the reactivity of primary amines. A 10–20 × *M* excess of dye is added to the sample tube dropwise, to prevent protein denaturation as dyes are typically dissolved in DMSO, and the labeling reaction is rotated at room temperature for approximately 2 h in the dark. A reducing agent (5 m*M* DTT) is then added to quench the reaction.
2. **a.** Bax and Bcl-X_L: Free dye is removed via gel filtration over a G-25 fine Sephadex column (10 mL bed volume, equilibrated with dialysis buffer). Bax- and Bcl-X_L-labeling reactions are applied to the column and approximately 12 0.5 mL fractions are collected.
   b. Bid: The Bid-labeling reaction is applied to a Ni-NTA column (0.2 mL bed volume) and passed through three to four times allowing labeled Bid to bind to the column. The column is washed with 50 mL of Bid-wash buffer and eluted with 5 mL of Bid-elution buffer, collecting 4 × 0.5 mL fractions.

3. Absorbance spectroscopy is used to determine protein containing fractions. Absorbance at both 280 nm (to detect protein) and at the absorbance peak of the dye used is determined. The fractions containing the highest amount of protein are pooled. Bid can now be cleaved (see Section 2.1.3, step 3). Bax, Bcl-X_L, and cBid are then dialyzed against 3×1 L dialysis buffer at 4 °C with stirring to remove any remaining free dye and detergent, and the protein is aliquoted and stored for later use.
4. After dialysis, labeling efficiency is calculated by first determining the concentration of your protein via absorbance at 280 (Bax), BCA assay (Bcl-X_L), or Bradford assay (cBid). Then the concentration of the dye in the protein sample is determined by the OD of the sample at the peak absorbance wavelength of the label, as outlined in the protocol supplied by the manufacturer. Assuming that the protein is only labeled at the single-cysteine residue and that there is no free dye in the sample, the concentration of the dye should equal that of the labeled protein. Labeling efficiency is the fraction of labeled protein to that of total protein.

Single-cysteine mutants of the purified recombinant protein need to be assayed functionally before and after labeling to determine if the mutation or the addition of the dye alters protein function. Ideally, we begin using mutants where one of the endogenous cysteines is present to minimize the amount of mutations introduced into the protein. If the protein does not contain any cysteines, choosing which residue to mutate to cysteine for efficient labeling and proper protein function is largely empirical. Typically, if the structure is known, one begins using solvent exposed residues, since those located in hydrophobic regions are difficult to label. Algorithms used to predict solvent exposure or antigenicity (antigenic sites tend to be both structured and solvent exposed) can often be useful in selecting a location if the structure of your protein is unknown.

2.3. Production of mitochondria-like liposomes

Large unilamellar vesicles (LUVs) are liposomes that have with a mean diameter of 120–140 nm and one lipid bilayer (Hope, Bally, Webb, & Cullis, 1985). OMM-like LUVs have been established as a valid biochemical model for membrane permeablization by Bcl-2 family members (Kuwana et al., 2002). These liposomes are assembled from lipids in fixed molar ratio similar to that of the OMM, based upon lipid composition studies from solvent extracted Xenopus mitochondria (Kuwana et al., 2002). Such liposome-based systems allow the analysis of Bcl-2 family proteins in a simple context while preserving their authentic functions. It is possible to more directly explore Bcl-2 family function in this kind of system because the protein and lipid components are well defined and tractable, unlike isolated mitochondria or proteoliposomes prepared from membranes.

2.3.1 *Preparing lipid films and generating liposomes*

1. Chloroform solublized lipids are added to a glass test tube to make a lipid mixture of a defined composition (Table 1.1) to a total of 1 mg lipid mass. The chloroform is evaporated off with nitrogen gas while rotating the tube to ensure an even distribution of lipids on the wall and then put under vacuum for 2 h at room temperature to remove any remaining chloroform. The dry lipid film is then either used immediately or can be stored for up to 2 weeks at −20 °C. To reduce lipid oxidation by atmospheric oxygen during storage, it is advisable to layer nitrogen or argon gas on top of the lipid film and seal the tube with parafilm.
2. The dry 1 mg lipid film is hydrated with 1 mL of assay buffer (10 m*M* HEPES, 200 m*M* KCl, 5 m*M* $MgCl_2$, 0.2 m*M* EDTA, pH 7). The lipids become suspended and spontaneously form lipid bilayer vesicles due to

Table 1.1 Mitochondria-like lipid film composition

Name	Company	Catalog #	Molar (%)	Molecular weight (g/mol)	Amount needed for 1 mg lipid film (mg)
"PC": L-α-phosphatidylcholine (egg, chicken)	Avanti	840051C	48	770.123	0.4596
"PE": L-α-phosphatidylethanolamine (egg, chicken)	Avanti	841118C	28	726.076	0.2528
"PI": L-α-phosphatidylinositol (liver, bovine)	Avanti	840042C	10	902.133	0.1122
"DOPS": 1,2-dioleoyl-*sn*-glycero-3-phospho-L-serine	Avanti	840035C	10	810.025	0.1007
"TOCL": 1,1′,2,2′-tetra-(9Z-octadecenoyl) cardiolipin	Avanti	710335C	4	1501.959	0.0747

the association of the hydrophobic tails, forming the center of the bilayer, and the grouping of the hydrophilic heads of the phospholipids, forming the edges of the bilayer. However, these vesicles are multilamellar as they contain more than one lipid bilayer and their size distribution is not homogeneous. To generate unilamellar liposomes, the lipid mixture is subjected to 8–10 freeze/thaw cycles by alternately placing the sample vial in liquid nitrogen and a warm water bath (Hope et al., 1985). The unilamellar liposomes are extruded 11 times through a filter with 0.1 μm pore size to produce liposomes of a uniform size, at a final concentration of 1 mg/mL lipid.

3. MEMBRANE PERMEABILIZATION ASSAY

The Bcl-2 family proteins play a pivotal role in regulating apoptosis by controlling the permeabilization of the OMM through the activation of Bax/Bak. Thus, a membrane permeablization assay is one crucial functional assay for the Bcl-2 family proteins. To assay liposome permeabilization, the liposomes are encapsulated with a polyanionic fluorophore, ANTS (8-aminonaphthalene-1,3,6-trisulfonic acid; Molecular Probes, Cat. #: A350), and cationic quencher, DPX (*p*-xylene-bis-pyridinium bromide; Molecular Probes, Cat. #: X1525). Due to the high local concentration of DPX, ANTS fluorescence is quenched when liposomes are still intact. Recombinant Bax and/or other Bcl-2 family proteins and/or reagents are added to the system in order to assay permeabilization. As the liposomes permeabilize, ANTS and DPX are released from the liposomes, greatly decreasing the local concentration of the quencher resulting in a gain of ANTS fluorescence. The kinetics and extent of membrane permeabilization can reveal crucial information for studying relationships between Bcl-2 family members and how they regulate membrane permeabilization.

3.1. ANTS/DPX release assay

1. A dry 1 mg lipid film is hydrated with 1 mL of assay buffer with the addition of ANTS (12.5 m*M*) and DPX (45 m*M*). The lipid suspension is vortexed until the ANTS and DPX dissolve, and liposomes are created as above via 10 freeze/thaw cycles and extrusion through a 0.1 μm pore size membrane.
2. Excess ANTS and DPX are removed by applying the extruded liposomes onto a CL2B size-exclusion column (10 mL bed volume), that separates the ANTS/DPX encapsulated liposomes from the free

ANTS/DPX in solution (Billen et al., 2008; Yethon et al., 2003). Fractions (1 mL each) are collected in glass tubes and the liposome containing fractions (typically fractions 3 and 4) are identified by an increase in cloudiness of the sample which occurs due to light scattering by the liposomes. The two liposome containing fractions are combined resulting in a final ANTS/DPX liposome concentration of approximately 0.5 mg/mL lipid. These liposomes can now be used to test the regulation of membrane permeabilization by the Bcl-2 family proteins.

3. The assay is set up in a low protein binding 96-well plate (Corning; Cat. #: 3686) and in each well to be measured, 8 μL of ANTS/DPX liposomes are added to 92 μL of assay buffer. Background measurements (F_0) are recorded at 30 °C using a fluorescence plate reader (Tecan M1000 pro) set to excite the sample at 355 nm (5 nm bandwidth) and collect emission at 520 nm (12 nm bandwidth).
4. Proteins are added to the desired concentrations and combinations in each well and fluorescence emission of ANTS (F) is recorded every minute for 3 h at 30 °C. Any increase in fluorescence emission is directly related to membrane permeabilization.
5. To normalize the data, 100% ANTS release is determined by the addition of Triton to each well at a final concentration of 0.2% (w/v) causing permeabilization of all liposomes and ANTS fluorescence is measured (F_{100}). This results in a slight overestimation of the intensity of 100% release due to the dye becoming trapped in detergent micelles. Nevertheless, the release percentage generally does not take this into account and is calculated as follows:

$$\text{ANTS release}\,(\%) = \frac{F - F_0}{F_{100} - F_0} \times 100\%$$

The ANTS/DPX release assay can be used to dissect exactly how the different classes of Bcl-2 family proteins affect permeabilization of the OMM. When cBid (20 n*M*), Bax (100 n*M*), or Bcl-X_L (40 n*M*) are added individually to liposomes they do not cause membrane permeabilization (Fig. 1.1A). Incubation of liposomes with cBid and Bax results in membrane permeabilization due to cBid binding to membranes causing separation of the two fragments of cBid with the p15 (tBid) fragment remaining membrane-bound and -activating Bax. Bcl-X_L inhibits this process by binding to and inhibiting both tBid and Bax (Billen et al., 2008; Lovell et al., 2008). Obviously, other techniques are needed to discern exactly how these

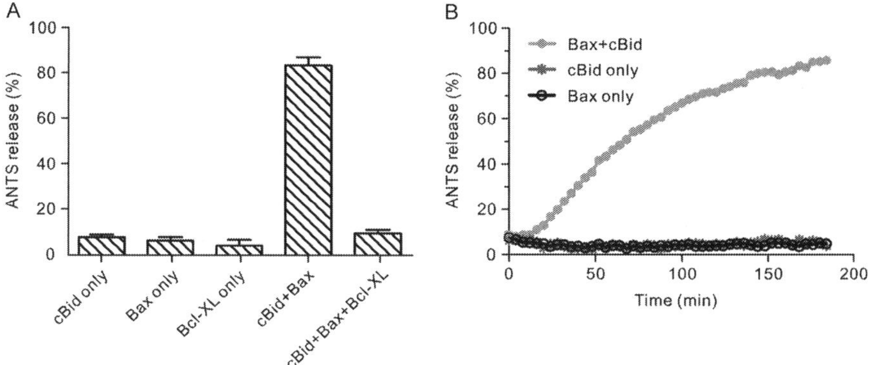

Figure 1.1 (A) Endpoint values of ANTS assay with 100 n*M* Bax, 20 n*M* cBid, 40 n*M* Bcl-X_L or both, or with 100 n*M* Bax, 20 n*M* tBid, and 40 n*M* Bcl-X_L. ($n=3$). (B) Liposomes encapsulated with ANTS and DPX were incubated with 100 n*M* Bax, 20 n*M* cBid, or both. Membrane permeabilization was assayed by an increase of ANTS fluorescence.

interactions occur (see Section 4); however, this dye release assay allows the functional consequence of the addition of any number of various combinations of Bcl-2 family members or small molecule effectors of the proteins to be determined. Furthermore, it provides information on how changes in relative concentrations of the proteins can vary the extent of permeabilization or how alterations in the parameters of the assay affect membrane permeabilization. For example, it is possible quantify how changes in liposome composition affect Bcl-2 family proteins functions to permeabilize membranes or test specific mutations that may inhibit/activate the protein of interest. Additionally, the kinetics of pore formation can be studied allowing the comparison of kinetics for Bax-mediated membrane permeabilization in response to various BH3-only activators (Fig. 1.1B).

4. FLUORESCENCE RESONANCE ENERGY TRANSFER

Here, fluorescence resonance energy transfer (FRET) will be used to detect binding between cBid and Bax, and Bax oligomerization. FRET is possible between fluorophores when the emission spectra of one fluorescent molecule, termed the donor, overlaps the excitation spectra of another fluorophore, the acceptor. When a donor fluorophore is excited by light, an electron moves to a higher energy state and, in the presence of an acceptor, the energy is transferred nonradiatively to the acceptor fluorophore via dipole–dipole interactions between the two probes. This transfer of energy results in a decrease of the donor emission, and it is the change in the light emitted by the donor that we track to measure FRET between two proteins.

One of the main advantages of FRET is that it requires both the donor and acceptor fluorophores to be in close proximity for the required dipole coupling to occur. As a result, FRET efficiency decreases to the sixth power of distance according to the formula for FRET efficiency (E) at a fixed donor acceptor distance:

$$E = \frac{R_0^6}{R_0^6 + r^6}$$

where R_0 is the Förster distance, the distance between a donor acceptor pair at which a 50% FRET efficiency is observed and r is the distance between the donor and acceptor. The distance dependence of FRET is illustrated in Fig. 1.2A where FRET efficiency is calculated for distances between a donor

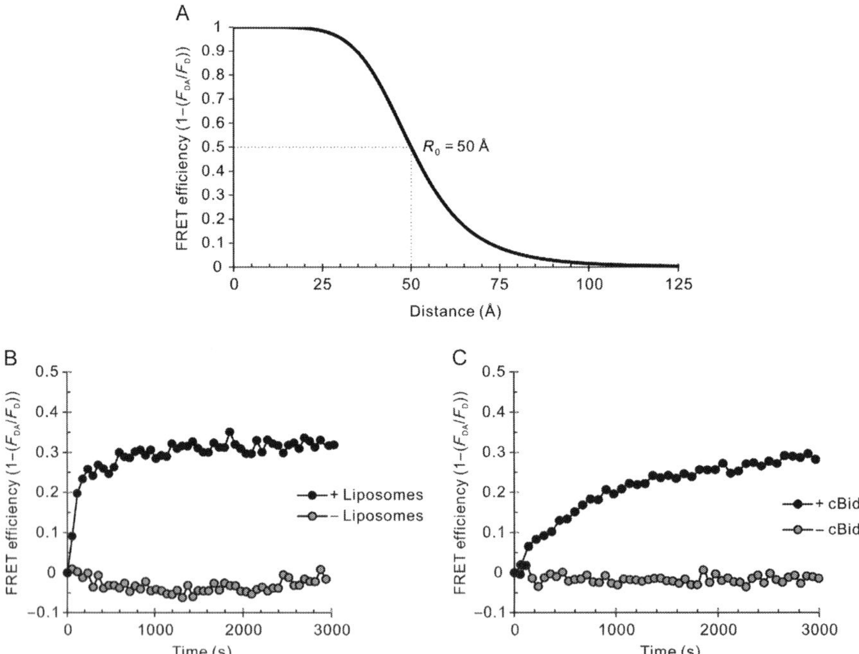

Figure 1.2 (A) FRET efficiency as a function of distance between a dye pair with a theoretical Förster distance of 50 Å. (B) FRET between cBid-DAC (20 n*M*) and Bax-NBD (100 n*M*) in the presence (black circles) and absence (gray circles) of liposomes (0.2 mg/mL). (C) FRET between Bax-DAC (20 n*M*) and Bax-NBD (100 n*M*) in samples containing liposomes (0.2 mg/mL) with (black circles) or without (gray circles) 20 n*M* cBid.

and acceptor pair with an R_0 of 50 Å (Lakowicz, 2006). For this dye pair, FRET will only be detected if the distance between the two fluorophores is 70 Å or less. Typical R_0 values for a donor and acceptor pair are between 30 and 60 Å, similar to the size of proteins; thus, if FRET between donor- and acceptor-labeled proteins is detected then they are bound to each other.

4.1. Detecting the interaction between two proteins using FRET

As mentioned in Section 1, the BH3-only protein cBid targets to and embeds within the OMM where it recruits and activates cytosolic Bax (Leber et al., 2007; Lovell et al., 2008). Active membrane-bound Bax oligomerizes within the OMM resulting in membrane permeabilization. Here, we are using DAC and NBD as the donor and acceptor molecules, respectively. We will be using FRET to detect (1) the binding between cBid and Bax and (2) the binding between Bax molecules during oligomerization.

1. Liposomes are made as above (Section 2.3.1) resulting in liposomes at a concentration of 1 mg/mL lipid.
2. The fluorimeter (Photon Technology International) is set to record the fluorescence of DAC (380 nm excitation, 2 nm slit width; 460 nm emission, 10 nm slit width) with stirring for 1 h at 37 °C. Either 200 μL of liposomes and 800 μL of assay buffer, or as a control, 1 mL of assay buffer is added to a quartz cuvette and the signal is read until it remains stable (~5 min). Two reactions are required to detect FRET. One that contains both the donor- and acceptor-labeled proteins and a control that contains the donor-labeled protein and unlabeled acceptor protein. This control accounts for any changes in the donor protein that occur due to binding interactions, conformational changes, or environment changes that may affect the spectral properties of the donor dye.
3. The donor-labeled protein is added to the cuvette at a concentration of 20 n*M* and DAC fluorescence is read until the signal is stable. At this point, the acceptor protein that is either labeled with NBD or unlabeled is added to the system at a concentration of 100 n*M*. It is important to keep the amount of acceptor higher (5–10 ×) than that of the donor. Keeping the donor protein in excess ensures, it will be saturated by the acceptor.
4. The DAC signal is recorded for 1 h at 37 °C. FRET efficiency (E) is measured by comparing the relative intensity of the donor in the presence of labeled (F_{DA}) and unlabeled (F_D) acceptor and is calculated by:

$$E = 1 - \frac{F_{\mathrm{DA}}}{F_{\mathrm{D}}}$$

Figure 1.2 illustrates two binding interactions between the Bcl-2 family proteins cBid and Bax. Donor (DAC)-labeled cBid (20 n*M*) is incubated with acceptor (NBD)-labeled Bax (100 n*M*), and only in the presence of liposomes do the two proteins interact (Fig. 1.2B). This underlines the point that many functional interactions of the Bcl-2 family proteins only occur in the presence of a lipid bilayer. Additionally, the activator protein cBid is required for Bax to oligomerize, since FRET between donor (DAC) and acceptor (NBD)-labeled Bax is only observed when cBid is added to the system (Fig. 1.2C). As we observe the interactions of two proteins in real time, kinetics of the reactions can be determined. Indeed, it is clear from the data shown that the cBid–Bax interaction occurs faster than Bax oligomerization, suggesting that cBid first binds to and activates Bax followed by Bax oligomerization. Additionally, it is possible to generate a binding curve where an affinity for the interaction can be determined as was done for the binding between cBid and Bax (Lovell et al., 2008). To do this multiple FRET measurements are obtained by titrating the amount of acceptor, while keeping the donor concentration fixed.

5. TRACKING THE CONFORMATION CHANGES OF A PROTEIN

NBD is an environment-sensitive low-molecular weight fluorescent dye that has been used to track environment changes of specific residues of proteins (Dattelbaum et al., 2005; Lin, Jongsma, Pool, & Johnson, 2011). The emission intensity and fluorescence lifetime increases and the emission peak of NBD blueshifts from 570 nm, in an aqueous environment, to 530 nm when it is in a hydrophobic environment due to a decrease in fluorescence quenching by water (Crowley, Reinhart, & Johnson, 1993). The small size of NBD allows specific labeling of single-cysteine mutants of proteins, with less potential perturbation of wild-type function. Importantly, NBD is uncharged but has sufficient polar characteristics that it remains stable in both polar and nonpolar environments such that it is less likely than other environment-sensitive dyes to change the membrane-binding characteristics and/or conformation of the protein being studied (Shepard et al., 1998). These properties of NBD make it particularly useful to study membrane-binding proteins such as Bax and cBid that transition from the aqueous environment and embed into a membrane bilayer (Lovell et al., 2008; Shamas-Din, Bindner, et al., 2013; Shamas-Din, Kale, et al., 2013).

5.1. NBD-emission assay

Real-time changes in the fluorescence of NBD can be measured to determine whether and when specific regions of Bax (labeled with NBD) insert into the membrane during the activation of Bax. It is known from chemical-labeling studies that Bax inserts helices 5, 6, and 9 into the membrane (Annis et al., 2005). By labeling Bax at residue 175 (helix 9), it is possible to track the conformational change of Bax as it transitions from a soluble monomer to membrane embedded oligomer.

1. The fluorimeter is set to record NBD fluorescence (475 nm excitation, 2 nm slit width; 530 nm emission, 10 nm slit width), and as in the FRET experiment above, 200 μL of 1 mg/mL liposomes are added to 800 μL of assay buffer in a quartz cuvette. Background signal (Bg) is recorded with stirring until stable at 37 °C.
2. NBD-labeled Bax (100 n*M*) is added to the cuvette. Since Bax does not insert into membranes in the absence of an activator (Hsu & Youle, 1998), Bax-NBD has a stable signal when incubated with liposomes and an initial fluorescence value can be recorded (F_0). Alternatively, the very first point upon addition of the protein can be used as the F_0 value if the protein insert into lipids too rapidly. This approach is useful for proteins that are unstable in the assay solution such as cBid which spontaneously targets to membranes (Shamas-Din, Bindner, et al., 2013; Shamas-Din, Kale, et al., 2013). In the absence of membranes, cBid has sufficient exposed hydrophobicity that it tends to aggregate and to stick to the walls of the cuvette.
3. In our example, the change in emission over time (ΔF) of the dye labeled Bax is collected once an activator, cBid, is added. Fluorescence intensity plateaus after the protein comes to equilibrium (1 h endpoint). By calculating the ΔF value, one can track the relative change in emission intensity of the labeled residue in real time:

$$\Delta F = \frac{F - \mathrm{Bg}}{F_0 - \mathrm{Bg}}$$

Both residues 3 and 175 of Bax transition to a more hydrophobic environment as indicated (Fig. 1.3A) by the relative change in emission (ΔF). As the environment change of the residue can be tracked over time, kinetics of membrane binding can be measured. Tracking the kinetics of the environment changes of various residues as a protein undergoes a conformational change has been used to order specific structural changes of the protein (Shamas-Din, Bindner, et al., 2013; Shamas-Din, Kale, et al., 2013). Here, the carboxyl terminus of Bax (residue 175) has slower kinetics compared to that of the amino terminus of Bax (residue 3) suggesting that Bax undergoes a conformational change at residue 3 before that of 175. Additionally, residue 175 moves to a more hydrophobic environment since Bax 175C-NBD has a larger change in NBD emission compared to Bax 3C-NBD. This

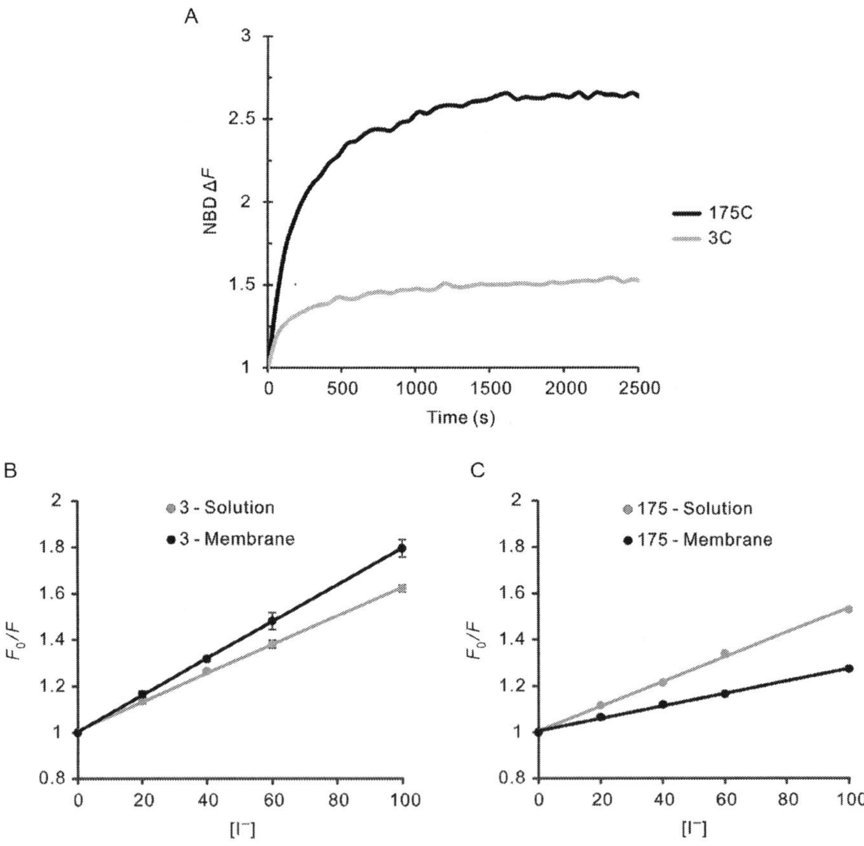

Figure 1.3 (A) NBD emission change for Bax 175C-NBD (100 n*M*) and Bax 3C-NBD (100 n*M*) upon addition of cBid (20 n*M*) in the presence of liposomes (0.2 mg/mL) (B and C) Iodide quenching data of 100 n*M* Bax 3C-NBD (B) or 175C-NBD (C) in solution (gray) or in the presence of liposomes (0.2 mg/mL) and cBid (20 n*M*) (black).

paired with the quenching data discussed below, suggests that residue 175 of Bax inserts into the phospholipid bilayer. As this residue is part of a larger hydrophobic sequence believed to span the bilayer, the kinetics for this residue likely represent insertion of the Bax carboxyl-terminal tail into membranes. This is in accordance with data that shows residue 175C is embedded within the mitochondrial membrane in cells (Annis et al., 2005). By using various activators of Bax or mutations known to perturb Bax function, it is possible to determine whether these changes affect the extent of or rate at which Bax helix 9 inserts into phospholipid bilayers.

6. DETERMINING THE TOPOLOGY OF PROTEINS WITHIN MEMBRANES

Fluorescence quenching by heavy atoms such as iodide can be used to determine how exposed a fluorescently tagged residue is to the solvent. This is due to collisional quenching that occurs when I^- collides with an excited fluorophore, resulting in a loss of energy back to ground state without emission of a photon. Typically, collisional quenching requires direct molecular interaction with the fluorophore such that the distance of quenching is <2 Å giving a very high resolution to detect solvent accessibility (Lakowicz, 2006).

Since I^- quenches NBD fluorescence (Crowley et al., 1993; Lin, Jongsma, Liao, & Johnson, 2011), this technique would be advantageous to look at the difference of residue solvent accessibility between soluble monomeric Bax, in the absence of activator, and membrane-bound oligomeric Bax, in the presence of an activator.

6.1. Iodide quenching of NBD-labeled Bax

1. As in the method for NBD emission change, the fluorimeter is set to record NBD fluorescence, and 200 μL of 1 mg/mL liposomes are added to 800 μL of assay buffer in a quartz cuvette. Background signal is read with stirring until stable at 37 °C.
2. NBD-labeled Bax (100 n*M*) and cBid (20 n*M*) are added to cuvettes containing either liposomes (200 μL liposomes, 800 μL assay buffer) for quenching of membrane bound Bax, or assay buffer only (1 mL assay buffer), for quenching of solution Bax. NBD emission (F_0) is recorded after incubation of the sample at 37 °C for 1 h.
3. Multiple quenching reactions are set up where aliquots of potassium iodide (2 *M*, supplemented with 2 m*M* sodium thiosulfate to prevent

oxidation) and potassium chloride (2 *M*) stock solutions are added to each sample so that the total ion concentration, and thus ionic strength, in samples is the same (typically 100 m*M*) (Table 1.2). The NBD emission for each concentration of KI is determined (F) and collisional quenching is calculated by the Stern–Volmer equation:

$$\frac{F_0}{F} = 1 + K_{sv}[\mathrm{Q}]$$

where F_0 is the fluorescence intensity in the absence of quencher, F is the fluorescence intensity at a specific quencher concentration [Q] and K_{sv} is the Stern–Volmer quenching constant.

As the concentration of I^- increases, so does the extent of quenching as determined by the Stern–Volmer equation, allowing the titration curve to be fit with a line where the slope is the Stern–Volmer constant (K_{sv}). The smaller the Stern–Volmer constant the more protected a residue is from the solvent. Quenching can then be used to compare the change in exposure of Bax residues to solvent upon the addition of an activator. Residue 3 shows a slight decrease in protection from quenching upon the addition of cBid (Fig. 1.3B), whereas residue 175C of Bax becomes more protected from quenching, in agreement with this region of Bax inserting into the bilayer (Fig. 1.3C) (Annis et al., 2005).

7. CONCLUSION

Here, four techniques have been highlighted to show how membrane proteins can be studied by fluorescence spectroscopy. The high sensitivity of fluorescence-based assays along with the ability to probe the dynamics of protein:protein and protein:membrane interactions in real time lends itself well to study a complex regulatory system such as the Bcl-2 family of proteins. The methods outlined use a simplified liposome system but can be extended to study Bcl-2 family regulation in isolated mitochondria (Shamas-Din, Bindner, et al., 2013; Shamas-Din, Kale, et al., 2013). Additionally, interactions between proteins can be detected via FRET in live cells using fluorescence lifetime imaging microscopy that recapitulate what we see *in vitro* (Aranovich et al., 2012). Thus, fluorescence is a very powerful technique that not only allows us to determine the core mechanism of Bcl-2 family regulation *in vitro* but also allows us to extend these studies to live cells and *in vivo*, bridging the gap between simplified and complex systems (Kale et al., 2012).

Table 1.2 Iodide quenching values of NBD-labeled Bax

		3C[a]		175C[a]	
KI (m*M*)	KCl (m*M*)	Solution (F_0/F)	Membrane (F_0/F)	Solution (F_0/F)	Membrane (F_0/F)
0	100	1	1	1	1
20	80	1.1351 ± 0.0105	1.1657 ± 0.0136	1.1158 ± 0.0002	1.0650 ± 0.0134
40	60	1.2645 ± 0.0068	1.3176 ± 0.0101	1.2140 ± 0.0029	1.1110 ± 0.0112
60	40	1.3812 ± 0.0180	1.4807 ± 0.0373	1.3397 ± 0.0019	1.1647 ± 0.0020
100	0	1.6224 ± 0.0156	1.7954 ± 0.0376	1.5294 ± 0.0002	1.2730 ± 0.0093

[a]Values for solution and membrane quenching are plotted in Fig. 1.3C.

ACKNOWLEDGMENTS

The work in the author's laboratory is supported by Grant FRN12517 from the Canadian Institutes of Health Research (CIHR) to D. W. A. and B. L. J. K. is recipient of a doctoral fellowship from the Canadian Breast Cancer Foundation, Ontario Division.

REFERENCES

Annis, M. G., Soucie, E. L., Dlugosz, P. J., Cruz-Aguado, J. A., Penn, L. Z., Leber, B., et al. (2005). Bax forms multispanning monomers that oligomerize to permeabilize membranes during apoptosis. *EMBO Journal*, *24*(12), 2096–2103.

Antonsson, B., Montessuit, S., Lauper, S., Eskes, R., & Martinou, J. C. (2000). Bax oligomerization is required for channel-forming activity in liposomes and to trigger cytochrome c release from mitochondria. *Biochemistry Journal*, *345*(Pt 2), 271–278.

Aranovich, A., Liu, Q., Collins, T., Geng, F., Dixit, S., Leber, B., et al. (2012). Differences in the mechanisms of proapoptotic BH3 proteins binding to Bcl-XL and Bcl-2 quantified in live MCF-7 cells. *Molecular Cell*, *45*(6), 754–763.

Billen, L. P., Kokoski, C. L., Lovell, J. F., Leber, B., & Andrews, D. W. (2008). Bcl-XL inhibits membrane permeabilization by competing with Bax. *PLoS Biology*, *6*(6), e147.

Bleicken, S., Classen, M., Padmavathi, P. V., Ishikawa, T., Zeth, K., Steinhoff, H. J., et al. (2010). Molecular details of Bax activation, oligomerization, and membrane insertion. *Journal of Biological Chemistry*, *285*(9), 6636–6647.

Bogner, C., Leber, B., & Andrews, D. W. (2010). Apoptosis: Embedded in membranes. *Current Opinion in Cell Biology*, *22*(6), 845–851.

Budd, S. L., Tenneti, L., Lishnak, T., & Lipton, S. A. (2000). Mitochondrial and extramitochondrial apoptotic signaling pathways in cerebrocortical neurons. *Proceedings of the National Academy of Sciences of the United States of America*, *97*(11), 6161–6166.

Crowley, K. S., Reinhart, G. D., & Johnson, A. E. (1993). The signal sequence moves through a ribosomal tunnel into a noncytoplasmic aqueous environment at the ER membrane early in translocation. *Cell*, *73*(6), 1101–1115.

Dattelbaum, J. D., Looger, L. L., Benson, D. E., Sali, K. M., Thompson, R. B., & Hellinga, H. W. (2005). Analysis of allosteric signal transduction mechanisms in an engineered fluorescent maltose biosensor. *Protein Sciences*, *14*(2), 284–291.

Hope, M. J., Bally, M. B., Webb, G., & Cullis, P. R. (1985). Production of large unilamellar vesicles by a rapid extrusion procedure: Characterization of size distribution, trapped volume and ability to maintain a membrane potential. *Biochimica et Biophysica Acta*, *812*(1), 55–65.

Hsu, Y. T., & Youle, R. J. (1997). Nonionic detergents induce dimerization among members of the Bcl-2 family. *Journal of Biological Chemistry*, *272*(21), 13829–13834.

Hsu, Y. T., & Youle, R. J. (1998). Bax in murine thymus is a soluble monomeric protein that displays differential detergent-induced conformations. *Journal of Biological Chemistry*, *273*(17), 10777–10783.

Kale, J., Liu, Q., Leber, B., & Andrews, D. W. (2012). Shedding light on apoptosis at subcellular membranes. *Cell*, *151*(6), 1179–1184.

Kuwana, T., Mackey, M. R., Perkins, G., Ellisman, M. H., Latterich, M., Schneiter, R., et al. (2002). Bid, Bax, and lipids cooperate to form supramolecular openings in the outer mitochondrial membrane. *Cell*, *111*(3), 331–342.

Lakowicz, J. R. (2006). *Principles of fluorescence spectroscopy* (3rd ed.). New York: Springer.

Landeta, O., Landajuela, A., Gil, D., Taneva, S., Diprimo, C., Sot, B., et al. (2011). Reconstitution of proapoptotic BAK function in liposomes reveals a dual role for mitochondrial lipids in the BAK-driven membrane permeabilization process. *Journal of Biological Chemistry*, *286*(10), 8213–8230.

Leber, B., Lin, J., & Andrews, D. W. (2007). Embedded together: The life and death consequences of interaction of the Bcl-2 family with membranes. *Apoptosis*, *12*(5), 897–911.

Lin, P. J., Jongsma, C. G., Liao, S., & Johnson, A. E. (2011). Transmembrane segments of nascent polytopic membrane proteins control cytosol/ER targeting during membrane integration. *Journal of Cell Biology*, *195*(1), 41–54.

Lin, P. J., Jongsma, C. G., Pool, M. R., & Johnson, A. E. (2011). Polytopic membrane protein folding at L17 in the ribosome tunnel initiates cyclical changes at the translocon. *Journal of Cell Biology*, *195*(1), 55–70.

Lovell, J. F., Billen, L. P., Bindner, S., Shamas-Din, A., Fradin, C., Leber, B., et al. (2008). Membrane binding by tBid initiates an ordered series of events culminating in membrane permeabilization by Bax. *Cell*, *135*(6), 1074–1084.

Malhotra, K., Sathappa, M., Landin, J. S., Johnson, A. E., & Alder, N. N. (2013). Structural changes in the mitochondrial Tim23 channel are coupled to the proton-motive force. *Nature Structural and Molecular Biology*, *20*(8), 965–972.

Perez-Lara, A., Egea-Jimenez, A. L., Ausili, A., Corbalan-Garcia, S., & Gomez-Fernandez, J. C. (2012). The membrane binding kinetics of full-length PKCalpha is determined by membrane lipid composition. *Biochimica et Biophysica Acta*, *1821*(11), 1434–1442.

Petros, A. M., Olejniczak, E. T., & Fesik, S. W. (2004). Structural biology of the Bcl-2 family of proteins. *Biochimica et Biophysica Acta*, *1644*(2–3), 83–94.

Ren, D., Tu, H. C., Kim, H., Wang, G. X., Bean, G. R., Takeuchi, O., et al. (2010). BID, BIM, and PUMA are essential for activation of the BAX- and BAK-dependent cell death program. *Science*, *330*(6009), 1390–1393.

Satsoura, D., Kucerka, N., Shivakumar, S., Pencer, J., Griffiths, C., Leber, B., et al. (2012). Interaction of the full-length Bax protein with biomimetic mitochondrial liposomes: A small-angle neutron scattering and fluorescence study. *Biochimica et Biophysica Acta*, *1818*(3), 384–401.

Shamas-Din, A., Bindner, S., Zhu, W., Zaltsman, Y., Campbell, C., Gross, A., et al. (2013). tBid undergoes multiple conformational changes at the membrane required for Bax activation. *Journal of Biological Chemistry*, *288*(30), 22111–22127.

Shamas-Din, A., Kale, J., Leber, B., & Andrews, D. W. (2013). Mechanisms of action of Bcl-2 family proteins. *Cold Spring Harbor Perspectives in Biology*, *5*(4), a008714.

Shepard, L. A., Heuck, A. P., Hamman, B. D., Rossjohn, J., Parker, M. W., Ryan, K. R., et al. (1998). Identification of a membrane-spanning domain of the thiol-activated pore-forming toxin Clostridium perfringens perfringolysin O: An alpha-helical to beta-sheet transition identified by fluorescence spectroscopy. *Biochemistry*, *37*(41), 14563–14574.

Yethon, J. A., Epand, R. F., Leber, B., Epand, R. M., & Andrews, D. W. (2003). Interaction with a membrane surface triggers a reversible conformational change in Bax normally associated with induction of apoptosis. *Journal of Biological Chemistry*, *278*(49), 48935–48941.

Zhang, Z., Zhu, W., Lapolla, S. M., Miao, Y., Shao, Y., Falcone, M., et al. (2010). Bax forms an oligomer via separate, yet interdependent, surfaces. *Journal of Biological Chemistry*, *285*(23), 17614–17627.

CHAPTER TWO

Photoreactive Stapled Peptides to Identify and Characterize BCL-2 Family Interaction Sites by Mass Spectrometry

Susan Lee[*,†], **Craig R. Braun**[*,†,‡], **Gregory H. Bird**[*,†], **Loren D. Walensky**[*,†,1]

[*]Department of Pediatric Oncology and the Linde Program in Cancer Chemical Biology, Dana-Farber Cancer Institute, Boston, Massachusetts, USA
[†]Department of Pediatrics, Children's Hospital Boston, Harvard Medical School, Boston, Massachusetts, USA
[‡]Department of Cell Biology, Harvard Medical School, Boston, Massachusetts, USA
[1]Corresponding author: e-mail address: loren_walensky@dfci.harvard.edu

Contents

Abstract

Protein interactions dictate a myriad of cellular activities that maintain health or cause disease. Dissecting these binding partnerships, and especially their sites of interaction, fuels the discovery of signaling pathways, disease mechanisms, and next-generation therapeutics. We previously applied all-hydrocarbon peptide stapling to chemically restore α-helical shape to bioactive motifs that become unfolded when taken out of context from native signaling proteins. For example, we developed stabilized alpha-helices of BCL-2 domains (SAHBs) to dissect and target protein interactions of the BCL-2 family, a critical network that regulates the apoptotic pathway. SAHBs are α-helical surrogates that bind both stable and transient physiologic interactors and have effectively uncovered novel sites of BCL-2 family protein interaction. To leverage stapled peptides for proteomic discovery, we describe our conversion of SAHBs into photoreactive agents that irreversibly capture their protein targets and facilitate rapid

Methods in Enzymology, Volume 544
ISSN 0076-6879
http://dx.doi.org/10.1016/B978-0-12-417158-9.00002-9

identification of the peptide helix binding sites. We envision that the development of photoreactive stapled peptides will accelerate the discovery of novel and unanticipated protein interactions and how they impact health and disease.

1. INTRODUCTION

Alpha-helical interactions are found throughout the cell and govern critical biological processes, such as infection (Weissenhorn, Dessen, Harrison, Skehel, & Wiley, 1997), the immune response (Blum, Stevens, & DeFranco, 1993), apoptosis (Sattler et al., 1997), and transcription (Kussie et al., 1996; Fig. 2.1). The natural complexity of peptide α-helices enables them to engage diverse cellular targets with high affinity and selectivity. Indeed, helical peptides can effectively discriminate among homologous proteins owing to the high fidelity key-in-lock specificity afforded by their amino acid composition. The conserved BCL-2 homology 3 (BH3) domain of BCL-2 family proteins is a critical interaction module that mediates

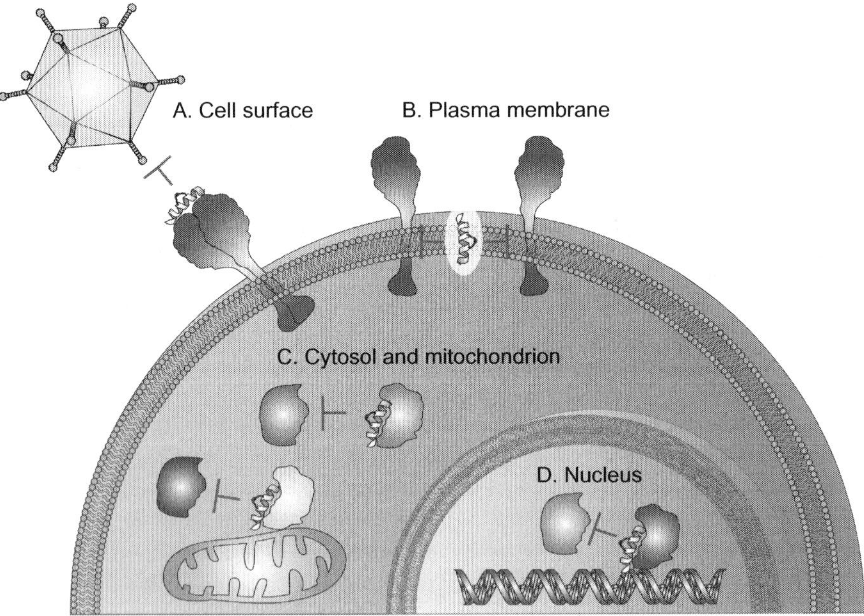

Figure 2.1 The peptide α-helix is a ubiquitous secondary structural motif that mediates a host of biomedically relevant protein interactions. Hydrocarbon-stapled peptides modeled after bioactive helices can be generated to target and modulate protein interactions from the surface to the inner nuclear core of the cell. (See the color plate.)

member-specific communication. The canonical complex between a proapoptotic BH3 α-helix and a surface groove on antiapoptotic members reflects the molecular wrestling match between pro- and antiapoptotic signals (Sattler et al., 1997). If antiapoptotic grooves are sufficient in number to bind and sequester the proapoptotic BH3 signals, cell survival prevails. In contrast, if the capacity to withstand proapoptotic assault is breached, cell death ensues (Fig. 2.2). Not only did the discovery of this protein interaction paradigm provide a mechanism for apoptotic regulation, but it also informed the development of drugs to reactivate cell death through targeted inhibition of antiapoptotic BH3-binding pockets (Oltersdorf et al., 2005).

Given the critical roles of amphipathic alpha-helices in mediating signal transduction, we advanced all-hydrocarbon stapling to refold bioactive

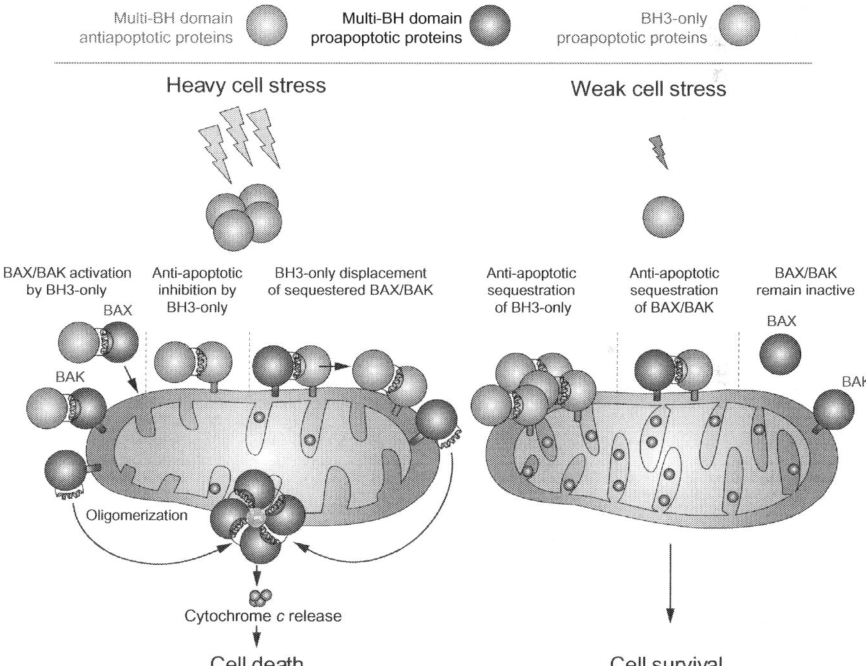

Figure 2.2 BCL-2 family proteins regulate the life and death decision of stressed cells. If mitochondrial antiapoptotic proteins (orange) can effectively harness their C-terminal binding pockets to trap and sequester the BH3-signaling helices of proapoptotic proteins, cell survival prevails. However, with increased cellular stress, proapoptotic signals overwhelm the antiapoptotic reserve. The BH3 domain helices of BH3-only proteins (blue) can directly activate the essential executioner proteins BAX and BAK (gray) and also release trapped forms from antiapoptotic inhibition by targeting the antiapoptotic groove. (See the color plate.)

alpha-helical peptides for use as research tools and prototype therapeutics (Fig. 2.3A). Depending on their amino acid composition and design, "stapled peptides" have proven to be structurally stable, protease resistant, and cell permeable agents capable of interrogating and modulating protein interactions *in vitro* and *in vivo* (Bernal et al., 2010; Kim et al., 2013; LaBelle et al., 2012; Takada et al., 2012; Walensky et al., 2004). Such stapled peptides have revealed unanticipated and functionally relevant helix/target interactions, including engagement of (1) antiapoptotic MCL-1 by its own MCL-1 BH3 helix (Stewart et al., 2010; Fig. 2.3B), (2) a novel "trigger site" on proapoptotic BAX by the BIM BH3 helix (Gavathiotis et al., 2008; Fig. 2.3C), and (3) glucokinase by a BH3-phosphorylated form of the BH3-only protein BAD (Danial et al., 2008). The capacity of these stabilized alpha-helices of BCL-2 domains (SAHBs) to access canonical and noncanonical protein interactors, and reveal sites of interaction by their use in NMR and X-ray crystallography studies, prompted us to consider whether they could be harnessed for higher throughput binding site discovery. Here, we describe our approach to converting stapled peptides into high fidelity photoaffinity reagents and the methodologies developed to rapidly identify their sites of target protein interaction.

2. OVERVIEW OF PHOTOREACTIVE STABILIZED ALPHA-HELICES METHODOLOGY

We generate photoreactive stabilized alpha-helices or pSAHs by substituting into the peptide sequence a nonnatural amino acid bearing a benzophenone moiety (4-benzoylphenylalanine, Bpa) (Fig. 2.4A), which covalently crosslinks to protein targets upon exposure to ultraviolet (UV) light (Dorman & Prestwich, 1994; Saghatelian, Jessani, Joseph, Humphrey, & Cravatt, 2004; Vodovozova, 2007). We then insert an (i, $i+4$) or (i, $i+7$) pair of nonnatural amino acids bearing olefin tethers, followed by ruthenium-catalyzed ring-closing metathesis (RCM), to generate a hydrocarbon-stapled peptide with reinforced α-helical structure (Blackwell et al., 2001; Schafmeister, Po, & Verdine, 2000; Fig. 2.4B). For binding site identification, pSAHs containing alternatively placed Bpa moieties are individually incubated with a protein-of-interest and UV irradiated (Fig. 2.4C). The mixture is subjected to electrophoresis and the crosslinked species excised, trypsinized, and prepared for mass spectrometry analysis, which is designed to identify peptidic fragments bearing Bpa crosslinks and thus the explicit sites of target protein intercalation (Fig. 2.4D).

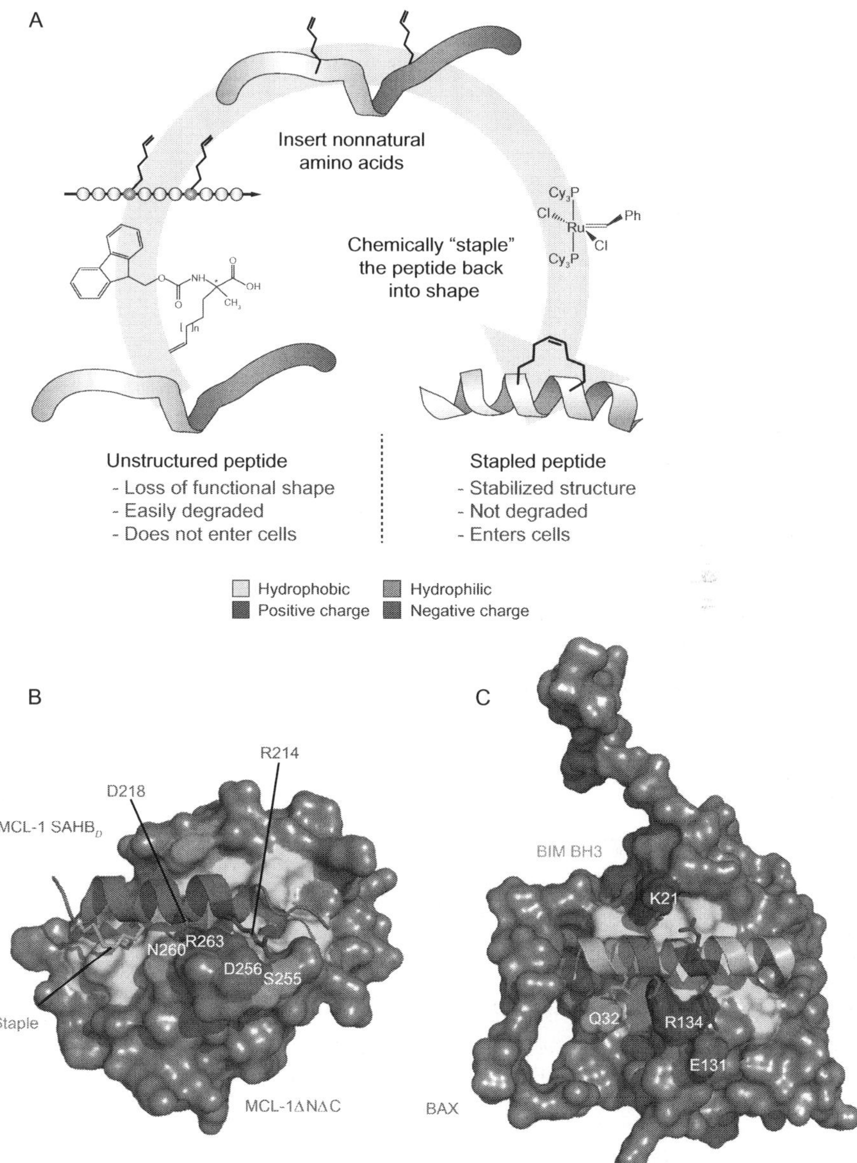

Figure 2.3 Synthetic overview of all-hydrocarbon stapling and its applications in developing α-helical peptides for structural studies of protein interactions. (A) Pairs of α,α-disubstituted nonnatural amino acids bearing olefin tethers are substituted into the peptide sequence at discrete locations (e.g., *i*, *i*+4), followed by ruthenium-catalyzed ring-closing metathesis (RCM) to generate "stapled peptides." (B) Crystal structure of a stapled MCL-1 BH3 helix in complex with antiapoptotic MCL-1 (Stewart, Fire, Keating, & Walensky, 2010). (C) Calculated model structure of a BIM BH3 helix engaging the N-terminal trigger site of full-length BAX, as derived from paramagnetic relaxation enhancement NMR analyses using full-length ^{15}N-BAX- and MTSL-labeled BIM SAHBs (aa 145–164) (Gavathiotis et al., 2008). (See the color plate.)

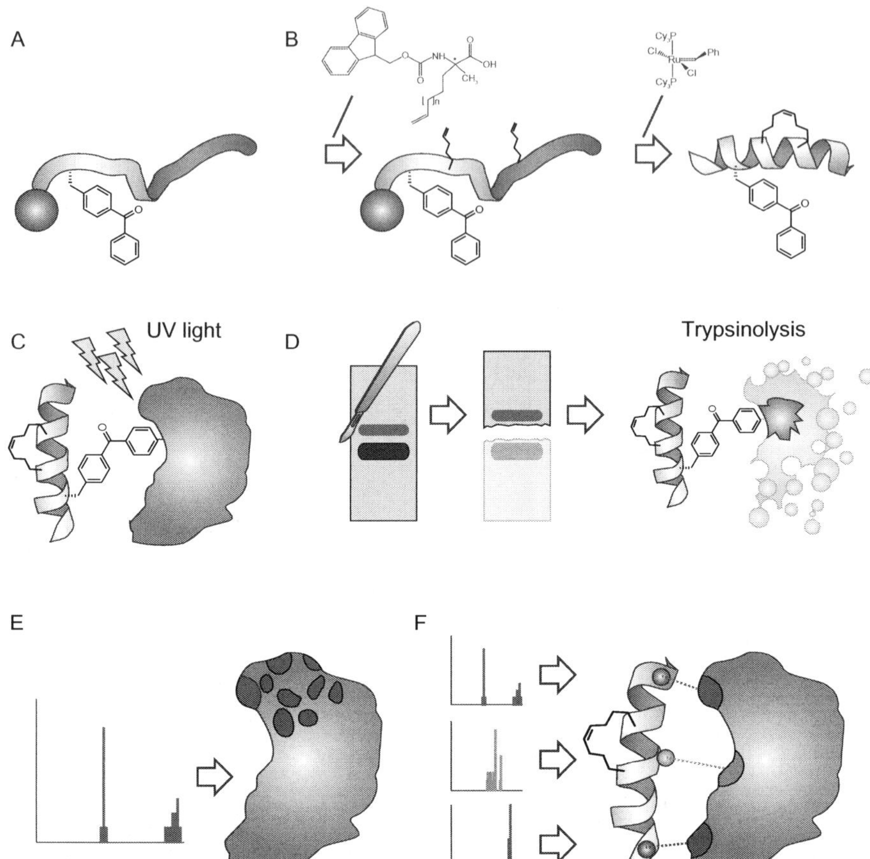

Figure 2.4 Design of photoreactive stabilized alpha-helices (pSAHs) for binding site identification by mass spectrometry. A photoreactive Bpa residue (A) and a pair of stapling amino acids (B) are inserted into the peptide template followed by RCM to generate a pSAH. Upon exposure to UV light, the bound pSAH covalently crosslinks to the target protein (C) and, following electrophoresis of the mixture, crosslinked protein is excised from the gel and subjected to in-gel digestion with trypsin (D). LC–MS/MS analysis identifies the explicit sites of covalent modification, which when mapped on to the protein structure reveal the region of pSAH interaction (E). Top scoring crosslinks from a series of experiments employing sequentially placed Bpa residues can provide interaction restraints for calculating model structures (F). (See the color plate.)

By mapping the crosslink sites of pSAHs bearing alternatively placed Bpa moieties onto the protein structure (if known), the location of the interaction site can be determined (Fig. 2.4E). Furthermore, if the structure of the protein target has been solved and a series of pSAHs with sequentially localized Bpa residues map sites of intercalation with sufficient resolution, computational docking can be applied to calculate a model structure of the complex for biochemical and functional validation (Fig. 2.4F).

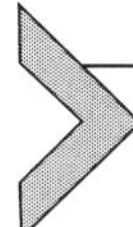

3. DESIGN AND SYNTHESIS OF PHOTOREACTIVE STAPLED PEPTIDES

The design of photoreactive stapled peptides is ideally based on a structure involving the template α-helix, so that the structurally reinforcing staples can be placed on the noninteracting surface and Bpa residues located at or near the putative interaction surface to facilitate UV crosslinking. Ideally, Bpa residues are substituted at positions of homologous, bulky, and hydrophobic residues (e.g., Phe) to minimize the effect of Bpa on the binding interface. Without structural information, a library of constructs can be generated by sampling alternative staple and Bpa positions along the length of the peptide template, selecting those constructs with successful α-helical induction (as determined by circular dichroism analysis; Bird, Bernal, Pitter, & Walensky, 2008) and preserved binding activity for experimental application. Of note, to simplify interpretation of the MS2-based identification of Bpa-crosslinked peptides, it is also preferable to locate Bpa residues in regions of the peptide such that trypsinization will yield short cleavage products. If necessary, a lysine or arginine residue can be strategically substituted into the peptide, ideally on the noninteracting surface or as a conserved substitution observed in other species (Leshchiner, Braun, Bird, & Walensky, 2013), to accomplish this goal.

Once designed, stapled peptides are synthesized as previously described in detail (Bird et al., 2008; Bird, Crannell, & Walensky, 2011) using Fmoc chemistry, stapling amino acids installed at $(i, i+4)$ or $(i, i+7)$ positions (to bridge one or two turns of the α-helix, respectively), and ruthenium-catalyzed olefin metathesis with Grubbs generation I catalyst (Fig. 2.5). Of note, methionine residues in the template sequence are typically replaced by norleucines, as the unprotected sulfur can decrease the efficiency of the Grubbs catalyst. Following peptide deprotection and cleavage, pSAHs are purified by reverse phase high performance liquid chromatography–mass spectrometry and quantified by amino acid analysis (AAA). Pure peptide is stored as a lyophilized powder at −20 °C until use.

In addition to the experimental conditions summarized in Fig. 2.5 and the step-by-step synthetic protocol with reagent list detailed in our recently updated methodologic review (Bird et al., 2011), our approach to stapled peptide synthesis incorporates the following general principles:

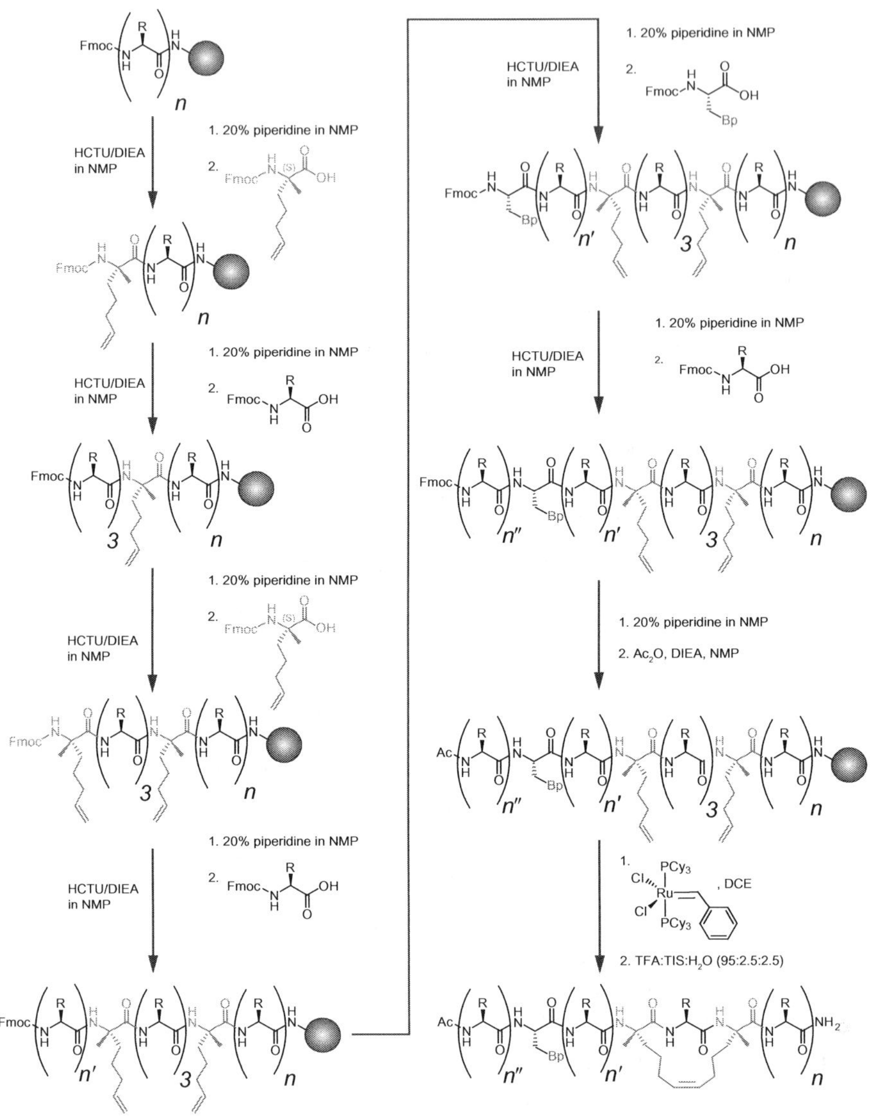

Figure 2.5 Synthetic steps for the automated production of pSAHs by Fmoc-based solid-phase peptide synthesis and ruthenium-catalyzed RCM. Alternatives to N-terminal acetylation include capping with FITC or biotin, depending on the desired application. Bp, 4-benzoylphenyl. (See the color plate.)

a. Create an automated synthesis program by employing standard double Fmoc deprotections, vigorous washes with DMF, and subsequent single 10-fold excess or double fivefold excess amino acid couplings. The precise timing and coupling excesses are optimized based on the specific coupling reagents, bases, and synthesizer employed. Particular attention is devoted to reactions involving the α,α-disubstituted amino acids, as the amine is linked to a quaternary carbon center, and therefore, can be recalcitrant to Fmoc removal and the subsequent acylation steps. For example, couplings that involve beta-branched amino acids and arginine can be especially challenging. Typical protocol modifications include multiple extended deprotections and acylations, and sometimes capping at these difficult steps with acetic acid (AcOH) or even decanoic acid in order to capture and terminate unreacted amine. In this manner, the production of full-length peptide species can be maximized, while also avoiding the challenge of chromatographic separation of truncated byproducts.

b. Prepare the peptide synthesis reagents and solutions by weighing out the resin and dissolving the calculated amount of each amino acid in anhydrous NMP. Although more expensive than DMF, NMP has greater stability and we have seen no deterioration in peptide quality using reagent solutions for up to 2 weeks. A secondary amine is used to remove the Fmoc protective group; appropriate choices include piperidine, piperazine, or 1,8-diazabicycloundec-7-ene. Likewise, there are multiple coupling reagents to select from. HCTU/DIEA and HOBt (or HOAt)/DIC nicely balance reactivity with solution stability, and more reactive coupling reagents range from uronium based (HATU) to the more exotic tris-pyrrolidinophosphonium derivatives.

c. Evaluate synthetic success by subjecting a small sample of beads to standard trifluoroacetic acid (TFA) cleavage (95:2.5:2.5 TFA:water: triisopropylsilane) and analysis by LC/MS. If the full-length species is identified in good yield, the next step is acylation with acetic anhydride or Fmoc-β-alanine, followed by stapling with Grubbs I catalyst (bis (tricyclohexylphosphine) benzylidine ruthenium (IV) dichloride) dissolved in dichloroethane. For most sequences and staple locations, a 2×2 h incubation using several milliliters of a 4 mg/mL solution will yield a completely metathesized product. If there is incomplete metathesis, the reaction can be repeated multiple times and with longer incubation periods.

d. After confirmation of successful stapling, the Fmoc group of Fmoc-β-alanine is removed and the N-terminus derivatized with biotin or fluorescein isothiocyanate (FITC), depending on the application (see later). The final product is cleaved off the resin in large scale, precipitated with ether:hexanes, and purified by HPLC or, if available, by LC/MS with mass-triggered fraction collection. The pure fractions are pooled, lyophilized, and peptide powder resuspended in a known volume, followed by submission of several dilutions for AAA. Although we typically store stapled peptides as dried powder (see above), the material can also be solubilized in 100% DMSO for storage at −20 °C.

4. PHOTOAFFINITY LABELING AND RETRIEVAL

With a panel of photoreactive stapled peptides in hand, the next step is to isolate pure protein-of-interest for photoaffinity labeling. We have typically expressed glutathione *S*-transferase, chitin-binding domain, or histidine-tagged fusion proteins followed by affinity purification, cleavage of the tag, and size exclusion chromatography. Protein purity and identity are confirmed by SDS-PAGE, protein staining (e.g., coomassie, silver), protein-specific western blot, and mass spectrometry analyses. Protein concentrations are determined using the Bradford assay (Biorad).

Photoreactive stapled peptide and pure recombinant protein (e.g., 10 μ*M*) are typically mixed at a 1:1 ratio, but higher pSAH concentrations can be used if necessary to increase the quantity of crosslinked product. The peptide/protein mixture is vortexed, incubated for 20 min in the dark, and then irradiated (365 nm) on ice for 1.5 h by use of a Spectroline handheld UV lamp (Model En280L, Spectronics Corporation). The duration of UV exposure can also be adjusted based on the efficiency of crosslinking for a particular peptide/protein complex. The mixtures are then diluted with 4× LDS loading buffer (Invitrogen), electrophoresed (e.g., 4–12% Bis–Tris gels [Invitrogen]), and analyzed by coomassie staining (SimplyBlue Safestain, Invitrogen). The appearance of a slower migrating band just above the protein starting material reflects the pSAH-crosslinked species, which is excised for MS analysis (see below).

Several supplementary experiments can be especially useful when piloting this approach. For example, to confirm that insertion of staples and Bpa residues at discrete sites within the peptide template do not disrupt target engagement, we recommend generating fluorescently labeled (FITC) pSAH derivatives for comparative fluorescence polarization (FP) binding analyses with a positive control ligand. In the case of FITC-BAD pSAHBs, the substitution of Bpa residues at four positions within the BAD BH3 sequence had no disruptive effect on binding to the antiapoptotic protein BCL-X_LΔC (Braun et al., 2010; Fig. 2.6A). The FITC-derivatized pSAHs can also be used for photoaffinity labeling, with conversion of recombinant protein into a fluorescent species that can readily be detected by fluorescence imaging (Gel Doc XR Molecular Imager and Quantity One software, Bio-Rad), providing further confirmation of effective crosslinking (Braun et al., 2010; Fig. 2.6B). Biotinylated pSAHs can also be used for high sensitivity screening of effective crosslinking, with electrophoresed protein analyzed by anti-biotin western blot (see below). Binding specificity experiments using bovine serum albumin (BSA), for example, are also advised. Crosslinking is performed as above, except that BSA is added in molar excess

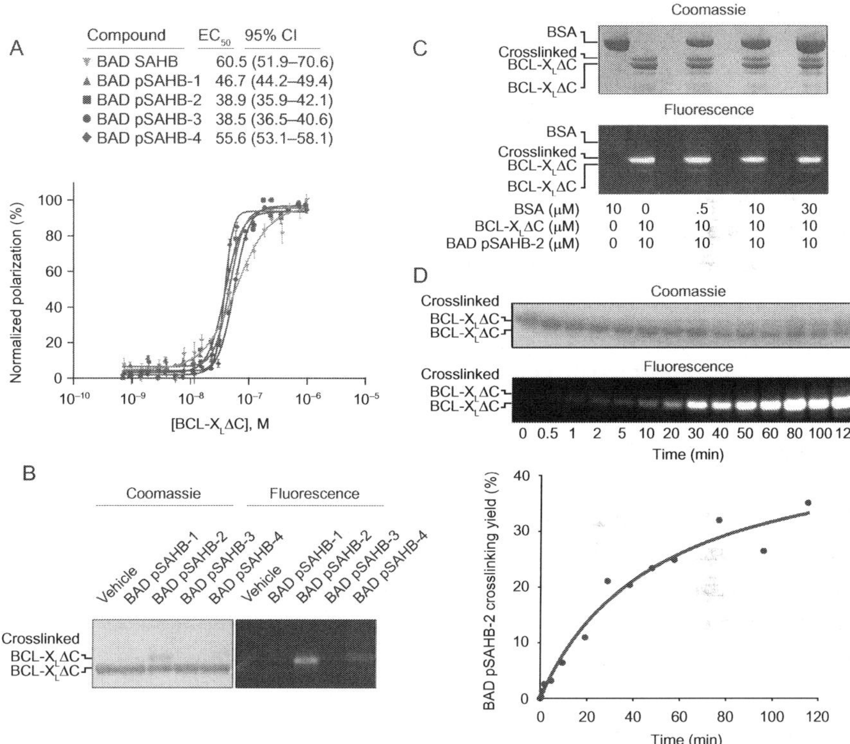

Figure 2.6 pSAHs bind and selectively crosslink to their protein targets, as exemplified by the photocrosslinking of BAD pSAHBs to antiapoptotic BCL-$X_L\Delta C$ (Braun et al., 2010). (A) A series of FITC-BAD pSAHBs retain nanomolar binding affinity to BCL-$X_L\Delta C$ as measured by FP analysis. (B) FITC-BAD pSAHBs exhibit a spectrum of BCL-$X_L\Delta C$-crosslinking efficiency upon UV exposure, with BAD pSAHB-2 demonstrating the greatest reactivity toward BCL-$X_L\Delta C$ (left, coomassie stain; right, fluorescence scan). (C) The selectivity of BAD pSAHB-2 is reflected by the absence of crosslinking to BSA and the lack of effect of added BSA on the BCL-$X_L\Delta C$-crosslinking efficiency of FITC-BAD pSAHB-2. (D) Time course for photocrosslinking of FITC-BAD pSAHB-2 to BCL-$X_L\Delta C$ as monitored by coomassie stain (top) and fluorescence scan (middle). A plot of crosslinking yield, calculated based on densitometry, demonstrates the time-dependent production of crosslinked BCL-$X_L\Delta C$ (bottom). (See the color plate.)

prior to addition of pSAH. In the case of FITC-BAD pSAHB-2/BCL-X-$_L\Delta$C crosslinking, the addition of molar excess BSA had no effect on the yield of pSAHB-crosslinked BCL-$X_L\Delta$C (Braun et al., 2010; Fig. 2.6C). Tracking the production of pSAH-crosslinked protein over time is useful for determining just how long to extend the reaction time to produce maximal reaction product for the MS analyses (Fig. 2.6D).

If the efficiency of pSAH crosslinking is suboptimal after increasing the peptide/protein ratio and extending the reaction time (i.e., crosslinked species not readily visualized by coomassie stain), the capacity to load sufficient material for electrophoresis and band excision can be compromised. Although we typically find that sampling a variety of pSAH constructs reveals lead photoaffinity labeling reagents (Fig. 2.6B), an alternative approach to enrich for crosslinked species entails the synthesis and application of pSAHs capped at the N-terminus with biotin. Photoaffinity labeling with biotinylated pSAHs is performed as above and then unreacted peptide is removed from the irradiated samples by overnight dialysis at 4 °C in 50 m*M* Tris pH 7.4, 200 m*M* NaCl buffer using 6–8 kDa molecular weight cut-off D-Tube dialyzers (EMD Biosciences). After addition of SDS to a final concentration of 0.2%, biotinylated protein is isolated from unreacted protein by incubation with high capacity streptavidin agarose (50 μL 50% slurry/reaction) for 2 h at room temperature. The streptavidin beads are successively washed at room temperature in 1% SDS in PBS, 1 *M* NaCl in PBS, and then 10% ethanol in PBS for 3 × 10 min each. Biotinylated proteins are eluted by boiling for 30 min in a 10% SDS solution (Promega) containing biotin (10 mg/mL), electrophoresed (e.g., 4–12% Bis–Tris gels [Invitrogen]), and then subjected to coomassie staining. Using this enrichment protocol for purifying biotinylated pSAH-crosslinked protein, isolating sufficient material for MS analyses of even low efficiency crosslinked complexes (e.g., weak or transient protein interactions) can typically be achieved (Edwards et al., 2013; Leshchiner et al., 2013).

5. MASS SPECTROMETRY ANALYSIS

To prepare pSAH-crosslinked protein for MS analysis, we perform an in-gel digest of the excised protein band using a standard protocol (Braun et al., 2010), mass spectrometry grade reagents, and due caution to avoid keratin contamination. Gels are subjected to coomassie or other MS-compatible stain (e.g., Pierce Silver Stain) and the excised gel slab cut into 1 mm cubes for processing as follows:

a. *Destain*: Add 500 μL of 50% acetonitrile (ACN)/50 m*M* ammonium bicarbonate (AB), pH 8 and incubate at 37 °C with shaking for 15 min; remove solution and repeat until the blue color is eliminated from the gel pieces.
b. *Dehydrate*: Add 500 μL of ACN, incubate at room temperature for 5 min, remove solution, and repeat.
c. *Reduce disulfides*: Add 100 μL of 10 m*M* dithiothreitol in AB, incubate at 56 °C for 30 min, and remove solution.
d. *Alkylate cysteines*: Add 100 μL of 50 m*M* iodoacetamide in AB, incubate at room temperature in the dark for 20 min, and then remove solution.
e. *Dehydrate*: As above.
f. *Digest*: Cover gel pieces with cold 12.5 ng/μL trypsin (Promega) in AB, incubate on ice for 45 min and then overnight at 37 °C. Peptides are extracted twice in 50% ACN/5% formic acid (FA) and dried in a speed-vac.

Samples are prepared for LC–MS/MS using home-made C18 stage tips or Agilent C18 Omix tips (Braun et al., 2010; Jakobsen, Schröder, Larsen, Lundberg, & Andersen, 2013). The sample preparation procedure below employs 100 μL Omix tips:
a. Place 6 μL of elution buffer (80% ACN/0.5% AcOH) in a clean elution tube for each sample.
b. Resuspend the dried peptides in 100 μL of 0.1% TFA.
c. Wet tips by pipetting 2 × 100 μL 40% ACN/0.5% AcOH. Be sure to prevent air from being introduced into the tips once wetted.
d. Equilibrate tips by pipetting 2 × 100 μL of 0.1% TFA.
e. Bind peptides in sample by carefully pipetting up and down 10 times.
f. Wash by pipetting 3 × 100 μL of 0.1% TFA.
g. Elute by carefully pipetting up and down in a prepared elution tube three times.
h. Dry the eluate by speed-vac.

We use a split flow LC system with in-house packed C4/C18 columns for MS analysis on a Thermo LTQ Orbitrap Discovery mass spectrometer. Samples are reconstituted in 5% ACN/5% FA and loaded onto the LC. We employ 30 cm of 100 μm inner diameter fused silica tips packed with <0.5 mm C4 (Sepax 5 μm, 120 Å) and then C18 slurry (Nest 3 μm, 200 Å). LC solutions are (A) 3.1% ACN/0.125% FA/water and (B) 0.125% FA/ACN. We typically employ a 2 h gradient from 10% to 40% B with a top 10 data-dependent method.

Analysis of the MS/MS spectra can be accomplished using the SEQUEST algorithm (Eng, McCormack, & Yates, 1994; Yates, Eng,

McCormack, & Schieltz, 1995), searching against a partially tryptic database containing the target protein, trypsin, and common keratin contaminants. Data analysis can also be accomplished using commercial software packages, such as Thermo Proteome Discoverer or XTandem! Analytical parameters include: enzyme, trypsin; missed cleavages, 2; parent tolerance, 10 ppm; fragment ion tolerance, 0.8 Da. Tryptic peptides containing Bpa crosslinks are identified by allowing for a variable modification corresponding to the mass of the Bpa-containing pSAH fragment on any amino acid of each protein in the database. Of note, the reversed protein sequences are also included in the database to generate an estimate of false-positive detection rate. For example, the subset of crosslinked peptides identified by a SEQUEST search are filtered based on tryptic state, charge state, mass accuracy, peptide length, and number of correct matches per protein, such that the maximum false-positive detection rate is 5%. Results containing more than one instance of photoaffinity labeling are excluded from the data set, as the crosslinking stoichiometry is assumed to be 1:1. pSAH modifications are further confirmed by manual examination of each MS/MS spectrum.

The resultant list of covalently modified sites is plotted by frequency of occurrence across the polypeptide sequence of the target protein. Fortunately, we have found that, if anything, the relative inefficiency of photoaffinity labeling with pSAHs results in the identification of too few rather than too many covalently modified sites, yielding reproducible and highly specific results—especially for relatively stable protein interactions. For example, our earliest crosslinking analysis of a photoreactive BAD pSAHB with antiapoptotic BCL-$X_L\Delta C$, identified a focal cluster of covalently modified sites with F105 emerging as the most frequently crosslinked residue (Fig. 2.7A). To depict the crosslinking "hot spots" on the protein structure, a color scale is assigned by converting crosslinking frequencies to percent of maximum occurrence and then colored by groups of 10 percentiles. When we compared the previously reported crystal structure of the BAD BH3/BCL-$X_L\Delta C$ complex (PDB ID 2BZW) with our crosslinking results, F105 indeed represented the nearest neighbor to Y110 of the BAD BH3 helix, the exact site of our Bpa substitution (Fig. 2.7B).

6. COMPUTATIONAL DOCKING

When the structure of a target protein-of-interest is known but the site of a putative peptide helix interaction site is not, our pSAH photoaffinity labeling/MS approach can be combined with computational docking to

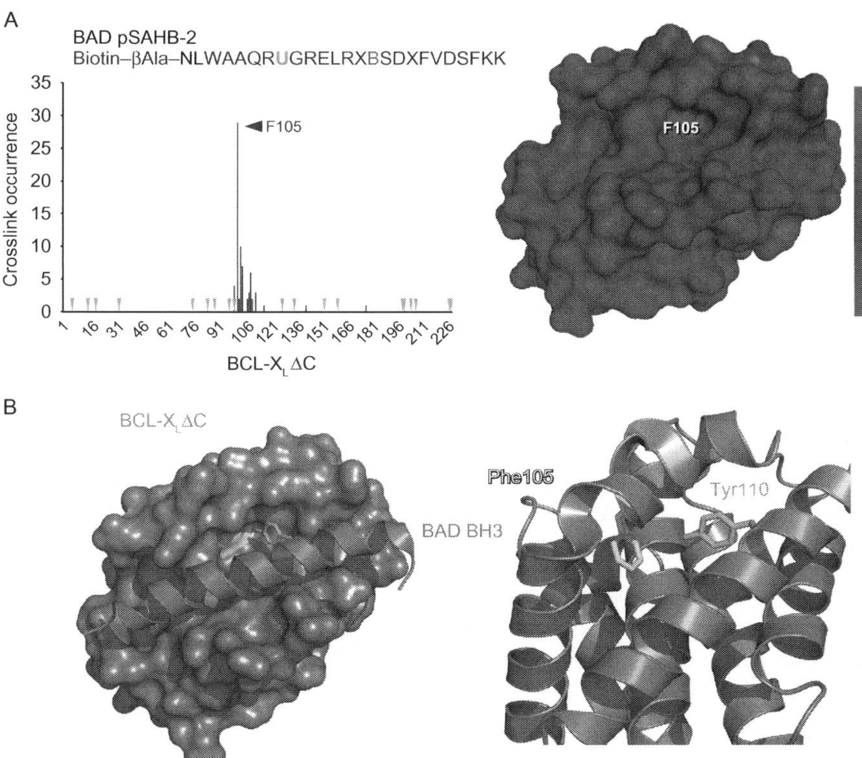

Figure 2.7 pSAHs localize helix/target interaction sites with high fidelity, as exemplified by the capacity of BAD pSAHB-2 to accurately map the BAD BH3 interaction site on BCL-X_LΔC (Braun et al., 2010). (A) The plot (left) depicts the frequency of crosslinked sites identified across the BCL-X_LΔC polypeptide sequence. Mapping of the crosslinked residues onto the BCL-X_LΔC structure (right) revealed their striking colocalization within a circumscribed region of the canonical BH3-binding pocket, with the frequency of occurrence reflected by the color scale. Green arrowheads, trypsin digestion sites. (B) The most abundant crosslink, located between BAD pSAHB-2 Bpa110 and BCL-X_LΔC F105, precisely matches the structurally defined interaction between BAD BH3 Y110 (cyan) and BCL-X_LΔC F105 (yellow) at the BH3-binding pocket of BCL-X_LΔC (PDB ID 2BZW). U, Bpa; X, stapling amino acid; B, norleucine. (See the color plate.)

calculate model structures of helix/target complexes for biochemical, structural, and functional validation. This approach is also useful for validating new pSAH reagents on established targets before exploring their putative binding interactions with novel targets. For example, to validate the binding specificity and utility of a BID pSAHB panel, we conducted crosslinking analyses with BCL-X_LΔC, employing pSAHBs with sequentially placed benzophenone moieties. Data analysis for each BID pSAHB construct revealed high regiospecificity of modification sites, enabling the calculation of a model structure (Fig. 2.8).

Computational docking analyses are performed using crystallography and NMR system solve (CNS) (Brunger et al., 1998) within HADDOCK 2.0 (Dominguez, Boelens, & Bonvin, 2003). Starting structures are based on reported structures of the helical ligand and protein target of interest, or their

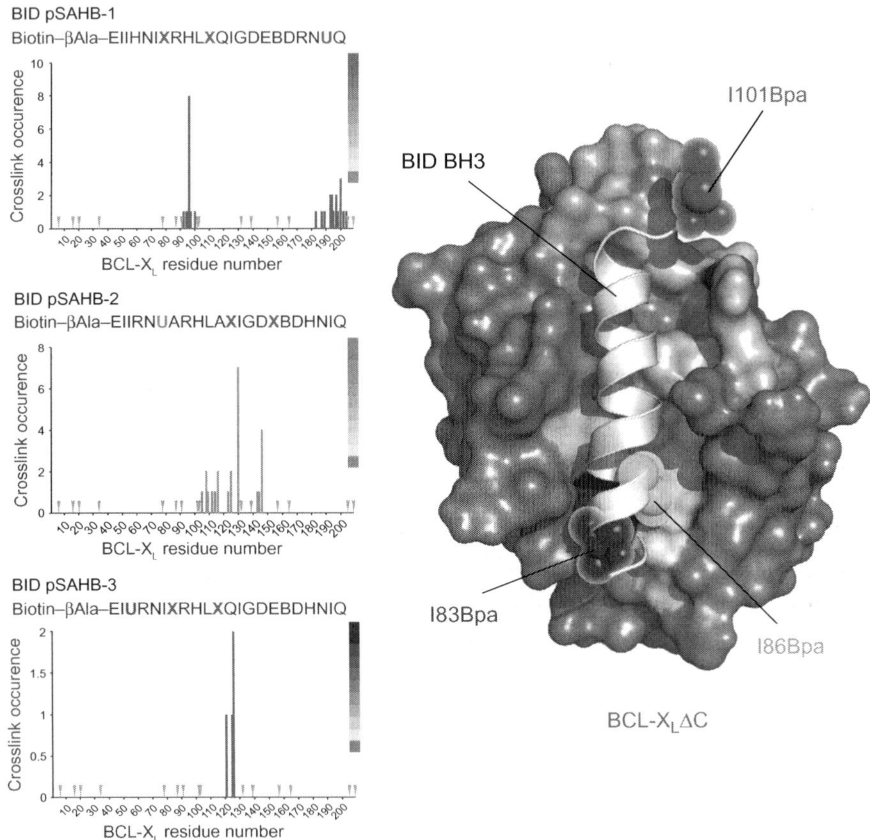

Figure 2.8 The high regiospecificity of pSAH crosslinking enables the calculation of model structures of helix/target complexes, as exemplified by a panel of BID pSAHBs crosslinked to BCL-X_LΔC (Leshchiner et al., 2013). The plots (left) and the corresponding mapping (right) reflect the frequency of crosslinked sites across the BCL-X_LΔC polypeptide sequence for three distinct BID pSAHBs. The covalently modified residues for each pSAHB construct maps to a highly circumscribed subregion of the canonical BH3-binding pocket of BCL-X_LΔC, facilitating the calculation of a model structure of the complex (right) using CNS within HADDOCK 2.0. Orange arrowheads, tryptic digestion sites; U, Bpa; X, stapling amino acid; B, norleucine. (See the color plate.)

close homologues. Models are generated by docking calculations that employ interaction restraints based on the top scoring crosslinks. For example, force constants for the restraints are set to 10 kcal mol^{-1} $Å^{-2}$ for the rigid-body docking stage and then set to a final value of 50 kcal mol^{-1} $Å^{-2}$ during the semiflexible simulated annealing stage. The restraints are set to be satisfied as soon as any one of the distances entering the sum average ($[\sum 1/r^6]^{-1/6}$), over all individual pairwise combinations, is within the defined cut-off distance of 2 Å. The docking calculations are performed with standard HADDOCK protocols (Arnesano, Banci, Bertini, & Bonvin, 2004). Initially, 2000 orientations/structures of the complex are generated by rigid-body docking energy minimization of the individual structures. At this stage, structures are docked based on the restraints, van der Waals, and electrostatic energy terms. The 100 better-scored structures are semiflexibly refined in torsion angle space and then refined in explicit water (Linge, Habeck, Rieping, & Nilges, 2003). Final structures are then calculated following water refinement and energy minimization. During the simulated annealing and water refinement, all amino acids (side chains and backbone) of the peptide helix are allowed to move to optimize the interface packing. In addition, at this stage, specific segments of the target protein structure that comprise the binding pocket are fully flexible (side chains and backbone). The 20 best-scored structures from the water refinement step are clustered into one group using pairwise backbone root mean square deviation (r.m.s.d.) of 3 Å. The final structure ensemble is evaluated for chirality and stereochemistry with the programs WHATCHECK (Hooft, Vriend, Sander, & Abola, 1996) and PROCHECK_NMR (Laskowski, Rullmannn, MacArthur, Kaptein, & Thornton, 1996), and then displayed and analyzed using PYMOL (DeLano, 2002).

7. MAPPING NOVEL BINDING SITES

In recent years, we have deployed the pSAH technology to identify novel BCL-2 family interaction sites, informing both canonical and noncanonical mechanisms of the cell death pathway. The complexes between a series of proapoptotic BH3 helices and a binding pocket at the C-terminal face of antiapoptotic proteins were the first BCL-2 family interactions to be structurally and functionally defined (Petros, Olejniczak, & Fesik, 2004). This work demonstrated how antiapoptotic proteins can intercept and neutralize proapoptotic BH3 signals and vice versa (Fig. 2.2). The question of how the essential executioner proteins, BAX and BAK, are triggered, however, remained an enigma. One theory was that proapoptotic BH3 helices could directly engage BAX and BAK to activate their pore-forming function (Wei et al., 2000; Fig. 2.2). To probe this hypothesis, we subsequently developed stapled peptides modeled after the BID and BIM BH3 helices (Walensky et al., 2006), identified a new binding site at the N-terminal face of full-length BAX by NMR spectroscopy (Gavathiotis et al., 2008), and correlated engagement of this site to structural and functional activation of BAX *in vitro* and in cells (Gavathiotis, Reyna, Davis, Bird, & Walensky, 2010; Gavathiotis et al., 2008).

Stimulated by these findings, we sought an approach that harnessed stapled peptides to rapidly and accurately map binding interfaces. Our goal in developing the pSAH technology was to bridge the gap between discovering protein interactions and defining biologically relevant interfaces by structural methods, which often takes years to accomplish. With pSAHs in hand, we turned our attention to BAK, a BAX homologue that had been refractory to protein expression and structural evaluation in full-length form, likely owing to its constitutive membrane disposition in the cell. After overcoming the challenge of producing pure, monomeric, and functionally active full-length BAK protein (Leshchiner et al., 2013), we applied BID pSAHBs and explicitly mapped the BID BH3-activating interaction to a binding pocket at the C-terminal face of the BAK protein, mirroring the location of BH3 interaction sites on antiapoptotic targets (Fig. 2.9A). The subsequent NMR analysis of a stapled BID BH3 peptide in complex with a BAK construct lacking its C-terminal helix (BAKΔC) provided a definitive structure that corroborated the BID pSAHB results (Moldoveanu et al., 2013). Interestingly, the interaction site analysis of full-length BAK protein did not detect BH3-engagement at the N-terminal face of BAK, yet the same BID pSAHBs peppered the N-terminal face of full-length BAX with covalent modification sites corresponding to the N-terminal "trigger site" that we previously identified by NMR analysis (Gavathiotis et al., 2008; Leshchiner et al., 2013). These data suggested that the N-terminal trigger site on BAX may reflect a unique and afferent direct activation mechanism required to translocate BAX from the cytosol to the mitochondria, a process not required for BAK given its constitutive localization at the mitochondrial outer membrane.

Further interrogation of this hypothesis using pSAHs revealed that both BID and PUMA BH3-based constructs predominantly crosslinked to the

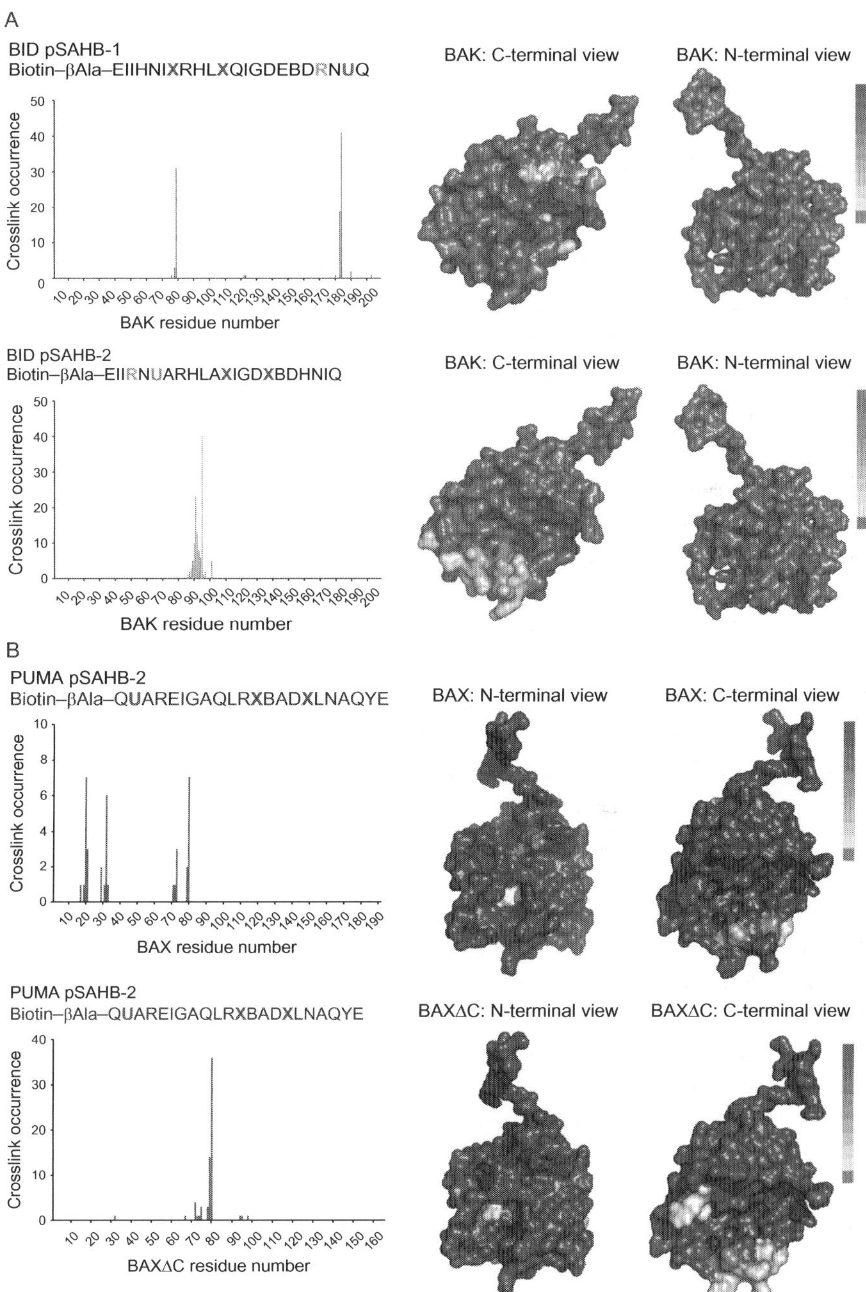

Figure 2.9 pSAHs identify novel sites of BCL-2 family protein interaction. BID, PUMA, and phospho-BAD pSAHBs revealed BH3 interaction sites on (A) full-length BAK

(Continued)

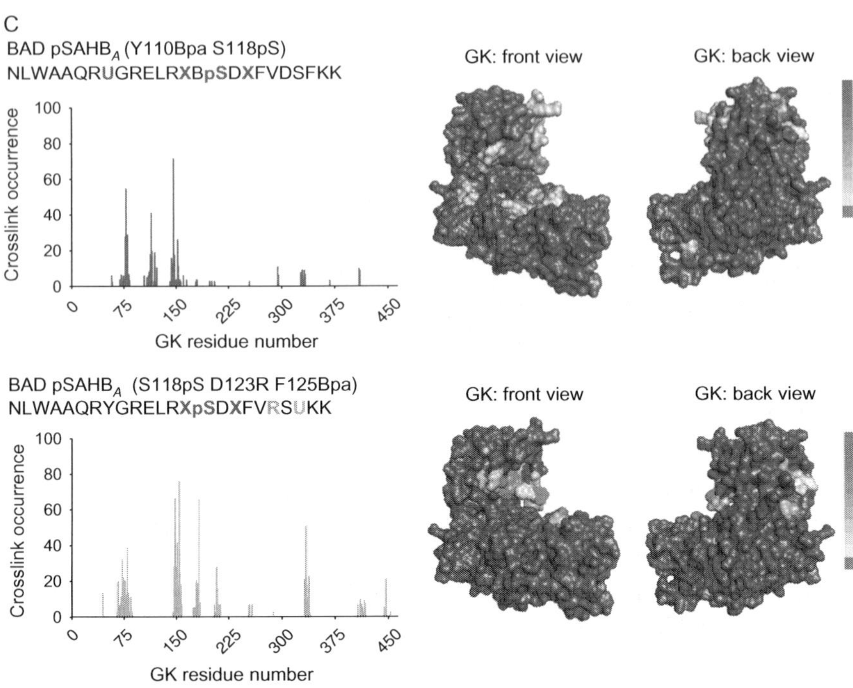

Figure 2.9—Cont'd (Leshchiner et al., 2013), (B) full-length versus C-terminally truncated BAX (i.e., BAX vs. BAXΔC) (Edwards et al., 2013), and (C) glucokinase (Szlyk et al., 2014), respectively. U, Bpa; X, stapling amino acid; B, norleucine; pS, phosphoserine; orange R, single Arg substitution to facilitate tryptic digestion of pSAHBs into shorter and more identifiable fragments by MS. (See the color plate.)

N-terminal BH3 interaction site on full-length BAX, with occasional covalent modification sites also detected at the C-terminal binding pocket (Edwards et al., 2013; Leshchiner et al., 2013; Fig. 2.9B). The corresponding NMR analyses consistently showed chemical shift changes at the N-terminal activation site upon BH3 ligand engagement (Edwards et al., 2013; Gavathiotis et al., 2008), with C-terminal helix changes becoming apparent at later time points and with increased ligand concentrations (Edwards et al., 2013; Gavathiotis et al., 2010). Thus, we hypothesized that allosteric exposure of the C-terminal pocket as a result of N-terminal site triggering reveals a secondary BH3-binding site, which may function to propel BAX activation at the mitochondrial membrane in a manner analogous to BAK. Indeed, we find that if the C-terminal helix of BAX is removed (i.e., BAXΔC), bypassing the N-terminal triggering step and providing ready exposure to the C-terminal binding pocket, the balance of PUMA pSAHB crosslinks shifts from the N-terminal to the C-terminal BH3-binding pocket (Edwards et al., 2013; Fig. 2.9B). The existence of such complexes between activating BH3 domain helices and the C-terminal pocket of BAXΔC has since been confirmed by X-ray crystallography (Czabotar et al., 2013). Thus, the development of pSAHs for interrogating the direct activation of BAX and BAK has yielded both timely and structurally validated mechanistic insights.

The pSAH approach has proven to be especially useful for exploring noncanonical interactions when definitive structures of the helix/target complexes are yet unknown or elusive. For example, the proapoptotic BH3-only protein BAD was discovered to have dual functions in regulating apoptosis and metabolism (Danial et al., 2003), with the molecular toggle between these biological activities regulated by the phosphorylation state of BAD's BH3 domain helix (Danial et al., 2008). Specifically, the BH3-phosphorylated form of BAD binds to and regulates the enzyme glucokinase, an essential mediator of the glucose-stimulated insulin secretion response and a *bona fide* drug target for diabetes. Small molecule glucokinase activators enhance the enzyme's glucose affinity via an allosteric mechanism but their potential benefits can be offset by hypoglycemia. Interestingly, a stapled peptide modeled after the phospho-BAD BH3 helix increases the enzyme catalytic rate without dramatically changing glucose affinity and stimulates cultured human islets to secrete more insulin (Szlyk et al., 2014). To explore just how this α-helical BAD BH3 phospho-mimic engages glucokinase, we generated a corresponding panel of phospho-BAD pSAHBs for crosslinking studies. MS analysis revealed that the predominant sites of covalent modification colocalized near the active site of glucokinase, a completely distinct region from the binding site for small molecule glucokinase activators (Szlyk et al., 2014; Fig. 2.9C). This work predicts that BAD BH3 phospho-mimics may serve as a novel class of GK activators by engaging this distinct binding site.

8. CONCLUSIONS AND FUTURE DIRECTIONS

We describe a strategy for rapidly localizing BH3 α-helix interaction sites on canonical and noncanonical targets of BCL-2 family proteins by combining peptide α-helix stabilization, photoaffinity labeling, proteomic analysis, and structure calculation. Indeed, the method can be broadly applied to interrogate the "α-helical interactome." All-hydrocarbon stapling reinforces the structure of α-helical peptide motifs, recapitulating their native biological activities. The incorporation of photoreactive moieties into stapled peptide α-helices adapts them for binding partner and interaction site identification, key steps in advancing mechanistic studies and drug-targeting efforts. We believe that future applications and enhancements of photoreactive stapled peptides will extend the potential for discovery of new and unforeseen protein interactions and how they impact health and disease. Importantly, stapled peptides developed to capture protein targets can be applied to validate them in a cellular context and also serve as templates for next-generation therapeutics.

ACKNOWLEDGMENTS

We are grateful to Walensky lab members, past and present, for their contributions to the development and application of pSAH technology, including Craig Braun, Amanda Edwards, Elizaveta Leshchiner, Michelle Stewart, Lauren Barclay, Evripidis Gavathiotis, Greg Bird, and Susan Lee. We are also indebted to our longstanding collaborators on this project, including Steven Gygi, Julian Mintseris, Nika Danial, and Ben Szlyk. We thank Eric Smith for figure preparation. This work was supported by NIH Grants 5R01GM090299 and 5R01CA050239, a Stand Up to Cancer Innovative Research Grant, and a William Lawrence and Blanche Hughes Foundation Award to L. D. W.

REFERENCES

Arnesano, F., Banci, L., Bertini, I., & Bonvin, A. M. (2004). A docking approach to the study of copper trafficking proteins; interaction between metallochaperones and soluble domains of copper ATPases. *Structure, 12*, 669–676.

Bernal, F., Wade, M., Godes, M., Davis, T. N., Whitehead, D. G., Kung, A. L., et al. (2010). A stapled p53 helix overcomes HDMX-mediated suppression of p53. *Cancer Cell, 18*, 411–422.

Bird, G. H., Bernal, F., Pitter, K., & Walensky, L. D. (2008). Synthesis and biophysical characterization of stabilized alpha-helices of BCL-2 domains. *Methods in Enzymology, 446*, 369–386.

Bird, G. H., Crannell, W. C., & Walensky, L. D. (2011). Chemical synthesis of hydrocarbon-stapled peptides for protein interaction research and therapeutic targeting. *Current Protocols in Chemical Biology, 3*, 99–117.

Blackwell, H. E., Sadowsky, J. D., Howard, R. J., Sampson, J. N., Chao, J. A., Steinmetz, W. E., et al. (2001). Ring-closing metathesis of olefinic peptides: Design, synthesis, and structural characterization of macrocyclic helical peptides. *Journal of Organic Chemistry, 66*, 5291–5302.

Blum, J. H., Stevens, T. L., & DeFranco, A. L. (1993). Role of the mu immunoglobulin heavy chain transmembrane and cytoplasmic domains in B cell antigen receptor expression and signal transduction. *Journal of Biological Chemistry, 268*, 27236–27245.

Braun, C. R., Mintseris, J., Gavathiotis, E., Bird, G. H., Gygi, S. P., & Walensky, L. D. (2010). Photoreactive stapled BH3 peptides to dissect the BCL-2 family interactome. *Chemistry & Biology, 17*, 1325–1333.

Brunger, A. T., Adams, P. D., Clore, G. M., DeLano, W. L., Gros, P., Grosse-Kunstleve, R. W., et al. (1998). Crystallography & NMR system: A new software suite for macromolecular structure determination. *Acta Crystallographica. Section D, Biological Crystallography, 54*, 905–921.

Czabotar, P. E., Westphal, D., Dewson, G., Ma, S., Hockings, C., Fairlie, W. D., et al. (2013). Bax crystal structures reveal how BH3 domains activate Bax and nucleate its oligomerization to induce apoptosis. *Cell, 152*, 519–531.

Danial, N. N., Gramm, C. F., Scorrano, L., Zhang, C. Y., Krauss, S., Ranger, A. M., et al. (2003). BAD and glucokinase reside in a mitochondrial complex that integrates glycolysis and apoptosis. *Nature, 424*, 952–956.

Danial, N. N., Walensky, L. D., Zhang, C. Y., Choi, C. S., Fisher, J. K., Molina, A. J., et al. (2008). Dual role of proapoptotic BAD in insulin secretion and beta cell survival. *Nature Medicine, 14*, 144–153.

DeLano, W. L. (2002). *The PyMOL molecular graphics system*. San Carlos: DeLano Scientific. http://www.pymol.org.

Dominguez, C., Boelens, R., & Bonvin, A. M. (2003). HADDOCK: A protein–protein docking approach based on biochemical or biophysical information. *Journal of the American Chemical Society, 125*, 1731–1737.

Dorman, G., & Prestwich, G. D. (1994). Benzophenone photophores in biochemistry. *Biochemistry, 33*, 5661–5673.

Edwards, A. L., Gavathiotis, E., LaBelle, J. L., Braun, C. R., Opoku-Nsiah, K. A., Bird, G. H., et al. (2013). Multimodal interaction with BCL-2 family proteins underlies the proapoptotic activity of PUMA BH3. *Chemistry & Biology, 20*, 888–902.

Eng, J. K., McCormack, A. L., & Yates, J. R. (1994). An approach to correlate tandem mass spectral data of peptides with amino acid sequences in a protein database. *Journal of the American Society for Mass Spectrometry, 5*, 976–989.

Gavathiotis, E., Reyna, D. E., Davis, M. L., Bird, G. H., & Walensky, L. D. (2010). BH3-triggered structural reorganization drives the activation of proapoptotic BAX. *Molecular Cell, 40*, 481–492.

Gavathiotis, E., Suzuki, M., Davis, M. L., Pitter, K., Bird, G. H., Katz, S. G., et al. (2008). BAX activation is initiated at a novel interaction site. *Nature, 455*, 1076–1081.

Hooft, R. W., Vriend, G., Sander, C., & Abola, E. E. (1996). Errors in protein structures. *Nature, 381*, 272.

Jakobsen, L., Schroder, J. M., Larsen, K. M., Lundberg, E., & Andersen, J. S. (2013). Centrosome isolation and analysis by mass spectrometry-based proteomics. *Methods in Enzymology, 525*, 371–393.

Kim, W., Bird, G. H., Neff, T., Guo, G., Kerenyi, M. A., Walensky, L. D., et al. (2013). Targeted disruption of the EZH2-EED complex inhibits EZH2-dependent cancer. *Nature Chemical Biology, 9*, 643–650.

Kussie, P. H., Gorina, S., Marechal, V., Elenbaas, B., Moreau, J., Levine, A. J., et al. (1996). Structure of the MDM2 oncoprotein bound to the p53 tumor suppressor transactivation domain. *Science, 274*, 948–953.

LaBelle, J. L., Katz, S. G., Bird, G. H., Gavathiotis, E., Stewart, M. L., Lawrence, C., et al. (2012). A stapled BIM peptide overcomes apoptotic resistance in hematologic cancers. *Journal of Clinical Investigation, 122*, 2018–2031.

Laskowski, R. A., Rullmannn, J. A., MacArthur, M. W., Kaptein, R., & Thornton, J. M. (1996). AQUA and PROCHECK-NMR: Programs for checking the quality of protein structures solved by NMR. *Journal of Biomolecular NMR, 8*, 477–486.

Leshchiner, E. S., Braun, C. R., Bird, G. H., & Walensky, L. D. (2013). Direct activation of full-length proapoptotic BAK. *Proceedings of the National Academy of Sciences of the United States of America, 110*, E986–E995.

Linge, J. P., Habeck, M., Rieping, W., & Nilges, M. (2003). ARIA: Automated NOE assignment and NMR structure calculation. *Bioinformatics, 19*, 315–316.

Moldoveanu, T., Grace, C. R., Llambi, F., Nourse, A., Fitzgerald, P., Gehring, K., et al. (2013). BID-induced structural changes in BAK promote apoptosis. *Nature Structural & Molecular Biology, 20*, 589–597.

Oltersdorf, T., Elmore, S. W., Shoemaker, A. R., Armstrong, R. C., Augeri, D. J., Belli, B. A., et al. (2005). An inhibitor of Bcl-2 family proteins induces regression of solid tumours. *Nature*, *435*, 677–681.

Petros, A. M., Olejniczak, E. T., & Fesik, S. W. (2004). Structural biology of the Bcl-2 family of proteins. *Biochimica et Biophysica Acta*, *1644*, 83–94.

Saghatelian, A., Jessani, N., Joseph, A., Humphrey, M., & Cravatt, B. F. (2004). Activity-based probes for the proteomic profiling of metalloproteases. *Proceedings of the National Academy of Sciences of the United States of America*, *101*, 10000–10005.

Sattler, M., Liang, H., Nettesheim, D., Meadows, R. P., Harlan, J. E., Eberstadt, M., et al. (1997). Structure of Bcl-xL-Bak peptide complex: Recognition between regulators of apoptosis. *Science*, *275*, 983–986.

Schafmeister, C., Po, J., & Verdine, G. (2000). An all-hydrocarbon cross-linking system for enhancing the helicity and metabolic stability of peptides. *Journal of the American Chemical Society*, *122*, 5891–5892.

Stewart, M. L., Fire, E., Keating, A. E., & Walensky, L. D. (2010). The MCL-1 BH3 helix is an exclusive MCL-1 inhibitor and apoptosis sensitizer. *Nature Chemical Biology*, *6*, 595–601.

Szlyk, B., Braun, C. R., Ljubicic, S., Patton, E., Bird, G. H., Osundiji, M. A., et al. (2014). A phospho-BAD BH3 helix activates glucokinase by a mechanism distinct from that of allosteric activators. *Nature Structural & Molecular Biology*, *21*, 36–42.

Takada, K., Zhu, D., Bird, G. H., Sukhdeo, K., Zhao, J. J., Mani, M., et al. (2012). Targeted disruption of the BCL9/beta-catenin complex inhibits oncogenic Wnt signaling. *Science Translational Medicine*, *4*, 148ra117.

Vodovozova, E. L. (2007). Photoaffinity labeling and its application in structural biology. *Biochemistry (Mosc)*, *72*, 1–20.

Walensky, L. D., Kung, A. L., Escher, I., Malia, T. J., Barbuto, S., Wright, R. D., et al. (2004). Activation of apoptosis *in vivo* by a hydrocarbon-stapled BH3 helix. *Science*, *305*, 1466–1470.

Walensky, L. D., Pitter, K., Morash, J., Oh, K. J., Barbuto, S., Fisher, J., et al. (2006). A stapled BID BH3 helix directly binds and activates BAX. *Molecular Cell*, *24*, 199–210.

Wei, M. C., Lindsten, T., Mootha, V. K., Weiler, S., Gross, A., Ashiya, M., et al. (2000). tBID, a membrane-targeted death ligand, oligomerizes BAK to release cytochrome c. *Genes & Development*, *14*, 2060–2071.

Weissenhorn, W., Dessen, A., Harrison, S. C., Skehel, J. J., & Wiley, D. C. (1997). Atomic structure of the ectodomain from HIV-1 gp41. *Nature*, *387*, 426–430.

Yates, J. R., 3rd., Eng, J. K., McCormack, A. L., & Schieltz, D. (1995). Method to correlate tandem mass spectra of modified peptides to amino acid sequences in the protein database. *Analytical Chemistry*, *67*, 1426–1436.

CHAPTER THREE

The Structural Biology of BH3-Only Proteins

Marc Kvansakul[*,1], **Mark G. Hinds**[†,‡,1]

[*]La Trobe Institute for Medical Science, La Trobe University, Bundoora, Victoria, Australia
[†]School of Chemistry, The University of Melbourne, Parkville, Victoria, Australia
[‡]Bio21 Molecular Science and Biotechnology Institute, The University of Melbourne, Parkville, Victoria, Australia
[1]Corresponding authors: e-mail addresses: m.kvansakul@latrobe.edu.au; mhinds@unimelb.edu.au

Contents

Abstract

B-cell lymphoma-2 (Bcl-2) homology-3 (BH3)-only proteins are considered members of the Bcl-2 family, though they bear little sequence or structural identity with the multi-BH motif prosurvival or proapoptotic Bcl-2 proteins like Bcl-2 or Bax. They are better considered a separate phylogenetic entity. In combination, results from biophysical, biochemical, cell biological, and animal studies in conjunction with structural investigations have elucidated the function and mechanism of action of these proteins. Either by antagonizing prosurvival Bcl-2 proteins or directly activating proapoptotic Bcl-2 proteins (Bax or Bak) they initiate apoptosis. BH3-only proteins are intrinsically disordered and fold and bind into a groove provided by their cognate receptor Bcl-2 family proteins. Our detailed molecular understanding of BH3-only protein action has aided the development of novel chemical entities that initiate cell death by mimicking the properties of BH3-only proteins.

Methods in Enzymology, Volume 544
ISSN 0076-6879
http://dx.doi.org/10.1016/B978-0-12-417158-9.00003-0

ABBREVIATIONS

Bcl-2 B-cell lymphoma-2
BH Bcl-2 homology
IDP intrinsically disordered protein
MOM mitochondrial outer membrane
MOMP MOM permeabilization

1. INTRODUCTION

BH3-only proteins integrate a diverse array of stimuli to initiate apoptosis, the programmed suicide of cells no longer required, infected, or dangerous to an organism (Strasser, Cory, & Adams, 2011). Eight BH3-only proteins (Bim, Bad, Bmf, Bid, Noxa, Bik, Puma, and Hrk) (Fig. 3.1) have been well characterized in mammals and two key roles of these members of

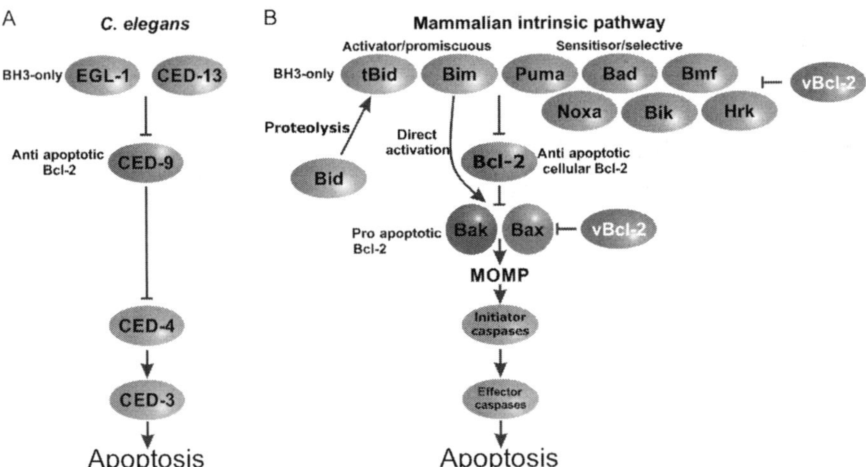

Figure 3.1 BH3-only protein initiation of apoptosis in *C. elegans* and mammals. Schemes for apoptosis in (A) *C. elegans* and (B) mammals. *C. elegans* expresses two BH3-only proteins that bind the prosurvival Bcl-2 protein CED-9 to release the caspase CED-4 that in turn activates the executioner caspase CED-3. In contrast, in mammals there are eight BH3-only proteins that either neutralize the six prosurvival Bcl-2 proteins (the sensitisors) and/or activate Bax and Bak (the activators) in "intrinsic" apoptosis. Activation of Bax/Bak releases apoptogenic factors from mitochondria via oligomerization and mitochondrial outer membrane permeabilization (MOMP) and release of caspase-activating factors. Viral Bcl-2 (vBcl-2) proteins either bind and inhibit BH3-only proteins or prevent Bax/Bak activation to inhibit apoptosis.

the B-cell lymphoma-2 (Bcl-2) family have been recognized. Selected BH3-only (Bim, Puma, and Bid) proteins may directly activate (Gavathiotis et al., 2008; Leshchiner, Braun, Bird, & Walensky, 2013) the two proapoptotic Bcl-2 proteins, Bax and Bak, to permeabilize the mitochondrial outer membrane (MOM) (Shamas-Din, Kale, Leber, & Andrews, 2013). Importantly, BH3-only proteins also inhibit the six members of the mammalian prosurvival Bcl-2 family (Bcl-2, Bcl-x_L, Bcl-w, Mcl-1, A1, and Bcl-B) that in turn regulate Bax and Bak. Though BH3-only proteins interact with the MOM (Wilfling et al., 2012), they do not directly cause its permeabilization, rather, it is their interaction with the multimotif Bcl-2 proteins to either antagonize prosurvival Bcl-2 interaction with Bax or Bak or direct interaction and activation of these two proteins that is significant. Oligomerization of Bax and Bak at the MOM and MOM permeabilization (MOMP) releases apoptogenic factors from the mitochondrial intermembrane space and is the key event in mammalian intrinsic apoptosis (Shamas-Din, Kale, et al., 2013). Release of cytochrome *c* and other factors from mitochondria activate the destructive proteolytic cascade of cysteine aspartyl proteases that is the hallmark of apoptosis (Fuentes-Prior & Salvesen, 2004). Finally, cells are packaged for recycling through phagocytosis. BH3-only or BH3-like proteins may also regulate autophagy (Kang, Zeh, Lotze, & Tang, 2011), though the crosstalk between autophagy and apoptosis is not yet well understood (Gump & Thorburn, 2011). Mechanisms of BH3-only initiated cell death are both highly regulated and conserved.

Not all proteins form well-folded entities and the intrinsically disordered proteins (IDP) that are defined by their lack of structure, including secondary structure, are an important class of macromolecular regulator (Uversky & Dunker, 2010). IDPs form the basis of many signaling networks, including those of apoptosis (Peng, Xue, Kurgan, & Uversky, 2013; Rautureau, Day, & Hinds, 2010), and are able to interact with multiple binding partners (Uversky & Dunker, 2010). The BH3-only proteins, with the exception of Bid (Chou, Li, Salvesen, Yuan, & Wagner, 1999; McDonnell, Fushman, Milliman, Korsmeyer, & Cowburn, 1999), are IDPs (Barrera-Vilarmau, Obregon, & de Alba, 2011; Hinds et al., 2007; Peng et al., 2013; Rautureau et al., 2010) and even in the case of Bid, its active state is nonordered as it requires both a proteolytic activation and an unfolding step prior to being a functional BH3-only protein (Shamas-Din, Bindner, et al., 2013; Yao, Bobkov, Plesniak, & Marassi, 2009). The results of multiple structural studies (Hinds & Day, 2005; Petros, Olejniczak, &

Fesik, 2004) have shown that BH3-only proteins fold and bind their multi-BH motif prosurvival Bcl-2 targets as an α-helix buried in a hydrophobic groove formed from the conserved regions (known as the BH1, BH2, and BH3 motifs) of these proteins (Rautureau et al., 2012; Sattler et al., 1997). In contrast, the multimotif Bcl-2 proteins are well-folded entities, though they bear intrinsically disordered regions (Rautureau et al., 2010). Combined, these sequence and structural differences between the multimotif Bcl-2 and BH3-only proteins lead to the conclusion they are phylogenetically separate entities (Hardwick, Chen, & Jonas, 2012).

BH3-motifs, the sequence signature, and interaction motif of BH3-only proteins are not unique to them, are also found on the multimotif Bcl-2 proteins and potentially other unrelated proteins. Two distinct roles are performed by BH3-motif: one as a ligand, the other as part of the receptor site groove on a multi-BH motif protein (Kvansakul & Hinds, 2013). In each case, different interactions are observed. The BH3-in-groove interaction is a key mechanistic aspect of prosurvival BH3-only protein molecular recognition that is now well understood biochemically, structurally, and cell biologically. The interaction of Bax and Bak with BH3-only proteins is less well understood structurally or thermodynamically, and involves relatively weak binding events compared to the prosurvival:BH3-only interactions. Exactly how Bax and Bak are activated is still being elucidated and two models have recently been hypothesized (Shamas-Din, Kale, et al., 2013) corresponding to two BH3-binding sites on Bak and Bax: the canonical groove (Leshchiner et al., 2013), as Bax and Bak have similar structures to prosurvival proteins, and a site on the opposite face termed the "α1-site" (Gavathiotis et al., 2008). The role of BH3-only proteins as the initiators of intrinsic apoptosis (Huang & Strasser, 2000) has led to the development of small molecules that mimic the behavior of BH3-only proteins to antagonize prosurvival Bcl-2 proteins for initiation of cell death in diseases where expression of these proteins prevents appropriate cell death (Lessene, Czabotar, & Colman, 2008).

2. WHAT CONSTITUTES A BH3-ONLY PROTEIN?

The Bcl-2 family, including the BH3-only proteins, is characterized by the presence of short (<20 residue) conserved sequence regions (Sato, Irie, Krajewski, & Reed, 1994), the BH motifs (frequently misnamed domains). The first two motifs identified, BH1 and BH2, were recognized in Bcl-2 (Yin, Oltvai, & Korsmeyer, 1994) and subsequently the BH3 motif designated in Bak (Chittenden et al., 1995) and recognized in Bik (Boyd et al., 1995), with the term "BH3 domain-only" protein coined for Bid (Wang, Yin, Chao, Milliman, & Korsmeyer, 1996). The central feature of BH3-only proteins is that in contrast with the other members of the Bcl-2 family they bear only a single BH motif, the BH3 that defines all proapoptotic Bcl-2 proteins (Huang & Strasser, 2000), while it may or may not be present in prosurvival proteins. Mouse Bad was the first BH3-only protein to be reported (however, its BH3 motif was not recognized at first), discovered in an expression screen using Bcl-2 and Bcl-x_L as probes (Yang et al., 1995), followed soon after by Bik (Boyd et al., 1995), Bid (Wang et al., 1996), Hrk (Imaizumi et al., 1997; Inohara, Ding, Chen, & Nunez, 1997), Bim (Hsu, Lin, & Hsueh, 1998; O'Connor et al., 1998), Noxa (Oda et al., 2000), Bmf (Puthalakath et al., 2001), and Puma (Han et al., 2001; Nakano & Vousden, 2001; Yu, Zhang, Hwang, Kinzler, & Vogelstein, 2001) which were established by similar means.

Sequence and structural analyses identified an interaction interface for BH3-only proteins with prosurvival Bcl-2 (Huang & Strasser, 2000). The identifying feature of BH3 sequences is the presence of an invariant aspartate four residues C-terminal to a highly conserved, though not invariant, leucine (Fig. 3.2). The Asp forms a key ionic interaction with a conserved arginine (Fig. 3.3), located in the BH1 motif of the prosurvival protein receptor (Hinds & Day, 2005). The geometry of this interaction favors Asp rather than Glu and its mutation to Glu substantially reduces the affinity (DeBartolo, Dutta, Reich, & Keating, 2012; Dutta, Chen, & Keating, 2013; Dutta et al., 2010) and its replacement with Ala abrogates binding (Boersma, Sadowsky, Tomita, & Gellman, 2008). The key residues for high-affinity interactions are contained within a 13-residue binding motif (Day et al., 2008), a length corresponding to approximately two heptads of an amphipathic helix. Four hydrophobic residues (including the Leu) generate its hydrophobic face (Fig. 3.2). Other sequence features of the BH3 include the presence of small residues (G, A, S, or C) immediately preceding the Asp and after the first hydrophobic residue of the core motif (Fig. 3.2) (Day et al., 2008).

Searches for BH3 motifs in numerous genomes identified many Bcl-2 proteins (Blaineau & Aouacheria, 2009; Lanave, Santamaria, & Saccone, 2004). Two BH3-only proteins have been recognized in *Caenorhabditis elegans*: Egl-1 (Conradt & Horvitz, 1998) and CED-13 (Schumacher et al., 2005). BH3-only proteins identified in hydra (Lasi et al., 2010) and Zebrafish (Eimon & Ashkenazi, 2010) have genes that mirror mammalian

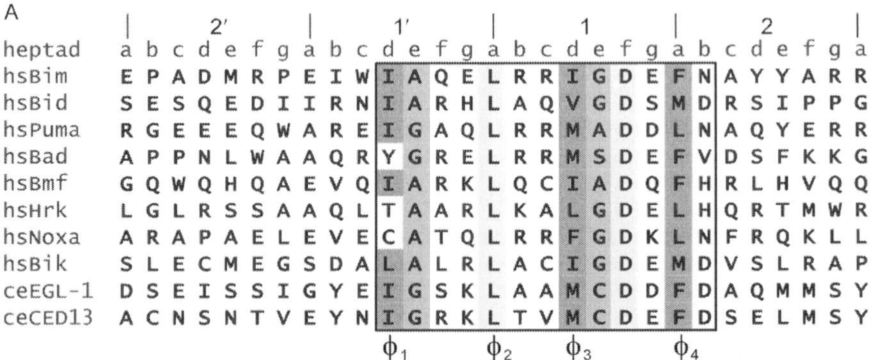

B

hsBnip1 L T C K I A I D N L E K A E L L Q G G D L L R Q R K T T K
hsBnip3 E E D D I E R R K E V E S I L K K N S D W I W D W S S R P
hsBnip3L S E E E V V E G E K E V E A L K K S A D W V S D W S S R P
hsBcn E A S D G G T M E N L S R R L K V T G D L F D I M S G Q T

Figure 3.2 Sequence alignment of BH3 regions of BH3-only proteins. (A) Core binding region of human and *C. elegans* BH3 motifs. The significant binding interactions occur within a 13-residue sequence (boxed) (Day et al., 2008). Conserved residues are indicated in shaded blocks with the two key BH3-motif defining residues Leu (ϕ_2) and Asp less intensely shaded. There is little sequence conservation outside the core residues that spans about two heptads of helix. Heptads are numbered above the protein sequences and the four conserved hydrophobic residues are indicated ϕ_1–ϕ_4. (B) Putative BH3-motifs of BNIP family proteins and Beclin-1 (Bcn). Hs, *Homo sapiens*; ce, *C. elegans*.

BH3-only proteins (Bid, Bim, Bad, Bik, Bmf, Noxa, and Puma) but an equivalent of Hrk is absent (Eimon & Ashkenazi, 2010). In contrast to mammalian, fish and worm apoptosis, insects are not reliant on BH3-only proteins for apoptosis induction though they have multimotif Bcl-2 proteins in their genome (Igaki & Miura, 2004). It is a moot point whether the BH3-only proteins should be considered part of the Bcl-2 family, as from a sequence and structure stand point they contain little similarity (Aouacheria, Rech de Laval, Combet, & Hardwick, 2013; Hardwick et al., 2012; Kvansakul & Hinds, 2013). BH3-only proteins have likely arisen through a combination of convergent and divergent evolution (Aouacheria et al., 2013). The relatively small number of identifying features and consequent lack of uniqueness have made it difficult to be certain that the protein is a "BH3-only" protein from sequence analysis alone. A discussion of what constitutes a BH3-only protein and how to identify them has been given by Aouacheria et al. (2013). Bcl-2 protein sequences have been accumulated into databases: BCL2DB (Blaineau & Aouacheria, 2009) (http://bcl2db.ibcp.fr/BCL2DB/); Deathbase (Doctor, Reed, Godzik, & Bourne, 2003)

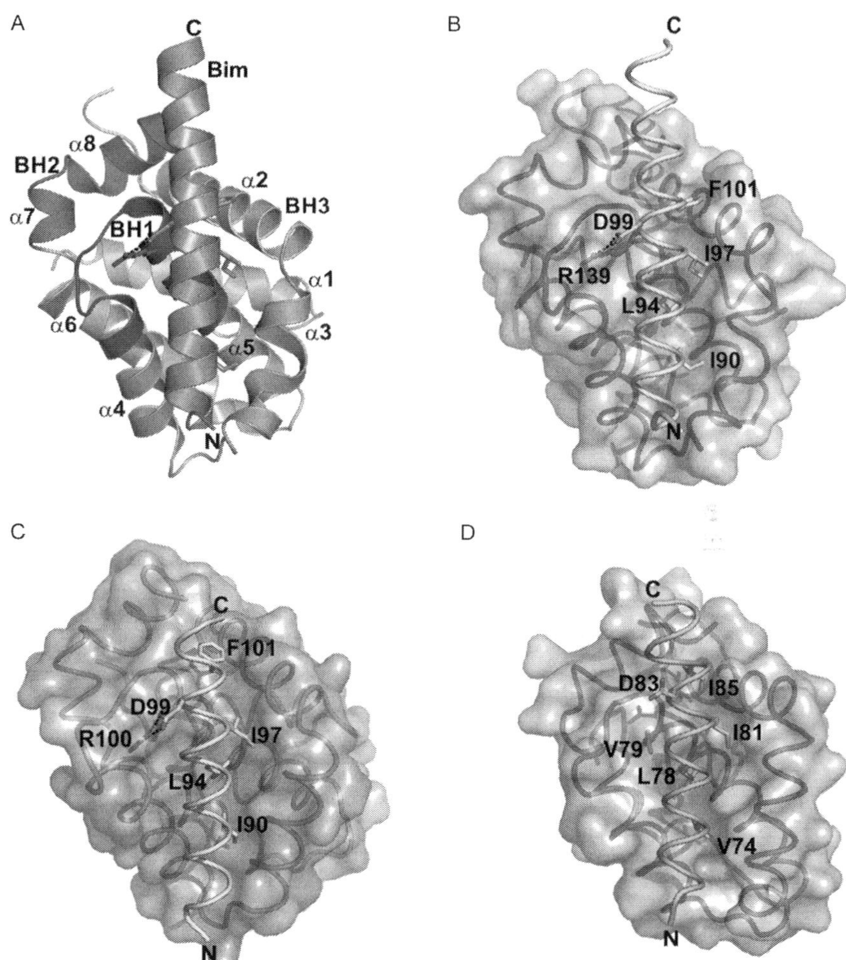

Figure 3.3 Structures of BH3-motifs bound to mammalian and viral prosurvival Bcl-2 proteins. (A) Ribbon representation of mouse Bcl-x_L:Bim BH3 complex (pdb: 1PQ1). Bim-BH3 (orange) binds as a helix in a groove provided by the BH1 (violet), BH2 (red); and BH3 (green) motifs of Bcl-x_L that corresponds to helices α2–α5 and α8. (B) Surface representation of Bcl-x_L:Bim BH3 in (A) with Bim depicted as a tube (green) and the highly conserved residues of Bim indicated. The conserved Asp–Arg interaction between Bim and Arg R139 in the BH1 motif of Bcl-x_L is also shown. (C) Interactions of BHRF1:Bim BH3 (pdb:2WH6) complex. (D) Interactions of M11L:Bak BH3 complex (pdb:1JBY). Molecular surfaces are shaded gray and the surface from residues within 4 Å from the key BH3-only residues in orange. (See the color plate.)

(http://deathbase.org); and consensus patterns based on sequence alignments derived on prosite (http://prosite.expasy.org).

Structural analysis of BH3-peptides bound to prosurvival proteins defined the contact residues of the BH3 motif (Fig. 3.2) (Day et al., 2008), but as an isolated constraint the exiguous sequence BH3 motif is of limited use in identifying BH3-only proteins (Aouacheria et al., 2013). The apparent simplicity of the BH3 motif has led to the proposal that it exists in many proteins (Beverly, 2012). For example, the commonly assumed BH3-only proteins of the BNIP family (BNIP1, BNIP3, and BNIP3L; Fig. 3.2) have a potential role in apoptosis, but there is little evidence for direct binding of prosurvival proteins (Certo et al., 2006). Notably, the putative BH3-motif of BNIPs lacks the positionally conserved hydrophobic residues of the known BH3-only proteins (Fig. 3.2) and is phylogenetically distinct from the BH3-only proteins (Aouacheria, Brunet, & Gouy, 2005). Likewise, the putative BH3-only protein Beclin-1 binds Bcl-x_L weakly (Oberstein, Jeffrey, & Shi, 2007), but does not fulfill the sequence motif criteria derived from the known BH3-only proteins (Fig. 3.2). We conclude that in order to be considered a BH3-only protein, a potential BH3-motif should bear the conserved residues, bind with reasonable affinity to prosurvival Bcl-2 proteins, and have some biological data supporting a direct role in Bcl-2 regulated apoptosis. While elements of the binding motif require their conservation, they do not need to be positionally conserved as the BH3-only proteins are disordered (discussed below). Using these criteria Beclin1 is not a "BH3-only" protein, as it does not fit the sequence motif well and in addition has an affinity, $K_d \sim 2\ \mu M$ for Bcl-x_L. There is little evidence of interaction of BNIP3L with prosurvival proteins (Certo et al., 2006) and other BNIP family members should be revaluated as BH3-only proteins in the absence of direct interaction with the multimotif Bcl-2 proteins. Other apparent BH3-bearing proteins have recently been disproved, such as Bcl-G (Giam, Okamoto, Mintern, Strasser, & Bouillet, 2012).

3. SPECIFICITY AND INTERACTIONS WITH THE BCL-2 FAMILY

The most well-understood aspect of BH3-only protein initiated apoptosis is their interaction with prosurvival Bcl-2 proteins (Sattler et al., 1997). Using a combination of biophysical techniques (SPR, ITC, fluorescence; Certo et al., 2006; Chen et al., 2005; Kuwana et al., 2005) and screening methodologies such as peptide libraries (DeBartolo et al., 2012; Dutta et al., 2013), bacterial surface display (Zhang & Link, 2011), yeast-two-hybrid screens, and phage display (Day et al., 2008) BH3-only protein specificity and affinities for their Bcl-2 targets were determined. Combined, these studies revealed a hierarchy of specificity for prosurvival proteins (Table 3.1). Generally less well understood are interactions of BH3-only proteins with Bax and Bak, but as with their interaction with prosurvival proteins there is a specificity of interaction observed. Experiments with mouse embryonic fibroblasts showed there is a differential response to Bim and Bid BH3-peptides. Bim activated Bax and Bid activated Bak to induce their respective oligomerization (Sarosiek et al., 2013), but in a nonstoichiometric manner (Leshchiner et al., 2013).

Understanding the optimal binding features of BH3-only proteins and the prosurvival proteins has been instrumental in the design of BH3-only peptides with novel specificities (Lee et al., 2008; Placzek et al., 2011). These reagents have been useful for delineating apoptotic pathways. For example, a potent selective inhibitor of Mcl-1 was found by randomized phage displayed peptide library of Bim (Lee et al., 2008). Enforced expression of this

Table 3.1 Bcl-2 binding profiles of the mammalian BH3-only proteins

BH3-only	Alternate name	Uniprot	Residues	Prosurvival Bcl-2 antagonists
Bim	Bod, Bcl2l11	O43521	198	Bcl-x_L, Bcl-2, Bcl-w, Mcl-1, A1, Bcl-B
Bid		P55957	195	Bcl-x_L, Bcl-2, Bcl-w, Mcl-1, A1
Bad	Bbc6, bcl2l8	Q92934	168	Bcl-x_L, Bcl-2, Bcl-w
Bmf		Q96LC9	184	Bcl-x_L, Bcl-2, Bcl-w, Mcl-1
Puma	Bbc3, Jfy-1	Q9BXH1	193	Bcl-x_L, Bcl-2, Bcl-w, Mcl-1, A1
Bik	Nbk, BIP1, BP4	Q13323	160	Bcl-x_L, Bcl-w, A1, Bcl-B
Hrk	DP5, Bid3	O00198	91	Bcl-x_L, Bcl-w, A1
Noxa	PMAIP1	Q13794	54	Mcl-1, A1

Uniprot accession codes and total number of residues for human BH3-only sequences.

inhibitor in acute myeloid leukemia derived cells that are dependent on Mcl-1 for survival initiates apoptosis (Glaser et al., 2012). The molecular determinants for Mcl-1 specificity over Bcl-x_L were determined using yeast surface display and peptide array analyses (Dutta et al., 2010) and a similar approach yielded selective A1 inhibitors (Dutta et al., 2013). In another approach, high-affinity interaction with Bid BH3 peptide and Bcl-x_L was achieved by stabilizing the helical BH3 sequence with a hydrocarbon bridge (Walensky et al., 2004) and a similarly conformationally constrained peptide based on the BH3 of Mcl-1 bound Mcl-1 tightly (Stewart, Fire, Keating, & Walensky, 2010).

4. STRUCTURES OF BH3-ONLY PROTEINS AND THEIR COMPLEXES

Contrasting with the well-ordered helical bundle structures of the multimotif Bcl-2 family (Muchmore et al., 1996; Sattler et al., 1997), with the exception of Bid that has a well-defined core structure similar to the Bcl-2 fold (Chou et al., 1999; McDonnell et al., 1999), BH3-only proteins are IDPs (Rautureau et al., 2010). Evidence for their unstructured nature derives from sequence analysis (Hinds et al., 2007; Peng et al., 2013), circular dichroism (Hinds et al., 2007), NMR data (Barrera-Vilarmau et al., 2011; Hinds et al., 2007), and proteolytic sensitivity (Hinds et al., 2007). While Bid has a structurally ordered hydrophobic helical core, it must be activated by proteolytic cleavage at sites in a disordered loop connecting helices α1 and α2. Multiple proteases, including caspase-8, Granzyme-B, calpains, and cathepsins cleave Bid at sites in this interhelical loop (Billen, Shamas-Din, & Andrews, 2008) but leave it stably folded (Chou et al., 1999). Fragmentation of cleaved Bid to releases the BH3-containing C-terminal fragment, truncated Bid (tBid), in the presence of a membrane environment (Lovell et al., 2008) releases the Bid BH3-motif for interactions. NMR studies have shown tBid to be dynamically disordered (Yao et al., 2009) and thus Bid is processed to become an IDP like the other BH3-only proteins. IDPs are common in eukaryotic genomes where they frequently form the basis of protein–protein interaction networks and undergo disorder to order transitions on binding (Hsu et al., 2013). Although the BH3-only proteins are both IDPs and promiscuous Bcl-2 binders (Table 3.1) they bind with high affinity, frequently in the low nanomolar range (Kvansakul & Hinds, 2013).

A number of structural studies have defined the principles of BH3-only protein interaction with their targets (Hinds & Day, 2005; Kvansakul & Hinds, 2013; Petros et al., 2004). Figure 3.3A shows the basic Bcl-2 fold, as exemplified by Bcl-x_L bound to a fragment of Bim (Liu, Dai, Zhu, Marrack, & Kappler, 2003). Remarkably, the same fold is shared by both the proapoptotic (Bax and Bak) and antiapoptotic multimotif Bcl-2 proteins. Structures of BH3-complexes of the prosurvival proteins show that the BH3 ligand binds in a hydrophobic groove formed from residues in the BH1, BH2, and BH3 motifs (or their structural equivalents where the BH motif is absent) of the prosurvival protein, corresponding to helices $\alpha 2$–$\alpha 5$ and $\alpha 8$. BH3 binding displaces the C-terminal residues from Bcl-x_L, Bcl-w, Mcl-1 and in the case of Bcl-w induces a structured to unstructured transition for these residues (Hinds et al., 2003) and is a prerequisite for membrane integration of Bcl-w (Wilson-Annan et al., 2003).

The BH3-only motif is a simple linear sequence element (discussed above, Fig. 3.2) and little outside the motif makes intermolecular contact with the Bcl-2 receptor (Day et al., 2008) or contributes to affinity (Wilson-Annan et al., 2003). A 26-mer Bim-BH3 spanning peptide binds with about equal affinity as longer constructs to Bcl-w (Wilson-Annan et al., 2003). BH3-only proteins target select Bcl-2 proteins from among the six prosurvival proteins and Bax and Bak with this simple molecular recognition feature. When folded into the binding groove of the Bcl-2 protein the four hydrophobic groups are buried and pack against conserved residues in the BH motifs from the multimotif Bcl-2 protein (Day et al., 2005) with a solvent exposed salt bridge between the Asp from the BH3 motif and the conserved Arg of the BH1 motif on helix $\alpha 5$. This folding and binding event buries about 1000 Å^2 of solvent accessible surface area (Fig. 3.3). No structures of full-length BH3-only proteins with their Bcl-2 partners have been determined; instead, peptides that span the BH3-region have been used as a surrogate for full-length protein. Attempts to crystallize full-length Bim in the presence of Bcl-x_L resulted in only a proteolytic fragment spanning the BH3-motif crystallizing as a Bcl-x_L complex (Liu et al., 2003). The first BH3-only protein solved as a complex was Bad with Bcl-x_L (Petros et al., 2000), which like earlier studies on a complex between Bak-BH3 and Bcl-x_L (Sattler et al., 1997) and others (Hinds & Day, 2005; Petros et al., 2004) showed the α-helical BH3-motif of Bad sitting in a groove provided by residues from the BH1, BH2, and BH3 motifs of Bcl-x_L, a configuration that now defines these interactions (Hinds & Day, 2005; Kvansakul & Hinds, 2013; Petros et al., 2004).

The role of BH3-only proteins varies between organisms but the essential structural features are maintained. While functionally worm BH3-only signaled developmental apoptosis is not equivalent to the mammalian Bcl-2 family (Fig. 3.1) and the sequence identity conserved between them is low (e.g., human Bim and *C. elegans* Egl-1 share approximately 15% identity), the structural aspects of Egl-1 binding by CED-9, including the key BH3 leucine and aspartate (Fig. 3.2), are conserved with those in mammals (Yan et al., 2004). CED-9 is isostructural with the multi-BH-motif Bcl-2 proteins and the 91-residue *C. elegans* BH3-only protein, Egl-1, binds CED-9 in a hydrophobic groove provided by surface helices analogously to that observed for mammalian complexes (Yan et al., 2004). The important functional difference with mammalian intrinsic apoptosis is transcription of Egl-1 does not result in MOMP and there is no equivalent to Bax or Bak in *C. elegans*. A further significant difference from mammalian Bcl-2 regulated apoptosis is that the Bcl-2-like CED-9 interacts with the adaptor protein CED-4 by binding and holding it in check, a function not apparent in the mammalian Bcl-2 family (Fig. 3.1). Upon binding, CED-9 in its BH3-binding groove Egl-1 elicits a conformational change in CED-9 releasing CED-4 (Yan et al., 2005, 2004) to assemble into a homo-oligomeric complex referred to as the apoptosome (Qi et al., 2010) that binds and activates the worm caspase CED-3 (Huang et al., 2013).

More recently it has been shown that there are alternate binding sites to the canonical hydrophobic groove in multidomain Bcl-2 proteins. For mammalian family members, such an alternative site has been proposed for Bax, where a constrained helical Bid BH3-peptide interacts at an interface remote to the BH3-binding groove, the "α1 site" (situated on the opposite face to the canonical groove of Fig. 3.2), with low affinity (Gavathiotis et al., 2008). However, Bid BH3-peptides bind Bak weakly in the canonical binding groove (Moldoveanu et al., 2013) but the structural evidence is currently weak. Notably, a complex of Bid BH3 with a detergent induced dimer of Bax has been solved (Czabotar et al., 2013). BH3-only proteins are also able to bind a number of alternative structural forms of prosurvival proteins, including homodimeric and higher order oligomers formed under the influence of heat, organic solvents, pH, surfactants, or sequence truncations (Denisov, Sprules, Fraser, Kozlov, & Gehring, 2007; Feng et al., 2008; Follis et al., 2013; Lee et al., 2011; O'Neill, Manion, Maguire, & Hockenbery, 2006) that retain a groove-like structure but have a domain swapped architecture. However, the functional role, if any, of these dimeric forms remains unclear.

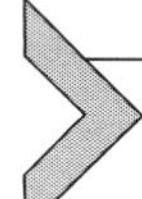

5. INTERACTION OF BH3-ONLY PROTEINS WITH VIRAL BCL-2 HOMOLOGUES

Viruses are obligate parasites reliant on host cells to provide the replicative machinery necessary for their survival. One response to impede viral invasion is to undergo apoptosis and viruses have evolved multiple strategies to subvert such host cell apoptosis. Among other apoptosis blocking strategies, many viruses have an ability to counter Bcl-2-regulated apoptosis in response to cellular infection (Taylor & Barry, 2006). Through expression of sequence, structural and functional mimics of Bcl-2 proteins (Cuconati & White, 2002) that sequester cellular BH3-only proteins apoptosis induction may be prevented in infected cells (Tait & Green, 2010). Viral Bcl-2 (vBcl-2) proteins retain key structural, but not necessarily sequence features, of their mammalian counterparts to inhibit both BH3-only and Bax-like death signals. The BH3-binding groove of mammalian Bcl-2 proteins is frequently retained. In general, the sequence divergence of vBcl-2 proteins from recognized Bcl-2 proteins has made them difficult to identify by sequence comparison and they frequently do not retain the BH sequence motifs. Structures of BH3-only proteins bound to vBcl-2 proteins show they interact in a similar way to complexes of mammalian family members (Kvansakul & Hinds, 2013) (Fig. 3.3).

Interaction studies revealed significant variations in the BH3-only binding profile of vBcl-2 proteins with respect to the range and nature of their ligands as well as their affinities. Yeast expression models have been invaluable in delineating the interactions between BH3-only proteins, prosurvival Bcl-2 and proapoptotic Bak and Bax, owing to the observation that overexpression of either Bak (Hanada, Aime-Sempe, Sato, & Reed, 1995) or Bax (Sato, Hanada, et al., 1994) in yeast arrests growth. This allowed reconstitution of some aspects of Bcl-2 mediated signaling in yeast for mammalian (Fletcher et al., 2008) worm (James, Gschmeissner, Fraser, & Evan, 1997) and viral (Banadyga et al., 2011; Juhasova et al., 2011) regulators of Bcl-2 signaling, where the adverse effects of Bak or Bax overexpression on yeast growth are rescued by coexpression of prosurvival Bcl-2 proteins. The rescue is countered by coexpressing BH3-only proteins as a third component of the reconstituted pathway, thus to some degree modeling events *in vivo* (Beaumont et al., 2013). A recent study has extended this by adapting reconstituted Bcl-2 signaling pathways in yeast to drug screening (Beaumont et al., 2013). Using these methods and both structural and other screening approaches specificity, affinities and structures have determined how BH3-only proteins engage vBcl-2 proteins.

Mechanistically the engagement of BH3-only proteins by viral Bcl-2 proteins is reminiscent of that observed for their cellular homologs. Key structural elements of cellular Bcl-2 proteins are retained by viral Bcl-2 proteins but sufficient differences exist to manifest themselves as differences in binding profiles and affinities for BH3-only proteins. Vaccinia virus F1L engages Bim, Bak, and Bax only and with affinities at least 20-fold lower compared to those of Bcl-x_L (IC_{50} 250 nM vs. 4.6 nM for Bcl-x_L) (Kvansakul et al., 2008). In contrast, sheep pox virus SPPV14 interacts with Bim, Bid, Bmf, Hrk, and Puma (Okamoto et al., 2012) with moderate to weak affinity (IC_{50} values 26, 341, 67, 63, 65 nM for Bim, Bid, Bmf, Hrk, and Puma, respectively) (Kvansakul & Hinds, 2013). Two of the most promiscuous BH3-only proteins Bid and Puma are only engaged by a limited number of viral Bcl-2 proteins and vBcl2 binding Noxa has not been observed. Immunoprecipitation analysis has indicated N1L associates with Bad and ORFV125 binds Noxa, but affinities determined with purified proteins have not been reported (Westphal et al., 2009). Most of the essential binding interactions between the mammalian BH3-only protein and the vBcl-2 are retained and they utilize the canonical in-groove mechanism (Kvansakul & Hinds, 2013; Kvansakul et al., 2007, 2010) (Fig. 3.3), however, structures of viral Bcl-2 proteins lacking detectable sequence identity to the family have revealed some unanticipated differences. Structures of viral Bcl-2 homologs BHRF1 and vBcl-2 in complex with mammalian BH3-only proteins revealed the four critical hydrophobic residues in the BH3 motif engage four pockets in the canonical hydrophobic groove in BHRF1 (Kvansakul et al., 2010) and γ-herpes virus Bcl-2 (Loh et al., 2005) (Fig. 3.3) as their mammalian counterparts do. Furthermore, the invariant Asp in the BH3 motif engages the conserved Arg in the BH1 domain, pointing to a high degree of conservation of the key features and mode of engagement between these obvious viral homologs and the endogenous mammalian proteins. However, structural studies of M11L and F1L that have sequences more divergent to the mammalian Bcl-2 proteins indicate that while the four hydrophobic interactions with BH3 motifs are retained, the ionic interaction between the Asp in the BH3 motif and Arg is absent. The M11L:Bak BH3 (Kvansakul et al., 2007) complex reveals that no Arg residue is found in the vicinity of the Asp in the BH3 motif, and its equivalent position in M11L is occupied by a Val, as is the case with F1L (Kvansakul et al., 2008), although final structural proof is missing as the structure of F1L was only determined in isolation and not bound to a BH3 protein. This suggests that important differences exist in the way certain viral Bcl-2 proteins engage BH3-only proteins when compared to their mammalian counterparts, and indicates that the Asp–Arg ionic interaction considered critical may not be as important as previously thought. Similar observations have been made in mammalian systems too, where an optimized A1 selective BH3-sequence derived from Bim can retain affinity and specificity without the ionic interaction where the key Asp was replaced by Lys (Dutta et al., 2013).

The change in ligand specificity for BH3-only proteins for certain vBcl-2 proteins compared to mammalian prosurvival Bcl-2 proteins reflects residue and conformational changes within the canonical binding groove. For example, M11L exhibits minimal conformational movement in the binding groove when it engages a BH3 ligand and displays high-affinity binding to a number of point-mutants of Bak-BH3 (Kvansakul et al., 2008). In contrast to the lack of conformational change in M11L on binding a BH3-motif, Bcl-x_L undergoes substantial changes, in the order of 9 Å, on binding (Liu et al., 2003; Sattler et al., 1997), and is unable to engage Bak harboring point mutations in the BH3 domain (Kvansakul et al., 2007). The reduced binding groove movement and robustness of ligand binding in the case of M11L could be a consequence of thermodynamic optimization to engage only a small subset of proapoptotic Bcl-2 members.

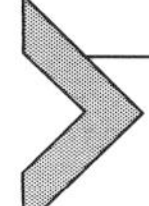

6. INTERACTION WITH NON-BCL-2 PROTEINS: BEYOND THE BH3-MOTIF

BH3-only proteins use regions outside of their BH3-motif to interact with cytoskeletal proteins and this modulates their apoptotic activity. Phosphorylation of Bad at serine sites sequesters it to cytosolic 14-3-3 proteins, preventing its interaction with Bcl-x_L and neutralizing its capacity to induce apoptosis (Zha, Harada, Yang, Jockel, & Korsmeyer, 1996), while Bim and Bmf are found associated with components of the dynein and myosin V actin motor complexes, respectively (Puthalakath, Huang, O'Reilly, King, & Strasser, 1999; Puthalakath et al., 2001). Bim and Bmf bind dynein light chains-1 (DYNLL1) (Puthalakath et al., 1999) and -2 (DYNLL2) (Puthalakath et al., 2001) components of the dynein and myosin V motor complexes, respectively, through a short linear recognition element. Mutation in the DYNLL1/2 binding sites on Bim or Bmf enhances apoptosis induction (Puthalakath et al., 1999, 2001). The structurally identical

DYNLL1 and DYNLL2 share >90% sequence identity and a single light chain residue appears responsible for Bim and Bmf localization to the appropriate light chain (Day et al., 2004) and may be a mechanism for their compartmentalization (Day et al., 2004). The sequence and structural basis for Bim binding has been determined by the solution structure of DYNLL1 with a fragment of Bim (Fan, Zhang, Tochio, Li, & Zhang, 2001). Recognition of DYNLL1/2 occurs through a short sequence motif, (K/R)XTQT (X is any residue), that forms a short β-strand that runs antiparallel to a strand on the five-stranded β-sheet of DYNLL1/2 at its dimer interface (Fan et al., 2001). In Bim, the binding residues are located on exon 4 and a splice variant that lacks this exon is a more potent apoptosis inducer (Puthalakath et al., 1999).

7. REGULATION

The potential finality of a death signal warrants its strict control and there are extensive translational and posttranslational controls on BH3-only proteins to prevent uncontrolled cell death (Puthalakath & Strasser, 2002). BH3-only proteins interact with posttranslational modifying enzymes for processes such as phosphorylation/dephosphorylation, ubiquitylation, or proteolysis and a key structural aspect is that they are IDPs or have intrinsically disordered regions capable of modification (Rautureau et al., 2010). IDPs are kept tightly regulated and in low concentrations in cells (Gsponer, Futschik, Teichmann, & Babu, 2008). BH3-only proteins have a low abundance in cells that have not been subjected to apoptotic stress. There are conceivably no free BH3-only proteins in a cell, for example, Bim is ordinarily found associated with a prosurvival protein (Gomez-Bougie, Bataille, & Amiot, 2005). Consistent with their IDP status, Noxa (Craxton et al., 2012) or Bim (Wiggins et al., 2011) can be degraded at proteasomes without ubiquitylation and this process can be blocked by Mcl-1. Also consistent with the IDP status of BH3-only proteins are the presence of many splice variants (Rautureau et al., 2010). Unlike folded proteins, IDPs can retain their functional relevance as the interaction motifs do not rely on epitopes arising from the 3D-fold, but rather on relatively short linear motifs. Splice variants that retain the BH3-motif could potentially activate apoptosis. For example, exon skipping in *mcl-1* generates a short Mcl-1 protein that retains the BH3-motif (Bingle et al., 2000) and inhibits Mcl-1 (Stewart et al., 2010). In another example, three main transcripts of Bim are known, Bim_S, Bim_L, and Bim_{EL} (short, long, and extra-long, respectively). Bim_S lacks the dynein binding motif and is the most potent apoptosis inhibitor, there also seems to be subtle differences in the activity between the short and longer Bim variants (Merino et al., 2009). Other splice variants of Bim have been observed and they can initiate apoptosis (Marani, Tenev, Hancock, Downward, & Lemoine, 2002). One posttranslational modification already pointed out above is the proteolytic activation of Bid that must undergo further disintegration to interact with Bcl-2 proteins. The eight BH3-only proteins present a diverse array of regulation methods for their induction and destruction, many of which rely on simple linear molecular recognition sequences compatible with their status as IDPs.

8. BH3-ONLY PROTEINS AND DISEASE

The multiple redundancies of the mammalian BH3-only proteins (Table 3.1) have required the use of multiple genetic targeting through knockouts and inhibitory RNA to disentangle their roles in tissue homeostasis and disease (Youle & Strasser, 2008). Dysregulation of apoptosis is thought to be a systemic underlying cause in cancer (Hanahan & Weinberg, 2011). BH3-only proteins are tumor suppressors initiating the apoptotic response to DNA damage or stress and their inactivation has been associated with human disease (Mestre-Escorihuela et al., 2007). Functional differences between the BH3-only proteins are also apparent, apart from targeting specific prosurvival proteins. For instance, Noxa but not other BH3-only proteins accelerate Mcl-1 degradation (Czabotar et al., 2007). Knock-in mice with the BH3 motif of Bim replaced by those from Bad, Noxa, or Puma were useful for investigating the *in vivo* functional role of BimBH3 (Merino et al., 2009). Despite binding all prosurvival proteins except Bcl-B (Table 3.1), Bim-PumaBH3 does not entirely replicate Bim and Bim-BadBH3 and Bim-NoxaBH3 have further reduced proapoptotic function due to their more selective binding profiles. Engaging the prosurvival proteins is a necessary but not a sufficient condition for Bim activation of apoptosis. The understanding of the molecular mechanisms of apoptosis induction by BH3-only proteins through genetic, cell biological, biophysical, and structural biological investigations have defined the network of interactions and their molecular basis for these proteins.

BH3-only proteins also play a role in therapeutic responses to disease, as their induction by chemotherapeutic agents or radiation initiates apoptosis. By probing plasma membrane permeabilized cancer cells with BH3-peptides (BH3-profiling) their dependence on particular survival Bcl-2 proteins can be determined for tailored cytotoxic therapies (Ni Chonghaile et al., 2011). Our understanding of the molecular structures has been exploited in the design of Bcl-2 family inhibitors, "BH3-mimics" that block the action of prosurvival Bcl-2 proteins by taking advantage of the molecular features that BH3-only proteins employ for high-affinity interactions (Oltersdorf et al., 2005). BH3-only mimics, such as ABT-737 (Oltersdorf et al., 2005) and its clinical homologs ABT-263 (Navitoclax) (Wilson et al., 2010) and ABT-199 (Souers et al., 2013) (Fig. 3.4), inhibit prosurvival proteins by binding in the BH3-binding groove (Fig. 3.4) to either directly initiate cell death or enhance the effects of cytotoxic drugs by reducing the burden of prosurvival Bcl-2 proteins. Considering the critical role of BH3-only proteins in apoptosis signaling, it appears likely that new BH3-only mimetics will eventuate to target diseases caused by dysfunctional apoptosis signaling.

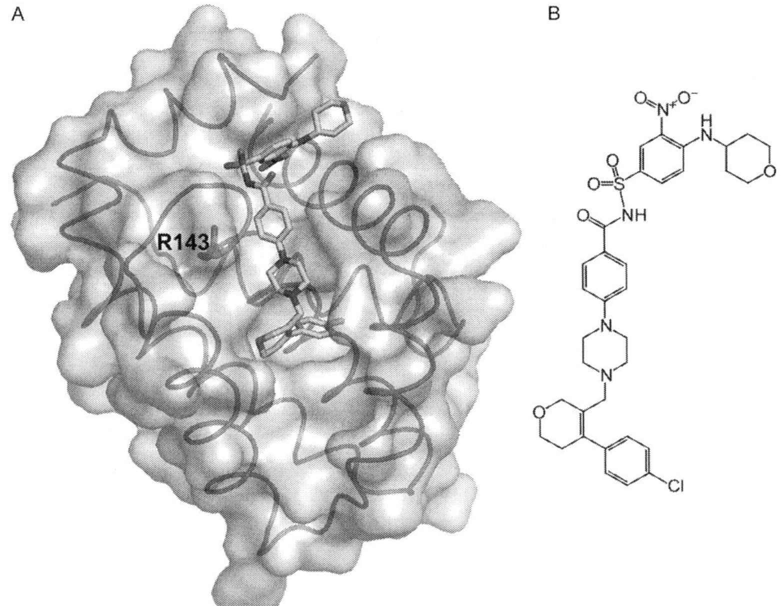

Figure 3.4 Mimics of BH3-only proteins bind in the BH3-binding groove. (A) Structure of an ABT-199 analog bound to Bcl-2 (pdb: 4LXD) (Souers et al., 2013). The drug binds in the same groove as BH3-only proteins but the conserved Arg in the BH1 motif of Bcl-2 does not make an ionic interaction with the acidic acylsulfonomide of the ABT-199 analog. The structure is aligned as those in Fig. 3.3 and Bcl-2 is depicted as a surface representation. (B) ABT-199 analog. (See the color plate.)

9. CONCLUSION

A diverse array of stimuli activate intrinsic apoptosis such as growth factor deprivation, viral invasion, UV or γ irradiation, glucocorticoids, or phorbol ester (Youle & Strasser, 2008). Although there are a number of BH3-only proteins proposed, only eight have compelling biological and structural evidence of interaction in the Bcl-2 family. Of these eight, Bim is the universal prosurvival protein antagonist, binding all six prosurvival proteins with high affinity while other BH3-only proteins have more selective binding profiles (Certo et al., 2006; Chen et al., 2005; Kuwana et al., 2005; Rautureau et al., 2012) (Table 3.1). Their role as cell sentinels and apoptosis initiators has enabled viruses to regulate the activation of Bak/Bak by sequestering BH3-only proteins through expression of prosurvival Bcl-2 mimics. These Bcl-2 mimics engage specific mammalian BH3-only proteins in a similar way to their mammalian counterparts. Structural and biophysical measurements have been invaluable in deciphering the role of BH3-only proteins in biology and disease, elucidating the molecular basis for their action and given us an appreciation of the network of interactions and complexity of Bcl-2 regulated apoptosis. These studies have paved the way for new anticancer therapies based on mimicking the properties of the BH3-only proteins. Gaps remain in our knowledge and the molecular details of BH3-only protein activation of Bax and Bak have yet to emerge. The BH3-only proteins also appear to be targeted to intracellular membranes, but it is not clear what their role is there. With their intrinsic disorder and apparently simple binding motifs the BH3-only proteins regulate Bcl-2 signaled apoptosis in a remarkable sophisticated manner.

ACKNOWLEDGMENTS

Our research is supported by the National Health and Medical Research Council of Australia (CDA fellowship 637372, and Project (APP1007918) to M. K.).

REFERENCES

Aouacheria, A., Brunet, F., & Gouy, M. (2005). Phylogenomics of life-or-death switches in multicellular animals: Bcl-2, BH3-Only, and BNip families of apoptotic regulators. *Molecular Biology and Evolution*, *22*, 2395–2416.

Aouacheria, A., Rech de Laval, V., Combet, C., & Hardwick, J. M. (2013). Evolution of Bcl-2 homology motifs: Homology versus homoplasy. *Trends in Cell Biology*, *23*, 103–111.

Banadyga, L., Lam, S. C., Okamoto, T., Kvansakul, M., Huang, D. C., & Barry, M. (2011). Deerpox virus encodes an inhibitor of apoptosis that regulates Bak and Bax. *Journal of Virology, 85*, 1922–1934.

Barrera-Vilarmau, S., Obregon, P., & de Alba, E. (2011). Intrinsic order and disorder in the bcl-2 member harakiri: Insights into its proapoptotic activity. *PLoS One, 6*, e21413.

Beaumont, T. E., Shekhar, T. M., Kaur, L., Pantaki-Eimany, D., Kvansakul, M., & Hawkins, C. J. (2013). Yeast techniques for modeling drugs targeting Bcl-2 and caspase family members. *Cell Death and Disease, 4*, e619.

Beverly, L. J. (2012). Regulation of anti-apoptotic BCL2-proteins by non-canonical interactions: The next step forward or two steps back? *Journal of Cellular Biochemistry, 113*, 3–12.

Billen, L. P., Shamas-Din, A., & Andrews, D. W. (2008). Bid: A Bax-like BH3 protein. *Oncogene, 27*(Suppl. 1), S93–S104.

Bingle, C. D., Craig, R. W., Swales, B. M., Singleton, V., Zhou, P., & Whyte, M. K. (2000). Exon skipping in Mcl-1 results in a bcl-2 homology domain 3 only gene product that promotes cell death. *Journal of Biological Chemistry, 275*, 22136–22146.

Blaineau, S. V., & Aouacheria, A. (2009). BCL2DB: Moving 'helix-bundled' BCL-2 family members to their database. *Apoptosis, 14*, 923–925.

Boersma, M. D., Sadowsky, J. D., Tomita, Y. A., & Gellman, S. H. (2008). Hydrophile scanning as a complement to alanine scanning for exploring and manipulating protein-protein recognition: Application to the Bim BH3 domain. *Protein Science, 17*, 1232–1240.

Boyd, J. M., Gallo, G. J., Elangovan, B., Houghton, A. B., Malstrom, S., Avery, B. J., et al. (1995). Bik, a novel death-inducing protein shares a distinct sequence motif with Bcl-2 family proteins and interacts with viral and cellular survival-promoting proteins. *Oncogene, 11*, 1921–1928.

Certo, M., Del Gaizo Moore, V., Nishino, M., Wei, G., Korsmeyer, S., Armstrong, S. A., et al. (2006). Mitochondria primed by death signals determine cellular addiction to antiapoptotic BCL-2 family members. *Cancer Cell, 9*, 351–365.

Chen, L., Willis, S. N., Wei, A., Smith, B. J., Fletcher, J. I., Hinds, M. G., et al. (2005). Differential targeting of prosurvival Bcl-2 proteins by their BH3-only ligands allows complementary apoptotic function. *Molecular Cell, 17*, 393–403.

Chittenden, T., Flemington, C., Houghton, A. B., Ebb, R. G., Gallo, G. J., Elangovan, B., et al. (1995). A conserved domain in Bak, distinct from BH1 and BH2, mediates cell death and protein binding functions. *EMBO Journal, 14*, 5589–5596.

Chou, J. J., Li, H., Salvesen, G. S., Yuan, J., & Wagner, G. (1999). Solution structure of BID, an intracellular amplifier of apoptotic signaling. *Cell, 96*, 615–624.

Conradt, B., & Horvitz, H. R. (1998). The *C. elegans* protein EGL-1 is required for programmed cell death and interacts with the Bcl-2-like protein CED-9. *Cell, 93*, 519–529.

Craxton, A., Butterworth, M., Harper, N., Fairall, L., Schwabe, J., Ciechanover, A., et al. (2012). NOXA, a sensor of proteasome integrity, is degraded by 26S proteasomes by an ubiquitin-independent pathway that is blocked by MCL-1. *Cell Death and Differentiation, 19*, 1424–1434.

Cuconati, A., & White, E. (2002). Viral homologs of BCL-2: Role of apoptosis in the regulation of virus infection. *Genes and Development, 16*, 2465–2478.

Czabotar, P. E., Lee, E. F., van Delft, M. F., Day, C. L., Smith, B. J., Huang, D. C., et al. (2007). Structural insights into the degradation of Mcl-1 induced by BH3 domains. *Proceedings of the National Academy of Sciences of the United States of America, 104*, 6217–6222.

Czabotar, P. E., Westphal, D., Dewson, G., Ma, S., Hockings, C., Fairlie, W. D., et al. (2013). Bax crystal structures reveal how BH3 domains activate Bax and nucleate its oligomerization to induce apoptosis. *Cell, 152*, 519–531.

Day, C. L., Chen, L., Richardson, S. J., Harrison, P. J., Huang, D. C., & Hinds, M. G. (2005). Solution structure of prosurvival Mcl-1 and characterization of its binding by proapoptotic BH3-only ligands. *Journal of Biological Chemistry, 280*, 4738–4744.

Day, C. L., Puthalakath, H., Skea, G., Strasser, A., Barsukov, I., Lian, L. Y., et al. (2004). Localization of dynein light chains 1 and 2 and their pro-apoptotic ligands. *Biochemical Journal, 377*, 597–605.

Day, C. L., Smits, C., Fan, F. C., Lee, E. F., Fairlie, W. D., & Hinds, M. G. (2008). Structure of the BH3 domains from the p53-inducible BH3-only proteins Noxa and Puma in complex with Mcl-1. *Journal of Molecular Biology, 380*, 958–971.

DeBartolo, J., Dutta, S., Reich, L., & Keating, A. E. (2012). Predictive Bcl-2 family binding models rooted in experiment or structure. *Journal of Molecular Biology, 422*, 124–144.

Denisov, A. Y., Sprules, T., Fraser, J., Kozlov, G., & Gehring, K. (2007). Heat-induced dimerization of BCL-xL through alpha-helix swapping. *Biochemistry, 46*, 734–740.

Doctor, K. S., Reed, J. C., Godzik, A., & Bourne, P. E. (2003). The apoptosis database. *Cell Death and Differentiation, 10*, 621–633.

Dutta, S., Chen, T. S., & Keating, A. E. (2013). Peptide ligands for pro-survival protein Bfl-1 from computationally guided library screening. *ACS Chemical Biology, 8*, 778–788.

Dutta, S., Gulla, S., Chen, T. S., Fire, E., Grant, R. A., & Keating, A. E. (2010). Determinants of BH3 binding specificity for Mcl-1 versus Bcl-xL. *Journal of Molecular Biology, 398*, 747–762.

Eimon, P. M., & Ashkenazi, A. (2010). The zebrafish as a model organism for the study of apoptosis. *Apoptosis, 15*, 331–349.

Fan, J., Zhang, Q., Tochio, H., Li, M., & Zhang, M. (2001). Structural basis of diverse sequence-dependent target recognition by the 8 kDa dynein light chain. *Journal of Molecular Biology, 306*, 97–108.

Feng, Y., Lin, Z., Shen, X., Chen, K., Jiang, H., & Liu, D. (2008). Bcl-xL forms two distinct homodimers at non-ionic detergents: Implications in the dimerization of Bcl-2 family proteins. *Journal of Biochemistry, 143*, 243–252.

Fletcher, J. I., Meusburger, S., Hawkins, C. J., Riglar, D. T., Lee, E. F., Fairlie, W. D., et al. (2008). Apoptosis is triggered when prosurvival Bcl-2 proteins cannot restrain Bax. Proceedings of the National Academy of Sciences of the United States of America. *105*, 18081–18087.

Follis, A. V., Chipuk, J. E., Fisher, J. C., Yun, M. K., Grace, C. R., Nourse, A., et al. (2013). PUMA binding induces partial unfolding within BCL-xL to disrupt p53 binding and promote apoptosis. *Nature Chemical Biology, 9*, 163–168.

Fuentes-Prior, P., & Salvesen, G. S. (2004). The protein structures that shape caspase activity, specificity, activation and inhibition. *Biochemical Journal, 384*, 201–232.

Gavathiotis, E., Suzuki, M., Davis, M. L., Pitter, K., Bird, G. H., Katz, S. G., et al. (2008). BAX activation is initiated at a novel interaction site. *Nature, 455*, 1076–1081.

Giam, M., Okamoto, T., Mintern, J. D., Strasser, A., & Bouillet, P. (2012). Bcl-2 family member Bcl-G is not a proapoptotic protein. *Cell Death and Disease, 3*, e404.

Glaser, S. P., Lee, E. F., Trounson, E., Bouillet, P., Wei, A., Fairlie, W. D., et al. (2012). Anti-apoptotic Mcl-1 is essential for the development and sustained growth of acute myeloid leukemia. *Genes and Development, 26*, 120–125.

Gomez-Bougie, P., Bataille, R., & Amiot, M. (2005). Endogenous association of Bim BH3-only protein with Mcl-1, Bcl-xL and Bcl-2 on mitochondria in human B cells. *European Journal of Immunology, 35*, 971–976.

Gsponer, J., Futschik, M. E., Teichmann, S. A., & Babu, M. M. (2008). Tight regulation of unstructured proteins: From transcript synthesis to protein degradation. *Science, 322*, 1365–1368.

Gump, J. M., & Thorburn, A. (2011). Autophagy and apoptosis: What is the connection? *Trends in Cell Biology, 21*, 387–392.

Han, J., Flemington, C., Houghton, A. B., Gu, Z., Zambetti, G. P., Lutz, R. J., et al. (2001). Expression of bbc3, a pro-apoptotic BH3-only gene, is regulated by diverse cell death and survival signals. *Proceedings of the National Academy of Sciences of the United States of America, 98*, 11318–11323.

Hanada, M., Aime-Sempe, C., Sato, T., & Reed, J. C. (1995). Structure-function analysis of Bcl-2 protein. Identification of conserved domains important for homodimerization with Bcl-2 and heterodimerization with Bax. *Journal of Biological Chemistry, 270*, 11962–11969.

Hanahan, D., & Weinberg, R. A. (2011). Hallmarks of cancer: The next generation. *Cell, 144*, 646–674.

Hardwick, J. M., Chen, Y. B., & Jonas, E. A. (2012). Multipolar functions of BCL-2 proteins link energetics to apoptosis. *Trends in Cell Biology, 22*, 318–328.

Hinds, M. G., & Day, C. L. (2005). Regulation of apoptosis: Uncovering the binding determinants. *Current Opinion in Structural Biology, 15*, 690–699.

Hinds, M. G., Lackmann, M., Skea, G. L., Harrison, P. J., Huang, D. C., & Day, C. L. (2003). The structure of Bcl-w reveals a role for the C-terminal residues in modulating biological activity. *EMBO Journal, 22*, 1497–1507.

Hinds, M. G., Smits, C., Fredericks-Short, R., Risk, J. M., Bailey, M., Huang, D. C., et al. (2007). Bim, Bad and Bmf: Intrinsically unstructured BH3-only proteins that undergo a localized conformational change upon binding to prosurvival Bcl-2 targets. *Cell Death and Differentiation, 14*, 128–136.

Hsu, S. Y., Lin, P., & Hsueh, A. J. (1998). BOD (Bcl-2-related ovarian death gene) is an ovarian BH3 domain-containing proapoptotic Bcl-2 protein capable of dimerization with diverse antiapoptotic Bcl-2 members. *Molecular Endocrinology, 12*, 1432–1440.

Hsu, W. L., Oldfield, C. J., Xue, B., Meng, J., Huang, F., Romero, P., et al. (2013). Exploring the binding diversity of intrinsically disordered proteins involved in one-to-many binding. *Protein Science, 22*, 258–273.

Huang, W., Jiang, T., Choi, W., Qi, S., Pang, Y., Hu, Q., et al. (2013). Mechanistic insights into CED-4-mediated activation of CED-3. *Genes and Development, 27*, 2039–2048.

Huang, D. C., & Strasser, A. (2000). BH3-Only proteins-essential initiators of apoptotic cell death. *Cell, 103*, 839–842.

Igaki, T., & Miura, M. (2004). Role of Bcl-2 family members in invertebrates. *Biochimica et Biophysica Acta, 1644*, 73–81.

Imaizumi, K., Tsuda, M., Imai, Y., Wanaka, A., Takagi, T., & Tohyama, M. (1997). Molecular cloning of a novel polypeptide, DP5, induced during programmed neuronal death. *Journal of Biological Chemistry, 272*, 18842–18848.

Inohara, N., Ding, L., Chen, S., & Nunez, G. (1997). harakiri, a novel regulator of cell death, encodes a protein that activates apoptosis and interacts selectively with survival-promoting proteins Bcl-2 and Bcl-X(L). *EMBO Journal, 16*, 1686–1694.

James, C., Gschmeissner, S., Fraser, A., & Evan, G. I. (1997). CED-4 induces chromatin condensation in Schizosaccharomyces pombe and is inhibited by direct physical association with CED-9. *Current Biology, 7*, 246–252.

Juhasova, B., Mentel, M., Bhatia-Kissova, I., Zeman, I., Kolarov, J., Forte, M., et al. (2011). BH3-only protein Bim inhibits activity of antiapoptotic members of Bcl-2 family when expressed in yeast. *FEBS Letters, 585*, 2709–2713.

Kang, R., Zeh, H. J., Lotze, M. T., & Tang, D. (2011). The Beclin 1 network regulates autophagy and apoptosis. *Cell Death and Differentiation, 18*, 571–580.

Kuwana, T., Bouchier-Hayes, L., Chipuk, J. E., Bonzon, C., Sullivan, B. A., Green, D. R., et al. (2005). BH3 domains of BH3-only proteins differentially regulate Bax-mediated mitochondrial membrane permeabilization both directly and indirectly. *Molecular Cell, 17*, 525–535.

Kvansakul, M., & Hinds, M. G. (2013). Structural biology of the Bcl-2 family and its mimicry by viral proteins. *Cell Death and Disease*, *4*, e909.
Kvansakul, M., van Delft, M. F., Lee, E. F., Gulbis, J. M., Fairlie, W. D., Huang, D. C., et al. (2007). A structural viral mimic of prosurvival Bcl-2: A pivotal role for sequestering proapoptotic Bax and Bak. *Molecular Cell*, *25*, 933–942.
Kvansakul, M., Wei, A. H., Fletcher, J. I., Willis, S. N., Chen, L., Roberts, A. W., et al. (2010). Structural basis for apoptosis inhibition by Epstein-Barr virus BHRF1. *PLoS Pathogens*, *6*, e1001236.
Kvansakul, M., Yang, H., Fairlie, W. D., Czabotar, P. E., Fischer, S. F., Perugini, M. A., et al. (2008). Vaccinia virus anti-apoptotic F1L is a novel Bcl-2-like domain-swapped dimer that binds a highly selective subset of BH3-containing death ligands. *Cell Death and Differentiation*, *15*, 1564–1571.
Lanave, C., Santamaria, M., & Saccone, C. (2004). Comparative genomics: The evolutionary history of the Bcl-2 family. *Gene*, *333*, 71–79.
Lasi, M., Pauly, B., Schmidt, N., Cikala, M., Stiening, B., Kasbauer, T., et al. (2010). The molecular cell death machinery in the simple cnidarian Hydra includes an expanded caspase family and pro- and anti-apoptotic Bcl-2 proteins. *Cell Research*, *20*, 812–825.
Lee, E. F., Czabotar, P. E., van Delft, M. F., Michalak, E. M., Boyle, M. J., Willis, S. N., et al. (2008). A novel BH3 ligand that selectively targets Mcl-1 reveals that apoptosis can proceed without Mcl-1 degradation. *The Journal of Cell Biology*, *180*, 341–355.
Lee, E. F., Dewson, G., Smith, B. J., Evangelista, M., Pettikiriarachchi, A., Dogovski, C., et al. (2011). Crystal structure of a BCL-W domain-swapped dimer: Implications for the function of BCL-2 family proteins. *Structure*, *19*, 1467–1476.
Leshchiner, E. S., Braun, C. R., Bird, G. H., & Walensky, L. D. (2013). Direct activation of full-length proapoptotic BAK. *Proceedings of the National Academy of Sciences of the United States of America*, *110*, 986–995.
Lessene, G., Czabotar, P. E., & Colman, P. M. (2008). BCL-2 family antagonists for cancer therapy. *Nature Reviews Drug Discovery*, 7, 989–1000.
Liu, X., Dai, S., Zhu, Y., Marrack, P., & Kappler, J. W. (2003). The structure of a Bcl-xL/Bim fragment complex: Implications for Bim function. *Immunity*, *19*, 341–352.
Loh, J., Huang, Q., Petros, A. M., Nettesheim, D., van Dyk, L. F., Labrada, L., et al. (2005). A surface groove essential for viral Bcl-2 function during chronic infection in vivo. *PLoS Pathogens*, *1*, e10.
Lovell, J. F., Billen, L. P., Bindner, S., Shamas-Din, A., Fradin, C., Leber, B., et al. (2008). Membrane binding by tBid initiates an ordered series of events culminating in membrane permeabilization by Bax. *Cell*, *135*, 1074–1084.
Marani, M., Tenev, T., Hancock, D., Downward, J., & Lemoine, N. R. (2002). Identification of novel isoforms of the BH3 domain protein Bim which directly activate Bax to trigger apoptosis. *Molecular and Cellular Biology*, *22*, 3577–3589.
McDonnell, J. M., Fushman, D., Milliman, C. L., Korsmeyer, S. J., & Cowburn, D. (1999). Solution structure of the proapoptotic molecule BID: A structural basis for apoptotic agonists and antagonists. *Cell*, *96*, 625–634.
Merino, D., Giam, M., Hughes, P. D., Siggs, O. M., Heger, K., O'Reilly, L. A., et al. (2009). The role of BH3-only protein Bim extends beyond inhibiting Bcl-2-like prosurvival proteins. *The Journal of Cell Biology*, *186*, 355–362.
Mestre-Escorihuela, C., Rubio-Moscardo, F., Richter, J. A., Siebert, R., Climent, J., Fresquet, V., et al. (2007). Homozygous deletions localize novel tumor suppressor genes in B-cell lymphomas. *Blood*, *109*, 271–280.
Moldoveanu, T., Grace, C. R., Llambi, F., Nourse, A., Fitzgerald, P., Gehring, K., et al. (2013). BID-induced structural changes in BAK promote apoptosis. *Nature Structural and Molecular Biology*, *20*, 589–597.

Muchmore, S. W., Sattler, M., Liang, H., Meadows, R. P., Harlan, J. E., Yoon, H. S., et al. (1996). X-ray and NMR structure of human Bcl-xL, an inhibitor of programmed cell death. *Nature, 381*, 335–341.

Nakano, K., & Vousden, K. H. (2001). PUMA, a novel proapoptotic gene, is induced by p53. *Molecular Cell, 7*, 683–694.

Ni Chonghaile, T., Sarosiek, K. A., Vo, T. T., Ryan, J. A., Tammareddi, A., Moore Vdel, G., et al. (2011). Pretreatment mitochondrial priming correlates with clinical response to cytotoxic chemotherapy. *Science, 334*, 1129–1133.

Oberstein, A., Jeffrey, P. D., & Shi, Y. (2007). Crystal structure of the Bcl-XL-Beclin 1 peptide complex: Beclin 1 is a novel BH3-only protein. *Journal of Biological Chemistry, 282*, 13123–13132.

O'Connor, L., Strasser, A., O'Reilly, L. A., Hausmann, G., Adams, J. M., Cory, S., et al. (1998). Bim: A novel member of the Bcl-2 family that promotes apoptosis. *EMBO Journal, 17*, 384–395.

Oda, E., Ohki, R., Murasawa, H., Nemoto, J., Shibue, T., Yamashita, T., et al. (2000). Noxa, a BH3-only member of the Bcl-2 family and candidate mediator of p53-induced apoptosis. *Science, 288*, 1053–1058.

Okamoto, T., Campbell, S., Mehta, N., Thibault, J., Colman, P. M., Barry, M., et al. (2012). Sheeppox virus SPPV14 encodes a Bcl-2-like cell death inhibitor that counters a distinct set of mammalian proapoptotic proteins. *Journal of Virology, 86*, 11501–11511.

Oltersdorf, T., Elmore, S. W., Shoemaker, A. R., Armstrong, R. C., Augeri, D. J., Belli, B. A., et al. (2005). An inhibitor of Bcl-2 family proteins induces regression of solid tumours. *Nature, 435*, 677–681.

O'Neill, J. W., Manion, M. K., Maguire, B., & Hockenbery, D. M. (2006). BCL-XL dimerization by three-dimensional domain swapping. *Journal of Molecular Biology, 356*, 367–381.

Peng, Z., Xue, B., Kurgan, L., & Uversky, V. N. (2013). Resilience of death: Intrinsic disorder in proteins involved in the programmed cell death. *Cell Death and Differentiation, 20*, 1257–1267.

Petros, A. M., Nettesheim, D. G., Wang, Y., Olejniczak, E. T., Meadows, R. P., Mack, J., et al. (2000). Rationale for Bcl-xL/Bad peptide complex formation from structure, mutagenesis, and biophysical studies. *Protein Science, 9*, 2528–2534.

Petros, A. M., Olejniczak, E. T., & Fesik, S. W. (2004). Structural biology of the Bcl-2 family of proteins. *Biochimica et Biophysica Acta, 1644*, 83–94.

Placzek, W. J., Sturlese, M., Wu, B., Cellitti, J. F., Wei, J., & Pellecchia, M. (2011). Identification of a novel Mcl-1 protein binding motif. *Journal of Biological Chemistry, 286*, 39829–39835.

Puthalakath, H., Huang, D. C., O'Reilly, L. A., King, S. M., & Strasser, A. (1999). The proapoptotic activity of the Bcl-2 family member Bim is regulated by interaction with the dynein motor complex. *Molecular Cell, 3*, 287–296.

Puthalakath, H., & Strasser, A. (2002). Keeping killers on a tight leash: Transcriptional and post-translational control of the pro-apoptotic activity of BH3-only proteins. *Cell Death and Differentiation, 9*, 505–512.

Puthalakath, H., Villunger, A., O'Reilly, L. A., Beaumont, J. G., Coultas, L., Cheney, R. E., et al. (2001). Bmf: A proapoptotic BH3-only protein regulated by interaction with the myosin V actin motor complex, activated by anoikis. *Science, 293*, 1829–1832.

Qi, S., Pang, Y., Hu, Q., Liu, Q., Li, H., Zhou, Y., et al. (2010). Crystal structure of the Caenorhabditis elegans apoptosome reveals an octameric assembly of CED-4. *Cell, 141*, 446–457.

Rautureau, G. J., Day, C. L., & Hinds, M. G. (2010). Intrinsically disordered proteins in bcl-2 regulated apoptosis. *International Journal of Molecular Sciences, 11*, 1808–1824.

Rautureau, G. J., Yabal, M., Yang, H., Huang, D. C., Kvansakul, M., & Hinds, M. G. (2012). The restricted binding repertoire of Bcl-B leaves Bim as the universal BH3-only prosurvival Bcl-2 protein antagonist. *Cell Death and Disease*, *3*, e443.

Sarosiek, K. A., Chi, X., Bachman, J. A., Sims, J. J., Montero, J., Patel, L., et al. (2013). BID preferentially activates BAK while BIM preferentially activates BAX, affecting chemotherapy response. *Molecular Cell*, *51*, 751–765.

Sato, T., Hanada, M., Bodrug, S., Irie, S., Iwama, N., Boise, L. H., et al. (1994). Interactions among members of the Bcl-2 protein family analyzed with a yeast two-hybrid system. *Proceedings of the National Academy of Sciences of the United States of America*, *91*, 9238–9242.

Sato, T., Irie, S., Krajewski, S., & Reed, J. C. (1994). Cloning and sequencing of a cDNA encoding the rat Bcl-2 protein. *Gene*, *140*, 291–292.

Sattler, M., Liang, H., Nettesheim, D., Meadows, R. P., Harlan, J. E., Eberstadt, M., et al. (1997). Structure of Bcl-xL-Bak peptide complex: Recognition between regulators of apoptosis. *Science*, *275*, 983–986.

Schumacher, B., Schertel, C., Wittenburg, N., Tuck, S., Mitani, S., Gartner, A., et al. (2005). C. elegans ced-13 can promote apoptosis and is induced in response to DNA damage. *Cell Death and Differentiation*, *12*, 153–161.

Shamas-Din, A., Bindner, S., Zhu, W., Zaltsman, Y., Campbell, C., Gross, A., et al. (2013). tBid undergoes multiple conformational changes at the membrane required for Bax activation. *Journal of Biological Chemistry*, *288*, 22111–22127.

Shamas-Din, A., Kale, J., Leber, B., & Andrews, D. W. (2013). Mechanisms of action of Bcl-2 family proteins. *Cold Spring Harbor Perspectives in Biology*, *5*, a008714.

Souers, A. J., Leverson, J. D., Boghaert, E. R., Ackler, S. L., Catron, N. D., Chen, J., et al. (2013). ABT-199, a potent and selective BCL-2 inhibitor, achieves antitumor activity while sparing platelets. *Nature Medicine*, *19*, 202–208.

Stewart, M. L., Fire, E., Keating, A. E., & Walensky, L. D. (2010). The MCL-1 BH3 helix is an exclusive MCL-1 inhibitor and apoptosis sensitizer. *Nature Chemical Biology*, *6*, 595–601.

Strasser, A., Cory, S., & Adams, J. M. (2011). Deciphering the rules of programmed cell death to improve therapy of cancer and other diseases. *EMBO Journal*, *30*, 3667–3683.

Tait, S. W., & Green, D. R. (2010). Mitochondria and cell death: Outer membrane permeabilization and beyond. *Nature Reviews Molecular Cell Biology*, *11*, 621–632.

Taylor, J. M., & Barry, M. (2006). Near death experiences: Poxvirus regulation of apoptotic death. *Virology*, *344*, 139–150.

Uversky, V. N., & Dunker, A. K. (2010). Understanding protein non-folding. *Biochimica et Biophysica Acta*, *1804*, 1231–1264.

Walensky, L. D., Kung, A. L., Escher, I., Malia, T. J., Barbuto, S., Wright, R. D., et al. (2004). Activation of apoptosis in vivo by a hydrocarbon-stapled BH3 helix. *Science*, *305*, 1466–1470.

Wang, K., Yin, X. M., Chao, D. T., Milliman, C. L., & Korsmeyer, S. J. (1996). BID: A novel BH3 domain-only death agonist. *Genes and Development*, *10*, 2859–2869.

Westphal, D., Ledgerwood, E. C., Tyndall, J. D., Hibma, M. H., Ueda, N., Fleming, S. B., et al. (2009). The orf virus inhibitor of apoptosis functions in a Bcl-2-like manner, binding and neutralizing a set of BH3-only proteins and active Bax. *Apoptosis*, *14*, 1317–1330.

Wiggins, C. M., Tsvetkov, P., Johnson, M., Joyce, C. L., Lamb, C. A., Bryant, N. J., et al. (2011). BIM(EL), an intrinsically disordered protein, is degraded by 20S proteasomes in the absence of poly-ubiquitylation. *Journal of Cell Science*, *124*, 969–977.

Wilfling, F., Weber, A., Potthoff, S., Vogtle, F. N., Meisinger, C., Paschen, S. A., et al. (2012). BH3-only proteins are tail-anchored in the outer mitochondrial membrane and can initiate the activation of Bax. *Cell Death and Differentiation*, *19*, 1328–1336.

Wilson, W. H., O'Connor, O. A., Czuczman, M. S., LaCasce, A. S., Gerecitano, J. F., Leonard, J. P., et al. (2010). Navitoclax, a targeted high-affinity inhibitor of BCL-2, in lymphoid malignancies: A phase 1 dose-escalation study of safety, pharmacokinetics, pharmacodynamics, and antitumour activity. *Lancet Oncology*, *11*, 1149–1159.

Wilson-Annan, J., O'Reilly, L. A., Crawford, S. A., Hausmann, G., Beaumont, J. G., Parma, L. P., et al. (2003). Proapoptotic BH3-only proteins trigger membrane integration of prosurvival Bcl-w and neutralize its activity. *The Journal of Cell Biology*, *162*, 877–887.

Yan, N., Chai, J., Lee, E. S., Gu, L., Liu, Q., He, J., et al. (2005). Structure of the CED-4-CED-9 complex provides insights into programmed cell death in *Caenorhabditis elegans*. *Nature*, *437*, 831–837.

Yan, N., Gu, L., Kokel, D., Chai, J., Li, W., Han, A., et al. (2004). Structural, biochemical, and functional analyses of CED-9 recognition by the proapoptotic proteins EGL-1 and CED-4. *Molecular Cell*, *15*, 999–1006.

Yang, E., Zha, J., Jockel, J., Boise, L. H., Thompson, C. B., & Korsmeyer, S. J. (1995). Bad, a heterodimeric partner for Bcl-XL and Bcl-2, displaces Bax and promotes cell death. *Cell*, *80*, 285–291.

Yao, Y., Bobkov, A. A., Plesniak, L. A., & Marassi, F. M. (2009). Mapping the interaction of pro-apoptotic tBID with pro-survival BCL-XL. *Biochemistry*, *48*, 8704–8711.

Yin, X. M., Oltvai, Z. N., & Korsmeyer, S. J. (1994). BH1 and BH2 domains of Bcl-2 are required for inhibition of apoptosis and heterodimerization with Bax. *Nature*, *369*, 321–323.

Youle, R. J., & Strasser, A. (2008). The BCL-2 protein family: Opposing activities that mediate cell death. *Nature Reviews Molecular Cell Biology*, *9*, 47–59.

Yu, J., Zhang, L., Hwang, P. M., Kinzler, K. W., & Vogelstein, B. (2001). PUMA induces the rapid apoptosis of colorectal cancer cells. *Molecular Cell*, *7*, 673–682.

Zha, J., Harada, H., Yang, E., Jockel, J., & Korsmeyer, S. J. (1996). Serine phosphorylation of death agonist BAD in response to survival factor results in binding to 14-3-3 not BCL-X(L). *Cell*, *87*, 619–628.

Zhang, S., & Link, A. J. (2011). Bcl-2 family interactome analysis using bacterial surface display. *Integrative Biology*, *3*, 823–831.

CHAPTER FOUR

How to Analyze Mitochondrial Morphology in Healthy Cells and Apoptotic Cells in *Caenorhabditis elegans*

Stéphane G. Rolland[1]
LMU Biocenter, Department Biology II, Ludwig-Maximilians-University, Munich, Germany
[1]Corresponding author: e-mail address: rolland@bio.lmu.de

Contents

Abstract

Mitochondria constantly undergo fusion and fission events. A proper balance of fusion and fission is essential in healthy cells, as disrupting this balance is associated with several neurodegenerative diseases. Mitochondrial fission has also been shown to play an important role during apoptosis. Hence, the machineries that control mitochondrial morphology have both nonapoptotic and apoptotic functions. Seminal work in yeast has identified some of the key components of these machineries. However, the list is certainly not complete and new factors that are specific to metazoans are being identified every year. In this review, we describe methodologies to test whether a particular candidate gene plays a role in the control of mitochondrial morphology in healthy cells and apoptotic cells using *Caenorhabditis elegans*.

Methods in Enzymology, Volume 544
ISSN 0076-6879
http://dx.doi.org/10.1016/B978-0-12-417158-9.00004-2

1. INTRODUCTION

Mitochondria are highly dynamic organelles that constantly fuse and divide (Chan, 2012; Nunnari & Suomalainen, 2012; Okamoto & Shaw, 2005; Westermann, 2010). The processes of mitochondrial fusion and fission are controlled by a conserved family of dynamin-related GTPases. DRP-1, the homolog of mammalian Drp1, is responsible for mitochondrial fission in *Caenorhabditis elegans* (Labrousse, Zappaterra, Rube, & van der Bliek, 1999). Conversely, *C. elegans* FZO-1 and EAT-3, the homologs of mammalian Mfn1,2 and Opa1, are responsible for the fusion of the outer mitochondrial membrane (OMM) and inner mitochondrial membrane (IMM), respectively (Ichishita et al., 2008; Kanazawa et al., 2008).

Mitochondrial morphology is defined by a balance of mitochondrial fusion and fission. This morphology has been shown to change in response to several cellular signals, such as metabolic, cell cycle, or apoptotic signals (Antico Arciuch, Elguero, Poderoso, & Carreras, 2012; Liesa & Shirihai, 2013; Wang & Youle, 2009). During apoptosis of mammalian cells, mitochondria undergo fission, hereafter referred to as "fragmentation" (Wang & Youle, 2009). In *C. elegans*, mitochondria also fragment during apoptosis and this fragmentation is required for efficient execution of the apoptotic process (Jagasia, Grote, Westermann, & Conradt, 2005). Whereas mitochondrial fragmentation during apoptosis in mammalian cells is associated with the release of proapoptotic factors such as cytochrome-*c* (Landes & Martinou, 2011), the role of mitochondrial fragmentation during *C. elegans* apoptosis remains elusive.

Understanding the role of mitochondrial morphology changes during apoptosis requires a better understanding of the machineries that control mitochondrial fusion and fission as well as how these machineries are regulated. In addition, these machineries play roles in both apoptotic and nonapoptotic cells. For example, mitochondrial fission is important during apoptosis as well as cell division (Antico Arciuch et al., 2012; Labrousse et al., 1999). The function and regulation of the mitochondrial fusion and fission machineries must therefore be understood in both apoptotic and nonapoptotic contexts.

Although the general components of these machineries (DRP-1, FZO-1, and EAT-3) are known in *C. elegans*, several additional factors have been recently identified and probably more will be identified in the future (Table 4.1) (Head et al., 2011; Ichishita et al., 2008; Lee et al., 2009). This

Table 4.1 Factors controlling mitochondrial morphology in *C. elegans*

Gene in *C. elegans*	Human homologs	Loss-of-function allele	Loss-of-function phenotype	Overexpression phenotype	Overall function
drp-1	Drp-1	*tm1108*	Mitochondrial elongation; mitochondrial distribution defect (Labrousse et al., 1999)	Mitochondrial fragmentation; ectopic apoptosis (Jagasia et al., 2005)	Mitochondrial fission
fzo-1	Mitofusin 1 and 2	*tm1133*	Mitochondrial fragmentation; no ectopic apoptosis (Breckenridge et al., 2008)	Mitochondrial clustering (Rolland, Lu, David, & Conradt, 2009)	OMM fusion
eat-3	Opa1	*tm1107*	Mitochondrial fragmentation; abnormal *cristae*; no ectopic apoptosis (Kanazawa et al., 2008)	Mitochondrial fragmentation (Rolland et al., 2009)	Inner mitochondrial membrane fusion
					Cristae maintenance
moma-1	my025 or APOOL	*tm1912*	Swollen mitochondria; abnormal *cristae* (Head, Zulaika, Ryazantsev, & van der Bliek, 2011)	ND	*Cristae* maintenance
immt-1	Mitofilins	*tm1730*		ND	*Cristae* maintenance
chch-3	CHCHD3	*tm2336*	Swollen mitochondria (Head et al., 2011)	ND	*Cristae* maintenance
immt-2	Mitofilins	*tm2366*	Interconnected mitochondria (Head et al., 2011)	ND	*Cristae* maintenance
dic-1	DICE1	*tm1615*	Reduced *cristae* number (Lee et al., 2009)	Increased *cristae* number (Lee et al., 2009)	*Cristae* maintenance

Continued

Table 4.1 Factors controlling mitochondrial morphology in *C. elegans*—cont'd

Gene in *C. elegans*	Human homologs	Loss-of-function allele	Loss-of-function phenotype	Overexpression phenotype	Overall function
ced-9	BCL-2 like proteins	*n2812*	No obvious phenotype (Rolland et al., 2009; Tan et al., 2008)	Mitochondrial hyperfusion (Rolland et al., 2009; Tan et al., 2008)	Regulation of mitochondrial morphology
egl-1	BH3-only proteins	*n3082*	Mitochondrial elongation (Lu, Rolland, & Conradt, 2011)	Mitochondrial fragmentation (Lu et al., 2011)	Regulation of mitochondrial morphology
mma-1	LRPPRC	*RNAi*	Mitochondrial hyperfusion (Rolland et al., 2013)	ND	Regulation of mitochondrial translation
fis-1	hFis1	*tm1867*	No obvious phenotype (Breckenridge et al., 2008)	ND	Unknown
fis-2		*gk363*			
mff-1	Mff	*tm2955*	ND	ND	Unknown
mff-2		*tm3041*	ND	ND	Unknown

ND, not determined.

review describes methodologies to test whether the inactivation and/or overexpression of a particular gene in *C. elegans* affects mitochondrial morphology in apoptotic and/or nonapoptotic cells.

2. WHY ADDRESS THESE QUESTIONS IN *C. ELEGANS*

Despite being a relatively simple metazoan with only 959 somatic cells, *C. elegans* is great model organism for cell biologists as it contains all major cell types (germ cells, muscle cells, neurons, and intestinal cells). Due to its essentially invariant cell lineage (Sulston, Schierenberg, White, & Thomson, 1983), it is also well suited for the study of developmentally regulated processes. Furthermore, with its several powerful genetic tools that have been developed over almost five decades, *C. elegans* has proven to be an excellent model for gene discovery (Brenner, 1974). More specifically concerning the study of the regulation of mitochondrial morphology, all the genes that control mitochondrial morphology in *C. elegans* are conserved in humans (Table 4.1). Hence, the new factors discovered in *C. elegans* are likely to be conserved and the understanding of the regulation of mitochondrial morphology in *C. elegans* will likely increase our understanding of these processes in humans as well.

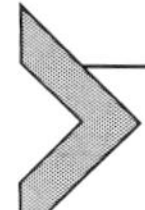

3. HOW TO INACTIVATE OR OVEREXPRESS A PARTICULAR CANDIDATE GENE IN *C. ELEGANS*

In this section, an overview of how to inactivate or overexpress a candidate gene is described, with an emphasis on genes that regulate mitochondrial morphology.

3.1. Inactivation of a particular candidate gene

3.1.1 Loss-of-function mutants

After the genome of *C. elegans* was published in 1998, efforts have been made to generate mutations in more than 20,000 genes of *C. elegans* (*C. elegans Consortium*, 1998). By 2012, with the effort of individual laboratories and the *C. elegans* Deletion Mutant Consortium, almost 30% of the *C. elegans* genes have been targeted (*C. elegans Consortium*, 2012). More recently, with the development of next-generation sequencing, a new library of mutants has been generated by the "million mutations project." For each *C. elegans* gene, on average eight different mutant alleles are available in this library (Thompson et al., 2013). The list of all these mutants or mutations can be found on Wormbase (http://www.wormbase.org/) and all these mutants are available from the *Caenorhabditis* Genetics Center (CGC) (http://www.cbs.umn.edu/cgc). Table 4.1 lists all the mutations (loss-of-function alleles) of genes controlling mitochondrial morphology in *C. elegans*.

To test the effect of a new candidate gene on mitochondrial morphology in embryos, animals carrying a mutation in this gene can be analyzed by TMRE staining (Section 4.2.2). To analyze the effect of such a mutation on mitochondrial morphology in other cell types, the mutant strain can be transformed with fluorescent reporters that label mitochondria (Section 4.3). (For detailed protocols to generate *C. elegans* transgenic animals; see Evans, 2006.)

3.1.2 Inactivation by RNA interference

RNA interference (RNAi) is a powerful tool for reverse genetics in *C. elegans*. Specifically, feeding *C. elegans* with an *Escherichia coli* strain expressing double-strand RNA (dsRNA) of a particular gene is often sufficient to knock-down its function (Timmons, Court, & Fire, 2001). Ahringer and coworkers have generated a RNAi feeding library that contains ~87% of the *C. elegans* genes (Kamath et al., 2003). In addition, Marc Vidal and coworkers have generated a RNAi-ORF feeding library (Rual et al., 2004). Whereas the Ahringer's library targets 500 bp conserved exon sequences, the RNAi-ORF library targets the entire ORF of a particular gene but only covers 55% of *C. elegans* genes (http://www.geneservice.co.uk and http://www.lifesciences.sourcebioscience.com/).

Even though RNAi-mediated inactivation of *drp-1*, *fzo-1*, and *eat-3* recapitulates the phenotypes observed in the loss-of-function mutants, their penetrances vary. For example, RNAi by feeding of *drp-1* induces mitochondrial elongation in 100% of treated animals (Rolland et al., 2009). In contrast, only 66% of animals have fragmented mitochondria upon *fzo-1* RNAi (Rolland et al., 2009). Injection of *in vitro* synthesized dsRNA into animals often results in a more penetrant phenotype compared to the feeding protocol. Hence, depending on the gene, injection of dsRNA may be more appropriate (Ahringer, 2006).

A protocol to analyze mitochondrial morphology in embryos or muscle cells in response to the inactivation of a candidate gene using RNAi by feeding is described below (Sections 4.2.2 and 4.3.4).

3.2. Overexpression of a particular candidate gene

3.2.1 In embryos, using heat-shock promoters

hsp-16.2 and *hsp-16.41* promoters are heat-inducible promoters that have been successfully used in embryos to overexpress genes that regulates mitochondrial morphology, such as *ced-9*, *egl-1*, *fzo-1*, *eat-3*, and *drp-1* (Table 4.1) (Jagasia et al., 2005; Lu et al., 2011; Rolland et al., 2013). *hsp-16.41* drives the expression of a transgene in the intestine and pharyngeal tissue, whereas *hsp-16.2* drives the expression of a transgene in neurons and hypodermal cells (Fire, Harrison, & Dixon, 1990; Stringham, Dixon, Jones, & Candido, 1992). Since these two promoters have a complementary pattern of expression, placing the candidate gene under the control of both promoters will ensure the expression of the transgene in the majority of the cells. (Subcloning can be performed into pPD49.78 and pPD49.83, which contain the *hsp-16.2* and *hsp-16.41* promoters, respectively; generated by A. Fire, Stanford School of Medicine and available at http://www.addgene.org/; plasmids 1447 and 1448.) Another important consideration is the presence of introns in the gene of interest, since introns are necessary to maximize its expression. Synthetic introns have been shown to increase expression of cDNA-based transgenes in *C. elegans* (Okkema, Harrison, Plunger, Aryana, & Fire, 1993). Such synthetic introns are present in pPD49.78 and pPD49.83 and were shown to be sufficient to properly drive the expression of *ced-9*, *egl-1*, *fzo-1*, and *eat-3* cDNA (Jagasia et al., 2005; Lu et al., 2011; Rolland et al., 2009).

To test the effect of the overexpression of a new candidate gene on mitochondrial morphology, transgenic animals expressing the cDNA of this gene and a mitochondrial matrix-targeted GFP (mitoGFP) under the control of the heat-shock promoters can be analyzed using the protocols described below (Section 4.3.3).

3.2.2 In specific cell types, using smg-inducible system

Different promoters can be used to drive the expression of a particular candidate gene in specific cell types (Section 4.3.2). However, constitutive expression of a particular gene may be toxic. In that case, a cell-type specific inducible expression system, such as the *smg*-inducible system, can be used.

The *smg*-inducible system takes advantage of a temperature-sensitive allele of the Smg RNA surveillance machinery (*smg-1(cc546*ts)) (Pulak & Anderson, 1993). The candidate gene is cloned under the control of a cell-type specific promoter and upstream of an abnormal 3′UTR (e.g., in pPD118.60; provided by A. Fire, Stanford School of Medicine, and available at http://www.addgene.org/; plasmid 1598). At permissive temperature (15 °C), the mRNA of the transgene is degraded due to its abnormal 3′UTR. At restrictive temperature (25 °C), the Smg RNA surveillance machinery is not functional, which results in the stabilization of the transgene mRNA and its translation. This system has been successfully used to drive the inducible expression of *ced-9* and *fzo-1* in body wall muscle cells (Rolland et al., 2009).

To analyze the role of a candidate gene on mitochondrial morphology in a specific cell-type, the *smg*-inducible system can be used in conjunction with the corresponding cell-type specific mitochondrial fluorescent reporter (Section 4.3).

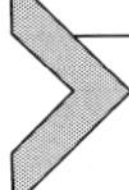

4. HOW TO VISUALIZE MITOCHONDRIA IN *C. ELEGANS*

4.1. General considerations for imaging

This review describes methodologies to visualize mitochondria using epifluorescence microscopy. Indeed, deconvolution of a *Z*-stack acquired with an epifluorescence microscope results in a quality that is sufficient to visualize and quantify mitochondrial morphology. Nonetheless, all the fluorescent imaging described below can also be performed using confocal microscopy.

4.2. Mitochondrial membrane potential-sensitive dyes

4.2.1 TMRE, Mitotracker, and others dyes

Several mitochondrial membrane potential-sensitive dyes have been used to visualize mitochondria in *C. elegans* (Table 4.2). Some of these dyes also allow one to measure the dynamics of mitochondrial membrane potential as they are rapidly and reversibly taken up by mitochondria (Loew, Tuft, Carrington, & Fay, 1993). For example, protocols using TMRE or DiS-$C_3(3)$ have been established in *C. elegans* to measure mitochondrial membrane potential at the organismal level (Gaskova et al., 2007; Yoneda et al., 2004).

The dyes described in this section have been used to stain mitochondria in several tissues (germline, embryonic cells, body wall muscle cells, hypodermis, neurons). However, the choice of the dye will depend on the tissue to be analyzed. For example, TMRE specifically stains mitochondria in embryos, whereas Mitotracker Red CMXRos does not (Rolland et al., 2009). In contrast, Mitotracker Red CMXRos stains specifically mitochondria in muscle cells (Kimura et al., 2007).

Table 4.2 *C. elegans* mitochondrial membrane potential-sensitive dyes

Dye	Concentration	Tissue analyzed	References
Rhodamine B hexyl ester	30 μ*M*	Embryonic cells	Jagasia et al. (2005)
TMRE	30 μ*M*	Embryonic cells	Breckenridge et al. (2008), Jagasia et al. (2005), and Rolland et al. (2009)
MitoTracker Red CMXRos	500 n*M*–5 μ*M*	Muscle, sperm	Al Rawi et al. (2011) and Kimura, Tanaka, Nakamura, Takano, and Ohkuma (2007)
Rhodamine 6G	5 μ*M*	Germline, embryonic cells	Badrinath & White (2003) and Labrousse et al. (1999)
DiS-C_3(3)	4 μ*M*	NA[a]	Gaskova, DeCorby, and Lemire (2007)
HRB and HR101	100 p*M*	Hypodermis, muscle, neurons, germline	Mottram, Forbes, Ackley, and Peterson (2012)

[a]Lemire and coworkers used DiS-C_3(3) to measure mitochondrial membrane potential (Gaskova et al., 2007).

In most protocols, mitochondrial dyes are simply incorporated in nematode growth medium (NGM) agar plates. *C. elegans* animals are fed with *E. coli* bacteria which are stained with the dye. An important issue with this approach is the background staining present in the animals' intestines. While it is not a problem to analyze mitochondrial morphology in embryos, the background staining considerably affects the mitochondria visualization in other tissues and requires several washing steps.

The concentration of the mitochondrial dye and the duration of exposure are important parameters to consider since extended period of staining or a high concentration of the dyes are toxic (Mottram et al., 2012). Recently, the development of new rhodamine derivatives (HRB and HR101) has improved both these parameters as these dyes are rapidly taken up (within 2 h) at low concentrations ($\geq$100 p*M*) and exhibit low toxicity (Mottram et al., 2012).

4.2.2 Protocol to visualize mitochondria in embryos using TMRE

4.2.2.1 Media and solutions

- Solutions for NGM agar plates (1 *M* $CaCl_2$, 1 *M* $MgSO_4$, 1 *M* potassium phosphate buffer, pH 6.0, and cholesterol solution (5 mg/ml in 100% ethanol)) (Stiernagle, 2006).

- NGM agar medium: prepare 50 ml aliquots using the following recipe:
  - 0.15 g NaCl
  - 0.37 g peptone
  - 0.85 g agar
  - qsp 50 ml with dH_2O

After autoclaving, the aliquots can be stored for several weeks at 4 °C.

- 0.1 *M* TMRE: dissolve 51.4 mg/ml of TMRE (molecular probes) in DMSO and store 50 μl aliquots at −20 °C in the dark. (Aliquots can be thawed and frozen three to four times.)
- *E. coli* OP50 culture: grow a *E. coli* OP50 culture as described (Stiernagle, 2006). (To ensure reproducibility, use a fresh culture inoculated the day before.)

4.2.2.2 Materials

- 32 mm Petri dishes
- Microscope slides and cover glasses
- 25G needles
- Epifluorescence microscope equipped with Nomarski optics and TRITC filter (e.g., Zeiss Axioskop 2 microscope with Zeiss Filter F41-032)

4.2.2.3 Protocol

Day 1:

1. Melt one aliquot of NGM agar medium using a microwave. When the medium has cooled down to 55 °C, add the following (in that order):
   - 50 μl 1 *M* $CaCl_2$
   - 50 μl 1 *M* $MgSO_4$
   - 1.25 ml 1 *M* potassium phosphate buffer pH 6.0
   - 100 μl cholesterol (5 mg/ml)
   - 15 μl 0.1 *M* TMRE
2. Pour 5 ml of medium per 32-mm Petri dish. Let the plates dry overnight in the dark. (Protect TMRE plates from light, since TMRE is light sensitive.)

Day 2:

1. Transfer 10 L4 larvae of the strain of interest and control strain on 60 mm NGM plates (without TMRE) and incubate at 20 °C for 24 h.
2. Seed TMRE plates with 50 μl of OP50 culture and incubate at 20 °C for 24 h in the dark.

Day 3:

Transfer 10 adults of each strain to the TMRE plates and incubate at 20 °C for 15 h (i.e., overnight).

Day 4:

1. Transfer 2–3 adults from the TMRE plate to a regular NGM plate (without TMRE). Let the animals crawl on the plate for 5 min. (This step is required to remove the excess of TMRE positive bacteria, which creates background signal.) (see Fig. 4.1).
2. Prepare a 2% agarose pad on a microscope slide using the setup described in Fig. 4.1.
3. Transfer the adults in 5 μl of M9 on a cover glass. (Alternatively, M9 containing 1 m*M* of levamisole can be used to paralyze the animals.)
4. Bisect adult animals using 25G needles to extract the embryos.
5. Transfer the cover glass up-side-down on the agarose pad.
6. Image mitochondria using an epifluorescence microscope.

The microscope slides can be analyzed for 10–15 min. To analyze the slides for an extended period of time, the cover glass can be sealed by brushing melted white vaseline along the edge of the cover glass.

To further process the images with a deconvolution software, it is necessary to acquire a *Z*-stack of the embryo. To define the top and bottom sections of the *Z*-stack, it is preferable to use Nomarski optics, in order to minimize the bleaching of the TMRE signal. For the same reason, during the acquisition of the TMRE signal, it is recommended to increase the imaging exposure time while using a neutral density filter to decrease the intensity of the excitation light. These values need to be adjusted depending on the microscope, excitation light, and camera setup.

As an indication, using a Zeiss Axioskop 2 equipped with a 100 × 1.3 NA oil lens, a TRITC filter (F41-032; Zeiss), a HAL-100 mercury lamp housing, a CCD camera (1300; Micromax), and a 50% neutral density filter, 100 ms exposure acquisition time results in sufficient quality for further processing.

Examples of TMRE stained embryos acquired with this setup are provided in Fig. 4.2.

4.2.2.4 Specific protocol for TMRE staining of animals treated with RNAi

RNAi is performed essentially as previously described (Ahringer, 2006). On Day 1, 32 mm RNAi plate supplemented with Carbenicillin (25 μg/ml) and 1 m*M* IPTG are inoculated with a fresh culture of the

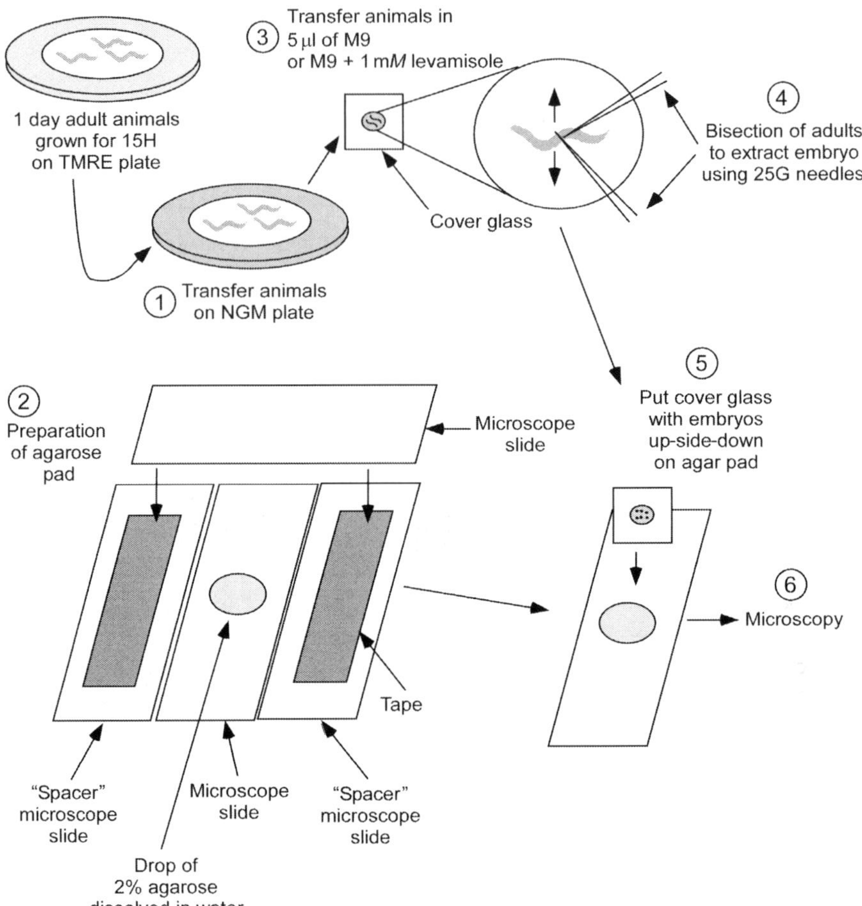

Figure 4.1 Flowchart of the protocol to visualize mitochondria in *C. elegans* embryos using TMRE staining. (1) 1-day post-L4 *C. elegans* adults grown for 15 h on TMRE medium are transferred for 5 min onto normal NGM medium. (2) 2% agarose thin pad is prepared between two microscope slides. The laboratory tape present on the two "spacer" microscope slides defines the thickness of the pad. (3) 2–3 adults are then transferred from the normal NGM medium to a microscope cover glass in 5 μl of M9 medium (that can be supplemented with 1 m*M* Levamisole). (4) Using two 25G needles, adults are bisected to extract the embryos. (5) The cover glass is then transferred up-side-down on the agarose pad. (6) The microscope slide is analyzed by Nomarski and fluorescent microcopy.

RNAi clone. On Day 2, 10 L4 larvae of the strain of interest are transferred onto the RNAi plate. On Day 3, adults exposed to RNAi for 24 h are transferred onto the TMRE plate. The rest of the protocol is performed as described above.

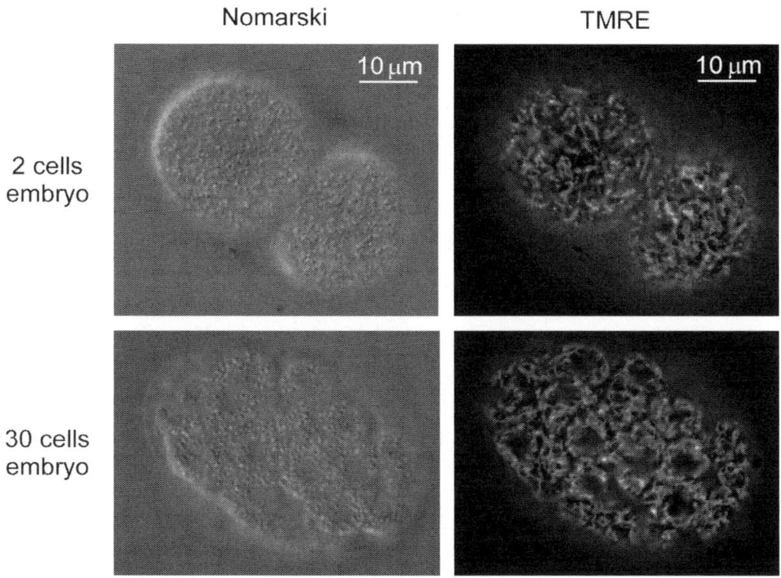

Figure 4.2 Representative images of TMRE stained embryos of *C. elegans*. 2 cells and 30 cells embryos were stained with TMRE and analyzed by Nomarski and fluorescent microscopy. The fluorescent *Z*-stacks were deconvolved using Autodeblur. A single plane image for each embryo is shown.

4.3. Mitochondrial-targeted fluorescent proteins

4.3.1 Which fluorescent proteins to use

As shown in Table 4.3, several fluorescent proteins have been used to label mitochondria in *C. elegans*: GFP, YFP, CFP, and the dsRedT3 variant. These proteins have been targeted to the mitochondrial matrix (Labrousse et al., 1999; Rolland et al., 2013). In addition, an OMM marker has been generated by fusing the N-terminal 30 amino acids of yeast TOM70 to YFP (TOM70::YFP) (Labrousse et al., 1999). Furthermore, a photoactivable mitoGFP (PAmitoGFP) has been recently generated (Rolland et al., 2013).

Unlike mitoGFP, mitoCFP, PAmitoGFP, and TOM70::YFP, the expression of mitodsRed seems to affect viability and growth of the transgenic animals. For this reason, mitodsRed reporters (such as pBC914) have to be used at lower concentrations (0.2 ng/μl) compared to other mitochondrial reporters such as mitoGFP (pVDB#1; 5 ng/μl) (Rolland et al., 2013).

Table 4.3 Plasmids encoding the *C. elegans* mitochondria fluorescent reporters

		Fluorescent proteins targeted to				
		Matrix				**OMM**
Tissue	**Promoter**	**mitoGFP**	**mitoCFP**	**mitodsRed**	**PAmitoGFP**	**TOM70::YFP**
Embryonic cells	*Phsp-16.2* and *Phsp-16.41*	pBC307 and pBC308 (Jagasia et al., 2005)				
Body wall muscle cells	*Pmyo*-3	pVDB#1 (Labrousse et al., 1999)	pVDB#2 (Labrousse et al., 1999)	pBC914 (Rolland et al., 2013)	pBC900 (Rolland et al., 2013)	pVDB#3 (Labrousse et al., 1999)
Apoptotic cells	*Pegl*-1	pBC306 (Jagasia et al., 2005)				

4.3.2 Which promoter to drive the expression of the mitochondrial marker

Since the heat-shock promoters *hsp-16.2* and *hsp-16.41* have a complementary pattern of expression, the two heat-shock plasmids expressing mitoGFP (Table 4.3) are usually cotransformed to maximize the number of mitoGFP-positive cells. These reporters become inducible at mid-gastrulation (~150 min after the first cell division at 20 °C) (Stringham et al., 1992). They are therefore suitable to study mitochondrial morphology in apoptotic cells (since the first cell death occurs at ~220 min at 20 °C) (Section 4.4). An example of an embryo carrying the P_{HS} *mitoGFP* transgene is shown in Fig. 4.3A and A′.

Van der Bliek and coworkers have pioneered the visualization of mitochondria in body wall muscle cells (muscles which are required for *C. elegans* locomotion) using the *myo-3* promoter to drive the expression of mitoGFP, mitoCFP, and TOM70::YFP (Table 4.3; Labrousse et al., 1999). More

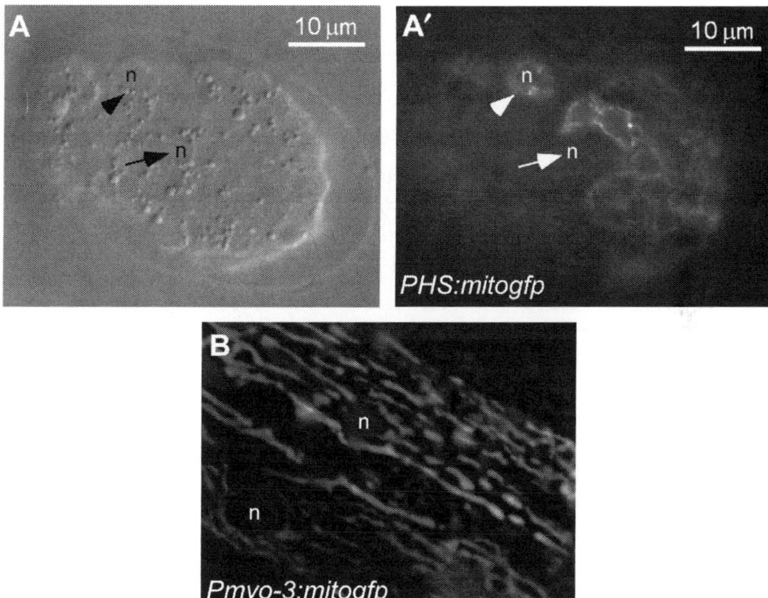

Figure 4.3 Representative images of different mitochondrial fluorescent reporters. MD2416 embryo analyzed by Nomarski (A) and fluorescent microscopy (A′). MD2416 animals are mosaic for the presence of the transgene and thus its expression. Only some cells (arrowhead) express the mitoGFP reporter whereas others do not (arrow) (n: nucleus). (B) MD2517 L4 larvae body wall muscle cells mitochondria labeled with mitoGFP and analyzed by fluorescent microscopy. Both fluorescent images are single plane images out of a *Z*-stack deconvolved using Autodeblur.

recently, mitodsRed and PAmitoGFP have been placed under the control of the *myo-3* promoter (Table 4.3). All these plasmids are often cotransformed in *C. elegans* with the pRF4 plasmid. pRF4 contains the *rol-6*(*su1006*dm) allele, which causes the transgenic animals to exhibit a dominant roller phenotype (Kramer, French, Park, & Johnson, 1990). This phenotype is accompanied by a helical twisting of the four muscle quadrants, which facilitates the visualization of the muscle cells and their mitochondria. An example of a transgenic animal carrying the $P_{myo\text{-}3}$ *mitoGFP* transgene is provided in Fig. 4.3B.

4.3.3 Visualizing embryonic mitochondria using the P_{HS} mitoGFP *strain*

4.3.3.1 Strain, media, and solutions

- MD2416 (P_{HS} *mitoGFP* strain; Table 4.4)
- 60 mm NGM plate seeded with *E. coli* OP50 strain (Stiernagle, 2006)
- M9 liquid media (Stiernagle, 2006)

4.3.3.2 Materials

- 20 and 32 °C incubators
- Microscope slides and cover glasses
- Epifluorescence microscope equipped with Nomarski optics and FITC filter (e.g., Zeiss Axioskop 2 microscope with Zeiss Filter F41-012)

Table 4.4 Transgenic lines expressing the *C. elegans* mitochondria fluorescent reporters

Strain	Genotype	References
MD2416	*lin-15*(*n765*ts); *bcEx555* [pL15EK (50 ng/μl) + pBC307 (1 ng/μl) + pBC308 (1 ng/μl)]	Jagasia et al. (2005)
MD2517	*bcEx620* [pRF4 (80 ng/μl) + pVDB#1 (5 ng/μl)]	Rolland et al. (2009)
MD2922	*bcEx856* [pRF4 (80 ng/μl) + pBC900 (2 ng/μl) + pBC914 (0.2 ng/μl)]	Rolland et al. (2013)
MD1948	*unc-76*(*e911*); *bcIs50* [pBC306 (5 ng/μl) + p76-16B (75 ng/μl)]	Jagasia et al. (2005)
MD1386	*unc-76*(*e911*); *bcEx383* [pBC307 (5 ng/μl) + pBC27 (5 ng/μl) + p76-16B (75 ng/μl)]	Jagasia et al. (2005)

4.3.3.3 Protocol

Day 1:

Transfer 20 MD2416 L4 larvae on 60 mm NGM plates seeded with OP50 and incubate at 20 °C for 24 h.

MD2416 carries *n765*ts, a temperature-sensitive mutation in the *lin-15AB* gene and *bcEx555*, an extra-chromosomal array composed of P_{HS} *mitoGFP* and pL15EK. At restrictive temperature (≥20 °C), nontransgenic animals exhibit a multivulva (Muv) phenotype, whereas transgenic animals are non-Muv or "wild-type" (due to the rescue of the *lin-15AB*(*n765*ts) mutation by the plasmid pL15EK) (Sternberg, 2005). At permissive temperature (15 °C), both transgenic and nontransgenic animals appear "wild-type." *The MD2416 strain must be therefore kept at* ≥20 °C. In addition, the Muv phenotype is only visible with a dissecting microscope at the adult stage. Therefore, it is essential to transfer 20 L4 larvae on Day 1, in order to have at least 10 transgenic adult animals on Day 2.

Day 2:

1. Transfer 10 transgenic (non-Muv) MD2416 adults animals onto a new 60 mm NGM plate seeded with OP50.
2. Incubate the plate at 20 °C for 2 h.
3. Incubate the plate at 32 °C for 45 min.
4. Incubate the plate at 20 °C for 1 h and 15 min.
5. Prepare a 2% agarose pad as described earlier (Fig. 4.1).
6. Transfer embryos that were laid on the plate in 5 μl of M9 on the agarose pad.
7. Cover the pad with a cover glass.
8. Image mitochondria using an epifluorescence microscope.

In order to further process the images with a deconvolution software, it is necessary to acquire a *Z*-stack of the embryo. Unlike for the TMRE staining, the top and bottom section of the *Z*-stack can be defined using fluorescence microscopy, since all the mitochondrial fluorescent reporters are stable. In addition, as mentioned earlier, MD2416 transgenic animals carry extra-chromosomal arrays. They are therefore mosaic for the presence of the transgene and consequently its expression. Hence, not all the cells of a transgenic embryo will express the mitoGFP reporter (Fig. 4.3A and A′).

4.3.4 Visualizing muscle cell mitochondria using the $P_{myo\text{-}3}$ mitoGFP *strain*

4.3.4.1 Strain, media, and solutions

– MD2517 ($P_{myo\text{-}3}$ *mitoGFP* strain; Table 4.4)

- 60 mm NGM plate seeded with *E. coli* OP50 strain (Stiernagle, 2006)
- M9 liquid media (Stiernagle, 2006) supplemented with 1 m*M* levamisole

4.3.4.2 Materials

- Microscope slides and cover glasses
- Epifluorescence microscope equipped with FITC filter (e.g., Zeiss Axioskop 2 microscope equipped with Zeiss Filter F41-012)

4.3.4.3 Protocol

1. Prepare a 2% agarose pad as described earlier (Fig. 4.1).
2. Transfer 10 transgenic (roller) L4 larvae animals in 5 μl of M9 supplemented with 1 m*M* levamisole on the agarose pad.
3. Cover the agarose pad with a cover glass.
4. Image mitochondria using an epifluorescence microscope.

It is critical to image mitochondria within 10–15 min. Longer incubations lead to the death of the animal and an artificial mitochondrial fragmentation. Recently, microfluidic devices have been developed to image *C. elegans* animals for an extended period of time (Mondal, Ahlawat, Rau, Venkataraman, & Koushika, 2011). An example of muscle cells mitochondria after deconvolution of a *Z*-stack is provided in Fig. 4.3C.

4.3.4.4 Specific protocol for RNAi treatment of $P_{myo\text{-}3}$ *mitoGFP* strain

RNAi is performed essentially as previously described (Ahringer, 2006). On Day 1, 32 mm RNAi plate supplemented with Carbenicillin (25 μg/ml) and 1 m*M* IPTG are inoculated with a fresh culture of the RNAi clone. On Day 2, two transgenic (roller) MD2517 L4 larvae are inoculated on the RNAi plate. At 20 °C, the next generation of L4 larvae can be analyzed using fluorescence microscopy as described above, 4–5 days later.

4.4. Visualizing mitochondria in apoptotic cells

One hundred and thirteen cells undergo programmed cell death during *C. elegans* embryogenesis (Sulston et al., 1983). The identity of each dying cell, the time at which it dies and its position in the embryo, is highly reproducible from individual to individual. This unique characteristic makes *C. elegans* an excellent model to study changes in mitochondrial morphology in response to apoptotic signals during development.

Apoptotic cells can be identified by Nomarski optics because they exhibit a characteristic "button-like" structure, appearing as flat, round disks. However, since mitochondrial fragmentation occurs before the apoptotic cell exhibits this characteristic structure (Jagasia et al., 2005), other tools have been developed to investigate mitochondrial morphology changes during apoptosis. Specifically, a reporter that expresses mitoGFP under the control of the promoter of the *C. elegans* BH3-only gene *egl-1* has been used to specifically label mitochondria in apoptotic cells (Jagasia et al., 2005) (P_{egl-1}*:mitoGFP*; Table 4.3). Since *egl-1* transcription is an early apoptotic event, this reporter allows visualizing mitochondrial morphology during the entire apoptotic process. Analysis by confocal time-lapse microscopy of a strain carrying this reporter revealed that mitochondrial fragmentation is an early feature of apoptosis in *C. elegans* (Jagasia et al., 2005).

As an alternative, overexpression of *egl-1*, using the heat-shock promoters described earlier, has been shown to induce ectopic apoptosis (Conradt & Horvitz, 1998). A *C. elegans* strain carrying the P_{HS} *egl-1* (pBC27) and P_{HS} *mitoGFP* (pBC307) transgenes (Table 4.4) can therefore be used to analyze changes in mitochondrial morphology during ectopic cell death (Jagasia et al., 2005).

4.5. Quantification of mitochondrial morphology

4.5.1 Qualitative analysis of mitochondrial morphology in embryos

Quantitative measurement of mitochondrial morphology in *C. elegans* embryos represents a technical challenge because of the three-dimensional structure of the mitochondrial network. In addition, during later stages of embryonic development, the small size of the cells makes the quantification even more challenging. For instance, at the stage when the first cell undergoes apoptosis, cells have a diameter of ~2 μm. In comparison, HeLa cells have a diameter of 20 μm and *C. elegans* body wall muscle cells in L4 larvae have a size of 60 μm × 8 μm.

To my knowledge, automated system for the quantification and classification of mitochondrial morphology has not yet been developed for *C. elegans*. As a consequence, the data are analyzed qualitatively using different mitochondrial morphology categories (such as fragmented, tubular, and highly elongated). To ensure unbiased analysis, the data are often analyzed blind by two independent investigators.

4.5.2 Quantitative morphometric analysis of mitochondrial morphology in larval muscle cells

Body wall muscle cells are well suited for quantitative measurement of mitochondrial morphology, because of the two-dimensional structure of their mitochondrial network. In order to increase the signal-to-noise ratio, it is preferable to acquire a *Z*-stack and to further process it using deconvolution. Mitochondria morphology can then be analyzed using a morphometric analysis software such as MetaMorph (MDS Analytical Technologies) or ImageJ (NIH). Among the parameters that can be measured, mitochondrial length is a good indicator of mitochondrial morphology (Lu et al., 2011; Rolland et al., 2009; Tan et al., 2008). Circularity or shape factor (whose values vary between 0 and 1; 0 indicating a perfect line and 1 indicating a perfect circle) is a good indicator of mitochondrial fragmentation (Bess, Crocker, Ryde, & Meyer, 2012). Finally, the mean area to perimeter ratio has been used as a measure of interconnectivity (Bess et al., 2012).

4.5.3 Quantification of mitochondrial morphology and mitochondrial fusion/fission rates using a photoactivable mitoGFP

Fluorescence loss in photobleaching with mitoGFP and more recently photoactivation of PAmitoGFP have been used to measure mitochondrial interconnectivity in body wall muscle cells (Labrousse et al., 1999; Rolland et al., 2013). The recent development of PAmitoGFP in *C. elegans* opens up the possibility to also measure mitochondrial fusion and fission rates, but this technique need to be further developed.

5. POTENTIAL OUTCOMES AND FUTURE EXPERIMENTS

Inactivation of a particular candidate gene or its overexpression may not lead to any obvious mitochondrial morphology changes. In order to conclude that this gene is not involved in the regulation of mitochondrial morphology, it is necessary to confirm that its inactivation or overexpression was successful. In the case of loss-of-function mutation, analysis of the DNA sequence can often predict if the mutation will affect protein function (e.g., deletion of an important domain of the protein or frame shift generating an early stop codon). However, if antibodies against the protein of interest are available, it is good to verify these predictions using western analysis. In the case of RNAi, before concluding on the absence of phenotype, it is important to confirm the efficiency of the inactivation (using RT-PCR or western analysis). Concerning the overexpression, no phenotype can simply mean that the transgene was not expressed or localized properly. This can be tested using antibodies against the protein of interest or alternatively by tagging this protein with GFP.

When the inactivation of a gene by RNAi leads to a change in mitochondrial morphology, it is recommended to confirm the phenotype using a strain carrying a loss-of-function mutation in this gene. In addition, loss-of-function mutations should always be outcrossed, as other mutations in the background may be responsible for the observed phenotype. Finally, if a different loss-of-function allele of the gene of interest is available, it should also be tested.

Once the phenotype has been confirmed, further experiments can be performed to gain a better understanding of how the newly discovered factor regulates mitochondrial morphology. For example, epistasis analysis can be performed to determine whether this new factor acts upstream or downstream of the known fusion and fission machineries or alternatively in a new pathway. The expression pattern and localization of this new factor can also be determined by using a GFP fusion protein. This can be performed with the help of new resources such as the TransgeneOme project, which has already generated a library of fosmid-based GFP transgene covering 70% of the *C. elegans* proteome (Sarov et al., 2012). Finally, if this factor is shown to regulate mitochondrial morphology in apoptotic cells, its role in the execution of the apoptotic process can be tested (Schwartz, 2007).

6. CONCLUDING REMARKS AND FUTURE CHALLENGES

This review describes tools for a candidate gene approach to test whether a particular factor control mitochondrial morphology in healthy and/or apoptotic cells. *C. elegans* being an excellent model for gene discovery, one could imagine a more unbiased approach such as a genetic screen for new factors regulating mitochondrial morphology. Some attempts have been made using RNAi (Ichishita et al., 2008) but with the development of next-generation sequencing, forward genetic screen can now be envisioned. The tools described in this review can be used to test the effect of new identified mutations on mitochondrial morphology in healthy and apoptotic cells.

ACKNOWLEDGMENTS

I thank Barbara Conradt, Saroj Regmi, and Nadin Memar for their comments on the manuscript. This work was supported by funding from the National Institutes of Health Grant GM076651 and the Center for Integrated Protein Science Munich to Barbara Conradt.

REFERENCES

Ahringer, J. (Ed.), (2006). Reverse genetics. In The C. elegans Research Community (Ed.), *WormBook*. http://dx.doi.org/10.1895/wormbook.1.47.1.

Al Rawi, S., Louvet-Vallee, S., Djeddi, A., Sachse, M., Culetto, E., Hajjar, C., et al. (2011). Postfertilization autophagy of sperm organelles prevents paternal mitochondrial DNA transmission. *Science*, *334*, 1144–1147.

Antico Arciuch, V. G., Elguero, M. E., Poderoso, J. J., & Carreras, M. C. (2012). Mitochondrial regulation of cell cycle and proliferation. *Antioxidants & Redox Signaling*, *16*, 1150–1180.

Badrinath, A. S., & White, J. G. (2003). Contrasting patterns of mitochondrial redistribution in the early lineages of *Caenorhabditis elegans* and *Acrobeloides* sp. PS1146. *Developmental Biology*, *258*, 70–75.

Bess, A. S., Crocker, T. L., Ryde, I. T., & Meyer, J. N. (2012). Mitochondrial dynamics and autophagy aid in removal of persistent mitochondrial DNA damage in *Caenorhabditis elegans*. *Nucleic Acids Research*, *40*, 7916–7931.

Breckenridge, D. G., Kang, B. H., Kokel, D., Mitani, S., Staehelin, L. A., & Xue, D. (2008). *Caenorhabditis elegans drp-1* and *fis-2* regulate distinct cell-death execution pathways downstream of *ced-3* and independent of *ced-9*. *Molecular Cell*, *31*, 586–597.

Brenner, S. (1974). The genetics of *Caenorhabditis elegans*. *Genetics*, 77, 71–94.

C. elegans Consortium. (1998). Genome sequence of the nematode *C. elegans*: A platform for investigating biology. *Science*, *282*, 2012–2018.

C. elegans Consortium. (2012). Large-scale screening for targeted knockouts in the *Caenorhabditis elegans* genome. *G3 (Bethesda)*, *2*, 1415–1425.

Chan, D. C. (2012). Fusion and fission: Interlinked processes critical for mitochondrial health. *Annual Review of Genetics*, *46*, 265–287.

Conradt, B., & Horvitz, H. R. (1998). The *C. elegans* protein EGL-1 is required for programmed cell death and interacts with the Bcl-2-like protein CED-9. *Cell*, *93*, 519–529.

Evans, T. C. (Ed.). (April 6, 2006). Transformation and microinjection. In The *C. elegans* Research Community (Ed.), *WormBook*. http://dx.doi.org/10.1895/wormbook.1.108.1.

Fire, A., Harrison, S. W., & Dixon, D. (1990). A modular set of lacZ fusion vectors for studying gene expression in *Caenorhabditis elegans*. *Gene*, *93*, 189–198.

Gaskova, D., DeCorby, A., & Lemire, B. D. (2007). DiS-C3(3) monitoring of in vivo mitochondrial membrane potential in *C. elegans*. *Biochemical and Biophysical Research Communications*, *354*, 814–819.

Head, B. P., Zulaika, M., Ryazantsev, S., & van der Bliek, A. M. (2011). A novel mitochondrial outer membrane protein, MOMA-1, that affects *cristae* morphology in *Caenorhabditis elegans*. *Molecular Biology of the Cell*, *22*, 831–841.

Ichishita, R., Tanaka, K., Sugiura, Y., Sayano, T., Mihara, K., & Oka, T. (2008). An RNAi screen for mitochondrial proteins required to maintain the morphology of the organelle in *Caenorhabditis elegans*. *Journal of Biochemistry*, *143*, 449–454.

Jagasia, R., Grote, P., Westermann, B., & Conradt, B. (2005). DRP-1-mediated mitochondrial fragmentation during EGL-1-induced cell death in *C. elegans*. *Nature*, *433*, 754–760.

Kamath, R. S., Fraser, A. G., Dong, Y., Poulin, G., Durbin, R., Gotta, M., et al. (2003). Systematic functional analysis of the *Caenorhabditis elegans* genome using RNAi. *Nature*, *421*, 231–237.

Kanazawa, T., Zappaterra, M. D., Hasegawa, A., Wright, A. P., Newman-Smith, E. D., Buttle, K. F., et al. (2008). The *C. elegans* Opa1 homologue EAT-3 is essential for resistance to free radicals. *PLoS Genetics*, *4*, e1000022.

Kimura, K., Tanaka, N., Nakamura, N., Takano, S., & Ohkuma, S. (2007). Knockdown of mitochondrial heat shock protein 70 promotes progeria-like phenotypes in *Caenorhabditis elegans*. *Journal of Biological Chemistry*, *282*, 5910–5918.

Kramer, J. M., French, R. P., Park, E. C., & Johnson, J. J. (1990). The *Caenorhabditis elegans rol-6* gene, which interacts with the *sqt-1* collagen gene to determine organismal morphology, encodes a collagen. *Molecular and Cellular Biology, 10*, 2081–2089.

Labrousse, A. M., Zappaterra, M. D., Rube, D. A., & van der Bliek, A. M. (1999). *C. elegans* dynamin-related protein DRP-1 controls severing of the mitochondrial outer membrane. *Molecular Cell, 4*, 815–826.

Landes, T., & Martinou, J. C. (2011). Mitochondrial outer membrane permeabilization during apoptosis: The role of mitochondrial fission. *Biochimica et Biophysica Acta, 1813*, 540–545.

Lee, T. H., Mun, J. Y., Han, S. M., Yoon, G., Han, S. S., & Koo, H. S. (2009). DIC-1 overexpression enhances respiratory activity in *Caenorhabditis elegans* by promoting mitochondrial formation. *Genes to Cells, 14*, 319–327.

Liesa, M., & Shirihai, O. S. (2013). Mitochondrial dynamics in the regulation of nutrient utilization and energy expenditure. *Cell Metabolism, 17*, 491–506.

Loew, L. M., Tuft, R. A., Carrington, W., & Fay, F. S. (1993). Imaging in five dimensions: Time-dependent membrane potentials in individual mitochondria. *Biophysical Journal, 65*, 2396–2407.

Lu, Y., Rolland, S. G., & Conradt, B. (2011). A molecular switch that governs mitochondrial fusion and fission mediated by the BCL2-like protein CED-9 of *Caenorhabditis elegans*. *Proceedings of the National Academy of Sciences of the United States of America, 108*, E813–E822.

Mondal, S., Ahlawat, S., Rau, K., Venkataraman, V., & Koushika, S. P. (2011). Imaging in vivo neuronal transport in genetic model organisms using microfluidic devices. *Traffic, 12*, 372–385.

Mottram, L. F., Forbes, S., Ackley, B. D., & Peterson, B. R. (2012). Hydrophobic analogues of rhodamine B and rhodamine 101: Potent fluorescent probes of mitochondria in living *C. elegans*. *Beilstein Journal of Organic Chemistry, 8*, 2156–2165.

Nunnari, J., & Suomalainen, A. (2012). Mitochondria: In sickness and in health. *Cell, 148*, 1145–1159.

Okamoto, K., & Shaw, J. M. (2005). Mitochondrial morphology and dynamics in yeast and multicellular eukaryotes. *Annual Review of Genetics, 39*, 503–536.

Okkema, P. G., Harrison, S. W., Plunger, V., Aryana, A., & Fire, A. (1993). Sequence requirements for myosin gene expression and regulation in *Caenorhabditis elegans*. *Genetics, 135*, 385–404.

Pulak, R., & Anderson, P. (1993). mRNA surveillance by the *Caenorhabditis elegans smg* genes. *Genes & Development, 7*, 1885–1897.

Rolland, S. G., Lu, Y., David, C. N., & Conradt, B. (2009). The BCL-2-like protein CED-9 of *C. elegans* promotes FZO-1/Mfn1,2- and EAT-3/Opa1-dependent mitochondrial fusion. *Journal of Cell Biology, 186*, 525–540.

Rolland, S. G., Motori, E., Memar, N., Hench, J., Frank, S., Winklhofer, K. F., et al. (2013). Impaired complex IV activity in response to loss of LRPPRC function can be compensated by mitochondrial hyperfusion. *Proceedings of the National Academy of Sciences of the United States of America, 110*, E2967–E2976.

Rual, J. F., Ceron, J., Koreth, J., Hao, T., Nicot, A. S., Hirozane-Kishikawa, T., et al. (2004). Toward improving *Caenorhabditis elegans* phenome mapping with an ORFeome-based RNAi library. *Genome Research, 14*, 2162–2168.

Sarov, M., Murray, J. I., Schanze, K., Pozniakovski, A., Niu, W., Angermann, K., et al. (2012). A genome-scale resource for in vivo tag-based protein function exploration in *C. elegans*. *Cell, 150*, 855–866.

Schwartz, H. T. (2007). A protocol describing pharynx counts and a review of other assays of apoptotic cell death in the nematode worm *Caenorhabditis elegans*. *Nature Protocols, 2*, 705–714.

Sternberg, P. W. (2005). Vulval development. In The C. elegans Research Community (Ed.), *WormBook*. http://dx.doi.org/10.1895/wormbook.1.6.1.

Stiernagle, T. (2006). Maintenance of C. elegans. In The C. elegans Research Community (Ed.), *WormBook*. http://dx.doi.org/10.1895/wormbook.1.101.1.

Stringham, E. G., Dixon, D. K., Jones, D., & Candido, E. P. (1992). Temporal and spatial expression patterns of the small heat shock (*hsp16*) genes in transgenic *Caenorhabditis elegans*. *Molecular Biology of the Cell*, *3*, 221–233.

Sulston, J. E., Schierenberg, E., White, J. G., & Thomson, J. N. (1983). The embryonic cell lineage of the nematode *Caenorhabditis elegans*. *Developmental Biology*, *100*, 64–119.

Tan, F. J., Husain, M., Manlandro, C. M., Koppenol, M., Fire, A. Z., & Hill, R. B. (2008). CED-9 and mitochondrial homeostasis in *C. elegans* muscle. *Journal of Cell Science*, *121*, 3373–3382.

Thompson, O., Edgley, M., Strasbourger, P., Flibotte, S., Ewing, B., Adair, R., et al. (2013). The million mutation project: A new approach to genetics in *Caenorhabditis elegans*. *Genome Research*, *23*, 1749–1762.

Timmons, L., Court, D. L., & Fire, A. (2001). Ingestion of bacterially expressed dsRNAs can produce specific and potent genetic interference in *Caenorhabditis elegans*. *Gene*, *263*, 103–112.

Wang, C., & Youle, R. J. (2009). The role of mitochondria in apoptosis*. *Annual Review of Genetics*, *43*, 95–118.

Westermann, B. (2010). Mitochondrial dynamics in model organisms: What yeasts, worms and flies have taught us about fusion and fission of mitochondria. *Seminars in Cell & Developmental Biology*, *21*, 542–549.

Yoneda, T., Benedetti, C., Urano, F., Clark, S. G., Harding, H. P., & Ron, D. (2004). Compartment-specific perturbation of protein handling activates genes encoding mitochondrial chaperones. *Journal of Cell Science*, *117*, 4055–4066.

CHAPTER FIVE

Apoptosis Initiation Through the Cell-Extrinsic Pathway

Pradeep Nair, Min Lu, Sean Petersen, Avi Ashkenazi[1]
Cancer Immunology, Genentech, Inc., San Francisco, California, USA
[1]Corresponding author: e-mail address: aa@gene.com

Contents

Abstract

Apoptosis is a tightly regulated cell suicide process used by metazoans to eliminate unwanted or damaged cells that pose a threat to the organism. Caspases—specialized proteolytic enzymes that are responsible for apoptosis initiation and execution—can be activated through two signaling mechanisms: (1) the cell-intrinsic pathway, consisting of Bcl-2 family proteins and initiated by internal sensors for severe cell distress and (2) the cell-extrinsic pathway, triggered by extracellular ligands through cognate death receptors at the surface of target cells. Proapoptotic ligands are often expressed on the surface of cytotoxic cells, for example, certain types of activated immune cells. Alternatively, these ligands can function in shed, soluble form. The mode of ligand presentation can substantially alter the cell response to receptor stimulation. Once receptor ligation on the target cell occurs, a number of intracellular signaling cascades may be initiated. These can lead to a variety of cellular outcomes, including

Methods in Enzymology, Volume 544
ISSN 0076-6879
http://dx.doi.org/10.1016/B978-0-12-417158-9.00005-4

caspase-mediated apoptosis, a distinct type of regulated cell death called necroptosis, or antiapoptotic or inflammatory responses. Death receptor signaling is kept tightly in check and plays critical homeostatic roles during embryonic development and throughout life.

1. INTRODUCTION

1.1. Apoptosis

Apoptosis is a process of cell suicide, executed by specialized protein-cleaving enzymes called caspases (Danial & Korsmeyer, 2004). Apoptosis enables multicellular organisms to eliminate unwanted or damaged cells, playing important roles in tissue sculpting during embryonic development and in tissue homeostasis throughout life. Aberrant apoptosis regulation contributes to a number of important diseases, including cancer, autoimmunity, diabetes, and neurodegeneration. Various types of cellular stress, for example, DNA damage or growth-factor deprivation, can trigger apoptotic cell elimination mediated either through the intrinsic or the extrinsic pathway.

The intrinsic apoptosis pathway is tightly regulated by the Bcl-2 gene family, made up of two major subclasses of proteins that share structural homology within Bcl-2 homology (BH) motifs (Cory & Adams, 2002). BH3-only proteins (e.g., Bid and Bad) contain a single BH motif and typically act as death agonists. Multi-BH motif proteins possess 3 or 4 BH regions and act, respectively, as agonists (e.g., Bax and Bak) or antagonists (e.g., Bcl-2 and Bcl-x_L) of apoptosis.

The extrinsic pathway is initiated by death receptors, a subset of the tumor necrosis factor receptor (TNFR) superfamily. TNFRs are typically type I transmembrane proteins possessing 2–4 similar cysteine-rich domains in the extracellular portion. Death receptors are further distinguished by the presence of a ~70 amino acid "death domain" in the cytoplasmic portion (Ashkenazi & Dixit, 1998; Nagata, 1997). Six death receptors have been identified that can regulate apoptosis either directly or indirectly: TNFR1 (TNFRSF1A), Fas/Apo1/CD95 (TNFRSF6), DR3 (TNFRSF25), DR4 (TNFRSF10A), DR5 (TNFRSF10B), and DR6 (TNFRSF21) (Wilson, Dixit, & Ashkenazi, 2009). Of these, Fas, DR4, and DR5 primarily mediate apoptosis, although they can trigger alternative signaling pathways such as the IKK/NF-κB or the JNK/c-Jun pathway in certain contexts in which apoptosis is circumvented. On the other hand, TNFR1 mainly controls IKK/NF-κB and JNK/c-Jun signaling, yet it can activate apoptosis if NF-κB activation is blocked, or in the presence of transcriptional or translational inhibitors such as actinomycin D or cycloheximide, respectively, or upon depletion of cellular inhibitor of apoptosis proteins (cIAPs), or in the context of unmitigated DNA damage (Vanlangenakker, Vanden Berghe, & Vandenabeele, 2012). DR3 mainly regulates noncanonical NF-κB signaling, yet it can activate apoptosis indirectly through TNFα (Schneider et al., 1999), by inducing TNFα via NF-κB activation while depleting cIAPs, thereby augmenting autocrine/paracrine apoptotic signaling through TNFR1 (Ikner & Ashkenazi, 2011). In addition to apoptosis, a number of death receptors are capable of inducing an alternative type of regulated cell death known as necroptosis, which is unmasked under circumstances in which apoptotic caspase activation is blocked (see Section 4).

Upon stimulation by proinflammatory mediators such as TNFα, interferon-α, -β, or -γ, or lipopolysaccharide, cytotoxic immune cells upregulate the expression of death ligands, such as TNFα, FasL/CD95L, and Apo2L/TRAIL (Cassatella et al., 2006; Ehrlich, Infante-Duarte, Seeger, & Zipp, 2003; Halaas, Vik, Ashkenazi, & Espevik, 2000; Sedger et al., 1999). This leads to higher levels of ligand expressed on the cell surface or released—often by proteolytic shedding—into the extracellular space. Consequently, these ligands may trigger apoptosis of target cells expressing cognate death receptors, to remove infected or oncogenically transformed cells, or to mediate immune modulatory functions (Fig. 5.1).

Death receptors that directly signal apoptosis activation, namely Fas, DR4, and DR5, trigger the cell-extrinsic apoptotic cascade by forming a death-inducing signaling complex (DISC), which most often contains the adaptor protein FADD and the apoptosis-initiating protease caspase-8 (Ashkenazi & Dixit, 1998). The DISC can be immunoprecipitated through epitope-tagged ligand, or using antireceptor antibodies that do not block ligand binding, or via FADD or caspase-8, while DISC components can be visualized by immunoblot (Kischkel et al., 1995, 2000; Sprick et al., 2000). Alternatively, DISC induction and immunoprecipitation can be performed using agonistic death receptor antibodies such as anti-Fas (Peter & Krammer, 2003) or anti-DR5 (Adams et al., 2008).

In contrast, TNFR1, which activates apoptosis indirectly, does so by inducing formation of a secondary, cytoplasmic signaling complex that most often contains receptor interacting protein kinase 1 (RIPK1) together with FADD and caspase-8 (Biton & Ashkenazi, 2011; Micheau & Tschopp, 2003), sometimes referred to as the ripoptosome (Bertrand & Vandenabeele, 2011). In cells resistant to the proapoptotic activity of Apo2L/TRAIL, stimulation with this ligand can trigger formation of a different cytosolic signaling complex after DISC formation (Varfolomeev et al., 2005). This secondary signaling complex is characterized by recruitment of factors such as RIPK1, TNFR-associated factor 2 (TRAF2), and NEMO, and leads to activation of NF-κB, JNK, and p38 MAPK, leading to prosurvival inflammatory responses normally observed in response to TNFR1 activation. Thus, all of the death receptors are capable of promoting either prosurvival activity or proapoptotic activity, and it is the precise biochemical and biophysical context in which these receptors interact with their extracellular ligands and intracellular signaling partners that governs cell fate.

2. LIGAND-INDUCED RECEPTOR CLUSTERING

2.1. Cell surface-displayed versus soluble ligands

Most of the ligands that comprise the TNF superfamily are synthesized as homotrimeric transmembrane proteins, and may function either as plasma membrane-anchored ligands at the surface of cells or secreted exosomes, or as soluble proteins shed into the extracellular space. The specific biophysical context in which these ligands interact with their cognate receptors affects which intracellular signaling cascades are activated. Clustering of the receptor in response to ligand can greatly amplify, and qualitatively alter, the cellular response to receptor ligation. In order to probe the signaling functions that are dependent on membrane localization, several variants of these ligands have been developed that display controllable cellular localization.

TNFα is synthesized by cells as a type II transmembrane protein of 26 kDa, with a stretch of hydrophobic amino acids at the amino terminus that anchors the protein in lipid bilayers while the carboxy terminus is extracellular (Cseh & Beutlers, 1989). Upon cleavage of this propeptide sequence by the metalloprotease TNF-converting enzyme, TACE/ADAM17 (Chang et al., 2002; Rosendahl et al., 1997), the protein is secreted as a soluble 51-kDa homotrimer. A noncleavable, surface-expressed TNFα mutant (mTNF; Table 5.1) was generated by deletion of the TACE cleavage site between Ala76 and Val77 (Perez et al., 1990). Soluble TNFα (sTNF; Table 5.1) was expressed by replacing the transmembrane domain of TNFα with the signal sequence for the secreted protein interferon-γ.

Table 5.1 TNF superfamily ligand variants, their subcellular localization, and their corresponding signaling activities

Mutant	Subcellular location	Outcome	References
mTNF	Cell membrane	Activates TNFR1/2 and cell death	Perez et al. (1990) and Grell et al. (1995)
sTNF	Solution	Activates TNFR1 and NF-κB signaling	Perez et al. (1990) and Grell et al. (1995)
CysTNF	Solution	Activates TNFR1/2 and cell death	Krippner-Heidenreich et al. (2002)
nanoTNF	Solution	Activates TNFR1/2 and cell death	Bryde et al. (2005)
mFasL	Cell membrane	Activates Fas, neutrophil recruitment, and apoptosis	Hohlbaum, Moe, and Marshak-Rothstein (2000) and O' Reilly et al. (2009)
sFasL	Solution	Dampened Fas effector function	Hohlbaum et al. (2000) and O' Reilly et al. (2009)
LZ-FasL	Solution	Activates Fas and apoptosis	Walczak et al. (1997)
sApo2L/TRAIL	Solution	Activates DR4/5 and apoptosis	Pitti et al. (1996) and Wiley et al. (1995)
LZ-TRAIL	Solution	Activates DR4/5 and enhanced apoptotic potency over sApo2L/TRAIL	Walczak et al. (1999)
Flag-Apo2L/TRAIL	Solution	Activates DR4/5 and enhanced apoptotic potency over sApo2L/TRAIL	Walczak et al. (1999) and Wilson et al. (2011)
His-Apo2L/TRAIL	Solution	Activates DR4/5 and enhanced apoptotic potency over sApo2L/TRAIL	Lawrence et al. (2001)

Using these genetically engineered TNFα variants, Grell and coworkers examined the signaling of TNFα as a function of its membrane localization (Grell et al., 1995). sTNF activates TNFR1 and in this context the ligand generally induces inflammatory, prosurvival NF-κB signaling. In contrast, activation of TNFR1 by membrane-displayed TNFα typically induces cell death via the cell-extrinsic apoptosis cascade.

The death ligands FasL and Apo2L/TRAIL are also natively expressed as type II transmembrane proteins upon inflammatory stimulation of immune cells, including cytotoxic T-lymphocytes, natural killer cells and dendritic cells (Wilson et al., 2009). FasL is synthesized as a 40-kDa protein and is proteolytically cleaved by either of the metalloproteases TACE or ADAM10, resulting in secretion of a 26-kDa fragment, soluble FasL (sFasL), that is, detected in serum as a homotrimer of ~70 kDa (Schneider et al., 1998). As in the case of TNFα, deletion of the metalloprotease-cleavage site in the FasL sequence, between Ser126 and Leu127 in the human protein, results in a noncleavable membrane-expressed variant of FasL (mFasL; Table 5.1). sFasL (Table 5.1) was generated by replacing the sequences for the intracellular and transmembrane domains of FasL with the signal peptide for the secreted cytokine, granulocyte colony stimulating factor (Hohlbaum et al., 2000).

Lymphoma cells expressing either wild-type FasL or mFasL induced an inflammatory response as well as activation of the cell-extrinsic apoptosis pathway. In contrast, lymphoma cells expressing only sFasL generated no inflammatory response and in fact soluble ligand protected against both inflammatory and cytotoxic Fas signaling in response to mFasL (Hohlbaum et al., 2000). Similarly, mice lacking sFasL but retaining mFasL had no apparent phenotype and their T cells were capable of inducing target cell death. However, mice lacking mFasL but retaining sFasL exhibited lymphadenopathy and splenomegaly, and displayed a systemic lupus erythematous phenotype, similar to mice lacking both forms of FasL (O' Reilly et al., 2009). Together, these results suggest that a critical element of death receptor stimulation is juxtacrine geometry, wherein both ligand and receptor are displayed on apposed cell membranes (Fig. 5.1.1).

2.1.1 Membrane receptor clusters associated with ligand response

In comparison to paracrine signaling—wherein soluble ligands are diffusing in the extracellular space—juxtacrine signaling—wherein ligands are membrane anchored—imparts unique biophysical parameters, including an increased likelihood of receptor clustering and greater resistance to ligand endocytosis (Dustin, Bromley, Davis, & Zhu, 2001). Both paracrine and juxtacrine modalities can either attenuate or amplify specific downstream signaling cascades in response to death ligands.

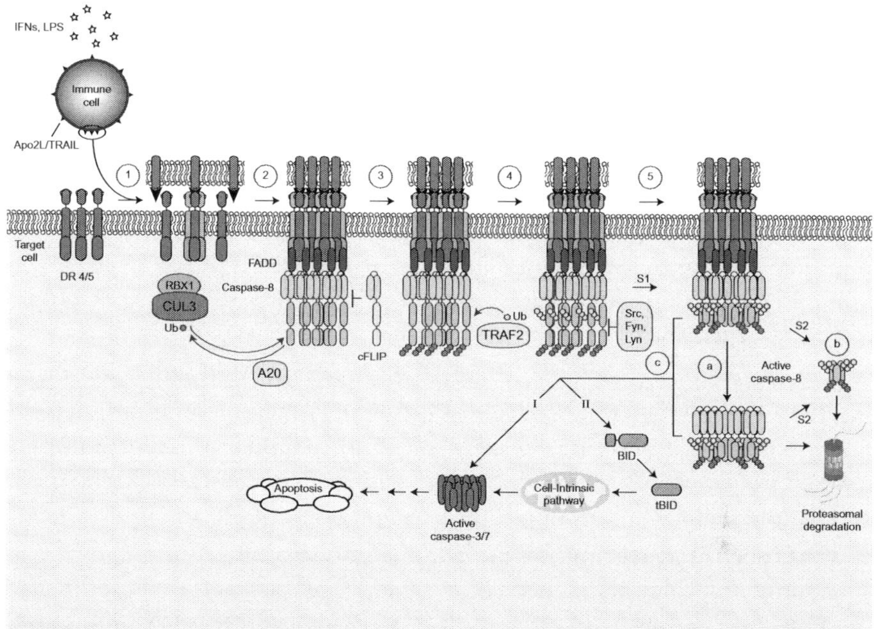

Figure 5.1

1. Death receptors 4 and 5, expressed on the surface of target cells, are ligated by Apo2L/TRAIL, presented on the surface of activated immune cells.
2. Multiple ligand–receptor complexes cluster together at the cell membrane, triggering recruitment of the adaptor protein FADD and the apoptosis-initiating protease caspase-8 to the DISC. An enzymatically inactive protein related to caspase-8, cFLIP, can also be recruited to the DISC, inhibiting caspase-8 activation.
3. In epithelial cells, the E3 ligase complex CUL3/RBX1 catalyzes polyubiquitination (likely via K63 chains) of lysine 461 on the p10 domain of caspase-8, leading to further clustering and activation of the caspase. The DUB enzyme A20 can reverse this ubiquitination.
4. The RING domain-containing adaptor protein TRAF2 mediates K48 ubiquitination on the p18 region of caspase-8. This targets caspase-8 for proteasomal degradation upon proteolytic processing and release from the plasma membrane into the cytosol, thereby setting a "shutoff timer" for caspase-8 activation.
5. S1 cleavage removes the linker between the p10 and the p18 domains of caspase-8, leading to the formation of a catalytically active $(p43/p10)_2$ heterodimer. S1 cleavage can be inhibited through phosphorylation on the linker region between the p10 and the p18 domains, catalyzed by members of the Src-family tyrosine kinases. Once S1 cleavage occurs, the pathway can proceed in three distinct directions:
   a. The $(p43/p10)_2$ heterodimer can dissociate from the membrane-anchored DISC, leading to a cytosolic pool of active caspase-8. The pre-added K48-ubiquitin on p18 (within p43) targets the p43 fragment for proteasomal degradation, thereby limiting the duration of caspase-8 activity.

(Continued)

Stimulation of Fas with agonistic antibodies or crosslinked sFasL triggers recruitment of acid sphingomyelinase (ASM) to the external leaflet of the plasma membrane (Grassmé, Jekle, et al., 2001), as well as palmitoylation of the membrane-proximal Cys199 of Fas (Feig, Tchikov, Schütze, & Peter, 2007). Upon extracellular translocation, ASM catalyzes the hydrolysis of sphingomyelin to C_{16}-ceramide, which self-associates within the lipid membrane to facilitate clustering of proteins such as Fas that are found in sphingomyelin-enriched membrane compartments (Grassmé, Schwarz, & Gulbins, 2001). This protein clustering results in formation of an SDS-stable, high-molecular weight DISC that is internalized and serves as the site of caspase-8 activation (Feig et al., 2007). Cells that lack ASM exhibit less sensitivity to FasL, which can be restored through exogenous addition of ASM or C_{16}-ceramide (Grassmé, Jekle, et al., 2001). Similarly, expression of a nonpalmitoylateable Cys199 mutant of Fas renders cells resistant to FasL stimulation (Feig et al., 2007). Thus, receptor clustering serves as a feed-forward mechanism to amplify initial Fas stimulation.

A similar mechanism has been observed in the case of another member of the TNF superfamily, CD40L, and its cognate receptor, CD40 (Grassmé, Bock, Kun, & Gulbins, 2002). Here, in addition to providing a scaffold for a greater number of ligand–receptor interactions, membrane receptor clustering also enhances recruitment of the downstream signaling molecules TRAF2 and TRAF3 to the receptor–ligand complex (Vidalain et al., 2000). In this manner, clustering of membrane receptors can enhance both the quantity and the signaling capacity of ligated receptors.

Figure 5.1—Cont'd

b. In either the membrane-associated or the cytosolic active caspase-8 pools, S2 cleavage can occur between the p18 and the N-terminal DEDs of caspase-8, releasing a $(p18/p10)_2$ heterodimer that no longer oligomerizes due to its separation from the DEDs. The pre-added K48-ubiquitin on p18 targets this fragment for proteasomal degradation, thereby limiting duration of activity.

c. Active caspase-8, either at the cell membrane or in the cytosol, can proceed to proteolytically cleave downstream substrates. In type I cells, caspase-8 activation is sufficient to directly cleave and activate executioner caspases, such as caspases-3 and -7 and commit the cell to apoptosis. In type II cells, less caspase-8 activity is generated and amplification of the signal is needed to drive sufficient effector caspase activity. This can be achieved through caspase-8-mediated cleavage of the BH3-only protein BID to form truncated BID (tBID). tBID triggers the cell-intrinsic (mitochondrial) pathway, augmenting the activation of executioner caspases and committing the cell to apoptotic demise. (See the color plate.)

2.1.2 Internalization of the DISC

Endocytosis is a mechanism of internalizing plasma membrane-associated proteins through highly regulated formation of membrane vesicles that encapsulate the cargo for internalization (Bonifacino & Traub, 2003). Stimulation with either soluble Apo2L/TRAIL (sApo2L/TRAIL; Table 5.1) or sFasL has been shown to inhibit clathrin-mediated endocytosis by caspase-dependent cleavage of the AP2 clathrin adaptor protein subunit as well as clathrin heavy chain (CHC) (Austin et al., 2006). In the case of sFasL stimulation, genetic silencing of AP2 or CHC prevents DISC formation and protects cells from Fas-dependent apoptosis (Lee et al., 2006). This suggests that ligand-induced Fas internalization critically supports its proapoptotic activity, and that caspase-induced cleavage of the endocytic machinery inhibits Fas-induced apoptosis.

In contrast, sApo2L/TRAIL-induced apoptosis proceeds even in the absence of receptor endocytosis (Austin et al., 2006; Kohlhaas, Craxton, Sun, Pinkoski, & Cohen, 2007). Moreover, internalization of the ligand dampens caspase activation, and inhibition of endocytosis augments apoptosis stimulation. The apoptosis-initiating protease caspase-8, activated upon stimulation with either sFasL or sApo2L/TRAIL, has been shown to cleave its substrates more effectively in proximity to the plasma membrane rather than in the cytoplasm or other compartments of the cell (Beaudouin, Liesche, Aschenbrenner, Hörner, & Eils, 2013). Taken together, these data suggest that in the case of sApo2L/TRAIL but not sFasL, the internalization of death receptors, as well as proximal DISC components, may act to inhibit activity of the cell-extrinsic apoptosis cascade. This internalization is likely itself inhibited in response to membrane-displayed ligand, since the cargo to be engulfed must either be first proteolytically cleaved or enlarged to include portions of the apposed cell membrane. If this does not occur, there will be significant energetic cost to exposing a hydrophobic transmembrane domain to the aqueous extracellular milieu.

2.2. Membrane receptor organization alters ligand sensitivity

The critical role for receptor clustering in death receptor activation raises the intriguing possibility that certain cell types may express partially preclustered receptors on their surface, rendering them better primed for stimulation by ligand. A number of studies have found evidence of ligand-independent preclustering of TNFR1, TNFR2, Fas, and DR4 (Chan, 2000; Papoff, 1999; Siegel, 2000). It should be noted, however that the functional relevance of such ligand-independent clusters has not yet been determined, as such clustering does not stimulate inflammatory or apoptotic pathways at physiological receptor concentrations in the absence of ligand.

In agreement with the hypothesis that receptor oligomerization on the membrane surface may contribute to ligand sensitivity, DR4 and DR5 have been shown to be O-glycosylated in a number of cancer cell lines sensitive to sApo2L/TRAIL (Wagner et al., 2007). Upon mutation of these posttranslational death receptor modification sites to prevent glycosylation, or genetic silencing of the glycosyltransferases required for these modifications, less DISC forms in response to sApo2L/TRAIL stimulation and cells become more resistant to sApo2L/TRAIL treatment. Importantly, this reduction in sensitivity is not accompanied by a commensurate decrease in expression of DR4/5 on the cell surface, since flow cytometry shows similar levels of receptors in all treatment conditions. Similarly, palmitoylation of both Fas and DR4 has been shown to play a critical role in organization of these receptors in the cell membrane, and subsequent proapoptotic activity in response to their cognate ligands (Chakrabandhu et al., 2007; Rossin, Derouet, Abdel-Sater, & Hueber, 2009). These findings highlight that not all death receptors on the cell surface are equivalent, and suggest that certain posttranslational modifications render death receptors more able to dynamically respond to ligand stimulation in a productive fashion.

2.3. Techniques to alter receptor clustering

The genetic approaches to altering ligand presentation described earlier allow for the direct observation of only membrane-displayed or soluble ligands interacting with their cognate receptors. The strength of these strategies is that they use the cell's own machinery to generate specifically localized ligands and to allow for the observation of subsequent downstream signaling. However, these strategies depend on the proper expression of an exogenous protein, and so unnatural splice isoforms or degradation products may occur, or the protein itself may not be expressed at physiological levels. Furthermore, these techniques are cell based and thus require coculture experiments, at least in the case of membrane-displayed mutants. While ligand-expressing cells can be fluorescently tagged, a heterogeneous cell population in the same flask introduces a multitude of experimental confounds due to intercellular interactions that may affect those being probed. Finally, the potential for therapeutic development is limited when potent ligands must be presented to target-cell populations on the surfaces of foreign cells.

Several techniques have been designed to specifically address these weaknesses, while still considering the role of membrane-ligand display in downstream signaling. Generally, these techniques involve variants of specific TNF superfamily ligands, or agonists targeting their cognate receptors that oligomerize through covalent or noncovalent interactions.

2.3.1 Generating multimeric ligands to stimulate receptor clustering

The signaling response of TNFR superfamily members is generally more potent, and sometimes qualitatively distinct, upon stimulation with membrane-displayed ligands compared to soluble ones. To address the functional requirement of clustered ligand, while still being able to add ligand in solution, several strategies have been developed to synthetically precluster soluble ligand.

TNFα derivatives have been formed that are either preclustered through an exogenous N-terminal cysteine (CysTNF; Table 5.1) (Krippner-Heidenreich et al., 2002) or through the use of nanoparticles that are covalently attached to a similar TNFα mutein (nanoTNF; Table 5.1) (Bryde et al., 2005). These variants of TNFα are thought to replicate behavior of the membrane-displayed ligand because they are capable of inducing a proapoptotic TNFR1 response, rather than the inflammatory response triggered by sTNF.

The addition of leucine zipper (LZ) motifs to the N-termini of TNF superfamily ligands induces controllable oligomerization that can enhance the signaling activity of the ligands (Harbury, Zhang, Kim, & Alber, 1993; Naito et al., 2013). This strategy has been used to develop LZ-FasL as well as LZ-TRAIL (Table 5.1; Walczak et al., 1997), both of which display proapoptotic signaling activity. In the case of Apo2L/TRAIL, addition of the LZ motif augments the proapoptotic potency of soluble ligand (Walczak et al., 1999).

Recombinant soluble Apo2L/TRAIL was initially purified by the addition of either polyhistidine tags (Pitti et al., 1996) or a Flag epitope (Wiley et al., 1995) to the amino terminus of sApo2L/TRAIL. These tags allowed for affinity purification using either a nickel column or immobilized antibody directed against the Flag epitope, respectively. However, both of these affinity tags were later used to induce preclustering of the death ligand. Addition of a polyhistidine tag causes a more heterogeneous mixture of Apo2L/TRAIL oligomers in solution, as well as insoluble aggregates at physiological temperatures (Lawrence et al., 2001), resulting in increased cytotoxicity toward cultured human hepatocytes (Jo et al., 2000). Use of the Flag epitope for ligand preclustering is more controllable and enhances proapoptotic activity of sApo2L/TRAIL, while still potentially allowing for a therapeutic window (Walczak et al., 1999; Wilson et al., 2012).

2.3.2 Using multivalent receptor agonists to induce signaling clusters

The crosslinking of death ligands aims to recapitulate the signaling-competent death receptor clusters formed in response to membrane-displayed ligand. A separate approach to regain this proapoptotic signaling involves the use of multimeric agonists targeted against DR4 or DR5 (Ashkenazi, 2008; Ashkenazi & Herbst, 2008; Johnstone, Frew, & Smyth, 2008). These proapoptotic receptor agonists (PARAs) typically have the advantage of being more stable *in vivo* and easier to manufacture than ligand variants, and they offer the potential to improve upon the apoptotic potency of the native ligand.

This class of PARAs includes not only agonistic antibodies directed against DR4 and DR5 (Ashkenazi & Herbst, 2008), but also antibody-derived biomolecules such as fusion proteins that induce higher order clustering of Apo2L/TRAIL homotrimers and thus enhance the potency of the ligand *in vivo* (Seifert et al., 2014). Another developing class of biomolecules is made up of single, high-affinity heavy chain antibodies, termed "nanobodies" that are chemically stable, facile to generate, and controllably crosslinkable (Cortez-Retamozo et al., 2004). One such nanobody, TAS266, is a tetrameric crosslinked agonist directed against DR5 and has shown improved proapoptotic activity compared to either the native ligand, or crosslinked conventional antibodies directed against DR5 (Huet et al., 2012).

In general, all of these PARAs show enhanced proapoptotic activity upon crosslinking, though the mechanism of crosslinking varies. In the case of conventional agonistic antibodies, their natural bivalency can be augmented through the addition of anti-Fc secondary antibodies, or through binding to Fc-receptors expressed by leukocytes *in vivo* (Wilson et al., 2011). Antibody fusions are designed to recruit multiple ligand homotrimers to a single cluster, while nanobodies are synthetically generated to present multiple death receptor binding sites separated by peptide linker sequences.

While these PARAs are all capable of activating the cell-extrinsic apoptosis cascade, their binding to the target death receptors, and likely the receptor clustering that follows, is significantly altered from that of the natural ligand (Adams et al., 2008). Nonetheless, they offer a chemically controllable approach to induce death receptor clustering in a manner that is at least phenotypically similar to that of membrane-displayed ligand.

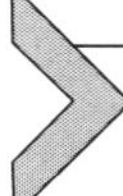

3. CONTROL OF CASPASE-8 ACTIVATION

3.1. Caspase-8 dimerization, processing, and activation

Ligation of death receptors such as DR4 and DR5 induces receptor clustering and subsequent recruitment of the FADD adaptor protein through interaction between the death domains of the receptors and the adaptor. FADD then recruits caspase-8 through homotypic death effector domain (DED) interactions (Fig. 5.1.2).

Disruption of caspase-8 expression leads to embryonic lethality in mice between embryonic days 10.5 and 11.5, with defects in cardiac development, growth retardation, and hematopoietic progenitor deficiency (Varfolomeev et al., 1998), unlike other caspase knockout mice which show only mild developmental defects. The physiological significance of caspase-8 is further highlighted by the observation that caspase-8 is mutated or deleted in some cancers (Lawrence et al., 2014). For example, *CASP8* is frequently silenced through DNA methylation as well as through gene deletion in neuroblastoma (Teitz et al., 2000), and a common sequence variant (D302H) of *CASP8* is associated with breast cancer risk (Cox et al., 2007). Thus, caspase-8 serves essential roles both in embryonic development and in proper tissue homeostasis.

The caspase-8 protein has several isoforms resulting from alternative mRNA splicing (Boldin, Goncharov, Goltsev, & Wallach, 1996; Muzio et al., 1996). The two major isoforms—p55 and p53—are functionally indistinguishable and often coexpressed. Each possesses two tandem DEDs in its N-terminal region, followed by the large (p18) and small (p10) catalytic subunits, separated by a short-intervening linker. Caspase-8 predominantly exists as a monomeric cytoplasmic protein and its recruitment to oligomeric platforms such as the DISC effectively increases its local concentration, which allows monomers to adapt a dimeric conformation (Boatright et al., 2003; Donepudi, Mac Sweeney, Briand, & Grutter, 2003). Dimerization is followed by two internal cleavage events: first between the large and small subunits, generating a p43 fragment (S1 processing), and subsequently, between the large subunit and the adjacent DED, producing p18 (S2 processing) (Van Raam & Salvesen, 2012; Fig. 5.1.5). Unlike executioner caspases (e.g., caspase-3), which are readily activated by interdomain cleavage alone, neither dimerization nor cleavage of caspase-8 alone is sufficient to achieve activation and trigger apoptosis (Oberst et al., 2010). S1 cleavage stabilizes the dimer and increases proteolytic activity (Keller, Mares, Zerbe, & Gruetter, 2009). Thus, coordinated dimerization and S1 processing are required for initial activation of caspase-8. However, additional aggregation events are often needed for maximal activation and apoptosis induction (see below). Human caspase-10 is structurally homologous to caspase-8 and—at least in some contexts—it may have similar activity (Kischkel et al., 2001). Antibodies to detect caspase-10 must be selected with care to avoid nonspecific crossreactivity (Kischkel et al., 2001). However, as compared to caspase-8, caspase-10 is poorly characterized, in part due to its absence in the mouse. Interestingly, the zebrafish has orthologues of both caspases-8 and -10 (Eimon et al., 2006).

3.2. Interaction with cFLIP

One of the major regulators of caspase-8 activation at the DISC is cFLIP. Multiple mRNA-splicing variants of cFLIP occur, but usually only two main cFLIP proteins are detected: cFLIP long (cFLIP$_L$, 55 kDa) and cFLIP short (cFLIP$_S$, 26 kDa) (Irmler et al., 1997; Scaffidi, Schmitz, Krammer, & Peter, 1999; Sharp, Lawrence, & Ashkenazi, 2005; Shu, Halpin, & Goeddel, 1997). Both cFLIP$_L$ and cFLIP$_S$ have two N-terminal DEDs similar to those of caspases-8 and -10. cFLIP$_L$ also has caspase-like p18 and p10 domains, but these are enzymatically inactive due to the lack of a critical cysteine residue in the catalytic triad (Irmler et al., 1997). cFLIP$_S$ has two DEDs and a short C-terminal tail lacking caspase-like catalytic domains (Scaffidi et al., 1999). Both cFLIP variants can be recruited to the DISC via homotypic DED interactions with FADD, and each is capable of dimerizing therein with itself or with caspase-8 or -10 (Krueger, Baumann, Krammer, & Kirchhoff, 2001; Thome & Tschopp, 2001). Formation of caspase-8:cFLIP$_S$ heterodimers inhibits caspase-8 homodimerization and processing, thereby blocking apoptosis activation (Krueger et al., 2001; Thome & Tschopp, 2001). When expressed at high concentrations, cFLIP$_L$ acts as an antiapoptotic molecule by competing for caspase-8 binding to FADD (Irmler et al., 1997). Furthermore, selective knockdown of cFLIP$_L$ augments caspase-8 activation and apoptosis (Sharp et al., 2005). Similarly, cFLIP$^{-/-}$ mouse embryo fibroblasts show enhanced sensitivity to extrinsic apoptosis induction (Yeh et al., 2000). However, at low levels, cFLIP$_L$ can facilitate caspase-8 activation at the DISC, thereby promoting apoptosis (Chang et al., 2002). Moreover, artificial formation of caspase-8:cFLIP$_L$ heterodimers readily triggers apoptosis; however, relative to caspase-8 homodimers, these heterodimers have attenuated activity toward many apoptotic substrates (Pop et al., 2011). Based on the mathematical modeling, it has been proposed that subtle differences in the expression of cFLIP isoforms determine whether cFLIP$_L$ exhibits pro- or antiapoptotic function (Fricker et al., 2010).

3.3. Posttranslational modifications of caspase-8

Caspase-8 is rarely regulated by transcriptional or translational events. However, it's activity can be modulated through specific posttranslational modification, including ubiquitination and phosphorylation.

3.3.1 Ubiquitination of caspase-8 in epithelial cells

The ubiquitination of substrate proteins is achieved by a series of reactions mediated by ubiquitin activating (E1), conjugating (E2), and ligating (E3) enzymes (Pickart, 2004). Specific types of ubiquitin modification direct the fate of conjugated substrate proteins: Lys48 (K48)-linked and some Lys11 (K11)-linked ubiquitin chains target proteins for degradation by the 26S proteasome. In contrast, Lys63 (K63)-linked chains, linear polyubiquitin chains, and some K11-linked chains usually provide scaffolding for recruitment and assembly of multiprotein complexes (Vucic, Dixit, & Wertz, 2011).

Death receptor ligation induces both K63- and K48-linked ubiquitin modification of caspase-8 (Gonzalvez et al., 2012; Jin et al., 2009). The E3 ligase cullin3 (CUL3), which belongs to the Cullin-RING ligase superfamily, associates with the DISC within a detergent-insoluble cellular compartment enriched in cytoskeletal proteins. The activated (NEDD8-modified) form of CUL3 is relatively abundant in this compartment and acts in conjunction with a substrate-adaptor called RBX1 to catalyze conjugation of ubiquitin chains (likely via K63-linkage) on lysine 461, within the p10 domain of caspase-8 (Fig. 5.1.3). This modification promotes further clustering and *en mass* activation of caspase-8 within ubiquitin-rich intracellular foci, which can be facilitated through interaction with the ubiquitin-binding protein p62/SEQSM1. CUL3-mediated ubiquitination of caspase-8 may be specific to cells of epithelial origin as compared with hematopoietic cells (Dickens et al., 2012).

In contrast to the relatively detailed understanding of caspase-8 activation, little is known about the mechanisms that extinguish activated caspase-8. Proteasome inhibition sensitizes various cancer cells to

DR-mediated apoptosis (Chen et al., 2010; Menke et al., 2011; Naumann, Kappler, von Schweinitz, Debatin, & Fulda, 2011), suggesting proteasomal degradation of key component(s) of the extrinsic apoptosis pathway. Recent work demonstrates that TRAF2, a protein involved in NF-κB signaling by several TNF superfamily members, directly mediates K48-polyubiquitination on the p18 subunit of caspase-8. Specifically, TRAF2 associates with the DISC downstream of CUL3 but upstream of caspase-8 processing, attaching a K48-ubiquitin "shutoff timer" to caspase-8 (Fig. 5.1.4). This ubiquitination process selectively targets the cytosolic pool of p43 and p18 for degradation upon S1 and S2 processing, thereby dampening caspase-8 activity. Consistent with this finding, caspase-8 is more active at the plasma membrane than within the cytosol (Beaudouin et al., 2013; Gonzalvez et al., 2012; Hughes et al., 2009; Oberst et al., 2010). Thus, TRAF2 serves a previously unrecognized role in setting a threshold for a cell's decision to commit to apoptosis via the extrinsic pathway, by regulating turnover of activated caspase-8.

Other components of the DISC are also subject to ubiquitin modification. Upon TNFα stimulation, the E3 ligase Itch, which is activated by JNK (Gao et al., 2004), ubiquitinates cFLIP, leading to its proteasomal degradation, and thus preventing it from inhibiting caspase-8 activation (Chang et al., 2006). Recently, it has been shown that FADD is regulated by Makorin Ring Finger Protein1 (MKRN1) E3 ligase-mediated ubiquitination and proteasomal degradation; downregulation of MKRN1 facilitates caspase-8 activation and suppresses tumor growth upon Apo2L/TRAIL treatment (Lee et al., 2012).

Ubiquitination is a reversible modification. The removal of ubiquitin residues from substrates is mediated by deubiquitinases (DUBs) (Vucic et al., 2011). A20, also known as TNFAIP3, associates with caspase-8 upon Apo2L/TRAIL stimulation and can reverse CUL3-mediated ubiquitination of caspase-8, which blocks the increase in caspase-8 activity driven by CUL3 (Jin et al., 2009). It is presently unknown whether A20 and/or other DUBs regulate K48-ubiquitination of caspase-8.

3.3.2 Phosphorylation of caspase-8

Phosphorylation of caspase-8 on tyrosine 380 (Y380) and 465 (Y465) by the Src-family tyrosine kinases Src, Fyn, and Lyn suppresses activation (Cursi et al., 2006; Jia, Parodo, Kapus, Rotstein, & Marshall, 2008; Senft, Helfer, & Frisch, 2007). Interestingly, Y380 lies in the linker sequence between the large and small subunits, a region that is removed upon caspase-8S1 processing and activation. Therefore, it is possible that Y380 inhibits caspase-8 activity by preventing caspase-8 processing, although this has not been experimentally confirmed. In some instances, phosphorylation of a signaling component is a prerequisite to its ubiquitination. For example, IKKβ phosphorylates the inhibitor of κB (IκB) protein, leading to its ubiquitination and proteasomal degradation, and permitting NF-κB activation (Vucic et al., 2011). Whether phosphorylation of caspase-8 regulates its ubiquitination requires further investigation.

3.4. Caspase-8 "chain" formation

Although it is well established that FADD, caspase-8, and cFLIP are core DISC components, the exact stoichiometry of these proteins in the complex has not been clearly defined. The conventional model has been that one ligand trimer binds one DR trimer. One DR trimer then binds two or three molecules of FADD, which in turn bind a similar number of caspase-8 and/or cFLIP monomers. Based on quantitative mass spectrometry and modeling, two groups recently proposed a more complex stoichiometric configuration of the core DISC components (Dickens et al., 2012; Schleich et al., 2012). Both studies showed that in hematopoietic cells caspase-8 is much more abundant at the DISC than FADD, although the exact ratio varied between the two studies, possibly due to differences in DISC isolation techniques. Furthermore, through structural or mathematical modeling, both groups proposed that caspase-8 binds FADD through its first DED and uses its second DED to recruit another caspase-8 molecule, thereby initiating caspase-8 "chain" assembly, which is crucial for activation. Schleich et al. also showed that the length of caspase-8 chains varied depending on stimulation strength and cellular amount of caspase-8, which could help explain why different types of cells vary in sensitivity to death ligands (Schleich et al., 2012). Caspase-8 chain assembly may require the actin cytoskeleton, the disruption of which perturbs cell-extrinsic apoptotic signaling, albeit in a cell type-dependent manner (Chaigne-Delalande, Moreau, & Legembre, 2008). Caspase-8 chains may serve a similar function as CUL3-mediated ubiquitination, by inducing further aggregation of caspase-8 to promote its full activation.

3.5. Techniques to alter caspase-8 activity

Caspase-8 plays a central role in activation of the cell-extrinsic apoptosis cascade, and subtle changes in its biochemical makeup can greatly alter the sensitivity of cells to death ligands. To study the function of caspase-8 and the mechanisms governing its activity, many tools have been developed that have greatly accelerated our understanding of the cell-extrinsic pathway.

3.5.1 Inhibitors

Fluoromethylketone (FMK)-derivatized peptides mimicking the cleavage sites of caspases act as effective irreversible caspase inhibitors, with few cytotoxic effects. Z-VAD-FMK is a commonly used pan-caspase inhibitor: it binds to the catalytic site of most caspases and can inhibit apoptosis induction. The addition of O-methyl side chains provides enhanced stability and increased cell permeability, thus facilitating the use of Z-VAD-FMK *in vitro* as well as *in vivo*. Another broad-spectrum caspase inhibitor, Q-VD-OPH, has significantly higher potency, greater selectivity for caspases (compared to other cysteine proteases), and decreased toxicity than the FMK derivatives (Chauvier, Ankri, Charriaut-Marlangue, Casimir, & Jacotot, 2007). Specific inhibitors for individual caspases have also been developed, including Z-DEVD-FMK, Z-IETD-FMK, and Z-LEHD-FMK, which inhibit caspases-3/7, caspase-8, and caspase-9, respectively. However, all such inhibitors have significant crossreactivity toward other caspases (Berger, Sexton, & Bogyo, 2006). Therefore, extreme care should be taken when performing experiments and when evaluating results based solely on these so-called selective inhibitors, for example, using RNA interference.

Cytokine response modifier A, a serpin-like protease inhibitor encoded by Cowpox virus (Ray et al., 1992), is able to bind and to inhibit caspases-1 and -8/-10 and thus prevent caspase-1-mediated processing of interleukin-1β as well as caspases-8/10-dependent apoptosis triggered by death receptor ligation (Muzio et al., 1996; Zhou et al., 1997).

Kosmotropic salts can be used to stabilize caspases in their active conformation by promoting dimerization (Boatright et al., 2003).

3.5.2 Mutations

Caspase-8 mutants can be classified into two major groups: (1) mutations that prevent dimerization and/or cleavage and (2) mutations that abrogate ubiquitination. Commonly used mutants are listed in Table 5.2.

Table 5.2 Caspase-8 mutants, the location of these mutations within the protein, and the biochemical outcome of these mutations

Mutant	**Location**	**Outcome**	**References**
C360A	p18 subunit	No cleavage and no activation	Boldin et al. (1996) and Hughes et al. (2009)
D374A/D384A	Interdomain linker	No dimerization, no S1 cleavage, and no activation	Gonzalvez et al. (2012), Hughes et al. (2009), and Keller et al. (2009)
E201A/ D210A/D216A	Prodomain linker	No S2 cleavage and enhanced activation	Gonzalvez et al. (2012) and Hughes et al. (2009)
D210A/ D216A/D223A	Prodomain linker	Enhanced activation	Oberst et al. (2010)
T467D	p10 subunit	No dimerization and no activation	Boatright et al. (2003) and Hughes et al. (2009)
F468A	p10 subunit	No dimerization and no activation	Keller et al. (2009)
K461R	p10 subunit	Decreased K63-ubiquitination and decreased caspase-8 activation	Jin et al. (2009)
K224;239;231R	p18 subunit	No K48-ubiquitination and enhanced caspase-8 activation	Gonzalvez et al. (2012)

4. DEATH RECEPTOR STIMULATION OF NECROPTOSIS

4.1. Alternative death pathways

Apoptosis was previously thought to be the only form of programmed cell death, both phenotypically and biochemically distinct from accidental necrosis. However, under conditions of cell stress in which apoptotic pathways are inhibited, a separate form of programmed cell death, termed necroptosis, can occur. Necroptosis shares certain features with apoptosis and other features with necrosis—passive unregulated cell death characterized by membrane swelling and leakage of cellular content into the extracellular space resulting in inflammation. While necroptosis is indistinguishable phenotypically from necrosis, it is a regulated process, as is apoptosis. Death receptor stimulation can promote this alternative form of programmed cell death, likely as a fail-safe mechanism that ensures elimination of unwanted cells (Vandenabeele, Galluzzi, Vanden Berghe, & Kroemer, 2010).

4.1.1 Caspase-8 suppresses necroptosis

Caspase-8 activity suppresses necroptosis, and its inhibition upon death receptor activation is required in order to unmask necroptotic death. In the absence of caspase-8 activity, death receptor ligands, such as TNFα, FasL, and Apo2L/TRAIL, can induce cell death in several cell lines, including murine L929 fibrosarcoma and human Jurkat T leukemia cells (Holler et al., 2000; Kawahara, Ohsawa, Matsumura, Uchiyama, & Nagata, 1998). This death phenotypically resembles necrosis, but is dependent on RIPK1 and FADD (Holler et al., 2000).

Necrostatin-1 prevents necroptosis activation in several cell lines and has also been shown to inhibit RIPK1 kinase activity (Degterev et al., 2008), suggesting that RIPK1 activity is required for necroptosis to proceed. The selectivity of necrostatin-1 for RIPK1 as compared to other kinases is remarkably high, but it does crossinhibit a small number of other kinases which must be examined before modulation by this inhibitor can be assigned specifically to RIPK1 inhibition (Biton & Ashkenazi, 2011).

4.1.2 RIPK3 and MLKL are key downstream components of necroptosis signaling

RIPK3 is structurally related to RIPK1 and recent studies demonstrate that deletion of RIPK3 also protects cells from necroptosis. Furthermore, only cells that express RIPK3 are able to undergo necroptosis and ectopic expression of RIPK3 in nonRIPK3-expressing cells enables necroptosis induction. These observations have led to the hypothesis that interaction between RIPK1 and RIPK3 regulates this cell death process (He et al., 2009; Zhang et al., 2009). Indeed, biochemical analysis has revealed a functional complex consisting of RIPK1 and RIPK3, dubbed the necrosome. Mechanistically, it appears that RIPK1 is deubiquitinated and then autophosphorylated (Degterev et al., 2008), which allows it to engage and phosphorylate RIPK3 (Cho et al., 2009) at the necrosome. It should be noted that RIPK1 autophosphorylation also can be required for apoptosis activation by autocrine feed-forward TNFα signaling in response to unmitigated DNA damage and ATM activation (Biton & Ashkenazi, 2011).

Screening of a small molecule library to further identify components of the necroptosis pathway led to the identification of mixed lineage kinase domain-like (MLKL), a pseudokinase, as a critical player downstream of the RIPK1/RIPK3 complex (Sun et al., 2012). Initially, MLKL was reported to be in a complex with RIPK1/RIPK3; however, the stability of this complex is uncertain (Murphy et al., 2013; Sun et al., 2012). Regardless, it is clear that RIPK1/RIPK3 activation leads to MLKL phosphorylation, which is dependent on RIPK3 kinase activity, itself dependent on phosphorylation by RIPK1 (Chen et al., 2013). Although MLKL is a pseudokinase and does not phosphorylate downstream targets, mutations in the pseudo-active site that mimick phosphorylation of MLKL relieve the dependence on RIPK3 for necroptosis activation, and deletion of MLKL blocks necroptosis activation. Taken together, these findings demonstrate that pseudo-activated MLKL is necessary and operate downstream of RIPK3 in the necroptosis cascade (Murphy et al., 2013). Recent work reveals that upon phosphorylation, MLKL is recruited to cellular membranes, where it aggregates and forms necrosis-promoting pores (Chen et al., 2014; Wang et al., 2014).

4.1.3 RIPK3 knockout rescues mouse embryonic lethality of caspase-8 knockout

Double caspase-8/RIPK3 KO mice are viable and show no overt negative phenotype (Kaiser et al., 2011; Oberst et al., 2011), in contrast to the caspase-8 single KO that displays embryonic lethality discussed earlier (Varfolomeev et al., 1998). Furthermore, mice with conditional KO of caspase-8 in the intestine develop necrotic lesions within the intestinal epithelium, and this phenotype is rescued by RIPK3 KO (Gunther et al., 2011). Thus, caspase-8 actively and critically suppresses necroptosis during normal development and in adult tissue homeostasis. Furthermore, suppression of necroptosis seems to be mediated by a functional complex containing FADD and heterodimerically associated caspase-8 and cFLIPL (Dillon et al., 2012; Oberst et al., 2011).

4.2. Necroptosis shares several signaling proteins with apoptosis

In vitro systems to analyze necroptosis often rely on the use of TNFα as the death ligand under conditions of caspase inhibition by Z-VAD-FMK or other caspase inhibitors, together with blockade of protein transcription or translation using actinomycin D or cycloheximide, respectively, or induced degradation of cIAPs using SMAC-mimetic cIAP antagonists (He et al., 2009; Wang, Du, & Wang, 2008).

Apo2L/TRAIL is also capable of inducing necroptosis, rather than apoptosis, under circumstances of cell stress coupled with caspase inhibition. Early studies showed that Apo2L/TRAIL could induce necroptotic cell death in the presence of Z-VAD-FMK and cycloheximide in human Jurkat cells (Holler et al., 2000) as well as in murine prostate adenocarcinoma TRAMP-C2 cells (Kemp, Kim, Crist, & Griffith, 2003). More recently, it has been shown that under low-extracellular pH conditions and in the presence of caspase inhibitors, Apo2L/TRAIL signaling can switch from triggering apoptosis to RIPK1- and RIPK3-dependent necroptotic death in HT-29 and HepG2 cells (Meurette et al., 2007). Intriguingly, PARP-1, a classical substrate of caspase-3 cleavage during apoptosis, is also required for necroptosis (Jouan-Lanhouet et al., 2012).

Several regulatory elements are shared between necroptosis and apoptosis, including FADD, RIPK1, and some RIPK1-modifying factors. The deubiquitinating (DUB) enzyme CYLD positively regulates necroptosis by deubiquitinating RIPK1, allowing it to interact with RIPK3 (Moquin, McQuade, & Chan, 2013; O'Donnell et al., 2011). A20, another DUB that regulates both TNFα signaling and caspase-8 activity upon death receptor ligation, acts as a negative regulator of necroptosis (Vanlangenakker, Bertrand, Bogaert, Vandenabeele, & Vanden Berghe, 2011). Deletion of the RIPK1 E3-ligases cIAP1 and cIAP2 promotes RIPK1 deubiquitination and allows RIPK1 to participate in either apoptotic or necroptotic signaling complexes, depending on the biochemical context (Biton & Ashkenazi, 2011; He et al., 2009; Wang et al., 2008).

5. CONCLUSION

In summary, death receptors control at least two types of regulated cell death—apoptosis and necroptosis. As the principal gatekeepers of these pathways, death receptors are precisely tuned to respond to subtle differences in ligand presentation by initiating distinct signaling pathways that can support either cell life or death. The balance between prosurvival and proapoptotic signals arising from death receptor ligation is tightly regulated through a series of biochemical checkpoints, some of which are only now becoming identified and experimentally accessible. Recent technical advances have significantly enhanced our understanding of the complex mechanisms by which extracellular cues, namely, binding of death receptor ligands, are ultimately transduced into cell fate decisions.

REFERENCES

Adams, C., Totpal, K., Lawrence, D., Marsters, S., Pitti, R., Yee, S., et al. (2008). Structural and functional analysis of the interaction between the agonistic monoclonal antibody Apomab and the proapoptotic receptor DR5. *Cell Death and Differentiation, 15,* 751–761.

Ashkenazi, A. (2008). Directing cancer cells to self-destruct with pro-apoptotic receptor agonists. *Nature Reviews. Drug Discovery,* 7, 1001–1012.

Ashkenazi, A., & Dixit, V. M. (1998). Death receptors: Signaling and modulation. *Science, 281,* 1305–1308.

Ashkenazi, A., & Herbst, R. S. (2008). To kill a tumor cell: The potential of proapoptotic receptor agonists. *The Journal of Clinical Investigation, 118,* 1979–1990.

Austin, C. D., Lawrence, D. A., Peden, A. A., Varfolomeev, E. E., Totpal, K., De Mazière, A. M., et al. (2006). Death-receptor activation halts clathrin-dependent endocytosis. *Proceedings of the National Academy of Sciences of the United States of America, 103,* 10283–10288.

Beaudouin, J., Liesche, C., Aschenbrenner, S., Hörner, M., & Eils, R. (2013). Caspase-8 cleaves its substrates from the plasma membrane upon CD95-induced apoptosis. *Cell Death and Differentiation, 20,* 599–610.

Berger, A. B., Sexton, K. B., & Bogyo, M. (2006). Commonly used caspase inhibitors designed based on substrate specificity profiles lack selectivity. *Cell Research, 16,* 961–963.

Bertrand, M. J. M., & Vandenabeele, P. (2011). The ripoptosome: Death decision in the cytosol. *Molecular Cell, 43,* 323–325.

Biton, S., & Ashkenazi, A. (2011). NEMO and RIP1 control cell fate in response to extensive DNA damage via TNF-α feedforward signaling. *Cell, 145,* 92–103.

Boatright, K. M., Renatus, M., Scott, F. L., Sperandio, S., Shin, H., Pedersen, I. M., et al. (2003). A unified model for apical caspase activation. *Molecular Cell, 11,* 529–541.

Boldin, M. P., Goncharov, T. M., Goltsev, Y. V., & Wallach, D. (1996). Involvement of MACH, a novel MORT1/FADD-interacting protease, in Fas/APO-1- and TNF receptor-induced cell death. *Cell, 85,* 803–815.

Bonifacino, J. S., & Traub, L. M. (2003). Signals for sorting of transmembrane proteins to endosomes and lysosomes. *Annual Review of Biochemistry, 72,* 395–447.

Bryde, S., Grunwald, I., Hammer, A., Krippner-Heidenreich, A., Schiestel, T., Brunner, H., et al. (2005). Tumor necrosis factor (TNF)-functionalized nanostructured particles for the stimulation of membrane TNF-specific cell responses. *Bioconjugate Chemistry, 16,* 1459–1467.

Cassatella, M. A., Huber, V., Calzetti, F., Margotto, D., Tamassia, N., Peri, G., et al. (2006). Interferon-activated neutrophils store a TNF-related apoptosis-inducing ligand (TRAIL/Apo-2 ligand) intracellular pool that is readily mobilizable following exposure to proinflammatory mediators. *Journal of Leukocyte Biology, 79,* 123–132.

Chaigne-Delalande, B., Moreau, J.-F., & Legembre, P. (2008). Rewinding the DISC. *Archivum Immunologiae et Therapiae Experimentalis, 56,* 9–14.

Chakrabandhu, K., Hérincs, Z., Huault, S., Dost, B., Peng, L., Conchonaud, F., et al. (2007). Palmitoylation is required for efficient Fas cell death signaling. *The EMBO Journal, 26,* 209–220.

Chan, F. K.-M. (2000). A domain in TNF receptors that mediates ligand-independent receptor assembly and signaling. *Science, 288,* 2351–2354.

Chang, L. F., Kamata, H., Solinas, G., Luo, J. L., Maeda, S., Venuprasad, K., et al. (2006). The E3 ubiquitin ligase itch couples JNK activation to TNF alpha-induced cell death by inducing c-FLIPL turnover. *Cell, 124,* 601–613.

Chang, D. W., Xing, Z., Pan, Y., Algeciras-Schimnich, A., Barnhart, B. C., Yaish-Ohad, S., et al. (2002). c-FLIPL is a dual function regulator for caspase-8 activation and CD95-mediated apoptosis. *The EMBO Journal, 21,* 3704–3714.

Chauvier, D., Ankri, S., Charriaut-Marlangue, C., Casimir, R., & Jacotot, E. (2007). Broad-spectrum caspase inhibitors: From myth to reality? *Cell Death and Differentiation, 14*, 387–391.

Chen, K. F., Liu, C. Y., Lin, Y. C., Yu, H. C., Liu, T. H., Hou, D. R., et al. (2010). CIP2A mediates effects of bortezomib on phospho-Akt and apoptosis in hepatocellular carcinoma cells. *Oncogene, 29*, 6257–6266.

Chen, W., Zhou, Z., Li, L., Zhong, C. Q., Zheng, X., Wu, X., et al. (2013). Diverse sequence determinants control human and mouse receptor interacting protein 3 (RIP3) and mixed lineage kinase domain-like (MLKL) interaction in necroptotic signaling. *The Journal of Biological Chemistry, 288*, 16247–16261.

Chen, X., Li, W., Ren, J., Huang, D., He, W., Song, Y., et al. (2014). Translocation of mixed lineage kinase domain-like protein to plasma membrane leads to necrotic cell death. *Cell Research, 24*, 105–121.

Cho, Y. S., Challa, S., Moquin, D., Genga, R., Ray, T. D., Guildford, M., et al. (2009). Phosphorylation-driven assembly of the RIP1–RIP3 complex regulates programmed necrosis and virus-induced inflammation. *Cell, 137*, 1112–1123.

Cortez-Retamozo, V., Backmann, N., Senter, P. D., Wernery, U., De Baetselier, P., Muyldermans, S., et al. (2004). Efficient cancer therapy with a nanobody-based conjugate. *Cancer Research, 64*, 2853–2857.

Cory, S., & Adams, J. M. (2002). The Bcl2 family: Regulators of the cellular life-or-death switch. *Nature Reviews. Cancer, 2*, 647–656.

Cox, A., Dunning, A. M., Garcia-Closas, M., Balasubramanian, S., Reed, M. W. R., Pooley, K. A., et al. (2007). A common coding variant in CASP8 is associated with breast cancer risk. *Nature Genetics, 39*, 352–358.

Cseh, K., & Beutlers, B. (1989). Alternative cleavage of the cachectin/tumor necrosis factor propeptide results in a larger, inactive form of secreted protein. *The Journal of Biological Chemistry, 264*, 16256–16260.

Cursi, S., Rufini, A., Stagni, V., Condo, I., Matafora, V., Bachi, A., et al. (2006). Src kinase phosphorylates caspase-8 on Tyr380: A novel mechanism of apoptosis suppression. *The EMBO Journal, 25*, 1895–1905.

Danial, N. N., & Korsmeyer, S. J. (2004). Cell death: Critical control points. *Cell, 116*, 205–219.

Degterev, A., Hitomi, J., Germscheid, M., Ch-en, I. L., Korkina, O., Teng, X., et al. (2008). Identification of RIP1 kinase as a specific cellular target of necrostatins. *Nature Chemical Biology, 4*, 313–321.

Dickens, L. S., Boyd, R. S., Jukes-Jones, R., Hughes, M. A., Robinson, G. L., Fairall, L., et al. (2012). A death effector domain chain DISC model reveals a crucial role for caspase-8 chain assembly in mediating apoptotic cell death. *Molecular Cell, 47*, 291–305.

Dillon, C. P., Oberst, A., Weinlich, R., Janke, L. J., Kang, T. B., Ben-Moshe, T., et al. (2012). Survival function of the FADD-caspase-8-cFLIP(L) complex. *Cell Reports, 1*, 401–407.

Donepudi, M., Mac Sweeney, A., Briand, C., & Grutter, M. G. (2003). Insights into the regulatory mechanism for caspase-8 activation. *Molecular Cell, 11*, 543–549.

Dustin, M. L., Bromley, S. K., Davis, M. M., & Zhu, C. (2001). Identification of self through two-dimensional chemistry and synapses. *Annual Review of Cell and Developmental Biology, 17*, 133–157.

Ehrlich, S., Infante-Duarte, C., Seeger, B., & Zipp, F. (2003). Regulation of soluble and surface-bound TRAIL in human T cells, B cells, and monocytes. *Cytokine, 24*, 244–253.

Eimon, P. M., Kratz, E., Varfolomeev, E., Hymowitz, S. G., Stern, H., Zha, J., et al. (2006). Delineation of the cell-extrinsic apoptosis pathway in the zebrafish. *Cell Death and Differentiation, 13*, 1619–1630.

Feig, C., Tchikov, V., Schütze, S., & Peter, M. E. (2007). Palmitoylation of CD95 facilitates formation of SDS-stable receptor aggregates that initiate apoptosis signaling. *The EMBO Journal, 26*, 221–231.

Fricker, N., Beaudouin, J., Richter, P., Eils, R., Krammer, P. H., & Lavrik, I. N. (2010). Model-based dissection of CD95 signaling dynamics reveals both a pro- and anti-apoptotic role of c-FLIPL. *The Journal of Cell Biology, 190*, 377–389.

Gao, M., Labuda, T., Xia, Y., Gallagher, E., Fang, D., Liu, Y. C., et al. (2004). Jun turnover is controlled through JNK-dependent phosphorylation of the E3 ligase Itch. *Science, 306*, 271–275.

Gonzalvez, F., Lawrence, D., Yang, B., Yee, S., Pitti, R., Marsters, S., et al. (2012). TRAF2 sets a threshold for extrinsic apoptosis by tagging caspase-8 with a ubiquitin shutoff timer. *Molecular Cell, 48*, 888–899.

Grassmé, H., Bock, J., Kun, J., & Gulbins, E. (2002). Clustering of CD40 ligand is required to form a functional contact with CD40. *The Journal of Biological Chemistry, 277*, 30289–30299.

Grassmé, H., Jekle, A., Riehle, A., Schwarz, H., Berger, J., Sandhoff, K., et al. (2001). CD95 signaling via ceramide-rich membrane rafts. *The Journal of Biological Chemistry, 276*, 20589–20596.

Grassmé, H., Schwarz, H., & Gulbins, E. (2001). Molecular mechanisms of ceramide-mediated CD95 clustering. *Biochemical and Biophysical Research Communications, 284*, 1016–1030.

Grell, M., Douni, E., Wajant, H., Löhden, M., Clauss, M., Maxeiner, B., et al. (1995). The transmembrane form of tumor necrosis factor is the prime activating ligand of the 80 kDa tumor necrosis factor receptor. *Cell, 83*, 793–802.

Gunther, C., Martini, E., Wittkopf, N., Amann, K., Weigmann, B., Neumann, H., et al. (2011). Caspase-8 regulates TNF-alpha-induced epithelial necroptosis and terminal ileitis. *Nature, 477*, 335–339.

Halaas, O., Vik, R., Ashkenazi, A., & Espevik, T. (2000). Lipopolysaccharide induces expression of APO2 ligand/TRAIL in human monocytes and macrophages. *Scandinavian Journal of Immunology, 51*, 244–250.

Harbury, P. B., Zhang, T., Kim, P. S., & Alber, T. (1993). A switch between two-, three-, and four-stranded coiled coils in GCN4 leucine zipper mutants. *Science, 262*, 1401–1407.

He, S., Wang, L., Miao, L., Wang, T., Du, F., Zhao, L., et al. (2009). Receptor interacting protein kinase-3 determines cellular necrotic response to TNF-alpha. *Cell, 137*, 1100–1111.

Hohlbaum, A. M., Moe, S., & Marshak-Rothstein, A. (2000). Opposing effects of transmembrane and soluble Fas ligand expression on inflammation and tumor cell survival. *The Journal of Experimental Medicine, 191*, 1209–1220.

Holler, N., Zaru, R., Micheau, O., Thome, M., Attinger, A., Valitutti, S., et al. (2000). Fas triggers an alternative, caspase-8-independent cell death pathway using the kinase RIP as effector molecule. *Nature Immunology, 1*, 489–495.

Huet, H., Schuller, A., Li, J., Johnson, J., Dombrecht, B., Meerschaert, K., et al. (2012). TAS266, a novel tetrameric nanobody agonist targeting death receptor 5 (DR5), elicits superior antitumor efficacy than conventional DR5-targeted approaches. *Cancer Research*, 72, Supplement 1: Abstract 3853.

Hughes, M. A., Harper, N., Butterworth, M., Cain, K., Cohen, G. M., & MacFarlane, M. (2009). Reconstitution of the death-inducing signaling complex reveals a substrate switch that determines CD95-mediated death or survival. *Molecular Cell, 35*, 265–279.

Ikner, A., & Ashkenazi, A. (2011). TWEAK induces apoptosis through a death-signaling complex comprising receptor-interacting protein 1 (RIP1), Fas-associated death domain (FADD), and caspase-8. *The Journal of Biological Chemistry, 286*, 21546–21554.

Irmler, M., Thome, M., Hahne, M., Schneider, P., Hofmann, B., Steiner, V., et al. (1997). Inhibition of death receptor signals by cellular FLIP. *Nature, 388*, 190–195.

Jia, S. H., Parodo, J., Kapus, A., Rotstein, O. D., & Marshall, J. C. (2008). Dynamic regulation of neutrophil survival through tyrosine phosphorylation or dephosphorylation of caspase-8. *The Journal of Biological Chemistry, 283*, 5402–5413.

Jin, Z., Li, Y., Pitti, R., Lawrence, D., Pham, V. C., Lill, J. R., et al. (2009). Cullin3-based polyubiquitination and p62-dependent aggregation of caspase-8 mediate extrinsic apoptosis signaling. *Cell*, *137*, 721–735.

Jo, M., Kim, T. H., Seol, D. W., Esplen, J. E., Dorko, K., Billiar, T. R., et al. (2000). Apoptosis induced in normal human hepatocytes by tumor necrosis factor-related apoptosis-inducing ligand. *Nature Medicine*, *6*, 564–567.

Johnstone, R. W., Frew, A. J., & Smyth, M. J. (2008). The TRAIL apoptotic pathway in cancer onset, progression and therapy. *Nature Reviews. Cancer*, *8*, 782–798.

Jouan-Lanhouet, S., Arshad, M. I., Piquet-Pellorce, C., Martin-Chouly, C., Le Moigne-Muller, G., Van Herreweghe, F., et al. (2012). TRAIL induces necroptosis involving RIPK1/RIPK3-dependent PARP-1 activation. *Cell Death and Differentiation*, *19*, 2003–2014.

Kaiser, W. J., Upton, J. W., Long, A. B., Livingston-Rosanoff, D., Daley-Bauer, L. P., Hakem, R., et al. (2011). RIP3 mediates the embryonic lethality of caspase-8-deficient mice. *Nature*, *471*, 368–372.

Kawahara, A., Ohsawa, Y., Matsumura, H., Uchiyama, Y., & Nagata, S. (1998). Caspase-independent cell killing by Fas-associated protein with death domain. *The Journal of Cell Biology*, *143*, 1353–1360.

Keller, N., Mares, J., Zerbe, O., & Gruetter, M. G. (2009). Structural and biochemical studies on procaspase-8: New insights on initiator caspase activation. *Structure*, *17*, 438–448.

Kemp, T. J., Kim, J. S., Crist, S. A., & Griffith, T. S. (2003). Induction of necrotic tumor cell death by TRAIL/Apo-2L. *Apoptosis*, *8*, 587–599.

Kischkel, F. C., Hellbardt, S., Behrmann, I., Germer, M., Pawlita, M., Krammer, P. H., et al. (1995). Cytotoxicity-dependent APO-1 (Fas/CD95)-associated proteins form a death-inducing signaling complex (DISC) with the receptor. *The EMBO Journal*, *14*, 5579–5588.

Kischkel, F. C., Lawrence, D. a., Chuntharapai, A., Schow, P., Kim, K. J., & Ashkenazi, A. (2000). Apo2L/TRAIL-dependent recruitment of endogenous FADD and caspase-8 to death receptors 4 and 5. *Immunity*, *12*, 611–620.

Kischkel, F. C., Lawrence, D. A., Tinel, A., LeBlanc, H., Virmani, A., Schow, P., et al. (2001). Death receptor recruitment of endogenous caspase-10 and apoptosis initiation in the absence of caspase-8. *The Journal of Biological Chemistry*, *276*, 46639–46646.

Kohlhaas, S. L., Craxton, A., Sun, X.-M., Pinkoski, M. J., & Cohen, G. M. (2007). Receptor-mediated endocytosis is not required for tumor necrosis factor-related apoptosis-inducing ligand (TRAIL)-induced apoptosis. *The Journal of Biological Chemistry*, *282*, 12831–12841.

Krippner-Heidenreich, A., Tübing, F., Bryde, S., Willi, S., Zimmermann, G., & Scheurich, P. (2002). Control of receptor-induced signaling complex formation by the kinetics of ligand/receptor interaction. *The Journal of Biological Chemistry*, *277*, 44155–44163.

Krueger, A., Baumann, S., Krammer, P. H., & Kirchhoff, S. (2001). FLICE-inhibitory proteins: Regulators of death receptor-mediated apoptosis. *Molecular and Cellular Biology*, *21*, 8247–8254.

Lawrence, D., Shahrokh, Z., Marsters, S., Achilles, K., Shih, D., Mounho, B., et al. (2001). Differential hepatocyte toxicity of recombinant Apo2L/TRAIL versions. *Nature Medicine*, *7*, 383–385.

Lawrence, M. S., Stojanov, P., Mermel, C. H., Robinson, J. T., Garraway, L. A., Golub, T. R., et al. (2014). Discovery and saturation analysis of cancer genes across 21 tumour types. *Nature*, *505*, 495–501.

Lee, K.-H., Feig, C., Tchikov, V., Schickel, R., Hallas, C., Schütze, S., et al. (2006). The role of receptor internalization in CD95 signaling. *The EMBO Journal*, *25*, 1009–1023.

Lee, E.-W., Kim, J.-H., Ahn, Y.-H., Seo, J., Ko, A., Jeong, M., et al. (2012). Ubiquitination and degradation of the FADD adaptor protein regulate death receptor-mediated apoptosis and necroptosis. *Nature Communications*, *3*, 1–12.

Menke, C., Bin, L., Thorburn, J., Behbakht, K., Ford, H. L., & Thorburn, A. (2011). Distinct TRAIL resistance mechanisms can be overcome by proteasome inhibition but not generally by synergizing agents. *Cancer Research*, *71*, 1883–1892.

Meurette, O., Rebillard, A., Huc, L., Le Moigne, G., Merino, D., Micheau, O., et al. (2007). TRAIL induces receptor-interacting protein 1-dependent and caspase-dependent necrosis-like cell death under acidic extracellular conditions. *Cancer Research*, *67*, 218–226.

Micheau, O., & Tschopp, J. (2003). Induction of TNF receptor I-mediated apoptosis via two sequential signaling complexes. *Cell*, *114*, 181–190.

Moquin, D. M., McQuade, T., & Chan, F. K. (2013). CYLD deubiquitinates RIP1 in the TNFalpha-induced necrosome to facilitate kinase activation and programmed necrosis. *PLoS One*, *8*, e76841.

Murphy, J. M., Czabotar, P. E., Hildebrand, J. M., Lucet, I. S., Zhang, J. G., Alvarez-Diaz, S., et al. (2013). The pseudokinase MLKL mediates necroptosis via a molecular switch mechanism. *Immunity*, *39*, 443–453.

Muzio, M., Chinnaiyan, A. M., Kischkel, F. C., O'Rourke, K., Shevchenko, A., Ni, J., et al. (1996). FLICE, a novel FADD-homologous ICE/CED-3-like protease, is recruited to the CD95 (Fas/APO-1) death-inducing signaling complex. *Cell*, *85*, 817–827.

Nagata, S. (1997). Apoptosis by death factor. *Cell*, *88*, 355–365.

Naito, M., Hainz, U., Burkhardt, U. E., Fu, B., Ahove, D., Stevenson, K. E., et al. (2013). CD40L-Tri, a novel formulation of recombinant human CD40L that effectively activates B cells. *Cancer Immunology, Immunotherapy*, *62*, 347–357.

Naumann, I., Kappler, R., von Schweinitz, D., Debatin, K.-M., & Fulda, S. (2011). Bortezomib primes neuroblastoma cells for TRAIL-induced apoptosis by linking the death receptor to the mitochondrial pathway. *Clinical Cancer Research*, *17*, 3204–3218.

O' Reilly, L. A., Tai, L., Lee, L., Kruse, E. A., Grabow, S., Fairlie, W. D., et al. (2009). Membrane-bound Fas ligand only is essential for Fas-induced apoptosis. *Nature*, *461*, 659–663.

O'Donnell, M. A., Perez-Jimenez, E., Oberst, A., Ng, A., Massoumi, R., Xavier, R., et al. (2011). Caspase 8 inhibits programmed necrosis by processing CYLD. *Nature Cell Biology*, *13*, 1437–1442.

Oberst, A., Dillon, C. P., Weinlich, R., McCormick, L. L., Fitzgerald, P., Pop, C., et al. (2011). Catalytic activity of the caspase-8-FLIP(L) complex inhibits RIPK3-dependent necrosis. *Nature*, *471*, 363–367.

Oberst, A., Pop, C., Tremblay, A. G., Blais, V., Denault, J.-B., Salvesen, G. S., et al. (2010). Inducible dimerization and inducible cleavage reveal a requirement for both processes in caspase-8 activation. *The Journal of Biological Chemistry*, *285*, 16632–16642.

Papoff, G. (1999). Identification and characterization of a ligand-independent oligomerization domain in the extracellular region of the CD95 death receptor. *The Journal of Biological Chemistry*, *274*, 38241–38250.

Perez, C., Albert, I., DeFay, K., Zachariades, N., Gooding, L., & Kriegler, M. (1990). A nonsecretable cell surface mutant of tumor necrosis factor (TNF) kills by cell-to-cell contact. *Cell*, *63*, 251–258.

Peter, M. E., & Krammer, P. H. (2003). The CD95(APO-1/Fas) DISC and beyond. *Cell Death and Differentiation*, *10*, 26–35.

Pickart, C. M. (2004). Back to the future with ubiquitin. *Cell*, *116*, 181–190.

Pitti, R. M., Marsters, S. A., Ruppert, S., Donahue, C. J., Moore, A., & Ashkenazi, A. (1996). Induction of apoptosis by Apo-2 ligand, a new member of the tumor necrosis factor cytokine family. *The Journal of Biological Chemistry*, *271*, 12687–12690.

Pop, C., Oberst, A., Drag, M., Van Raam, B. J., Riedl, S. J., Green, D. R., et al. (2011). FLIPL induces caspase 8 activity in the absence of interdomain caspase 8 cleavage and alters substrate specificity. *The Biochemical Journal*, *433*, 447–457.

Ray, C. A., Black, R. A., Kronheim, S. R., Greenstreet, T. A., Sleath, P. R., Salvesen, G. S., et al. (1992). Viral inhibition of inflammation: Cowpox virus encodes an inhibitor of the interleukin-1 beta converting enzyme. *Cell*, *69*, 597–604.

Rosendahl, M. S., Ko, S. C., Long, D. L., Brewer, M. T., Rosenzweig, B., Hedl, E., et al. (1997). Identification and characterization of a pro-tumor necrosis factor-alpha-processing enzyme from the ADAM family of zinc metalloproteases. *The Journal of Biological Chemistry*, *272*, 24588–24593.

Rossin, A., Derouet, M., Abdel-Sater, F., & Hueber, A.-O. (2009). Palmitoylation of the TRAIL receptor DR4 confers an efficient TRAIL-induced cell death signalling. *The Biochemical Journal*, *419*, 185–192.

Scaffidi, C., Schmitz, I., Krammer, P. H., & Peter, M. E. (1999). The role of c-FLIP in modulation of CD95-induced apoptosis. *The Journal of Biological Chemistry*, *274*, 1541–1548.

Schleich, K., Warnken, U., Fricker, N., Oeztuerk, S., Richter, P., Kammerer, K., et al. (2012). Stoichiometry of the CD95 death-inducing signaling complex: Experimental and modeling evidence for a death effector domain chain model. *Molecular Cell*, *47*, 306–319.

Schneider, P., Holler, N., Bodmer, J., Hahne, M., Frei, K., Fontana, A., et al. (1998). Conversion of membrane-bound Fas(CD95) ligand to its soluble form is associated with downregulation of its proapoptotic activity and loss of liver toxicity. *Nature*, *187*, 1205–1213.

Schneider, P., Schwenzer, R., Haas, E., Mühlenbeck, F., Schubert, G., Scheurich, P., et al. (1999). TWEAK can induce cell death via endogenous TNF and TNF receptor 1. *European Journal of Immunology*, *29*, 1785–1792.

Sedger, L. M., Shows, D. M., Blanton, R. a., Peschon, J. J., Goodwin, R. G., Cosman, D., et al. (1999). IFN-gamma mediates a novel antiviral activity through dynamic modulation of TRAIL and TRAIL receptor expression. *Journal of Immunology*, *163*, 920–926.

Seifert, O., Plappert, A., Fellermeier, S., Siegemund, M., Pfizenmaier, K., & Kontermann, R. E. (2014). Tetravalent antibody-scTRAIL fusion proteins with improved properties. *Molecular Cancer Therapeutics*, *13*, 101–111.

Senft, J., Helfer, B., & Frisch, S. M. (2007). Caspase-8 interacts with the p85 subunit of phosphatidylinositol 3-kinase to regulate cell adhesion and motility. *Cancer Research*, *67*, 11505–11509.

Sharp, D. A., Lawrence, D. A., & Ashkenazi, A. (2005). Selective knockdown of the long variant of cellular FLICE inhibitory protein augments death receptor-mediated caspase-8 activation and apoptosis. *The Journal of Biological Chemistry*, *280*, 19401–19409.

Shu, H. B., Halpin, D. R., & Goeddel, D. V. (1997). Casper is a FADD- and caspase-related inducer of apoptosis. *Immunity*, *6*, 751–763.

Siegel, R. M. (2000). Fas preassociation required for apoptosis signaling and dominant inhibition by pathogenic mutations. *Science*, *288*, 2354–2357.

Sprick, M. R., Weigand, M. a., Rieser, E., Rauch, C. T., Juo, P., Blenis, J., et al. (2000). FADD/MORT1 and caspase-8 are recruited to TRAIL receptors 1 and 2 and are essential for apoptosis mediated by TRAIL receptor 2. *Immunity*, *12*, 599–609.

Sun, L., Wang, H., Wang, Z., He, S., Chen, S., Liao, D., et al. (2012). Mixed lineage kinase domain-like protein mediates necrosis signaling downstream of RIP3 kinase. *Cell*, *148*, 213–227.

Teitz, T., Wei, T., Valentine, M. B., Vanin, E. F., Grenet, J., Valentine, V. A., et al. (2000). Caspase 8 is deleted or silenced preferentially in childhood neuroblastomas with amplification of MYCN. *Nature Medicine*, *6*, 529–535.

Thome, M., & Tschopp, J. (2001). Regulation of lymphocyte proliferation and death by flip. *Nature Reviews. Immunology, 1*, 50–58.

Van Raam, B. J., & Salvesen, G. S. (2012). Proliferative versus apoptotic functions of caspase-8 hetero or homo: The caspase-8 dimer controls cell fate. *Biochimica et Biophysica Acta, 1824*, 113–122.

Vandenabeele, P., Galluzzi, L., Vanden Berghe, T., & Kroemer, G. (2010). Molecular mechanisms of necroptosis: An ordered cellular explosion. *Nature Reviews Molecular Cell Biology, 11*, 700–714.

Vanlangenakker, N., Bertrand, M. J., Bogaert, P., Vandenabeele, P., & Vanden Berghe, T. (2011). TNF-induced necroptosis in L929 cells is tightly regulated by multiple TNFR1 complex I and II members. *Cell Death & Disease, 2*, e230.

Vanlangenakker, N., Vanden Berghe, T., & Vandenabeele, P. (2012). Many stimuli pull the necrotic trigger, an overview. *Cell Death and Differentiation, 19*, 75–86.

Varfolomeev, E., Maecker, H., Sharp, D., Lawrence, D., Renz, M., Vucic, D., et al. (2005). Molecular determinants of kinase pathway activation by Apo2 ligand/tumor necrosis factor-related apoptosis-inducing ligand. *The Journal of Biological Chemistry, 280*, 40599–40608.

Varfolomeev, E. E., Schuchmann, M., Luria, V., Chiannilkulchai, N., Beckmann, J. S., Mett, I. L., et al. (1998). Targeted disruption of the mouse caspase 8 gene ablates cell death induction by the TNF receptors, Fas/Apo1, and DR3 and is lethal prenatally. *Immunity, 9*, 267–276.

Vidalain, P. O., Azocar, O., Servet-Delprat, C., Rabourdin-Combe, C., Gerlier, D., & Manié, S. (2000). CD40 signaling in human dendritic cells is initiated within membrane rafts. *The EMBO Journal, 19*, 3304–3313.

Vucic, D., Dixit, V. M., & Wertz, I. E. (2011). Ubiquitylation in apoptosis: A post-translational modification at the edge of life and death. *Nature Reviews. Molecular Cell Biology, 12*, 439–452.

Wagner, K. W., Punnoose, E. A., Januario, T., Lawrence, D. A., Pitti, R. M., Lancaster, K., et al. (2007). Death-receptor O-glycosylation controls tumor-cell sensitivity to the proapoptotic ligand Apo2L/TRAIL. *Nature Medicine, 13*, 1070–1077.

Walczak, H., Degli-Esposti, M. A., Johnson, R. S., Smolak, P. J., Waugh, J. Y., Boiani, N., et al. (1997). TRAIL-R2: A novel apoptosis-mediating receptor for TRAIL. *The EMBO Journal, 16*, 5386–5397.

Walczak, H., Miller, R. E., Ariail, K., Gliniak, B., Griffith, T. S., Kubin, M., et al. (1999). Tumoricidal activity of tumor necrosis factor-related apoptosis-inducing ligand in vivo. *Nature Medicine, 5*, 157–163.

Wang, H., Sun, L., Su, L., Rizo, J., Liu, L., Wang, F., et al. (2014). Mixed lineage kinase domain-like protein MLKL causes necrotic membrane disruption upon phosphorylation by RIP3. *Molecular Cell, 54*, 133–146.

Wang, L., Du, F., & Wang, X. (2008). TNF-alpha induces two distinct caspase-8 activation pathways. *Cell, 133*, 693–703.

Wiley, S. R., Schooley, K., Smolak, P. J., Din, W. S., Huang, C. P., Nicholl, J. K., et al. (1995). Identification and characterization of a new member of the TNF family that induces apoptosis. *Immunity, 3*, 673–682.

Wilson, N. S., Dixit, V., & Ashkenazi, A. (2009). Death receptor signal transducers: Nodes of coordination in immune signaling networks. *Nature Immunology, 10*, 348–355.

Wilson, N. S., Yang, A., Yang, B., Couto, S., Stern, H., Gogineni, A., et al. (2012). Proapoptotic activation of death receptor 5 on tumor endothelial cells disrupts the vasculature and reduces tumor growth. *Cancer Cell, 22*, 80–90.

Wilson, N. S., Yang, B., Yang, A., Loeser, S., Marsters, S., Lawrence, D., et al. (2011). An Fcγ receptor-dependent mechanism drives antibody-mediated target-receptor signaling in cancer cells. *Cancer Cell, 19*, 101–113.

Yeh, W. C., Itie, A., Elia, A. J., Ng, M., Shu, H. B., Wakeham, A., et al. (2000). Requirement for casper (c-FLIP) in regulation of death receptor-induced apoptosis and embryonic development. *Immunity*, *12*, 633–642.

Zhang, D. W., Shao, J., Lin, J., Zhang, N., Lu, B. J., Lin, S. C., et al. (2009). RIP3, an energy metabolism regulator that switches TNF-induced cell death from apoptosis to necrosis. *Science*, *325*, 332–336.

Zhou, Q., Snipas, S., Orth, K., Muzio, M., Dixit, V. M., & Salvesen, G. S. (1997). Target protease specificity of the viral serpin CrmA—Analysis of five caspases. *The Journal of Biological Chemistry*, *272*, 7797–7800.

CHAPTER SIX

Using RNAi Screening Technologies to Interrogate the Extrinsic Apoptosis Pathway

Jean Philippe Stephan
Protein Chemistry Department/Discovery Oncology Department, Genentech Inc., South San Francisco, California, USA

Contents

Abstract

Despite the knowledge accumulated during the last two decades about programmed cell death, further investigations of the complex regulatory network of apoptosis, including the extrinsic pathways, are still needed to gain an exhaustive and comprehensive understanding of this critical biological process. In addition, the identification of novel modulators of apoptosis may represent a good opportunity for making new paths into an otherwise heavily investigated area, therefore providing a molecular basis for new therapeutic strategies. In the last decade, RNA interference has become the technology of choice for discovering genes that encode molecules with previously unknown functions in biological pathways of interest. Various RNAi reagents and library formats have been developed and harnessed for high-throughput screening technologies to enable almost limitless investigation to uncover gene functions and networks in the context of basic biology and biomedical research including cancer biology. Although RNAi screening has been demonstrated to be a very powerful tool, various caveats and pitfalls have been progressively uncovered, including, but not limited to the enduring off-target effects. As the novelty of its bells and whistles have begun to diminish, functional genomic screens have morphed into a specialized field within the high-throughput screening community, where expert investigators progressively establish rigorous strategies to mitigate most of its possible flaws. Using various examples of RNAi screens conducted to further understand the extrinsic apoptosis pathway, this chapter describes the different RNAi tools and screening formats available and reviews the parameters one has to critically consider in order to be successful in implementing this technology.

Methods in Enzymology, Volume 544
ISSN 0076-6879
http://dx.doi.org/10.1016/B978-0-12-417158-9.00006-6

1. INTRODUCTION

Apoptosis, a form of programmed cell death is essential for multicellular organisms to maintain tissue integrity (Jacobson, Weil, & Raff, 1997; Steller, 1995). It normally occurs during development and aging to maintain cell population homeostasis by mediating the equilibrium between cell proliferation and cell death. As a critical biological process, apoptosis is precisely regulated and controlled by a diverse range of cell signals leading to two distinct signaling pathways that ultimately result in cell death: (1) the extrinsic, or extracellularly activated, pathway also called the death receptor pathway (Ashkenazi & Dixit, 1999) and (2) the intrinsic, or mitochondrial-mediated, pathway (Fulda & Debatin, 2006; Tait & Green, 2010). Both pathways activate a cascade of cysteine proteinases, the caspases that act as death effector molecules to mediate cellular dismantling events (Degterev, Boyce, & Yuan, 2003). Caspase activation will result in rapid morphological changes that involve the three major compartments of the cell: the plasma membrane, the cytoplasm, and the nucleus and are associated with a distinct set of biochemical and physical changes such as cell membrane blebbing, cell shrinkage, chromatin condensation, and DNA fragmentation. Not surprisingly, deregulation of apoptosis has been recognized for some time as a key process in cancer development and progression (Hanhan & Weinberg, 2000) and as such signaling events leading to apoptosis have been highly investigated from both fundamental and applied aspects. Despite the understanding of apoptosis that has been gained in the last two decades and the resulting clinical trials currently ongoing (Fischer & Schulze-Osthoff, 2005; Wong, 2011), there is still a need to better understand the complex regulatory network of apoptosis that might provide a molecular basis for new therapeutic strategies targeting cell death pathways.

In this context, the availability of genomic sequencing as well as genetic tools have provided unprecedented abilities to systematically query the involvement of thousand of genes of unknown function in a specific biological process. In the last decade, RNA interference (RNAi) has arisen as a powerful investigation tool. RNAi is an endogenous, sequence-specific, post-transcriptional process whereby messenger RNAs are targeted for degradation by long double-stranded RNAs (dsRNAs) of identical sequence, leading to gene silencing (Chapman & Carrington, 2007; Fire et al., 1998). Since its discovery in *Caenorhabditis elegans* (Fire et al., 1998), RNAi provides to the cell biologist a powerful technique for the derivation and analysis of loss-of-function phenotypes in vertebrate cells, where alternative approaches are either arduous or frequently ineffective. RNAi techniques can be applied in a high-throughput fashion on both the cell or organism scale and therefore offer the opportunity to investigate the role of a large number of genes in a specific biological process. This chapter describes the different RNAi tools and screening formats available and reviews the critical parameters one has to consider in order to be successful when implementing the technology.

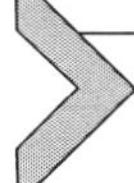

2. RNAi-MEDIATED GENE KNOCKDOWN IN MAMMALIAN CELLS

2.1. RNAi tools

RNAi leads to sequence-specific knockdown of gene function and therefore the technology enables functional genomics investigations in various cell types, including cell lines and primary cells as well as whole organisms. Combined with high-throughput screening (HTS) technologies, a large number of genes may be investigated concurrently within a specific defined biological context (e.g., apoptosis pathway) such that one could theoretically begin to isolate multiple members of a functional pathway as well as discover new regulatory nodes in a given biological function, process, complex, or cellular behavior. RNAi screening relies on the availability of libraries of various sizes with one or more unique RNAi reagent directed against each target gene. Unlike in *Drosophila* cells, the interferon response prevents the use of *in vitro* synthesized dsRNA in mammalian cells (Mohr, Bakal, & Perrimon, 2010). In mammalian cells, RNAi is generally induced by reagents in the form of either small interfering RNAs (siRNAs) duplex or short hairpin RNAs (shRNAs). There are critical differences between these reagents that need to be carefully considered when designing a screen. As discussed elsewhere in more detail (Cullen, 2004), siRNA are 19 to 25-base-pair synthetic dsRNA bearing 2-nucleotides 3′-overhang at each end that mimic the structure of microRNA (miRNA) duplex intermediates formed during the processing of endogenous miRNA transcripts. In the case of both exogenously introduced siRNA duplex and endogenously transcribed miRNA duplexes, one strand of the duplex is taken up by the RNA-induced silencing complex (RISC), where it then acts as a "guide" RNA to direct RISC to complementary mRNAs, while the second, "passenger" strand is degraded (Hammond, Bernstein, Beach, & Hannon, 2000; Khvorova, Reynolds, & Jayasena, 2003; Schwarz et al., 2003;

Schwarz, Hutvágner, Haley, & Zamore, 2002). One caveat to the siRNA technology is that since siRNA produces transient transfection, the gene knockdown effect is not carried from one generation to the next and will only persist for 4–5 days, regardless of the transfection efficiency. Moreover, only certain cells, such as "immortalized" cancer cell lines physically accept siRNA transfection. The advantage of siRNA, however, is that it is relatively facile, and the reagents are easy to work with.

To overcome the limitation of transient transfection, various methods for the stable induction of RNAi using expression plasmids or viral vectors have been developed (Brummelkamp, Bernards, & Agami, 2002; Paddison, Caudy, Bernstein, Hannon, & Conklin, 2002; Rutz & Scheffold, 2004; Stegmeier, Hu, Rickles, Hannon, & Elledge, 2005; Zeng, Wagner, & Cullen, 2002). These alternative approaches generally rely on the expression of RNAs that mimic earlier intermediates in the biogenesis of cellular miRNAs. The most common method relies on shRNAs, approximately 50 nucleotides long RNA hairpins consisting of a stem of 19 to 26-base-pair linked to a terminal loop and bearing a 2-nucleotides 3′-overhang (Brummelkamp et al., 2002; Paddison et al., 2002). shRNAs mimic the pre-miRNA hairpins that result from the processing of endogenous primary miRNA transcripts by the RNase III enzyme Drosha (Cullen, 2004, Lee et al., 2003). Subsequently, artificial shRNAs are processed similarly to endogenous pre-miRNAs through a second RNase III enzyme called Dicer to produce siRNA/miRNA duplex intermediates (Bernstein, Caudy, Hammond, & Hannon, 2001; Cullen, 2004). The use of endogenously transcribed artificial shRNA to generate siRNAs that are then loaded into RISC has several advantages over direct transfection of siRNA duplexes, including a more stable RNAi phenotype and substantially lower cost. shRNAs are the reagents of choice when the phenotype takes longer than about 2 weeks to develop. However, shRNA is disadvantageous in that it requires use of an expression vector that can pose safety concerns.

Other technologies have been developed during the last couple of years as genome-editing tools and could be applied to functional genomic HTS applications. These include complexes known as transcription activator-like effector nucleases (TALENs) (Bogdanove & Voytas, 2011; Joung & Sander, 2013; Sun & Zhao, 2013) as well as clustered regulatory interspaced short palindromic repeat (CRISPR)/Cas-9 based RNA-guided DNA endonucleases (Cong et al., 2013; Gilbert et al., 2013; Qi et al., 2013) that cut the genome in specific locations resulting in gene knockout (Gaj, Gersbach, & Barbas, 2013). Although TALENs could be expensive and difficult to assemble, some investigators assembled large libraries that could be used for screening efforts (Kim et al., 2013). Also the CRISPR/Cas-9 system could represent a significant advancement in the field of genome-editing, its specificity has yet to be fully determined and possible application to functional genomic screens demonstrated.

2.2. RNAi screening formats

siRNA and shRNA libraries come in two main formats that have significant advantages and disadvantages. Both of these formats have been successfully applied to investigate various biological questions including further dissection of the extrinsic apoptosis pathway. Multiple parameters need to be considered when choosing an RNAi screening format (Fig. 6.1) and are reviewed in the following section.

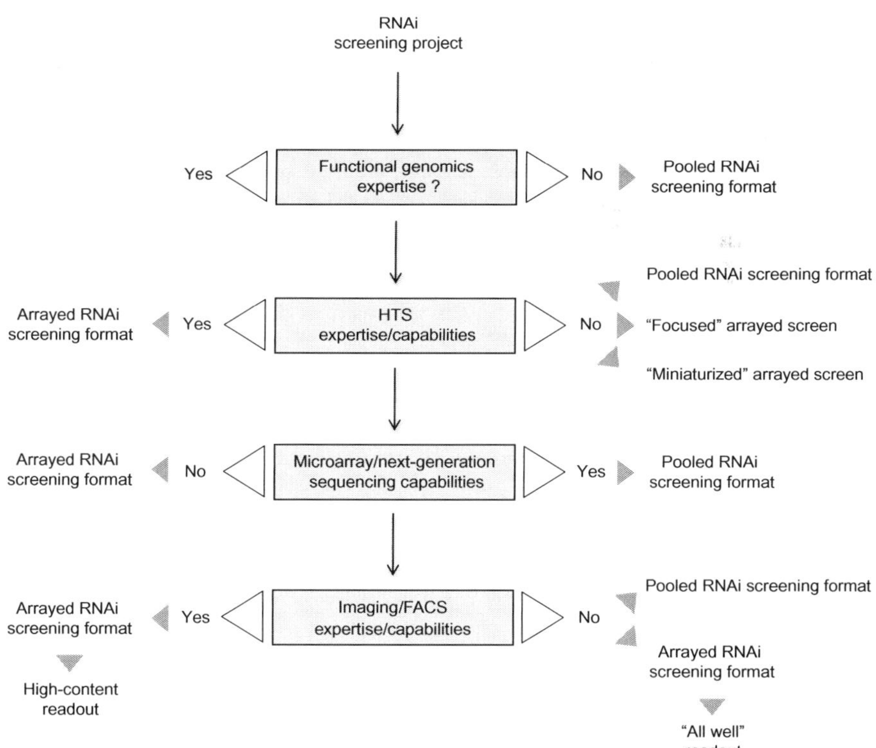

Figure 6.1 Parameters to be considered when choosing an RNAi screening format.

2.2.1 Arrayed RNAi screening format

The arrayed format uses a unique RNAi reagent (si or shRNA) per well in microtiter plates (96- or 384-well plate). Although this format could physically limit to some extent the number of genes and/or conditions that could be simultaneously queried within a single screen, it provides more flexibility in term of screen readouts. Due to concerns over off-target effects (Jackson & Linsley, 2010; Sigoillot & King, 2011), investigators typically test multiple RNAi (typically three or four) per target. These reagents are delivered and analyzed either individually or as a "mini-pool" (also called "small pool" or "smart pool"). Although the mini-pool approach sounds attractive to mitigate the size of the screen while increasing the number of RNAi tested per gene, in our experience it could result in a significant rate of false positive/negative identifications. In many cases the rational for the "mini-pool," where the off-target effect of each individual reagent would be diluted while the on-target effect would be the main phenotypic driver, does not hold true since one or two reagents within the pool could ultimately drive the phenotype specifically or not. However, with a sufficiently complex siRNA pool (more than 30 siRNAs), Hannus and colleagues suggest that the on-target gene knockdown effect will remain strong while the off-target effects of each individual siRNA are diluted with an increasing number of siRNA designs contained by the pool (Smith-Vikos, 2013). By spiking in a control siRNA previously published as demonstrating strong off-target effect (Hubner et al., 2010), Hannus and his collaborators demonstrated that the number of deregulated genes was dramatically reduced by increasing the complexity of the siRNA pool, as measured on an Affimetrix genomic microarray. Furthermore, the researchers discovered that a pool of approximately 30 siRNAs provided sufficient complexity necessary to substantially dilute any off-target effects. This technology, termed "siPools," provides complex but defined siRNA pools. Using an algorithm, optimal sequences are designed and subsequently produced using a novel enzymatic approach combining *in vitro* transcription and a specific nuclease. All siRNAs are precisely 21 nucleotides long to provide maximum efficiency and avoid any possible effect from the interferon response. Furthermore, as the enzymatic approach is far cheaper than chemically synthesized siRNA, large amounts of "siPools" can be produced at a low cost, offering an attractive alternative for *in vivo* RNAi applications. Ultimately, the reliability of the complex siRNA pools will have to be experimentally demonstrated more widely.

The use of individual reagents rather than "mini-pools" enables investigators to call out the "true" hits more rapidly avoiding the need for the pool to be subsequently deconvoluted. Depending on the size of the arrayed library and the screen conditions (e.g., number of cell lines tested, multiple treatment conditions), a large number of individual assay plates might have to be processed. Therefore, arrayed screens are usually handled in specialized groups that master various HTS techniques and take advantage of automation systems. However, there are several approaches that smaller laboratories could consider when planning to perform an arrayed si/shRNA screens. One very simple and effective strategy is to reduce the number of genes to be screened and select a focused pathway collection (e.g., PI3K pathway), gene family cassettes (e.g., GPCRs or kinases), disease-specific library (e.g., cancer), or subsets of the genome (e.g., druggable genome) based on some scientific rational. Other approaches are definitely worth considering such as the implementation of miniaturized arrays that use microarray slides on which thousands of siRNA or other RNAi reagents have been printed, facilitating reverse transcription of reagents into cells (Erfle et al., 2007; Erfle, Simpson, Bastiaens, & Pepperkok, 2004; Wheeler et al., 2004). This type of approach has been successfully implemented by various investigators (Neumann et al., 2006; Simpson et al., 2007) and shows several advantages over microplate-based screens: (1) reduced costs due to the small amount of RNAi and other reagents needed, (2) faster data acquisition due to the high number of experiments per array, and (3) low heterogeneity due to the absence of physical barriers between experiments, thus increasing screening data quality.

2.2.2 Pooled RNAi screening format

Pooled formats provide a more convenient approach to query large RNAi reagents collections, such as genome-wide mammalian shRNA libraries. With a pooled screen, a large population of cells is infected or transfected "en masse" and at random with a pool of different shRNA vectors, such that any given cells within the treated population will receive on-gene-specific RNAi reagent. After viral integration and antibiotic selection (e.g., puromycin) are complete, cells could be divided into different sets (e.g., treated vs. non-treated), depending on the experimental design. All pooled screens come in two flavors depending on the type of selection applied. In screens where a positive selection is applied, cells are infected with the pooled library and those that acquire the phenotype of interest are separated from those that do not. This separation could be applied based on, for example, survival in the presence of a drug (Burgess et al., 2008), ability to migrate through a chamber (Smolen et al., 2010), change in the expression of a specific marker or fluorescent transcriptional reporter that can be sorted and collected by flow cytometry (Mak et al., 2011). Using a positive selection approach, Zhi Sheng identified a molecular pathway that regulates the transcription of activating transcription factor 5 (ATF5), one of the key regulators of glioma cell survival. This work showed that ATF5 antagonizes apoptosis and increases the survival of the glioma cells through upregulation of MCL1, a member of the anti-apoptotic BCL2 family (Sheng et al., 2010). In screens where a negative selection is applied, also called dropout screens, the cells that acquired the phenotype of interest are lost (Schlabach et al., 2008). In that case, comparing the cell population before and after selection enables the identification of the hits. A great example of a negative selection screen would be the sensitization of formerly drug resistant cells in response to specific gene knockdowns (Fredbohm, Wolf, Hoheisel, & Boettcher, 2013; Mills et al., 2013). In general, positive selection screens have less noise and are easier to perform, but negative selection screens can be quite powerful if properly optimized and controlled. Subsequently, a molecular method, such as PCR amplification/next-generation sequencing or microarray analysis, is used to detect which RNAi reagents are present in each set (via detection of the RNAi-inducing sequence itself or a unique molecular "barcode" that identifies each reagent). Regardless of the design of a particular screen, the manner in which the viral transduction and subsequent cell culture are performed is critical to the success of the screen. The maintenance of hairpin representation (the number of cells infected with each shRNA) and logarithmic cell growth are of particular importance (Bassik et al., 2009; Silva et al., 2008; Zuber et al., 2011). Pooled format offers improvements in speed and scale compared to plate-based screening and therefore do not require the use of automated systems and specialized assay readout instruments making it an attractive approach for standard laboratories. Another advantage of the pooled shRNA format is the possibility to include an *in vivo* step to perform what is known as *ex vivo* screens (Bric et al., 2009; Meacham, Ho, Dubrovsky, Gertler, & Hemann, 2009). However, the pooled format requires specific approaches (e.g., microarrays, next-generation sequencing) to identify the subset of RNAi reagents that are enriched and/or depleted in the experimental versus starting or control condition. In addition, the pooled screen approach significantly limits the type of readouts that could be implemented such as, for example, high-content image-based cell assays.

2.3. Critical considerations when designing an RNAi screen

The following section reviews the different steps (Fig. 6.2) one should consider from a practical standpoint when developing and implementing an RNAi screen both in general and also more specifically for the investigation of the extrinsic apoptosis pathway.

2.3.1 Defining the goal(s) of the project

This first step is to thoroughly define the goal of the project as this will be absolutely critical in influencing many aspects of the screen and subsequent validation steps from the type of RNAi technology to use (e.g., siRNA vs. shRNA), the nature and size of the RNAi library (e.g., arrayed vs. pooled library, focused pathway collection vs. entire genome), the different arms of the screen (e.g., treated vs. nontreated conditions), the readout of the primary and secondary screens, and finally the type of follow-up investigations one would want to implement to fully validate the hits. Depending on what is known about the biological process or pathway to be investigated, one might chose a well-defined set of genes to identified suspected missing components. For example, Gonzalvez et al. (2012) used a smart pool RNAi library targeting most of the known E3-ligases (705 genes) to identify genes that augmented caspase-8 activation and sensitized cells to Apo2L/TRAIL. Others investigated a broader question and decided to screen the entire human genome. This was the case of Dompe et al. (2011) who employed a genome-wide pooled shRNA screen to identify regulators of death receptor-induced apoptosis. One key aspect to consider when defining the project goal and the experimental approach is to remain flexible in case possible adjustments are needed depending on the data generated during the screen optimization phase or the pilot screen.

2.3.2 Selecting the controls and optimizing the screen

2.3.2.1 Controls

Regardless of the approach chosen for the project, controls will be critical throughout the entire process from the optimization of the screen, the screen itself, the data processing, and interpretation. Therefore, positive and negative controls as well as transfection and untreated controls should be selected. Positive controls provide a means to optimize the transfection conditions and to ensure that an efficient siRNA delivery is achieved. Ideally, positive controls should encompass a wide range of effects on the studied phenotype in order to ensure that the readout used for the screen could identify both weak and strong hits. For the investigation of the extrinsic apoptosis pathway, investigators used known key components of this pathway

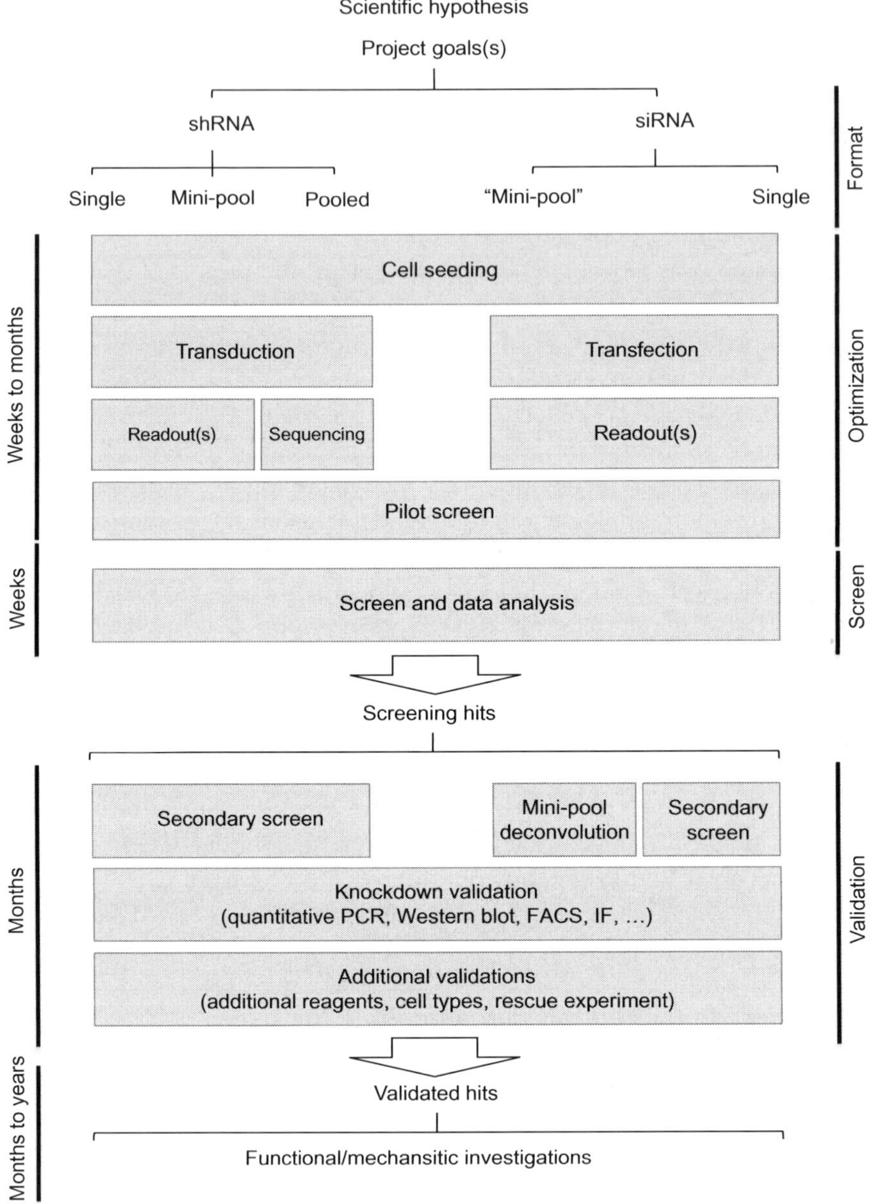

Figure 6.2 Critical steps when designing and implementing an RNAi screen. Description of the main steps involved with the development and implementation of an RNAi screen depending on the reagents (si/shRNA) and formats (single, mini-pool, or pooled library) used. Estimate of time commitment for each step is also provided.

such as CASP8, DR4, and BID as positive controls (Aza-Blanc et al., 2003; Gonzalvez et al., 2012). Negative controls distinguish sequence-specific silencing from nonspecific effects. These are designed to not target any known gene in the cell, and therefore produce minimal effects on the cell viability and phenotype. Standard negative controls are commercially available (e.g., nontarget controls (NTC) from Dharmacon, Lafayette, CO, USA). Transfection controls are not specific to the pathway or biological process investigated and they provide the investigator an additional way to check that an efficient RNAi delivery was achieved. Readouts for transfection controls are usually simple such as cell viability (e.g., siTOX from Dharmacon) or nuclear fluorescence (e.g., siGLO Green and siGLO Red from Dharmacon).

2.3.2.2 Optimizing the screen conditions

Once the controls are selected, various transfection conditions for the siRNA and transduction conditions for the shRNA need to be investigated. Many investigators underestimate the critical aspect of this step in the overall performance of the screen. It is important to perform a very thorough optimization since the time invested in this step will contribute to reduce false positives in the screen and subsequently speed up the hit validation work.

Cell seeding conditions—One of the first parameters that need to be defined are the cell-seeding conditions depending on the screen format (96-, 384-well plate), the desired duration of the screen, and the expected positive phenotype. This could be rapidly optimized by testing various cell-seeding densities and evaluating cell confluence and cell viability at multiple time points. For maximal cell viability during transfection, cells must be healthy at the beginning of the experiment. As a rule, cells should never be allowed to cover the entire surface area of their culture dish. Instead, cells should ideally be between 20% and 80% confluent. Subculturing cells before they become overcrowded minimizes instability in continuous cell lines and reduces variability from experiment to experiment. Here, the implementation of real-time readouts enabling, for example, measurement of the cellular confluence or numeration of the fluorescent events is clearly an advantage. In this area, platforms such as the ESSEN Incucytes revealed themselves as being particularly adaptable for this need (www.essenbioscience.com).

Transfection conditions—Once the optimal cell-seeding conditions have been defined, the RNAi delivery method can be optimized. Nucleic acids, including siRNA, carry a net negative charge on the sugar–phosphate backbone under normal physiological conditions. In order to enter the cell, siRNA must come into contact with the lipid bilayer of the cell membrane, whose head groups are also negatively charged. Three major technologies have been established to introduce siRNA into target cells: transfection, electroporation, and virus-mediated RNAi transfer or transduction. The most routinely employed transfection protocol, where cells are seeded a day prior to transfection, is referred to as "forward transfection." Forward transfection methods work well for most adherent cell types that are seeded a day prior to transfection in order to achieve an actively dividing cell population at the time of transfection. For suspension cells and/or high-throughput applications, a "reverse transfection" protocol, where freshly passaged cells are added to pre-plated transfection complexes, is often employed. Transfection typically involves the use of cationic lipids. The siRNA–lipid complexes offer protection to siRNA from enzymatic degradation and efficient endocytosis of siRNA by the cell (Kim, Chang, Lee, & Kim, 2007; Love, Moffett, & Novina, 2008). As an alternative to cationic lipid-based delivery, polymer-based vehicles have also been developed for delivery of siRNAs into specific tissues or cell types of interest to promote efficient and stable knockdown (Wu, Chien, Lui, Yan, & Lin, 2012). In the case of a lipid-based transfection, multiple lipids at various concentrations need to be evaluated. Here, the main goal is to achieve the strongest effect in the presence of a positive control (e.g., siTOX) while avoiding any impact of the negative control (e.g., NTC) on the cells. Although the absence of an effect of either the lipid alone or the lipid in the presence of the negative control is an absolute requirement, this criterion is not sufficient to insure the best transfection conditions. Indeed, investigators have to select a lipid providing the best transfection efficiency at the lowest concentration, thus reducing the chance of any off-target effect(s) triggered by an excessive amount of lipid. Although the delivery of the RNAi reagents through lipid-based transfection is a straightforward experimental approach, it does not work for every cell type. Electroporation has been demonstrated to be a possible alternative to siRNA deliver into hard-to-transfect cells, such as primary cells and suspension cells. During a brief but powerful electric pulse, hydrophilic pores and localized perturbations in the integrity of the cell membrane are induced. The temporary loss of the semipermeability of cell membranes leads to the escape of intracellular contents, such as ions and metabolites, and simultaneous uptake of drugs, molecular probes, and nucleic acids. Despite the high efficiency of nucleic acid transfer, electroporation can induce high cell mortality (Tsong, 1991). Therefore, the individual parameters of electroporation, including the voltage, length of the electric pulse, and number of pulses, must be optimized for different cell types so that a careful balance is established between delivery efficiency and cell mortality. Both transfection and electroporation approaches also require a careful optimization of the amount of RNAi to be delivered. The use of a minimum effective dose of siRNA significantly reduces the risk of potential off-target effects (Caffrey et al., 2011).

Transduction conditions—The silencing effects of transfected synthetic siRNAs are usually transient and not amenable to every cell type. Thus, the delivery of siRNA from DNA templates can be carried out using viral vectors. These viral vectors have been engineered and optimized to facilitate the entry of RNAi into cells that are not readily amendable to transfection. Different viral delivery systems have been developed using adenoviral, retroviral, and lentiviral platforms to deliver shRNA silencing constructs, each of which has its own set of advantages. Pseudotyped lentiviral vectors have proven to be among the most versatile delivery systems by providing delivery of large genomic payloads, broad tropism, and long-term transgene expression by integration of the viral payload into the host genome. In addition, lentiviral vectors mediate sustained, long-term expression in both dividing and nondividing cells, *in vitro* and *in vivo* (Shi, 2003). Regardless of the screen format, arrayed or pooled, the transduction optimization includes testing several critical parameters: the cell-seeding density, the amount of lentivirus, the cross linker and antibiotic concentrations, and the incubation time postinfection. Infection conditions should be optimized for each cell line and cell-based assay. For adherent cells, seeding conditions should be optimized such that the cells are approximately 40% confluent at the time of transductions. For suspension cell lines in 96-well culture plate, the ideal number of cells per well is in the 10,000–40,000 range and 50,000–200,000 for primary immune cells. In addition, it has been shown that the presence of cross linker (positively charged polycations) reduces the electrostatic repulsion forces between a negatively charged cell and an approaching enveloped lentiviral particle resulting in an increase in the transduction efficiency in a wide range of cell types (Andreadis & Palsson, 1997; Cornetta & Anderson, 1989; Hodgson & Solaiman, 1996; Le Doux, Landazuri, Yarmush, & Morgan, 2001; Seitz et al., 1998; Swaney, Sorgi, Bahnson, & Barranger, 1997; Themis et al., 1998; Toyoshima & Vogt, 1969). Polybrene (hexadimethrine bromide) is the most commonly used and seems to be most effective in a wide range of cell types. Protamine sulfate, diethylaminoethyl (DEAE)-dextran, poly-L-lysine, or cationic liposomes can also be effectively used (Denning et al., 2013). Transducing cells in serum-free media has also been shown to enhance lentiviral transduction efficiency. However, both low serum levels and the presence of cross linker can be toxic to certain cells and therefore these two parameters need to be carefully optimized in order to provide the best transduction efficiency while preserving cell viability. To effectively eliminate cells with no shRNA insert, the concentration of antibiotic (puromycin/blasticidin) needs to be optimized as well. This can be accomplished by performing a kill curve to determine the optimal concentration of antibiotic needed to eliminate untransduced cells. The lowest concentration of antibiotic that kills all untransduced cells should be utilized for subsequent experiments. Utilizing higher concentrations of antibiotic may lead to unacceptably high cytotoxicity and potential off-target effects. In addition, the amount of lentivirus to be used in the screen also needs to be carefully optimized. The expression of an shRNA in a given cell is influenced by its site of integration and the nearby host genetic elements. This problem can be mitigated by generating a sufficiently high number of independent infections (i.e., the representation rate) for each shRNA in the library. A high representation rate for each shRNA is also necessary to prevent its spurious loss from the pool (Hu & Luo, 2012). To properly sample the library and successfully identify shRNA sequences of interest, it is critical to maintain the representation of an shRNA (number of cells expressing that specific shRNA) at 200–2000 and 1000–2000 for positive and negative selection screens (Schlabach et al., 2008), respectively, and that the majority of cells are infected with only one shRNA. Such representation must be carefully maintained throughout all steps in the screen from shRNA delivery to shRNA recovery. Any single step in the screen that significantly reduces the representation of the library would bottleneck the pool and introduces additional, random fluctuations in library composition and thus contributes to the noise in the screen. This is achieved by carefully optimizing the ratio of infectious agents based on the virus titer to the number of cells being infected. This ratio is called multiplicity of infection (MOI). By using a low MOI, the probability of multiple integrants per cell is greatly decreased, hence reducing the likelihood of the emergence of combinatorial phenotypes. The actual number of viruses that will enter any given cell is a statistical process defined by a Poisson distribution. With the Poisson distribution one can calculate how many cells are transduced with 0, 1, 2, and more viruses at a given MOI. While MOI cannot be measured directly, the percentage of infected cells can be determined comparing cell viability with or without antibiotic. Typically, an infection rate of 30–50% is recommended, such that 70–80% of viable cells have only one shRNA. For example, in the case of a library of 100,000 shRNAs, 40–65 million cells must be initially infected in order to achieve representation of each hairpin in at least 200 cells, assuming an infection efficiency of 30–50%. However, transduction efficiencies, and therefore desired MOIs, depend strongly on the target cell type. Therefore, it is imperative that determination of the optimal MOI is carried out before starting the screen in a new cell type.

2.3.3 Selecting the screen readout

When selecting the readouts for an RNAi screen and the subsequent hit validation, careful consideration must be given. The biological relevance of the readout both from a mechanistic and a specificity point of view as well as the ease of implementation should guide the selection of the approach. The following section reviews multiple readout options for RNAi screens specifically designed toward interrogating the extrinsic apoptosis pathway and discusses critical considerations for RNAi screen readouts, in general.

In view of the diversity of manifestations of the apoptotic process including changes in plasma membrane (such as loss of membrane symmetry and attachment), a decrease in cell size with condensation of the cytoplasm and nucleus, alteration in mitochondrial structure and function (including loss of mitochondrial membrane potential), altered cellular redox status, activation of caspases, and associated protein cleavage, as well as ordered double-strand DNA fragmentation ultimately resulting in loss of cell viability, numerous techniques to analyze apoptosis have been developed (Table 6.1) (Brunet et al., 1998; Gavrieli, Sherman, & Ben-Sasson, 1992; Häcker, 2000; Kerr, Wyllie, & Currie, 1972; Lazebnik, Cole, Cooke, Nelson, & Earnshaw, 1993; McCarthy, Whyte, Gilbert, & Evan, 1997; Messam & Pittman, 1998; Solary, Bertrand, Kohn, & Pommier, 1993; Wyllie, 1980). A basic understanding of the changes that occur during apoptosis will help decide which endpoint to choose (Riss & Moravec, 2004). In addition, the exact apoptotic phenotype manifested appears to depend both on the cell type and the nature of the apoptotic trigger. Routine methods for detecting cell number, viability, and/or basic metabolic readouts in an automated fashion are well established. Thus, it is perhaps not surprising that cell viability/proliferation is one of the most widely used parameters to indirectly measure cell death in the context of RNAi screens. Typically, cell viability assays are carried out in a homogeneous fashion with simple "add, mix, and measure" protocols and rely on the measurement of critical cellular parameters such as metabolic activity or membrane integrity. Investigators used several of

Table 6.1 Example of assay readouts based on various apoptotic events using different modalities (HTS readouts, flow cytometry or high-content imaging)

Apoptotic event	HTS readout (e.g., homogenous assays)	Flow cytometry	High-content imaging
Upstream caspase activation	✓ (e.g., Caspase-Glo® 8 assays)	✓ (e.g., Vybrant® FAM Caspase-8 assay)	✓ (e.g. Image-iT® LIVE Caspase-8 detection)
Downstream caspase activation	✓ (e.g., Caspase-Glo® 3/7 assay)	✓ (e.g., CellEvent® Caspase 3/7 assay)	✓ (e.g., CellEvent® Caspase 3/7 assay)
Loss of mitochondrial membrane potential	✓ (e.g., Mitochondrial ToxGlo® Assay)	✓ (e.g., MitoProbe® JC-1 imaging)	✓ (e.g., MitoProbe™ JC-1, MitoTracker®, Mito-ID® MP kits)
Phosphatidylserine exposure	✓ (e.g., TR-FRET assay[a])	✓ (e.g., Annexin V assay)	✓ (e.g., Annexin V imaging)
Chromatin condensation	–	✓ (e.g., Hoechst 33342, Vybrant® DyeCycle® Violet stains staining)	✓ (e.g., Hoechst 33342 staining, Nuclear-ID® detection kit)
Increase cell membrane permeability	✓ (e.g., LDH release assay)	✓ (e.g., Ethidium bromide, YO-PRO-1, Hoechst 33342 dye)	✓ (e.g., Ethidium bromide, YO-PRO-1, Hoechst 33342 dye)
Increase reactive oxygen species	✓ (e.g., ROS Glo assay)	✓ (e.g., 2′,7′-Dichlorodihydrofluorescin diacetate (DCFH-DA) assay)	✓ (e.g. H2-DCF assay, Image-iT® ROS assay)

DNA fragmentation	✓ (e.g., ELISA histone-complexed DNA fragments)	✓ (e.g., TUNEL assay)	✓ (e.g., TUNEL assay, Hoechst 33342)
Formation of apoptotic bodies	–	–	✓ (e.g., Ethidium bromide, YO-PRO-1, Hoechst
Cell viability	✓ (e.g., CellTiter-Glo® assay)	✓ (e.g., Ethidium bromide, YO-PRO-1, Hoechst 33342 dye)	✓ (e.g., Ethidium bromide, YO-PRO-1, Hoechst

[a]Gasser, Hehl and Millward (2009).

these simple assay modalities such as resazurin/resorufin conversion (Alamar Blue assay) (Aza-Blanc et al., 2003; MacKeigan, Murphy, & Blenis, 2005), ATP quantitation (Gonzalvez et al., 2012), or lactate dehydrogenase release (Jiang et al., 2009). However, general assays such as those that measure cellular ATP levels (e.g., CellTiter Glo®) as surrogates for the cell viability are significantly more prone to off-target effects than specific assays addressing cause–effect relationship more directly. Therefore, screening strategies tend to combine multiple readouts including assays more specific to the apoptotic pathway. In this context, various apoptosis assays were developed based on protease activity especially caspases that comprise a key component of the apoptosis machinery of the cell, participating in an enzyme cascade that results in cellular disassembly. Multiple assays are available to measure the engagement of various critical caspases such as caspase 3/7 and 8. For example, Stec and collaborators (Stec et al., 2012) determined caspase 3/7 activities to identify genes in the poly ADP ribose polymerase (PARP) pathway, while Gonzalvez and colleagues (Gonzalvez et al., 2012) used caspase-8 activity to uncover the critical role of TRAF2 in the extrinsic apoptosis. Markers of apoptosis such as caspase activity may be present only transiently. Therefore, understanding the kinetics of the cell death process in the biological context used for the screen including cell type and treatment is critical. Whereas many of the investigations mentioned above have used simple viability assays alone or in combination with apoptosis-specific homogeneous readouts, additional studies have used fluorescence-assisted cell sorting (FACS) or imaged-based assay readouts to uncover specific phenotypes related to cell death. Although screening strategies relying on these types of readouts are usually more challenging to implement, they have the significant advantage to provide data about cell culture heterogeneity. Indeed, resolution at the individual cell level is essential for some screens, especially in the context of apoptosis-related investigations where individual variability could impact sensitivity to treatment.

Flow cytometry based readouts have been used in arrayed-based RNAi screens to determine DNA content or the relative levels of different markers at the individual cell level. This was the case for Ren and colleagues (Ren et al., 2004) who used a FACS-based readout to monitor DR4, DR5, TNFSFR6, and TNFSFR1 expression in response to siRNA-mediated gene knockdown in HCT15 colon carcinoma cells. Other investigators performed a high-throughput cell-survival analysis after transfection of an shRNA library with genome-wide coverage to detect cells that escaped from cell death, which were further analyzed by terminal deoxynucleotidyl transferase dUTP nick end labeling assay (Kimura et al., 2008). Using flow cytometry readout for RNAi screening campaign is particularly useful when dealing with low number of cells especially when nonadherent cells such as cells representing the immune system are used. However, because of the relatively slow speed of FACS analysis in microtiter plate formats, researchers more often turn to high-content imaging approaches.

High-content screening (HCS) combines the efficiency of high-throughput techniques with the ability of cellular imaging to collect quantitative data from complex biological systems. Image-based screening found an initial foothold in oncology research, due to early applications for measuring apoptosis (Cumming et al., 2008; Inglefield, Larson, Gibson, Lebrec, & Miller, 2006; Lovbor, Gullbo, & Larsson, 2005; To et al., 2007; Vogt et al., 2008). The combination of RNAi technology with image-based screening capabilities has opened up the possibility to perform cellular imaging in functional genomics HTS applications (Conrad & Gerlich, 2010) and provides a great opportunity to manage the detection of off-target effects due to the intrinsic multiparametric nature of the image-based readouts. Although HCS approaches enable the investigation of multiple death signals like caspase activation, mitochondrial membrane potential loss, chromatin condensation, and mitochondrial superoxide generation simultaneously in a HTS fashion (Joseph, Seervi, Sobhan, & Retnabai, 2011), examples of multiparametric high-content RNAi screens to investigate the apoptotic process are relatively limited. Eifert and collaborators carried out a large-scale loss-of-function analysis of more than 80 human tyrosine kinases using shRNA to identify novel survival factors for breast cancer cells using a cleaved caspase-3 antibody to detect apoptosis along with Hoechst nucleic acid stain (Eifert et al., 2013). Others took advantage of the miniaturized cell spot microarray method expanding the cell-based functional genetic screens to include more RNAi constructs, allow combinatorial RNAi analyses, multiparametric phenotypic readouts, or comparative analysis of many different cell types (Rantala et al., 2011). In this study, the impacts of the knockdown of GPCR coding genes on cell proliferation and induction of apoptosis were evaluated through the detection of nuclear Ki-67 and cleaved PARP, respectively, in human prostate cancer cells. In addition to HCS, live imaging also represents a very attractive approach to investigate programmed cell death using RNAi technology. For example, Antczak and collaborators reported the successful adaptation of a high-content assay method using the caspase-activated

DEVDNucView488™ fluorogenic substrate (Antczak, Takagi, Ramirez, Radu, & Djaballah, 2009). For the first time, these investigators showed caspase activation in live cells induced by siRNA, demonstrating the potential of this approach to identify previously uncovered enhancers and suppressors of the apoptotic machinery.

Finally, although imaging often dramatically increases the content of a screening assay, it poses new challenges to achieve accurate quantitative annotation and therefore needs to be carefully adjusted to the specific needs of individual screening applications (Conrad & Gerlich, 2010).

2.3.4 From the pilot screen to the primary screen and data analysis

Once the screen optimization is completed, it is generally recommended to perform a pilot screen on tens to hundreds of random genes including the positive and negative controls, especially when considering a large screening effort (e.g., whole-genome screen). The main purpose of such a small-scale preliminary screen is to make sure that every step could be carried out as planned and without any time delays. This could be particularly important when time sensitive procedures are included and therefore screening plates might have to be processed by batches in the primary screen in order to avoid a drift in the dataset. Another key benefit of running a pilot screen is to allow the opportunity of evaluating the hit rate. This will be defined by both the screening procedure itself as well as the downstream analysis method and the criteria used to select positive signal over background. If the hit rate is too high, this could either be due to the fact that the assay is not specific enough or that the hit selection criteria are too lenient. In this case the screen might need to be reoptimized especially the readout portion or the analysis methods reconsidered. On the other hand if the hit rate is too low, this could either be due to the fact that the screen is not sensitive enough or the hit selection criteria are too stringent. In this case the transfection/infection efficiency might be too low and should be further optimized or the screen readout is not sensitive enough and should be improved or reconsidered. Once the pilot screen is completed the investigator should be highly confident that the entire screen procedure, including the data analysis step, is as robust as possible and will generate meaningful hits. One can never emphasize enough the fact that the assay optimization and the screen itself are only a relatively small fraction of the work involved and that significant resources will have to be spent on the hit validation and follow-up mechanistic investigations in order to truly be successful. Therefore, time spent ensuring optimum performance of the screen is a wise investment.

Depending on the size of the library or the number of conditions to be tested, the primary screen is usually performed in duplicate or triplicate. The amount of data generated could be significant with files from the megabyte to terabyte size. Therefore, it is critical to establish in advance the computational tools, including hardware and software, required to reliably and efficiently handle the mass of data generated. The end result of the primary screen will be a list of numbers that will initially need to be associated with the specific RNAi reagents. Specific databases need to be in place to enable that step. Subsequently, the data will have to be processed including averaging, normalization, calculation of statistical scores, and setting appropriate cutoff values for significant results. The statistical techniques used to analyze RNAi screens are frequently borrowed directly from small-molecule screening; however, small-molecule and RNAi data characteristics differ in meaningful ways and have been extensively described (Birmingham et al., 2009; Boutros, Brás, & Huber, 2006; Malo, Hanley, Cerquozzi, Pelletier, & Nadon, 2006; Stone et al., 2007; Wang, Tu, & Sun, 2009; Wiles, Ravi, Bhavani, & Bishop, 2008; Zhang, Chung, & Oldenburg, 1999; Zhang et al., 2008; Zhang & Heyse, 2009; Zhang et al., 2009; Zimmermann et al., 2006). For example, independent siRNA targeting the same transcript could trigger phenotypes of different strengths. Weak but genuine phenotypes can be caused either by partial knockdown of a gene with a critical role in the biology investigated or a strong knockdown of a gene with a marginal role. This clearly represents a challenge when defining hits and therefore rigorous statistical methods have to be implemented to ultimately help to discriminate true from false hits (Birmingham et al., 2009; König et al., 2007; Zhang et al., 2007). In this respect, screens that are performed in various conditions and rely on multiparametric readouts have a clear advantage. For example, Gonzalvez et al. (2012) performed an siRNA screen in HCT116 Bax$^{-/-}$ in the presence or absence of Apo2L/TRAIL and used both caspase-8 activity monitored by caspase-8 Glo assay (Promega, Madison, WI, USA) and cell viability measurements through Cell-Titer Glo assay (Promega) to uncover the critical role of TRAF2 in the extrinsic apoptosis. Others like Stec and colleagues (Stec et al., 2012) employed HeLa cells treated with or without a PARP inhibitor to screen siRNA pools targeting 23886 unique genes using the Alamar Blue assay as cell viability readout and Caspase 3/7 assay to measure apoptosis.

Several analytical tools have been developed for the analysis of RNAi screening data specifically such as the cellHTS software package based on Bioconductor/R (http://www.dkfz.de/signaling/cellHTS) (Boutros et al.,

2006). A more recent system called ScreenSifter (Kumar, Goh, Wongphayak, Moreau, & Bard, 2013) associates the data analysis capabilities with the ability to visualize associated gene information in an interactive fashion. Users can upload their raw signal intensities and will reveal complete quality control of the screen, hit selection, plotting of hit genes with gene ontology, replicates comparisons, or channels comparison. In addition, ScreenSifter has visualization tools to plot specific genes or gene groups in the screen data. ScreenSifter also provides Gene Set Enrichment Analysis, protein–protein interaction directly on the plot.

Ultimately, investigators have to implement a data analysis strategy enabling them to weed out false positives and select true hits also they might have a relatively weak phenotype. In a well-designed and well-controlled screen, the investigator will be able to call out many of the hits with a very high level of confidence. However, others hit candidates could be more ambiguous either because of a weak reproducibility between replicate wells despite some of these presenting a strong phenotype or because of a relatively weak phenotype while consistent across replicate wells. Ideally, being able to repeat a completely independent run of the screen and selecting hits that were only reproduced in both runs significantly increases the confidence level for the final hit selection. Although this strategy might sounds cumbersome, it is good scientific practice and worth considering whenever possible. However, the size and the cost of some screens could be prohibitive and therefore prevent the investigator on performing two independent run for the hit selection. In that case it could be useful to define high- and low-confidence hits list to prioritize the validation work. In addition, the investigator's expert knowledge of the biological system can be used to make an informed decision.

2.3.5 Hit validation and follow-up investigations

Confirming the specificity, and hence the validity, of data obtained using RNAi is an essential aspect of the experiments, including screens, that rely on this technique. Although early reports in the literature described si/shRNAi as specific and nonimmunogenic (Amarzguioui, Holen, Babaie, & Prydz, 2003; Caplen, Parrish, Imani, Fire, & Morgan, 2001; Chi et al., 2003; Elbashir, Martinez, Patkaniowska, Lendeckel, & Tuschl, 2001; Holen, Amarzguioui, Wiiger, Babaie, & Prydz, 2002; Hutvagner & Zamore, 2002; Semizarov et al., 2003; Tushi, Zamore, Lehman, Bartel, & Sharp, 1999), recent studies such as unbiased, genome-scale expression profiling studies have revealed that si/shRNA-mediated silencing is less specific than was originally believed. It is now recognized that certain properties of si/shRNA and/or technologies used to deliver them in cells could lead to suboptimal specificity (Jackson & Linsley, 2010; Sigoillot & King, 2011). Off-target activity of si/shRNA can trigger a hit in a screen (false positive) and therefore compromise the biological interpretation of the data set. The off-target effects associated with the RNAi technology fall into mainly three categories: siRNA-induced sequence-dependent regulation of unintended transcripts through partial sequence complementarity to their 3′-UTR also called miRNA-like off-target effects (Jackson et al., 2003; Jackson & Linsley, 2004; Scacheri et al., 2004; Semizarov et al., 2003; Snove & Holen, 2004); an inflammatory response through activation of Toll-like receptors triggered by siRNAs (Sledz, Holko, de Veer, Silverman, & Willimas, 2003) and/or their delivery vehicle such as cationic lipids (lipid-mediated response) (Fedorov et al., 2005); and widespread effects on miRNA processing and functionality through saturation of the endogenous RNAi machinery by exogenous siRNA. Understanding the off-target effects triggered by the RNAi technology has led to the development of reagents and experimental approaches that in some extent mitigate these effects. However, off-target effect and subsequent false discovery remains a significant and difficult problem with si/shRNA screening and therefore hits have to be carefully validated before engaging in additional investigations and drawing conclusions.

Simple experimental approaches can be used to increase the confidence in the hits. First of all, it is a good practice to retest the reagents with the same assay, if possible comparing the screening reagent(s) with a new batch or a resynthesis. It is important to keep in mind especially when working with large libraries that errors could happen and it is always possible that a specific reagent end up in the wrong well. As an initial validation step, single reagents should also be tested in an arrayed format assuming the primary screen was run in a pooled format. Depending on the number of individual sequences scoring in the screen, it is also critical to expand the sequence diversity testing reagents designed to target different regions of the gene. For example, vendors generally provide siRNA libraries with three to four sequences per genes. A minimum of three sequences should score in the assay before investing more time in the target. Testing reagents using a different chemistry can also be beneficial. Once the hit has been reproduced with both the original and additional reagents, the specific knockdown has to be confirmed at the mRNA or protein level, comparing cells transfected with the specific si/shRNA versus negative controls. If antibodies against the specific protein corresponding to the hits are available, the knockdown validation at the protein level is preferred since the protein ultimately drives the cellular phenotype. Western blot, FACS, or immunofluorescence experiments are often commonly employed for this purpose. If specific reagents are not available to validate knockdown at the protein level, quantitative PCR can be used to validate knockdown at the mRNA level. Another critical step to validate the hits is to test the si/shRNA reagents in different cell types as well as use a different readout. In case small molecule(s) targeting the specific protein of interest exist, one should definitely take advantage of this opportunity to validate the phenotype. Finally, the key control for an RNAi experiment, which should be provided whenever technically feasible, is rescue of the observed RNAi phenotype through artificial expression of a resistant form of the targeted mRNA (genomic fragment, cDNA or open reading frame construct) that evades RNAi knockdown (Cullen, 2006). This approach is usually considered the "gold standard" validation to verification of an RNAi results at the gene level. However, these types of experiments could be challenging to perform and do not always provide conclusive data.

3. CONCLUDING REMARKS AND PERSPECTIVES

One of the major challenges in the post-genomic era is the functional analysis of all human genes. In barely a decade, RNA-mediated interference has evolved from a fascinating biological phenomenon into an experimental tool ideal for discovering genes that encode molecules with previously unknown functions in biological pathways of interest. Rapidly, the RNAi technology was harnessed to massive screening efforts through multiple arrayed and pooled si/shRNA libraries to uncover gene involved in discrete parts of complex signaling cascades such as the extrinsic apoptosis pathway. While the application of the RNAi HTS technology resulted in an impressive amount of data contributing to the understanding of gene functions and networks in multiple basic and applied biological fields (Neumüller & Perrimon, 2010), the occurrence of unintended off-target effects raised serious and valid concerns about the gene targets identified through more than 580 published RNAi screens (Bhinder & Djaballah, 2013; Djaballah, 2012). Today many of the major caveats and pitfalls of the RNAi technology are known and strategies to mitigate them were developed both at the individual and collective level within the expert community. Although accepted standards and guidelines have not yet been established globally, investigators new to the field should particularly pay attention to: (1) carefully consider the design of the siRNA and shRNA expression vector along with library format with an eye to minimizing the potential for off-target effects through either specific or nonspecific mechanisms; (2) the screen strategy and feasibility, emphasizing the use of multiple conditions and multiparametric readouts whenever possible in order to possibly weed out false positives; (3) the RNAi delivery conditions, balancing efficiency versus viability and potential off-target effect; (4) the data analysis and hit selection parameters, focusing on meaningful and reproducible biological effects; (5) the careful validation of the RNAi reagents correlated to the phenotypic observation. In this respect, gene silencing by RNAi should be monitored on the protein levels, which in most cases is the relevant readout for the interpretation of observed phenotypes. Finally, one should not forget that the RNAi screen by itself is only the first step of a long and comprehensive experimental process geared toward a deeper understanding of the hit significance at both a molecular and a system level.

REFERENCES

Amarzguioui, M., Holen, T., Babaie, E., & Prydz, H. (2003). Tolerance for mutations and chemical modifications in a siRNA. *Nucleic Acid Research*, *31*(2), 589–595.

Andreadis, S., & Palsson, B. O. (1997). Coupled effects of polybrene and calf serum on the efficiency of retroviral transduction and the stability of retroviral vectors. *Human Gene Therapy*, *8*(3), 285–291.

Antczak, C., Takagi, T., Ramirez, C. N., Radu, C., & Djaballah, H. (2009). Live-cell imaging of caspase activation for high-content screening. *Journal of Biomolecular Screening*, *14*(8), 956–969.

Ashkenazi, A., & Dixit, V. M. (1999). Apoptosis control by death and decoy receptors. *Current Opinion in Cell Biology*, *11*(2), 255–260.

Aza-Blanc, P., Cooper, C. L., Wagner, K., Batalov, S., Deveraux, Q. L., & Cooke, M. P. (2003). Identification of modulators of TRAIL-induced apoptosis via RNAi-based phenotypic screening. *Molecular Cell*, *12*(3), 627–637.

Bassik, M. C., Lebbink, R. J., Churchman, L. S., Ingolia, N. T., Patena, W., LeProust, E., et al. (2009). Rapid creation and quantification monitoring of high coverage shRNA libraries. *Nature Methods*, *6*(6), 443–445.

Bernstein, E., Caudy, A. A., Hammond, S. M., & Hannon, G. J. (2001). Role for a bidentate ribonuclease in the initiation step of RNA interference. *Nature*, *409*(6818), 363–366.

Bhinder, B., & Djaballah, H. (2013). Systematic analysis of RNAi reports identifies dismal commonality at gene-level & reveals an unprecedented enrichment in pooled shRNA screens. *Combinatorial Chemistry & High Throughput Screening*, *16*(9), 665–681.

Birmingham, A., Selfors, L. M., Forster, T., Wrobel, D., Kennedy, C. J., Shanks, E., et al. (2009). Statistical methods for analysis of high-throughput RNA interference screens. *Nature Methods*, *6*(8), 569–575.

Bogdanove, A. J., & Voytas, D. F. (2011). TAL effectors: Customizable proteins for DNA targeting. *Science*, *333*(6051), 1843–1846.

Boutros, M., Brás, L. P., & Huber, W. (2006). Analysis of cell-based RNAi screens. *Genome Biology*, 7(7), R66.

Bric, A., Miething, C., Bialucha, C. U., Scuoppo, C., Zender, L., Krasnitz, A., et al. (2009). Functional identification of tumor-suppressor genes through an in vivo RNA interference screen in a mouse lymphoma model. *Cancer Cell, 16*, 324–335.

Brummelkamp, T. R., Bernards, R., & Agami, R. (2002). Stable suppression of tumorigenicity by virus-mediated RNA interference. *Cancer Cell, 2*(3), 243–247.

Brunet, C. L., Gunby, R. H., Benson, R. S., Hickman, J. A., Watson, A. J., & Brady, G. (1998). Commitment to cell death measured by loss of clonogenicity is separable from the appearance of apoptotic markers. *Cell Death and Differentiation, 5*(1), 107–115.

Burgess, D. J., Doles, J., Zender, L., Xue, W., Ma, B., McCombie, W. R., et al. (2008). Topoisomerase levels determine chemotherapy response in vitro and in vivo. *Proceedings of the National Academy of Sciences of the United States of America, 105*(26), 9053–9058.

Caffrey, D. R., Zhao, J., Song, Z., Schaffer, M. E., Haney, S. A., & Subramanian, R. R. (2011). siRNA off-target effects can be reduced at concentrations that match their individual potency. *PLoS One, 6*(7), e21503.

Caplen, N. J., Parrish, S., Imani, F., Fire, A., & Morgan, R. A. (2001). Specific inhibition of gene expression by small double-stranded RNAs in invertebrate and vertebrate systems. *Proceedings of the National Academy of Sciences of the United States of America, 98*(17), 9742–9747.

Chapman, E. J., & Carrington, J. C. (2007). Specialization and evolution of endogenous small RNA pathways. *Nature Reviews Genetics, 8*(11), 884–896.

Chi, J. T., Chang, H. Y., Wang, N. N., Chang, D. S., Dunphy, N., & Brown, P. O. (2003). Genomewide view of gene silencing by small interfering RNAs. *Proceedings of the National Academy of Sciences of the United States of America, 100*(11), 6343–6346.

Cong, L., Ran, F. A., Cox, D., Lin, S., Barretto, R., Habib, N., et al. (2013). Multiplex genome engineering using CRISPR/Cas systems. *Science, 339*(6121), 819–823.

Conrad, C., & Gerlich, D. W. (2010). Automated microscopy for high-content RNAi screening. *The Journal of Cell Biology, 188*(4), 453–461.

Cornetta, K., & Anderson, W. F. (1989). Protamine sulfate as an effective alternative to polybrene in retroviral-mediated gene-transfer: Implications for human gene therapy. *Journal of Virological Methods, 23*(2), 187–194.

Cullen, B. R. (2004). Derivation and function of small interfering RNAs and microRNAs. *Virus Research, 102*, 3–9.

Cullen, B. R. (2006). Enhancing and confirming the specificity of RNAi experiments. *Nature Methods, 3*(9), 677–681.

Cumming, J., Hodgkinson, C., Odedra, R., Sini, P., Heaton, S. P., Mundt, K. E., et al. (2008). Preclinical evaluation of M30 and M65 ELISAs as biomarkers of drug induced tumor cell death and antitumor activity. *Molecular Cancer Therapeutics*, 7(3), 455–463.

Degterev, A., Boyce, M., & Yuan, J. (2003). A decade of caspases. *Oncogene, 22*(53), 8543–8567.

Denning, W., Das, S., Guo, S., Xu, J., Kappes, J. C., & Hel, Z. (2013). Optimization of the transductional efficiency of lentiviral vectors: Effect of sera and polycations. *Molecular Biotechnology, 53*(3), 308–314.

Djaballah, H. (2012). Random RNAi screening data analysis: A call for standardization. *Combinatorial Chemistry & High Throughput Screening, 15*(9), 685.

Dompe, N., Sanchez Rivers, C., Li, L., Cordes, S., Schwickart, M., Punnoose, E. A., et al. (2011). A whole-genome RNAi screen identifies an 8q22 gene cluster that inhibits death receptor-mediated apoptosis. *Proceedings of the National Academy of Sciences of the United States of America, 108*(43), E943–E951.

Eifert, C., Wang, X., Kokabee, L., Kourtidis, A., Jain, R., Gerdes, M. J., et al. (2013). A novel isoform of the B cell tyrosine kinase BTK protects breast cancer cells from apoptosis. *Genes, Chromosomes and Cancer, 52*(10), 961–975.

Elbashir, S. M., Martinez, J., Patkaniowska, A., Lendeckel, W., & Tuschl, T. (2001). Functional anatomy of siRNA for mediating efficient RNAi in Drosophila melanogaster embryo lysate. *The EMBO Journal*, *20*(23), 6877–6888.

Erfle, H., Neumann, B., Liebel, U., Rogers, P., Held, M., Walter, T., et al. (2007). Reverse transfection on cell arrays for high content screening microscopy. *Nature Protocols*, *2*(2), 392–399.

Erfle, H., Simpson, J. C., Bastiaens, P. I., & Pepperkok, R. (2004). siRNA cell arrays for high-content screening microscopy. *Biotechniques*, *37*(3), 454–462.

Fedorov, Y., King, A., Anderson, E., Karpillow, J., Llsley, D., Marshall, W., et al. (2005). Different delivery methods-different expression profiles. *Nature Methods*, *2*(4), 241.

Fire, A., Xu, S., Montgomery, M. K., Kostas, S. A., Driver, S. E., & Mello, C. C. (1998). Potent and specific genetic interference by double-stranded RNA in Caenorhabditis elegans. *Nature*, *391*(6669), 806–811.

Fischer, U., & Schulze-Osthoff, K. (2005). New approaches and therapeutics targeting apoptosis in disease. *Pharmacological Reviews*, *57*(2), 187–215.

Fredbohm, J., Wolf, J., Hoheisel, J. D., & Boettcher, M. (2013). Depletion of RAD17 sensitizes pancreatic cancer cells to gemcitabine. *Journal of Cell Science*, *126*(Pt. 15), 3380–3389.

Fulda, S., & Debatin, K. M. (2006). Extrinsic versus intrinsic apoptosis pathways in anticancer chemotherapy. *Oncogene*, *25*(34), 4798–4811.

Gaj, T., Gersbach, C. A., & Barbas, C. F. (2013). ZFN, TALEN, and CRISPR/Cas-based methods for genome engineering. *Trends in Biotechnology*, *31*(7), 397–405.

Gasser, J., Hehl, M., & Millward, T. A. (2009). A homogeneous time-resolved fluorescence resonance energy transfer assay for phosphatidylserine exposure on apoptotic cells. *Analytical Biochemistry*, *384*, 49–55.

Gavrieli, Y., Sherman, Y., & Ben-Sasson, S. A. (1992). Identification of programmed cell death in situ via specific labeling of nuclear DNA fragmentation. *The Journal of Cell Biology*, *119*(3), 493–501.

Gilbert, L. A., Larson, M. H., Morsut, L., Liu, Z., Brar, G. A., Torres, S. E., et al. (2013). CRISPR-mediated modular RNA-guided regulation of transcription in eukaryotes. *Cell*, *154*(2), 442–451.

Gonzalvez, F., Lawrence, D., Yang, B., Yee, S., Pitti, R., Marsters, S., et al. (2012). TRAF2 sets a threshold for extrinsic apoptosis by tagging caspase-8 with a ubiquitin shutoff timer. *Molecular Cell*, *48*(6), 888–899.

Häcker, G. (2000). The morphology of apoptosis. *Cell and Tissue Research*, *301*(1), 5–17.

Hammond, S. M., Bernstein, E., Beach, D., & Hannon, G. J. (2000). An RNA-directed nuclease mediates post-transcriptional gene silencing in Drosophila cells. *Nature*, *404*(6775), 293–296.

Hanhan, D., & Weinberg, R. A. (2000). The hallmarks of cancer. *Cell*, *100*(1), 57–70.

Hodgson, C. P., & Solaiman, F. (1996). Virosomes: Cationic liposomes enhance retroviral transduction. *Nature Biotechnology*, *14*, 339–342.

Holen, T., Amarzguioui, M., Wiiger, M. T., Babaie, E., & Prydz, H. (2002). Positional effects of short interfering RNAs targeting the human coagulation trigger Tissue Factor. *Nucleic Acids Research*, *30*(8), 1757–1766.

Hu, G., & Luo, J. (2012). A primer on using pooled shRNA libraries for functional genomic screens. *Acta Biochimica et Biophysica Sinica*, *44*(2), 103–112.

Hubner, N. C., Wang, L. H. C., Kaulich, M., Descombes, P., Poser, I., & Nigg, E. A. (2010). Re-examination of siRNA specificity questions role of PICH and Tao1 in the spindle checkpoint and identifies Mad2 as a sensitive target for small RNAs. *Chromosoma*, *119*(2), 149–165.

Hutvagner, G., & Zamore, P. D. (2002). A microRNA in a multiple-turnover RNAi enzyme complex. *Science*, *297*(5589), 2056–2060.

Inglefield, J. R., Larson, C. J., Gibson, S. J., Lebrec, H., & Miller, R. L. (2006). Apoptotic response in squamous carcinoma and epithelial cells to small-molecule toll-like receptor agonist evaluated with automated cytometry. *Journal of Biomolecular Screening*, *11*(6), 575–585.

Jackson, A. L., Bartz, S. R., Schelter, J., Kobayashi, S. V., Burchard, J., Mao, M., et al. (2003). Expression profiling reveals off-target gene regulation by RNAi. *Nature Biotechnology*, *21*(6), 635–637.

Jackson, A. L., & Linsley, P. S. (2004). Noise amidst the silence: Off-target effects of siRNAs? *Trends in Genetics*, *20*(11), 521–524.

Jackson, A. L., & Linsley, P. S. (2010). Recognizing and avoiding siRNA off-target effects for target identification and therapeutic application. *Nature Reviews Drug Discovery*, *9*(1), 57–67.

Jacobson, M. D., Weil, M., & Raff, M. C. (1997). Programmed cell death in animal. *Cell*, *88*(3), 347–354.

Jiang, J., McDonald, P. R., Dixon, T. M., Franicola, D., Zhang, X., Nie, S., et al. (2009). Synthetic protection short interfering RNA screen reveals glyburide as a novel radioprotector. *Radiation Research*, *172*(4), 414–422.

Joseph, J., Seervi, M., Sobhan, P. K., & Retnabai, S. T. (2011). High throughput ratio imaging to profile caspase activity: Potential application in multiparameter high content apoptosis analysis and drug screening. *PLoS One*, *6*(5), e20114.

Joung, J. K., & Sander, J. D. (2013). TALENs: A widely applicable technology for targeted genome editing. *Nature Reviews Molecular Cell Biology*, *14*(1), 49–55.

Kerr, J. F., Wyllie, A. H., & Currie, A. R. (1972). Apoptosis: A basic biological phenomenon with wide-ranging implications in tissue kinetics. *British Journal of Cancer*, *24*(4), 239–257.

Khvorova, A., Reynolds, A., & Jayasena, S. D. (2003). Functional siRNAs and miRNAs exhibit strand bias. *Cell*, *115*(2), 209–216.

Kim, W. J., Chang, C. W., Lee, M., & Kim, S. W. (2007). Efficient siRNA delivery using water soluble lipopolymer for anti-angiogenic gene therapy. *Journal of Controlled Release*, *118*(3), 357–363.

Kim, Y., Kweon, J., Kim, A., Chon, J. K., Yoo, J. K., Kim, H. J., et al. (2013). A library of TAL effector nucleases spanning the human genome. *Nature Biotechnology*, *31*(3), 251–258.

Kimura, J., Nguyen, S. T., Liu, H., Taira, N., Miki, Y., & Yoshida, K. (2008). A functional genome-wide RNAi screen identifies TAF1 as a regulator for apoptosis in response to genotoxic stress. *Nucleic Acids Research*, *36*(16), 5250–5259.

König, R., Chiang, C. Y., Tu, B. P., Yan, S. F., DeJesus, P. D., Romero, A., et al. (2007). A probability-based approach for the analysis of large-scale RNAi screens. *Nature Methods*, *4*(10), 847–849.

Kumar, P., Goh, G., Wongphayak, S., Moreau, D., & Bard, F. (2013). ScreenSifter; analysis and visualization of RNAi screening data. *BMC Bioinformatics*, *14*, 290.

Lazebnik, Y. A., Cole, S., Cooke, C. A., Nelson, W. G., & Earnshaw, W. C. (1993). Nuclear events of apoptosis in vitro in cell-free mitotic extracts: A model system for analysis of the active phase of apoptosis. *The Journal of Cell Biology*, *123*(1), 7–22.

Le Doux, J. M., Landazuri, N., Yarmush, M. L., & Morgan, J. R. (2001). Complexation of retrovirus with cationic and anionic polymers increase the efficiency of gene transfer. *Human Gene Therapy*, *12*(13), 1611–1621.

Lee, Y., Ahn, C., Choi, H., Kim, J., Yim, J., Lee, J., et al. (2003). The nuclear RNase III Drosha initiates microRNA processing. *Nature*, *425*(6956), 415–419.

Lovbor, H., Gullbo, J., & Larsson, R. (2005). Screening for apoptosis classical and emerging techniques. *Anti-Cancer Drugs*, *16*, 593–599.

Love, T. M., Moffett, H. F., & Novina, C. D. (2008). Not miR-ly small RNAs: Big potential for microRNAs in therapy. *The Journal of Allergy and Clinical Immunology*, *121*(2), 309–319.

MacKeigan, J. P., Murphy, L. O., & Blenis, J. (2005). Sensitized RNAi screen of human kinases and phosphatases identifies new regulators of apoptosis and chemoresistance. *Nature Cell Biology*, 7(6), 591–600.

Mak, A. B., Blakely, K. M., Williams, R. A., Penttilä, P. A., Shukalyuk, A. I., Osman, K. T., et al. (2011). CD133 protein N-glycosylation processing contributes to cell surface recognition of the primitive cell marker AC133 epitope. *The Journal of Biological Chemistry*, *286*(47), 41046–41056.

Malo, N., Hanley, J. A., Cerquozzi, S., Pelletier, J., & Nadon, R. (2006). Statistical practice in high-throughput screening data analysis. *Nature Biotechnology*, *24*(2), 167–175.

McCarthy, N. J., Whyte, M. K. B., Gilbert, C. S., & Evan, G. I. (1997). Inhibition of Ced-3/ICE-related proteases does not prevent cell death induced by oncogenes, DNA damage, or the Bcl-2 homologue Bak. *The Journal of Cell Biology*, *136*(1), 215–227.

Meacham, C. E., Ho, E. E., Dubrovsky, E., Gertler, F. B., & Hemann, M. T. (2009). In vivo RNAi screening identifies regulators of actin dynamics as key determinants of lymphoma progression. *Nature Genetics*, *41*(10), 1133–1137.

Messam, C. A., & Pittman, R. N. (1998). Asynchrony and commitment to die during apoptosis. *Experimental Cell Research*, *238*(2), 389–398.

Mills, J., Malina, A., Lee, T., Di Paola, D., Larsson, O., Miething, C., et al. (2013). RNAi screening uncovers Dhx9 as a modifier of ABT-737 resistance in an Eμ-*myc*/Bcl-2 mouse model. *Blood*, *121*, 3402–3412.

Mohr, S., Bakal, C., & Perrimon, N. (2010). Genomic screening with RNAi: Results and challenges. *Annual Review of Biochemistry*, *79*, 37–64.

Neumann, B., Held, M., Liebel, U., Erfle, H., Rogers, P., Pepperkok, R., et al. (2006). High-throughput RNAi screening by time-lapse imaging of live human cells. *Nature Methods*, *3*(5), 385–390.

Neumüller, R. A., & Perrimon, N. (2010). Where gene discovery turns into systems biology: Genome-scale RNAi screens in Drosophila. *Wiley Interdisciplinary Reviews: Systems Biology and Medicine*, *3*(4), 471–478.

Paddison, P. J., Caudy, A. A., Bernstein, E., Hannon, G. J., & Conklin, D. S. (2002). Short hairpin RNAs (shRNAs) induce sequence-specific silencing in mammalian cells. *Genes & Development*, *16*(8), 948–958.

Qi, L. S., Larson, M. H., Gilbert, L. A., Doudna, J. A., Weissman, J. S., Arkin, A. P., et al. (2013). Repurposing CRISPR as an RNA-guided platform for sequence-specific control of gene expression. *Cell*, *152*(5), 1173–1183.

Rantala, J. K., Mäkelä, R., Aaltola, A. R., Laasola, P., Mpindi, J. P., Nees, M., et al. (2011). A cell spot microarray method for production of high density siRNA transfection microarrays. *BMC Genomics*, *12*, 162.

Ren, Y. G., Wagner, K. W., Knee, D. A., Aza-Blanc, P., Nassof, M., & Deveraux, Q. L. (2004). Differential regulation of the TRAIL death receptors DR4 and DR5 by the signaling recognition particle. *Molecular Biology of the Cell*, *15*, 5064–5074.

Riss, T. L., & Moravec, R. A. (2004). Use of multiple assay endpoints to investigate the effects of incubation time, dose of toxin, and plating density in cell-based cytotoxicity assays. *Assay and Drug Development Techologies*, *2*(1), 51–62.

Rutz, S., & Scheffold, A. (2004). Towards in vivo application of RNA interference—New toys, old problems. *Arthritis Research and Therapy*, *6*(2), 78–85.

Scacheri, P. C., Rozenblatt-Rosen, O., Caplen, N. J., Wolfsberg, T. G., Umayan, L., Lee, J. C., et al. (2004). Short interfering RNAs can induce unexpected and divergent changes in the levels of untargeted proteins in mammalian cells. *Proceedings of the National Academy of Sciences of the United States of America*, *101*(7), 1892–1897.

Schlabach, M., Luo, J., Solimini, N. L., Hu, G., Xu, Q., Li, M. Z., et al. (2008). Cancer proliferation gene discovery through functional genomics. *Science*, *319*(5863), 620–624.

Schwarz, D. S., Hutvágner, G., Du, T., Xu, Z., Aronin, N., & Zamore, P. D. (2003). Asymmetry in the assembly of the RNAi enzyme complex. *Cell*, *115*(2), 199–208.

Schwarz, D. S., Hutvágner, G., Haley, B., & Zamore, P. D. (2002). Evidence that siRNAs function as guides, not primers, in the Drosophila and human RNAi pathways. *Molecular Cell*, *10*(3), 537–548.

Seitz, B., Baktanian, E., Gordon, E. M., Anderson, W. F., LaBree, L., & McDonnell, P. J. (1998). Retroviral vector-mediated gene transfer into keratocytes: In vitro effects of polybrene and protamine sulfate. *Graefe's Archive for Clinical and Experimental Ophthalmology*, *236*(8), 602–612.

Semizarov, D., Frost, L., Sarthy, A., Kroeger, P., Halbert, D. N., & Fesik, S. W. (2003). Specificity of short interfering RNA determined through gene expression signatures. *Proceedings of the National Academy of Sciences of the United States of America*, *100*(11), 6447–6452.

Sheng, Z., Li, L., Zhu, L. J., Smith, T. W., Demers, A., et al. (2010). A genome-wide RNA interference screen reveals an essential CREB3L2-ATF5-MCL1 survival pathway in malignant glioma with therapeutic implications. *Nature Medicine*, *16*, 671–677.

Shi, Y. (2003). Mammalian RNAi for the masses. *Trends in Genetics*, *19*, 9–12.

Sigoillot, F. D., & King, R. W. (2011). Vigilance and validation: Keys to success in RNAi screening. *ACS Chemical Biology*, *6*(1), 47–60.

Silva, J. M., Marran, K., Parker, J. S., Silva, J., Golding, M., Schlabach, M. R., et al. (2008). Profiling essential genes in human mammary cells by multiplex RNAi screening. *Science*, *319*(5863), 617–620.

Simpson, J. C., Cetin, C., Erfle, H., Joggerst, B., Liebel, U., Ellenberg, J., et al. (2007). An RNAi screening platform to identify secretion machinery in mammalian cells. *Journal of Biotechnology*, *129*(2), 352–365.

Sledz, C. A., Holko, M., de Veer, M. J., Silverman, R. H., & Willimas, B. R. (2003). Activation of the interferon system by short-interfering RNAs. *Nature Cell Biology*, *5*(9), 834–839.

Smith-Vikos, T. (2013). Advances in RNAi tools and technologies. *Genetic Engineering & Biotechnologies News*, *33*, 20–25.

Smolen, G. A., Zhang, J., Zubrowski, M. J., Edelman, E. J., Luo, B., Yu, M., et al. (2010). A genome-wide RNAi screen identifies multiple RSK-dependent regulators of cell migration. *Genes & Development*, *24*(23), 2654–2665.

Snove, O., & Holen, T. (2004). Many commonly used siRNAs risk off-target activity. *Biochemical and Biophysical Research Communications*, *319*(1), 256–263.

Solary, E., Bertrand, R., Kohn, K. W., & Pommier, Y. (1993). Differential induction of apoptosis in undifferentiated and differentiated HL-60 cells by DNA topoisomerase I and II inhibitors. *Blood*, *81*(5), 1359–1368.

Stec, E., Locco, L., Szymanski, S., Bartz, S. R., Toniatti, C., Needham, R. H., et al. (2012). A multiplexed siRNA screening strategy to identify genes in the PARP pathway. *Journal of Biomolecular Screening*, *17*(10), 1316–1328.

Stegmeier, F., Hu, G., Rickles, R. J., Hannon, G. J., & Elledge, S. J. (2005). A lentiviral microRNA-based system for single-copy polymerase II-regulated RNA interference in mammalian cells. *Proceedings of the National Academy of Sciences of the United States of America*, *102*(37), 13212–13217.

Steller, H. (1995). Mechanism and genes of cellular suicide. *Science*, *2767*(5203), 1445–1449.

Stone, J. D., Marine, S., Majercak, J., Ray, J. W., Espeseth, A., et al. (2007). High-throughput screening by RNA interference: Control of two distinct types of variance. *Cell Cycle*, *6*, P898–P901.

Sun, N., & Zhao, H. (2013). Transcription activator-like effector nucleases (TALENs): A highly efficient and versatile tool for genome editing. *Biotechnology and Bioengineering*, *110*(7), 1811–1821.

Swaney, W. P., Sorgi, F. L., Bahnson, A. B., & Barranger, J. A. (1997). The effect of cationic liposome pretreatment and centrifugation on retrovirus-mediated gene transfer. *Gene Therapy, 4*(12), 1379–1386.
Tait, S. W., & Green, D. R. (2010). Mitochondria and cell death: Outer membrane permeabilization and beyond. *Nature Reviews Molecular Cell Biology, 11*(9), 621–632.
Themis, M., Forbes, S. J., Chan, L., Cooper, R. G., Etheridge, C. J., & Miller, A. D. (1998). Enhanced in vitro and in vivo gene delivery using cationic agent complexed retrovirus vectors. *Gene Therapy, 5*(9), 1180–1186.
To, K., Zhao, Y., Jiang, K., Hu, M., Wang, J., Wu, C., et al. (2007). The phosphoinositide-dependent kinase-1 inhibitor 2-amino-*N*-[4-[5-(2-phenanthrenyl)-3-(trifluoromethyl)-1*H*-pyrazol-1-yl]phenyl]-acetamide (OSU-03012) prevents Y-box binding protein-1 from inducing epidermal growth factor receptor. *Molecular Pharmacology, 72*, 641–652.
Toyoshima, K., & Vogt, P. K. (1969). Temperature sensitive mutants of an ovarian sarcoma virus. *Virology, 39*(4), 930–931.
Tsong, T. Y. (1991). Electroporation of cell membranes. *Biophysical Journal, 60*(2), 297–306.
Tushi, T., Zamore, P. D., Lehman, R., Bartel, D. P., & Sharp, P. A. (1999). Targeted mRNA degradation by double-stranded RNA in vitro. *Genes & Development, 13*(24), 3191–3197.
Vogt, A., McDonald, P. R., Tamewitz, A., Sikorski, R. P., Wipf, P., Skoko, J. J., et al. (2008). A cell-active inhibitor of mitogen-activated protein kinase phosphatases restores paclitaxel-induced apoptosis in dexamethasone-protected cancer cells. *Molecular Cancer Therapeutics*, 7, 330–340.
Wang, L., Tu, Z., & Sun, F. (2009). A network-based integrative approach to prioritize reliable hits from multiple genome-wide RNAi screens in Drosophila. *BMC Genomics, 10*(1), 220.
Wheeler, D. B., Bailey, S. N., Guertin, D. A., Carpenter, A. E., Higgins, C. O., & Sabatini, D. M. (2004). RNAi living-cell microarrays for loss-of-function screens in Drosophila melanogaster cells. *Nature Methods, 1*(2), 127–132.
Wiles, A. M., Ravi, D., Bhavani, S., & Bishop, A. J. (2008). An analysis of normalization methods for Drosophila RNAi genomic screens and development of a robust validation scheme. *Journal of Biomolecular Screening, 13*(8), 777–784.
Wong, R. S. Y. (2011). Apoptosis in cancer: From pathogenesis to treatment. *Journal of Experimental & Clinical Cancer Research, 30*(1), 87–101.
Wu, Z. W., Chien, C. T., Lui, C. Y., Yan, J. Y., & Lin, S. Y. (2012). Recent progress in copolymer-mediated siRNA delivery. *Journal of Drug Targeting, 20*(7), 551–560.
Wyllie, A. H. (1980). Glucocorticoid-induced thymocyte apoptosis is associated with endogenous endonuclease activation. *Nature, 285*, 555–556.
Zeng, Y., Wagner, E. J., & Cullen, B. R. (2002). Both natural and designed micro RNAs can inhibit the expression of cognate mRNAs when expressed in human cells. *Molecular Cell, 9*(6), 1327–1333.
Zhang, J. H., Chung, T. D., & Oldenburg, K. R. (1999). A simple statistical parameter for use in evaluation and validation of high throughput screening assays. *Journal of Biomolecular Screening, 4*(2), 67–73.
Zhang, X. H. D., Espeseth, A. S., Johnson, E. N., Chin, J., Gates, A., et al. (2008). Integrating experimental and analytic approaches to improve data quality in genome-scale RNAi screens. *Journal of Biomolecular Screening, 13*(5), 378–389.
Zhang, X. H. D., Ferrer, M., Espeseth, A. S., Marine, S. D., Stec, E. M., Crackower, M. A., et al. (2007). The use of strictly standardized mean difference for hit selection in primary RNA interference high-throughput screening experiments. *Journal of Biomolecular Screening, 12*, 645–655.
Zhang, X. H. D., & Heyse, J. F. (2009). Determination of sample size in genome-scale RNAi screens. *Bioinformatics, 25*, 841–844.

Zhang, X. H. D., Kuan, P. F., Ferrer, M., Shu, X., Liu, Y. C., Gates, A. T., et al. (2009). Hit selection with false discovery rate control in genome-scale RNAi screens. *Nucleic Acids Research*, *36*, 4667–4679.

Zimmermann, T. S., Lee, A. C. H., Akinc, A., Bramlage, B., Bumcrot, B., et al. (2006). RNAi-mediated gene silencing in non-human primates. *Nature*, *441*, 111–114.

Zuber, J., McJunkin, K., Fellmann, C., Dow, L. E., Taylor, M. J., Hannon, G. J., et al. (2011). Toolkit for evaluating genes required for proliferation and survival using tetracycline-regulated RNAi. *Nature Biotechnology*, *29*(1), 79–83.

CHAPTER SEVEN

Caspase Enzymology and Activation Mechanisms

Peter D. Mace*, Stefan J. Riedl†, Guy S. Salvesen†,[1]
*Biochemistry Department, University of Otago, Dunedin, New Zealand
†Program in Cell Death and Survival Networks, NCI Designated Cancer Center, Sanford-Burnham Medical Research Institute, La Jolla, California, USA
[1]Corresponding author: e-mail address: gsalvesen@sanfordburnham.org

Contents

Abstract

Apical caspases 8, 9, and 10 are only active as dimers. These dimers are unstable, and to characterize their activity they need to be maintained *in vitro* in a dimeric state. We provide updated methods for those looking to characterize various aspects of caspase function. We describe full methods for those looking to activate caspases *in vitro* using kosmotropic reagents, an essential step in characterizing upstream (apical) caspases. We detail methods for fusion of caspase domains to engineered dimerization domains as an alternative method to trigger regulated dimerization of caspases. We also describe methods to determine caspase activity profiles in cells and provide methods for studying the ability of SMAC-mimetic reagents to release inhibition of caspases by IAPs.

1. INTRODUCTION

Caspases are a family of proteases that play a key role in cellular processes such as apoptosis and inflammation—as recently reviewed in Crawford and Wells (2011), McIlwain, Berger, and Mak (2013), and Pop

Methods in Enzymology, Volume 544
ISSN 0076-6879
http://dx.doi.org/10.1016/B978-0-12-417158-9.00007-8

and Salvesen (2009). They act by cleaving substrates following aspartate residues, with additional specificity determined by the sequence immediately surrounding this aspartate. Although there is some functional overlap, caspases 3, 6, 7, 8, 9, and 10 primarily act to regulate apoptosis, whereas caspases 1, 4, and 5 process inflammatory cytokines during inflammation. Other roles have been ascribed to caspase 14 in keratinocyte differentiation, and a related class of "paracaspases" have recently come to light that share some structural homology with caspases but cleave following an arginine rather than an aspartate residue (Hachmann et al., 2012).

The apoptotic caspases function in a hierarchical manner where initiator members (2, 8, 9, and 10) are first activated and subsequently cleave effector caspases (3, 6, and 7), which go on to cleave a more diverse range of substrates (Crawford & Wells, 2011). Because of these distinct roles, activation mechanisms of initiator and effector caspases are different (Pop & Salvesen, 2009). Considerable work has established that upstream caspases require proximity-based dimerization for activation, as well as cleavage within their intradomain linkers for stabilization. Various platforms for the activation of apical caspases exist and differ depending on the pathway in which a particular caspase functions. For instance, caspase 9 is activated during intrinsically triggered apoptosis at the apoptosome (Bratton & Salvesen, 2010; Yuan & Akey, 2013), caspase 2 is activated by cytotoxic stress at the PIDDosome (Tinel & Tschopp, 2004), and extrinsically triggered apoptosis activates caspase 8 at the death-inducing signaling complex (DISC) (Mace & Riedl, 2010). In an analogous manner, caspase 1 is activated by self-association as part of the "inflammasome," which forms in response to inflammatory stimuli (Martinon & Tschopp, 2007).

In contrast, effector caspases in apoptotic signaling cascades preexist as inactive dimers and are activated following cleavage by upstream caspases (Pop & Salvesen, 2009). In such a manner, various apoptotic stimuli can converge upon effector caspases to amplify proteolytic signaling and apoptosis. As one may expect, various points of regulation exist in such a crucial signaling cascade that proceeds toward such an irreversible outcome as cell death. The fact that upstream caspases require both dimerization and cleavage for avid signaling is itself an important factor in preventing spurious activation of death signaling. Another crucial regulator the apoptotic caspase pathway is X-linked inhibitor of apoptosis (XIAP), which directly binds and inhibits the activity of the effector caspases 3, 7, and the initiator caspase 9—reviewed in Lopez and Meier (2010), Eckelman, Salvesen, and Scott (2006) and thus can counteract an apoptotic signal by blocking two levels of this caspase cascade. Remarkably, the regulator protein SMAC (second mitochondrial derived activator of caspases) can counteract this blockage restoring the activity (Vaux & Silke, 2003).

Caspase activity is often studied in cells using fluorimetric or colorimetric reagents that act as active-site directed enzyme activity reporters. While these reagents can be very useful, they have several large downsides. The largest of these drawbacks is that even though caspases have different optimal sequences (e.g., DEVD for caspase 3, LEHD for caspase 9, IETD for caspase 8, VEID for caspase 6, and VDVAD for caspase 2), there is considerable overlap in substrates that they can actually cleave. For this reason, and as has been strongly demonstrated previously (McStay, Salvesen, & Green, 2008; Pop & Salvesen, 2009), significant care must be used when interpreting results of experiments that employ these reagents. These can often be skewed depending on which caspase is the most prevalent or active in a particular sample. Thus, thorough biochemical characterization of a caspase–substrate pair, as well as orthogonal methods that allow the specificity of this protease–substrate pair to be investigated in a cellular setting, remain key steps in determining functional relevance.

Here, we provide an updated reference for those looking to characterize various aspects of caspase function. We describe full methods for those looking to activate caspases *in vitro* using kosmotropic reagents, which is a vital step in characterizing a caspase–substrate cleavage event for upstream caspases. We detail methods for fusion of caspase domains to dimerization domains as an alternative method to trigger regulated dimerization of caspases in the presence of small molecules. We also describe methods to determine caspase activity profiles in cells and provide methods for studying the ability of SMAC-mimetic reagents to release inhibition of caspases by IAPs.

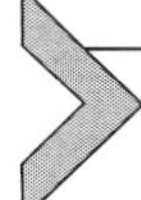

2. ACTIVATING INITIATOR CASPASES

2.1. Kosmotrope-based activation

Recombinantly expressed initiator caspases exist as a mixture of dimers and monomers (Boatright et al., 2003). This is expected given the characteristics of their dimerization: typical purification of these proteins from *E. coli* result in a protein concentration of approximately 10 μ*M*, which is slightly below the measured dissociation constants of caspase 8 and 9 (50 and 100 μ*M*, respectively) (Donepudi, Mac Sweeney, Briand, & Gruetter, 2003; Renatus, Stennicke, Scott, Liddington, & Salvesen, 2001). This led us to develop a general method using kosmotropic salts to drive dimerization and fully activate caspases 8, 9, and 10 (Fig. 7.1A). Kosmotropes stabilize proteins in an aqueous environment. They act either by ordering the structure of water or by compensating for the loss of hydrogen bonding in

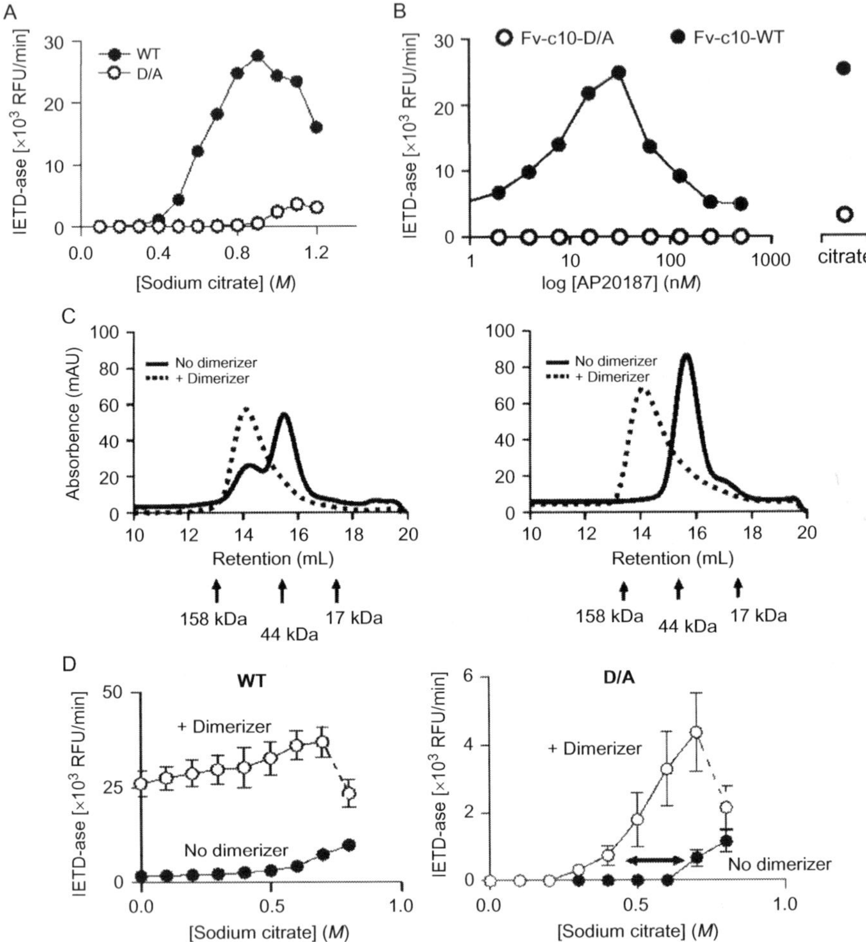

Figure 7.1 (A) Induction of caspase 10 activity by sodium citrate. IETDase activity of caspase 10 wild-type, and cleavage site mutant D297A (D/A) at various concentrations of sodium citrate (B) Activity of dimerizable caspase 10 wild-type (Fv-c10-WT) and cleavage site mutant (Fv-c10-D/A) at various concentrations of AP20187. The citrate data to the right of panel B reflects the amount of activity elicited at optimal citrate concentration (1 *M*). (C) Size-exclusion chromatography of Fv-caspase 10 (WT or D/A) in the presence and absence of equimolar AP20187. (D) IETDase activity of Fv-caspase 10 (WT or D/A) at various concentrations of sodium citrate. *Figure adapted with permission from Wachmann et al. (2010). Copyright (2010) American Chemical Society.*

proteins at high salt concentration (Collins & Washabaugh, 1985); either way, the overall effect is "order" or less entropy. Examples of ionic kosmotropic salts are citrate and related organic acid salts (malic, malonic, and succinic), phosphates, ammonium, and sulfates. Examples of nonionic kosmotropes are polyethylene glycol, proline, glycine, betaine, and trehalose.

At concentrations ranging from 0.5 to 1.5 *M* kosmotropic salts can have two effects: inducing dimerization of caspase monomers, which effectively increases k_{cat} by increasing the number of active sites in solution, and promoting order in the loops flanking the caspase active site to enhance substrate binding (i.e., decreasing K_M; Gouvea et al., 2006; Schmidt & Darke, 1997). In reality, both of these effects occur simultaneously for the initiator caspases.

The most effective kosmotrope for initiator caspase activation is sodium citrate. We have illustrated this for purified caspase 8, 9, and 10, where monomers isolated by gel filtration can be activated in excess of 100-fold by 1 *M* sodium citrate. Mixed samples purified by affinity but not size-exclusion chromatography show activation that is more modest (between 10 and 100-fold) owing to the fact that they are already a mixture of monomer and dimer (Boatright et al., 2003). In contrast, for recombinant effector caspases 3 and 7 that are already 100% dimeric in solution, sodium citrate induces only a two to threefold increase in activity, likely by stabilizing the catalytic loops (Pop, Fitzgerald, Green, & Salvesen, 2007). We observe that malonic, malic, and succinic acids can also efficiently activate apical caspases, whereas aspartic and maleic acid salts are less efficient activators (unpublished).

2.1.1 Reagents

Recombinant caspases 3, 7, 8, 9, and 10: expressed in *E. coli* and purified as previously described (Stennicke & Salvesen, 1999). Expression plasmids for these caspases are available from www.addgene.com.

Sodium citrate stock: 1.4 *M* sodium citrate, 50 m*M* Tris or 50 m*M* NaH_2PO_4/Na_2HPO_4, pH 7.4. The stock is made by dissolving the buffer reagents in distilled water, followed by slow addition of solid sodium citrate or citric acid and then adjustment of the pH with concentrated HCl or NaOH, respectively. *Note*: the solubility limit of sodium citrate in water is 1.7 *M*—so correct adjustment of the solution volume is important to avoid citrate crystallization. The solution is sterile filtered to prevent bacterial growth and is stable for several months at room temperature.

DTT: 1 *M* in water, stored in aliquots at −20 °C.

CHAPS stock: 10% (w/v) in water, stable at room temperature for several months.

Low salt buffer: 50 m*M* Tris, pH 7.4, or 50 mM NaH_2PO_4/Na_2HPO_4, pH 7.4.

Caspase assay buffer: 20 m*M* PIPES, 0.1 *M* NaCl, 5% (w/v) sucrose, 0.1% (w/v) CHAPS, 10 m*M* DTT (freshly added), pH 7.4. This buffer is the general optimal assay buffer for effector caspases (Stennicke & Salvesen, 1999).

Caspase substrates: 10 m*M* stocks in DMSO. Acetyl-Ile-Glu-Thr-Asp-7-amino-4-trifluoromethyl coumarin (Ac-IETD-AFC) for caspases 8 and 10, Ac-LEHD-AFC for caspase 9, and Ac-DEVD-AFC for caspase 3, 6, and 7. Either fluorogenic substrates or colorimetric substrates are compatible with this procedure.

Other kosmotrope stocks: 1.4 *M* sodium malonate, 1.4 *M* sodium malate, 1.4 *M* sodium succinate, 1.1 *M* sodium aspartate in 25 m*M* NaH_2PO_4/Na_2HPO_4, pH 7.4. (These should have similar long-term stability as sodium citrate, although this has not been rigorously tested.)

2.1.2 Procedure

Caspase activity assays are carried out in 96-well plates.

1. From the stock solutions, prepare 1 *M* sodium citrate, 50 m*M* Tris, pH 7.4, containing 10 m*M* DTT and 0.05% CHAPS, by diluting in low salt buffer. (*Note*: CHAPS precipitates in 1 *M* sodium citrate; therefore, use concentrations less than 0.1%. CHAPS can be omitted if the experiments last less than 2 h.)
2. Add kosmotrope containing buffer to the caspase of interest, to give final concentrations of 10–20 n*M* for caspases 8, 9, 10, or 0.5 to 1 n*M* for caspase 3 and 7. Mix by pipetting.
3. Incubate at 37 °C. The activation process for effector caspases 3 and 7, proceeds rapidly, so the incubation time should not exceed 15–20 min. Activation kinetics for the initiator caspases are slower and depend on both the enzyme concentration and temperature. Higher enzyme concentrations require less time for full activation than lower concentrations. A suitable incubation time should be chosen for the particular application, considering aspects such as substrate stability (see note below).
4. For determination of the catalytic activity, add the appropriate fluorogenic substrate to a final concentration of 100 μ*M*. The initial velocity

is monitored by use of a plate reader equipped with filters that excite at 405 nm and detect emission at 510 nm. As a control, the experiment is repeated in caspase assay buffer.

Note: Cleavage of natural substrates, either purified or in complex mixtures, can also be monitored in the presence of kosmotropes, as long as they do not precipitate under the assay conditions (Pop, Timmer, Sperandio, & Salvesen, 2006). This is primarily achieved by monitoring cleavage of the protein substrate using SDS-PAGE. However, high concentrations of kosmotropic salts interfere with SDS-PAGE, so proteins must first be TCA precipitated if PAGE is the chosen detection method. Sodium citrate generally increases the long-term stability of initiator caspases compared with caspase assay buffer. Nanomolar concentrations of caspases 8 and 10 maintain their activity overnight at ambient temperature, in the presence of 0.05% CHAPS and kosmotrope. We do not observe precipitation of either caspase 8 or 10 in 1 *M* sodium citrate, even at micromolar concentrations. However, caspases 8 and 10 become unstable at sodium citrate concentrations above 1 *M*. Caspase 9 at submicromolar concentration loses approximately 20% activity after overnight incubation in sodium citrate and at concentrations above 1 m*M* it precipitates in 1 *M* sodium citrate. In sodium citrate, caspase 3 and 7 are only stable for a few hours, whereas caspase 6 loses approximately 50% activity because of precipitation. Caspase 7 precipitates at micromolar concentrations in sodium citrate.

When preparing the working solution containing kosmotropes, it is important to adjust the pH to 7.4 with compatible buffers that are stable in the neutral range.

The concentration of "active" initiator caspase by active-site titration in kosmotrope is only true for the conditions containing the same amount of kosmotrope.

Activation by kosmotropes is reversible. Removal of kosmotrope by dilution or dialysis is sufficient to inactivate apical caspases. Also, addition of chaotropic salts, like $NaClO_4$, $NaNO_3$, or $MgCl_2$, to a mixture of monomer–dimer will shift the equilibrium to the inactive species, because of dimer dissociation (Pop et al., 2007).

2.2. Activation by dimerization reagents

As emphasized above, dimerization is a key step in activation of apical caspases. While kosmotropes can be used to promote dimerization and activation of purified caspases in a purified system, inducible dimerization through chimeric caspase/dimerization domain constructs provides an orthogonal tool to investigate activation of caspases either purified from recombinant sources, or in cells. The most convenient method to achieve this is to fuse the relevant caspase catalytic domain to the dimerization domain of the 12 kDa FK506 binding protein (FKBP12 or FKBP), in a method deriving from Muzio, Stockwell, Stennicke, Salvesen, and Dixit (1998). In their natural context, FKBP domains are monomeric. However, upon addition of FK506 (a rapamycin-like molecule), residues from two constituent domains bind FK506 (or derivatives thereof) at the dimer interface. Addition of FK506 thus has the ability to induce dimerization of the monomeric domain, either alone or in the context of fusion proteins such as FKBP-caspase hybrid proteins. A further variation of this system uses the rapamycin binding domain from mTor kinase (Frb) fused to one protein, in conjunction with the FKBP domain fused to another protein to induce heterodimerization of the two respective fusion partners. Inducible dimerization has proven to be a powerful method for studying a range of systems that are regulated by dimerization and protein–protein interactions, especially activation of caspases as well as downstream caspase signaling.

While kosmotropes are convenient if your aim is to simply activate a caspase, if one wishes to study the activation mechanism it is useful to be able to separate the dimerization and cleavage events. Studies have used inducible dimerization to dissect the requirement for both dimerization and cleavage to activate caspase 8 (Oberst et al., 2010), and in combination with kosmotropes to probe the activation mechanism of caspase 10 (Wachmann et al., 2010). This is achieved by combining dimerization domains with noncleavable mutant variants of both caspases 8 and 10, allowing observation of the activating potential of dimerization alone, and providing a closer mimic of the natural activation mechanism. For example, kosmotropes activate caspase 8 variants even when they have noncleavable sequences in the linker between the large and small subunits. In contrast, using dimerization domains only caspase 8 with a cleavable linker (i.e., the native linker) between the large and small subunits is activated fully upon addition of dimerization reagent (Oberst et al., 2010).

Before employing chemically inducible dimerization one must consider the most appropriate junction for in-frame cloning of the FKBP (or variant thereof) dimerization domain. With regard to caspases, this consideration is made more straightforward as the aim is generally to replace the natural recruitment domain(s) (i.e., the CARD domain of caspase 9 and the DED domains of caspase 8 or caspase 10) with the FKBP domain. However, given that it is not precisely known how CARDs or DEDs of caspases self-associate, the inclusion of a small additional linker provides some leeway in this regard. Here, we describe how we applied this system to study the activation mechanism of caspase 10 (Wachmann et al., 2010). However, the general principles may be applied to any apical caspase or other proteases that are believed to be activated by dimerization.

2.2.1 Reagents

Recombinant caspase 10: was expressed in *E. coli* as described in Wachmann et al. (2010). Our FKBP domain was derived from the pC4-Fv1E expression vector (previously available from ARIAD Pharmaceuticals as the ARGENT Regulated Homodimerization Kit, now available as pHom-1 from Clontech Inc.).

Caspase assay buffer: 20 m*M* PIPES, 0.1 *M* NaCl, 5% (w/v) sucrose, 0.1% (w/v) CHAPS, 10 m*M* DTT (freshly added), pH 7.4.

Caspase substrate: Ac-IETD-AFC.

Size-exclusion chromatography buffer: 50 m*M* Tris and 100 m*M* NaCl buffer, pH 7.4.

Dimerization reagent: The dimerization reagent used is determined by the precise dimerization domain fused to the protein of interest. We use AP20187, which is paired with the dimerization domain from pC4-Fv1E/pHom-1 and is available from Clontech Inc.

2.2.2 Procedure

2.2.2.1 Preparation of recombinant proteins

ΔDED caspase 10 and the ΔDED caspase 10 D297A were cloned into a pET-28b vector containing an FKBP domain. We used vectors for expressing recombinant FKBP fusions of caspase 10 in *E. coli* incorporating an N-terminal His_6 tag to facilitate straightforward affinity purification. Immobilized nickel affinity chromatography was performed using standard methods equivalent to that used for wild-type catalytic domains. Precise concentrations were measured by active-site titration with benzoxycarbonyl-Val-Ala-Asp-fluoromethyl ketone (zVAD-fmk) as previously described (Stennicke & Salvesen, 2000).

Like caspase 8, wild-type caspase 10 expressed in *E. coli* is efficiently autoproteolytically processed in the linker between its large and small subunits. Both caspase 8 and caspase 10 contain two potential cleavage sites in this linker. We find that while caspase 8 can be processed at either of these sites, a single mutation of caspase 10 at aspartate 297 (D297A) completely abrogated autoprocessing, resulting in an intact polypeptide of 35 kDa. The same phenomenon was observed for the equivalent FKBP fusion proteins.

2.2.2.2 Establishing efficiency of dimerization reagent

Catalytic activity of FKBP fused caspase 8 or 10 was measured in 96-well plates using Ac-IETD-AFC in caspase assay buffer, applying a similar procedure to that described above.

1. Incubate recombinant FKBP-caspase 10 at 30 n*M* with a serial dilution of dimerization reagent for 30 min at room temperature.
2. Add Ac-IETD-AFC (100 μ*M*). We observed maximal activity at an equimolar concentration of dimerization reagent to caspase (Fig. 7.1B). This is as expected, because at concentrations beyond equimolar the FKBP domains can become saturated with AP20187 while still monomeric, whereas submicromolar concentrations are not sufficient to drive full dimerization.

With this established, subsequent experiments were carried out using equimolar concentrations of caspase and dimerizing reagent. Caspase 10 catalytic domain bearing the D297A mutation could not be activated at any concentration of dimerizing reagent, reflecting the essential nature of processing at this site for activation in the absence of kosmotropic salts.

Size-exclusion chromatography was performed on both wild-type and D297A caspase 10-FKBP fusion proteins to establish that AP20187 treatment was indeed leading to dimerization. Using a properly calibrated size-exclusion column and sensitive UV detection at 280 nm meant that this experiment could be carried out on a very small-scale, limited only by the detection sensitivity of the FPLC system.

1. Generally, calibrate a size-exclusion column (Superdex 200 or Superdex 75 depending on the size of the protein of interest) using proteins of verified size.
2. Incubate recombinant caspase with an equimolar concentration of AP20187 for 30 min prior to loading onto a Superdex 200 column (which is typically the column of choice for the expected apparent molecular weight range).

In the absence of AP20187, we observed that wild-type caspase 10 existed as a mixture of monomer and dimer, consistent with our previous observations for caspase 8 (Boatright et al., 2003). Treatment with AP20187 efficiently shifted the population almost entirely to an apparent molecular weight of a dimer (Fig. 7.1C). Without dimerizing reagent, the D297A caspase 10-FKBP fusion protein eluted almost entirely as a monomer. This reflects the fact that dimerization and cleavage are coupled events, with cleavage sufficient to stabilize but not to drive caspase activation. Like the wild-type protein, addition of dimerizing reagent to D297A caspase 10-FKBP caused a complete shift of elution time to that expected for a dimeric species.

2.2.2.3 Combined effects of kosmotropes and inducible dimerization

Having verified that dimerization of our recombinant caspase 10 was indeed inducible, one can proceed to analyze the effects of dimerization on activity in the context of kosmotrope treatment.

1. Incubate either wild-type or D297A caspase 10-FKBP (30 n*M*) with or without AP20187 at an equal ratio for 30 min at room temperature.
2. Add sodium citrate buffer (as outlined above), followed by incubation for another 30 min at room temperature.
3. Measure substrate hydrolysis of Ac-IETD-AFC (100 μ*M*) in 96-well plates.

Our experiments showed that the wild-type protein in the presence of dimerization reagent effectively did not need sodium citrate for activation, that is, having already been cleaved, dimerization is sufficient for full activation of the recombinant wild-type protease (Fig. 7.1D). Overall activity of the cleavage site mutant was substantially lower than the wild-type protein. However, presence of the dimerizing reagent significantly increased the activity of the cleavage site mutant and induced this activity at much lower citrate concentrations than the monomeric species without AP20187. Having characterized the effect of dimerization domains on caspase activity *in vitro*, one may also use similar constructs (recloned into the appropriate expression vectors) for testing the effects of induced dimerization in cell lines.

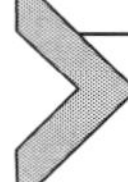

3. INVESTIGATING CASPASE ACTIVITY AND INHIBITION

3.1. Determining caspase activity in cells

In many instances, initiator caspases are cleaved in cells as a bystander event during apoptosis. For example, caspase 8 is cleaved in cells undergoing granzyme B–induced apoptosis, yet no ligand-stimulated DISC was formed, and caspase 8 was never activated (Boatright et al., 2003). In contrast, for effector caspases 3, 6, and 7, cleavage is the activating event. Therefore, demonstrating "active" caspase 3, 6, or 7 is relatively straightforward, with detection of the cleaved caspase by simple immunoblot being the technique of choice. Unfortunately, it is not so easy to demonstrate "active" caspase

8 and 9 in whole cells, and immunoblotting alone is never conclusive. To make matters worse, as described earlier, small peptidic substrates are not specific enough to be used in complex samples such as cell lysates (McStay et al., 2008).

To address this technical problem, a number of groups have developed activity-based probes to trap the active rather than the latent form of a particular caspase. The inhibitor usually consists of a "handle" to allow purification and/or detection (e.g., biotin), the recognition sequence to guide specificity, and a "warhead" (e.g., a fluoromethyl ketone, chloromethyl ketone, or acyloxymethyl ketone; reviewed in Berger, Vitorino, & Bogyo, 2004). Nucleophilic attack leads to the catalytic cysteine becoming covalently bound to the inhibitor. The "handle" of the probe then facilitates purification from the complex sample by way of the biotin group (Denault & Salvesen, 2003a; Faleiro, Kobayashi, Fearnhead, & Lazebnik, 1997; Tu et al., 2006). For a broad range of caspases, the most commonly used probe is biotinylated VAD-fmk (bVAD-fmk). To detect active caspases in whole cells, it is important to use O-methylated bVAD-fmk (bVAD(Ome)-fmk), because this version has increased cell permeability. For labeling active caspases in cell lysates, non-methylated bVAD-fmk is adequate. Thus, the principle of this approach is to use the "warhead" to achieve labeling of only the active caspases, the "handle" for pulling down the now labeled caspase and subsequent caspase-specific immune blotting to unambiguously identify the labeled caspase.

Here, we describe an optimized procedure for *in vitro* trapping of effector caspases. We have included a preincubation step with a proteasome inhibitor (MG132) to stabilize caspases at the protein level during apoptosis. Active caspase 3 is readily precipitated from Jurkat cells undergoing caspase 8–dependent apoptosis and active caspase 8 can be detected in mouse T-cells activated *in vitro* (Leverrier, Salvesen, & Walsh, 2011). When bVAD(Ome)-fmk is preincubated with cells before induction of apoptosis, autocatalytic removal of caspase 3N-terminal peptide is prevented (p20 large subunit). If bVAD(Ome)-fmk is added after apoptosis is induced, both the p20 and p17 (N-peptide removed) forms of caspase 3 are detected.

3.1.1 Reagents

Biotinyl-Val-Ala-(O-methyl)Asp-fluoromethyl ketone ((bVAD(Ome)-fmk), MP Bioscience).

MG132 (Calbiochem)

Agonistic anti-Fas antibody, clone CH11 (Millipore).

PBS: 136 m*M* NaCl, 2.6 m*M* KCl, 10 m*M* Na_2HPO_4, 1.76 m*M* KH_2PO_4, pH 7.4.
CHAPS buffer: 50 m*M* HEPES, 150 m*M* KCl, 0.1% (w/v) CHAPS, pH 7.4, CHAPS buffer containing protease inhibitors. To prevent adventitious proteolysis, we use CHAPS buffer with 10 μ*M* zVAD-fmk, 10 μ*M* E-64, 10 μ*M* leupeptin, 10 μ*M* MG132, 150 μg/ml PMSF, and 1 μ*M* pepstatin, but commercial protease inhibitor cocktails and tablets are also available.
Streptavidin agarose beads (Pierce).
Antibodies: monoclonal caspase 8 antibody (clone C15, generous gift from Dr. Markus Peter, University of Chicago, Chicago, IL) and monoclonal caspase 3 antibody (BD Pharmingen).
Horseradish peroxidase-conjugated streptavidin (Sigma-Aldrich).

3.1.2 Procedure

1. Wash 5×10^7 Jurkat cells in fresh RPMI media, resuspend the cells in 1 ml of RPMI containing 10 μ*M* MG132 and 50 μ*M* bVAD(Ome)-fmk and incubate at 37 °C for 1 h.
2. Add agonistic anti-Fas antibody (CH11) to 100 ng/ml and incubate cells for a further 4 h.
3. Wash cells twice with PBS, resuspend in 500 ml of CHAPS buffer containing protease inhibitors, and freeze–thaw once to lyse cells. (Lysates can be stored at −20 °C for as long as 1 week, and at −70 °C for longer periods of time.)
4. Add 30 μl of streptavidin agarose beads (washed once in 1 ml of CHAPS buffer) to the lysate and incubate at 4 °C for 4 h on a rocking platform (samples can be incubated overnight).
5. Wash beads three times with 1 ml CHAPS buffer, once with 1 ml CHAPS buffer containing 0.5 *M* NaCl and once more in 1 ml CHAPS buffer.
6. Resuspend the beads in 50 ml SDS loading buffer, electrophorese on an 8–18% SDS-PAGE gel, transfer to PVDF membrane, and immunoblot with antibodies specific for the caspase in question as previously described (Denault & Salvesen, 2003b).

Note: This procedure is most useful for suspension cells, because they can be treated in a small volume at high density, cutting down on the amount of apoptotic agent required.

3.2. Inhibition and derepression of caspase activity by IAPs

Inhibitor of apoptosis proteins (IAPs) are a major regulator of caspase activity in a cellular context. Various functions of IAPs are reviewed elsewhere (Fulda & Vucic, 2012; Mace & Riedl, 2010; Salvesen & Duckett, 2002), however, a key facet of the ability of IAPs to regulate caspases and apoptosis stems from their baculoviral IAP repeat (BIR) domains. Among the well-characterized IAP proteins, XIAP but not cIAP1/2 are able to directly inhibit caspases (Eckelman & Salvesen, 2006).

A groove on the surface of some BIR domains binds to a tetrapeptide motif, canonically consisting of Ala-Val-Pro-Ile. This tetrapeptide motif and variations thereof are found in apoptotic regulators and effectors including SMAC and processed caspase 9 (Shiozaki et al., 2003; Srinivasula et al., 2001). The SMAC tetrapeptide motif and its derivatives have given rise to a promising new class of therapeutic agents termed "SMAC mimetics" or "IAP antagonists" (Flygare et al., 2012; Fulda & Vucic, 2012). As the name implies, SMAC mimetics work in a similar manner to SMAC, displacing caspases from the inhibitory presence of the BIR domain. Most SMAC mimetics cross-react to some extent with different BIR domains from XIAP and cIAPs, among others, which is an important consideration given that BIR domains can have distinct functions. These distinct functions include recruiting substrates for ubiquitylation as well as inhibition of caspases (Scott et al., 2005; Shiozaki et al., 2003; Zhuang, Guan, Wang, Burlingame, & Wells, 2013). Inhibition of caspase 3 and 7 is mediated through the BIR2 domain of XIAP, while inhibition of caspase 9 is achieved by BIR3.

We have recently developed assays to measure the ability of SMAC mimetics to both directly compete with caspases for binding by IAP proteins, and also to "derepress" preformed IAP–caspase complexes (Vamos et al., 2013). Derepression assays may more realistically reflect the state of IAP-inhibited caspases in cells, and also provide deeper insight into the dynamics of both caspase and inhibitor binding by BIR domains. Overall, these assays allow a more information-rich analysis of the functionality of different SMAC-mimetic compounds (Fig. 7.2).

3.2.1 Reagents

Recombinant caspases 3, 7, and 9: were expressed in *E. coli* and purified as previously described (Stennicke & Salvesen, 1999).

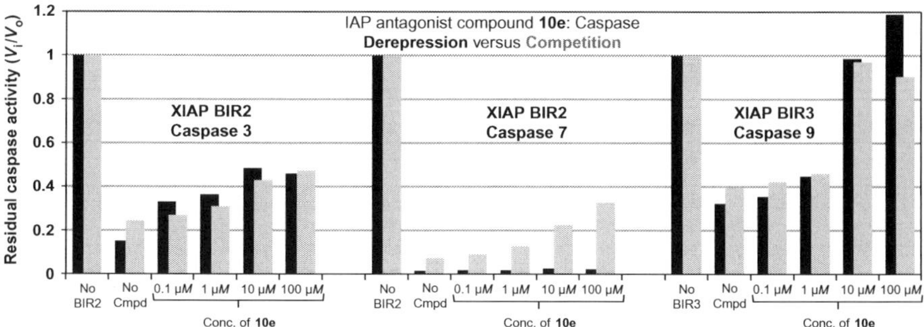

Figure 7.2 Caspase rescue ability of SMAC mimetics. Purified recombinant human caspases 3, 7, or 9 were used in combination with XIAP BIR2 or BIR3. Derepression assays are shown in black and competition assays are shown in grey, for increasing concentrations of SMAC-mimetic compound 10e (Vamos et al., 2013). The ratio of inhibited to uninhibited caspase activity (V_i/V_o) is obtained by monitoring the release of fluorescence from the fluorogenic substrates Ac-DEVD-AFC for caspases 3 and 7 or Ac-LEHD-AFC for caspase 9. *Figure adapted with permission from Vamos et al. (2013). Copyright (2013) American Chemical Society.*

BIR domains: were expressed as His_6 fusion proteins and purified as described previously (Vamos et al., 2013).

Caspase substrates: 10 m*M* stocks in DMSO. Ac-LEHD-AFC for caspase 9, and Ac-DEVD-AFC for caspases 3 and 7.

SMAC mimetics: SMAC mimetics (synthesized as in Vamos et al., 2013) were dissolved in DMSO and diluted to the indicated concentrations directly before incubating with caspase.

3.2.2 Procedure

Samples are mixed in 96-well plates and monitored in the same manner as caspase assays described above.

1. Purified recombinant caspases were used at 0.1 n*M* for caspase 3, 1 n*M* for caspase 7 and 2.2 μ*M* for caspase 9. Citrate is required for the assay of caspase 9 (see above).
2. (a) For derepression assays, mix BIR domains with relevant caspases (BIR2 for caspase 3 or 7, BIR3 for caspase 9) and incubate at room temperature for 30 min. (b) For competition assays omit this step.
3. Prepare a serial dilution of relevant SMAC mimetic. Combine serially diluted compound with caspase–BIR mixture, or for competition assays first mix compound and BIR domain, and initiate by adding this mixture to caspase.

4. Monitor the release of fluorescence from the fluorogenic substrates Ac-DEVD-AFC for caspases 3 and 7 or Ac-LEHD-AFC for caspase 9.
5. Obtain the ratio of inhibited to uninhibited caspase activity (V_i/V_o). These values should be interpreted as a proportion of the BIR free sample.

Note: Controls containing no BIR domain should be included to determine the full catalytic potential of the caspase preparation, as well as a control with BIR domain and 0.1% DMSO (or applicable compound-free buffer matched to compound solute) to show full potential inhibition.

4. CONCLUSIONS

While the diversity of cell death pathways is still coming into focus, caspases undoubtedly remain a central player. We hope that the methods described here provide useful guidance for those wishing to more fully explore the diverse roles of caspases in cell death, as well as other signalling pathways.

ACKNOWLEDGMENTS

Supported by NIH grants GM09040 and CA16374.

REFERENCES

Berger, A. B., Vitorino, P. M., & Bogyo, M. (2004). Activity-based protein profiling: Applications to biomarker discovery, in vivo imaging and drug discovery. *American Journal of Pharmacogenomics*, *4*, 371.

Boatright, K. M., Renatus, M., Scott, F. L., Sperandio, S., Shin, H., Pedersen, I. M., et al. (2003). A unified model for apical caspase activation. *Molecular Cell*, *11*, 529–541.

Bratton, S. B., & Salvesen, G. S. (2010). Regulation of the Apaf-1-caspase-9 apoptosome. *Journal of Cell Science*, *123*, 3209.

Collins, K. D., & Washabaugh, M. W. (1985). The Hofmeister effect and the behaviour of water at interfaces. *Quarterly Reviews of Biophysics*, *18*, 323.

Crawford, E. D., & Wells, J. A. (2011). Caspase substrates and cellular remodeling. *Annual Review of Biochemistry*, *80*, 1055.

Denault, J. B., & Salvesen, G. S. (2003a). Human caspase-7 activity and regulation by its N-terminal peptide. *The Journal of Biological Chemistry*, *278*, 34042.

Denault, J. B., & Salvesen, G. S. (2003b). Expression, purification, and characterization of caspases. *Current Protocols in Protein Science*, Chapter 21, Unit 21.13.

Donepudi, M., Mac Sweeney, A., Briand, C., & Gruetter, M. G. (2003). Insights into the regulatory mechanism for caspase-8 activation. *Molecular Cell*, *11*, 543.

Eckelman, B. P., & Salvesen, G. S. (2006). The human anti-apoptotic proteins cIAP1 and cIAP2 bind but do not inhibit caspases. *The Journal of Biological Chemistry*, *281*, 3254.

Eckelman, B. P., Salvesen, G. S., & Scott, F. L. (2006). Human inhibitor of apoptosis proteins: Why XIAP is the black sheep of the family. *EMBO Reports*, *7*, 988.

Faleiro, L., Kobayashi, R., Fearnhead, H., & Lazebnik, Y. (1997). Multiple species of CPP32 and Mch2 are the major active caspases present in apoptotic cells. *The EMBO Journal, 16*, 2271.

Flygare, J. A., Beresini, M., Budha, N., Chan, H., Chan, I. T., Cheeti, S., et al. (2012). Discovery of a potent small-molecule antagonist of inhibitor of apoptosis (IAP) proteins and clinical candidate for the treatment of cancer (GDC-0152). *Journal of Medicinal Chemistry, 55*, 4101.

Fulda, S., & Vucic, D. (2012). Targeting IAP proteins for therapeutic intervention in cancer. *Nature Reviews Drug Discovery, 11*, 109.

Gouvea, I. E., Judice, W. A., Cezari, M. H., Juliano, M. A., Juhász, T., Szeltner, Z., et al. (2006). Kosmotropic salt activation and substrate specificity of poliovirus protease 3C. *Biochemistry, 45*, 12083–12089.

Hachmann, J., Snipas, S. J., van Raam, B. J., Cancino, E. M., Houlihan, E. J., Poreba, M., et al. (2012). Mechanism and specificity of the human paracaspase MALT1. *The Biochemical Journal, 443*, 287–295.

Leverrier, S., Salvesen, G. S., & Walsh, C. M. (2011). Enzymatically active single chain caspase-8 maintains T-cell survival during clonal expansion. *Cell Death and Differentiation, 18*, 90.

Lopez, J., & Meier, P. (2010). To fight or die—Inhibitor of apoptosis proteins at the crossroad of innate immunity and death. *Current Opinion in Cell Biology, 22*, 872.

Mace, P. D., & Riedl, S. J. (2010). Molecular cell death platforms and assemblies. *Current Opinion in Cell Biology, 22*, 828.

Martinon, F., & Tschopp, J. (2007). Inflammatory caspases and inflammasomes: Master switches of inflammation. *Cell Death and Differentiation, 14*, 10.

McIlwain, D. R., Berger, T., & Mak, T. W. (2013). Caspase functions in cell death and disease. *Cold Spring Harbor Perspectives in Biology, 5*, a008656.

McStay, G. P., Salvesen, G. S., & Green, D. R. (2008). Overlapping cleavage motif selectivity of caspases: Implications for analysis of apoptotic pathways. *Cell Death and Differentiation, 15*, 322.

Muzio, M., Stockwell, B. R., Stennicke, H. R., Salvesen, G. S., & Dixit, V. M. (1998). An induced proximity model for caspase-8 activation. *The Journal of Biological Chemistry, 273*, 2926.

Oberst, A., Pop, C., Tremblay, A. G., Blais, V., Denault, J. B., Salvesen, G. S., et al. (2010). Inducible dimerization and inducible cleavage reveal a requirement for both processes in caspase-8 activation. *The Journal of Biological Chemistry, 285*, 16632–16642.

Pop, C., Fitzgerald, P., Green, D. R., & Salvesen, G. S. (2007). Role of proteolysis in caspase-8 activation and stabilization. *Biochemistry, 46*, 4398.

Pop, C., & Salvesen, G. S. (2009). Human caspases: Activation, specificity, and regulation. *The Journal of Biological Chemistry, 284*, 21777.

Pop, C., Timmer, J., Sperandio, S., & Salvesen, G. S. (2006). The apoptosome activates caspase-9 by dimerization. *Molecular Cell, 22*, 269.

Renatus, M., Stennicke, H. R., Scott, F. L., Liddington, R. C., & Salvesen, G. S. (2001). Dimer formation drives the activation of the cell death protease caspase 9. *Proceedings of the National Academy of Sciences of the United States of America, 98*, 14250.

Salvesen, G. S., & Duckett, C. S. (2002). IAP proteins: Blocking the road to death's door. *Nature Reviews Molecular Cell Biology, 3*, 401.

Schmidt, U., & Darke, P. L. (1997). Dimerization and activation of the herpes simplex virus type 1 protease. *The Journal of Biological Chemistry, 272*, 7732.

Scott, F. L., Denault, J. B., Riedl, S. J., Shin, H., Renatus, M., & Salvesen, G. S. (2005). XIAP inhibits caspase-3 and −7 using two binding sites: Evolutionarily conserved mechanism of IAPs. *The EMBO Journal, 24*, 645–655.

Shiozaki, E. N., Chai, J., Rigotti, D. J., Riedl, S. J., Li, P., Srinivasula, S. M., et al. (2003). Mechanism of XIAP-mediated inhibition of caspase-9. *Molecular Cell*, *11*, 519–527.

Srinivasula, S. M., Hegde, R., Saleh, A., Datta, P., Shiozaki, E., Chai, J., et al. (2001). A conserved XIAP-interaction motif in caspase-9 and Smac/DIABLO regulates caspase activity and apoptosis. *Nature*, *410*, 112–116.

Stennicke, H. R., & Salvesen, G. S. (1999). Caspases: Preparation and characterization. *Methods*, *17*, 313.

Stennicke, H. R., & Salvesen, G. S. (2000). Caspase assays. *Methods in Enzymology*, *322*, 91.

Tinel, A., & Tschopp, J. (2004). The PIDDosome, a protein complex implicated in activation of caspase-2 in response to genotoxic stress. *Science*, *304*, 843.

Tu, S., McStay, G. P., Boucher, L. M., Mak, T., Beere, H. M., & Green, D. R. (2006). In situ trapping of activated initiator caspases reveals a role for caspase-2 in heat shock-induced apoptosis. *Nature Cell Biology*, *8*, 72–77.

Vamos, M., Welsh, K., Finlay, D., Lee, P. S., Mace, P. D., Snipas, S. J., et al. (2013). Expedient synthesis of highly potent antagonists of inhibitor of apoptosis proteins (IAPs) with unique selectivity for ML-IAP. *ACS Chemical Biology*, *8*, 725–732.

Vaux, D. L., & Silke, J. (2003). Mammalian mitochondrial IAP binding proteins. *Biochemical and Biophysical Research Communications*, *304*, 499.

Wachmann, K., Pop, C., van Raam, B. J., Drag, M., Mace, P. D., Snipas, S. J., et al. (2010). Activation and specificity of human caspase-10. *Biochemistry*, *49*, 8307–8315.

Yuan, S., & Akey, C. W. (2013). Apoptosome structure, assembly, and procaspase activation. *Structure*, *21*, 501.

Zhuang, M., Guan, S., Wang, H., Burlingame, A. L., & Wells, J. A. (2013). Substrates of IAP ubiquitin ligases identified with a designed orthogonal E3 ligase, the NEDDylator. *Molecular Cell*, *49*, 273.

CHAPTER EIGHT

Turning ON Caspases with Genetics and Small Molecules

Charles W. Morgan[*,†,1], **Olivier Julien**[*,1], **Elizabeth K. Unger**[‡,§,1], **Nirao M. Shah**[‡,2], **James A. Wells**[*,¶,2]

[*]Department of Pharmaceutical Chemistry, University of California, San Francisco, California, USA
[†]Graduate Group in Chemistry and Chemical Biology, University of California, San Francisco, California, USA
[‡]Department of Anatomy, University of California, San Francisco, California, USA
[§]Program in Biomedical Sciences, University of California, San Francisco, California, USA
[¶]Department of Cellular and Molecular Pharmacology, University of California, San Francisco, California, USA
[1]These authors contributed equally to this work.
[2]Corresponding authors: e-mail address: nms@ucsf.edu; jim.wells@ucsf.edu

Contents

Methods in Enzymology, Volume 544
ISSN 0076-6879
http://dx.doi.org/10.1016/B978-0-12-417158-9.00008-X

Abstract

Caspases, aspartate-specific cysteine proteases, have fate-determining roles in many cellular processes including apoptosis, differentiation, neuronal remodeling, and inflammation (for review, see Yuan & Kroemer, 2010). There are a dozen caspases in humans alone, yet their individual contributions toward these phenotypes are not well understood. Thus, there has been considerable interest in activating individual caspases or using their activity to drive these processes in cells and animals. We envision that such experimental control of caspase activity can not only afford novel insights into fundamental biological problems but may also enable new models for disease and suggest possible routes to therapeutic intervention. In particular, localized, genetic, and small-molecule-controlled caspase activation has the potential to target the desired cell type in a tissue. Suppression of caspase activation is one of the hallmarks of cancer and thus there has been significant enthusiasm for generating selective small-molecule activators that could bypass upstream mutational events that prevent apoptosis. Here, we provide a practical guide that investigators have devised, using genetics or small molecules, to activate specific caspases in cells or animals. Additionally, we show genetically controlled activation of an executioner caspase to target the function of a defined group of neurons in the adult mammalian brain.

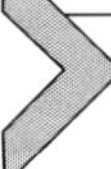

1. USE OF CHEMICAL-INDUCED DIMERIZERS TO ACTIVATE CASPASES

1.1. Controlling protein–protein interactions leads to selective activation of caspases

The elucidation of cell signaling has been empowered by chemical genetic approaches. In particular, protein-fragment complementation assays and chemical-induced dimerization (CID) have been very powerful approaches (Fig. 8.1A) (Fegan, White, Carlson, & Wagner, 2010; Remy & Michnick, 2007; Spencer, Wandless, Schreiber, & Crabtree, 1993). While the ability to control transcription-based circuits has been known for decades, only recently have signaling biologists learned to harness and engineer orthogonal circuits to selectively activate proteases. Proteolysis is an irreversible posttranslational-modification that directs the fate of cells in apoptosis, protein degradation, blood coagulation, and many other processes. Here, we discuss practical approaches that we and others have recently employed to selectively activate caspases using CID strategies.

There are two classes of apoptotic caspases: initiators (caspase-8, -9, -10) which are activated by oligomerization, and the primary executioners (caspase-3 and -7) that are activated by proteolysis mediated by upstream

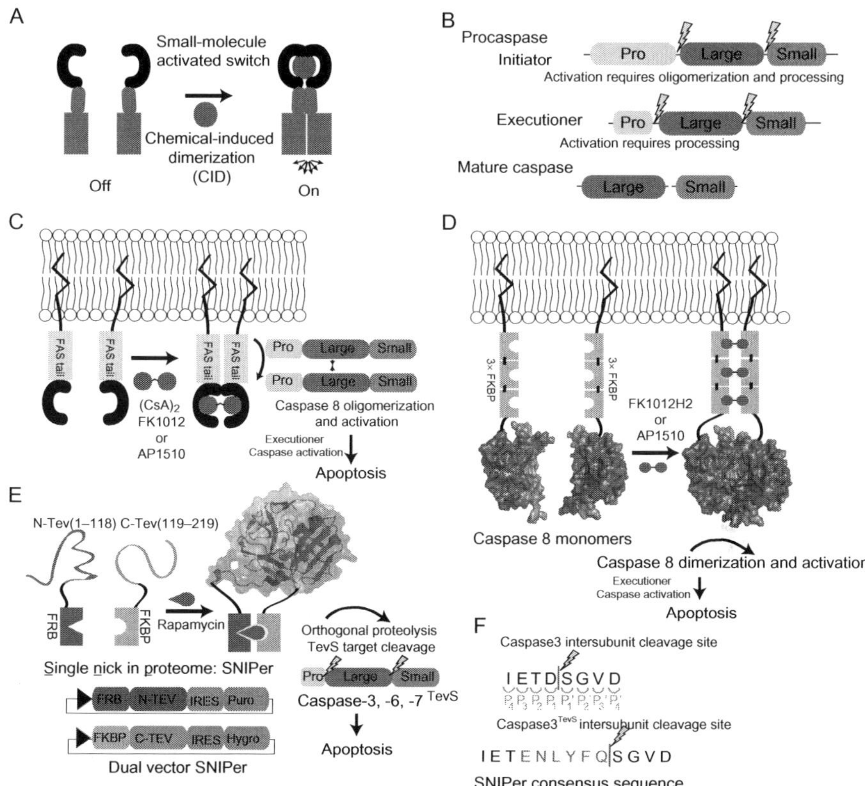

Figure 8.1 Schematic overview of chemical genetic strategies for selective caspase activation. (A) Chemical-induced dimerization (CID) utilizes dimerization domains, dependent on cell-permeable small molecules, to control the proximity of signal transduction domains (blue). The chemical genetic strategy of CID creates an orthogonal circuit in the cell, thereby enabling the control of the ON/OFF state of signal transduction. (B) Procaspases are composed of three domains; a prodomain, a large subunit, and a small subunit. Initiators have a long prodomain and require both oligomerization and processing for activation. The executioners are predimerized and they have a short prodomain and proteolysis at the large–small domain junction is sufficient for activation. (C) Utilizing a CID-based approach, FAS tails are dimerized by the addition of the symmetric small molecule, AP1510 or FK1012, thereby reconstituting the caspase-8 (PDB ID 1QTN) activation scaffold and ultimately resulting in apoptosis. (D) Direct caspase-8 activation via CID. (E) The engineered split-tobacco etch viral (TEV) variant, single nick in proteome (SNIPer), used for inducible and selective cleavage of the executioner caspase isoforms (TEV structure PDB ID 1LVM). Each SNIPer half is expressed from a single plasmid that coexpresses an IRES-driven drug-resistance marker for stable cell line engineering. The addition of rapamycin causes the heterodimerization of FKBP and FRB, thereby rescuing TEV protease activity. (F) The TEV protease consensus sequence (green) is inserted into the caspase-processing site to generate a TevS allele, susceptible to SNIPer-mediated proteolysis. (See the color plate.)

proteases (initiator caspases or granzyme B) (Fig. 8.1B). The purpose of this first section is to (1) briefly review previous approaches and research highlights that employed chemical genetic strategies used to activate caspases, primarily caspase-8, (2) summarize the progress that our laboratories have made in selectively turning on apoptosis via orthogonal executioner caspase activation, and (3) provide a detailed method for selective activation of executioner caspases employing this chemical genetic strategy.

1.2. Oligomerization strategies for selective caspase activation

1.2.1 Selective caspase-8 oligomerization and activation

The first target for CID activation of caspases was caspase-8, a classic initiator caspase involved in the extrinsic apoptosis pathway (Muzio, Stockwell, Stennicke, Salvesen, & Dixit, 1998; Yang, Chang, & Baltimore, 1998). CID activation of caspase-8 facilitated elegant experimentation of the induced-proximity hypothesis (Salvesen & Dixit, 1999), which is that the monomeric proenzyme is activated through dimerization. Previous studies have shown that heterologous expression of the inactive zymogens retained modest catalytic activity prior to their proteolytic processing. Initial tests of the caspase-8 cellular activation mechanism were described in two simultaneous papers that demonstrated the role of receptor oligomerization in caspase-8 activation (Belshaw, Spencer, Crabtree, & Schreiber, 1996; Spencer et al., 1996). This work showed that it was possible to activate procaspase-8 through CID of the intracellular FAS receptors using either a homodimerization analog of FK506, FK1012, a cyclosporin derivative $(CsA)_2$ (Fig. 8.1C). A year later, the symmetric small molecule, AP1510, was used to oligomerize FAS tails (Amara et al., 1997). The CID approach was then applied directly to caspase-8. Competing caspase-8 chimeras, containing trimeric homodimerization domains demonstrated that oligomerization was sufficient for robust caspase-8 activation (Muzio et al., 1998; Yang et al., 1998) (Fig. 8.1D). Other investigators showed that conditional dimerization of known monomeric initiator caspases-8, -9, or 10 could result in activation yet the predimerized executioner caspase-3 was not activated by oligomerization (Chen, Orozco, Spencer, & Wang, 2002). CIDs have now been applied to several caspases and together constitute a collection of artificial death switches (Steller, 1998).

Further work showed that during activation, procaspase-8 both dimerizes and undergoes proteolysis (Chang, Xing, Capacio, Peter, & Yang, 2003; Hughes et al., 2009; Kang et al., 2008; Murphy, Creagh, & Martin, 2004; Pop, Fitzgerald, Green, & Salvesen, 2007; Sohn, Schulze-Osthoff, &

Jänicke, 2005). Utilizing the unique specificity of tobacco etch viral (TEV) protease in mammalian cells and the previously described CID approaches for oligomerization, both methods of activation have been beautifully tested simultaneously (Oberst et al., 2010). Oberst et al. showed that dimerization alone in the absence of proteolysis of the intersubunit linker or prodomain failed to lead to activation, as detected by IETDase activity. Similarly, selective cleavage by TEV protease, in mutated linker regions with the TEV consensus sequences failed to activate the caspase-8. Only the combination of induced dimerization and proteolytic processing resulted in full activation of caspase-8. These experiments highlight the utility of engineered inducible and orthogonal activation strategies for the caspases, thereby permitting a more complete understanding of their function and regulation.

1.2.2 In vivo *strategies for selective caspase activation*

Inducible caspase activation has been utilized in several *in vivo* strategies that include conditional induction of apoptosis during development in *C. elegans*, mouse adipocytes, and a caspase-9-based artificial death switch for T-cell therapies. In order to reconstitute Ced-3/caspase-3 activity in *C. elegans* during development, individual subunits of caspase-3 chimeras were fused to antiparallel leucine-zippers and expressed by separate promoters, and selective apoptosis was observed only when promoters overlapped in cells (Chelur & Chalfie, 2007). Pajvani and colleagues demonstrated the power in reversible CID-based approaches in mice, upon removal of the dimerizer, thereby leading to the inactivation of the caspase-8 death switch in adipocytes; mice regained weight lost during treatment (Pajvani et al., 2005). Cell-based treatments for cancer and regenerative-based medicine hold promise, yet in the event of an adverse effect, elimination of the engineered cells would be critical. Toward this end, several artificial death switches, utilizing similar CID strategies discussed above, have demonstrated their usefulness in selectively killing engineered T-cells, demonstrated in mice and later in human patients (Di Stasi et al., 2011; Straathof et al., 2005).

1.3. Design of orthogonal and selective proteolysis for individual caspase activation

1.3.1 Inducible and selective proteolysis

We have recently shown one can selectively activate the executioner caspases through an orthogonal and selective CID approach of an initiating protease. This utilized a reengineered version of the split-TEV protease (Wehr et al., 2006), we call the SNIPer (single nick in proteome). The

SNIPer was optimized by truncating the C-terminus of TEV and swapping the order of the heterodimerization domains of FKBP/FRB (Gray, Mahrus, & Wells, 2010). This new design greatly reduced background of TEV activity caused by small amounts of constitutive TEV dimer formation. The SNIPer leads to the rescue of the TEV activity by the addition of the cell-permeable small-molecule, rapamycin, which was shown using a genetically encoded TEV FRET reporter (Gray et al., 2010). Selective and exclusive proteolysis of each target caspase by TEV is achieved by inserting the TEV consensus cleavage motif into the endogenous caspase-processing sites (Fig. 8.1E). Once each caspase is activated, it is able to process endogenous targets, including wild-type caspases, but cannot activate the TEV-sensitive caspase alleles (caspase^{TevS}) since they are missing the caspase consensus site.

TEV cleavage sequences were introduced into the two major processing sites in caspases-3, -6, and -7. Selective cleavage of the site between the prodomain and the large subunit did not lead to activation, but all three caspases were activated upon SNIPer-mediated proteolysis between the large and small subunit. When these alleles were stably transfected into human 293 cells, only activation of caspases-3 and -7 lead to apoptosis, whereas caspase-6 did not, therefore suggesting it is not a typical executioner. The executioner caspases were shown to be ubiquitinated and degraded by the ubiquitin proteasome system after activation, and proteasome inhibitors greatly accelerate SNIPer-mediated caspase^{TevS}-driven apoptosis. Interestingly, caspase-6^{TevS} activation by SNIPer is lethal in the presence of proteasome inhibitors.

1.3.2 Generating SNIPer substrates: TEVs alleles

In order to generate SNIPer-cleavable substrates, we employ an insertion strategy using standard molecular biology techniques. Briefly, we replace the P1 residue of the endogenous cleavage site with the six residues required for TEV recognition and cleavage (ENLYFQ↓). This insertion yields an orthogonal substrate (i.e., no longer cleavable by caspases) because the substrate lacks the required aspartic residue followed by a small amino acid (G/S/A). We also verify the P1′ residue as TEV prefers G/S at P1′ and depending on the endogenous sequence, this may also need to be mutated. A variety of PCR techniques can be used to insert the six residues required for TEV cleavage, such as standard overlap extension PCR or kunkel mutagenesis for simultaneous insertions (Tonikian, Zhang, Boone, & Sidhu, 2007). We always test our constructs using transient transfections prior to making stable cell lines.

1.4. Cell engineering

Direct caspase dimerization or orthogonal upstream protease activation both require substantial cellular engineering. The requirements for each strategy vary and each has its own set of advantages and disadvantages. Here, we discuss in detail our approach to cellular engineering stable cell lines that express the SNIPer and subsequent screening for caspase^{TevS}-sensitive proteolysis. It should be noted that since the SNIPer is a highly selective and orthogonal mechanism capable of inducing caspase processing, the target substrate does not need to be a caspase but the processing site needs to be accessible to TEV protease. Our first approach consisted of engineering cells to constitutively express the SNIPer using traditional MMLV retrovirus transduction (Ory, Neugeboren, & Mulligan, 1996). However, we recently have had better success using a lentiviral-based approach that enables for a wider range of engineered target cell lines (Cooray, Howe, & Thrasher, 2012).

1.4.1 Protocol for SNIPer stable cell line engineering (inducible and selective proteolysis)

Our strategy for stable expression of both halves of SNIPer has been described previously (Gray et al., 2010). The purpose of this section is to highlight some specific details, technical challenges, and potential solutions that can be anticipated during cellular engineering with the SNIPer (Fig. 8.2).

1. Culture HEK293/GPG cells in DMEM supplemented with 10% FBS, nonessential amino acids, sodium pyruvate, L-glutamine, penicillin–streptomycin, 300 μg/mL G418, 2 μg/mL puromycin, and 1 μg/mL doxycycline.
2. A modified protocol for lipofectamine2000 (Life Technologies) is used to transfect GPG cells. We find robust transfection efficiency using only ¼ of the recommended amount of lipid mixture and that the decreased amount also minimizes cellular toxicity. We co-transfect both N- and C-terminal SNIPer plasmids together to increase successful generation of double-positive clones. The media is replaced after 5 h with fresh DMEM without doxycycline and antibiotics. Transfection efficiency is visually inspected at 24 h posttransfection by examining a control well that has been transfected in parallel with a fluorescent protein (EGFP or mCherry). We generally observe >80% transfection efficiency.
3. Viral-containing supernatant is collected at 48 and 72 h posttransfection.

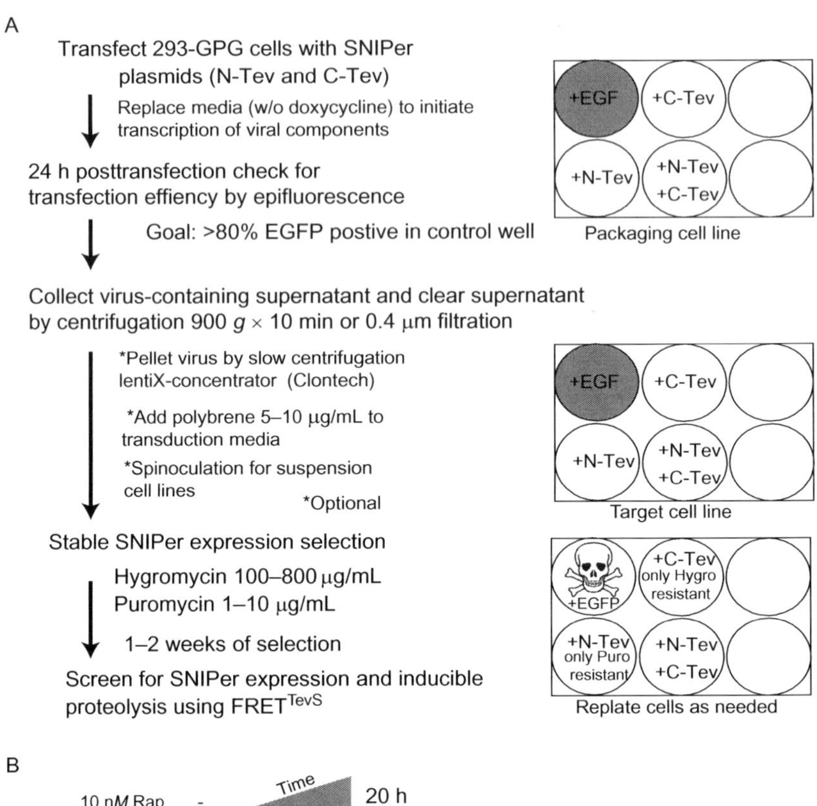

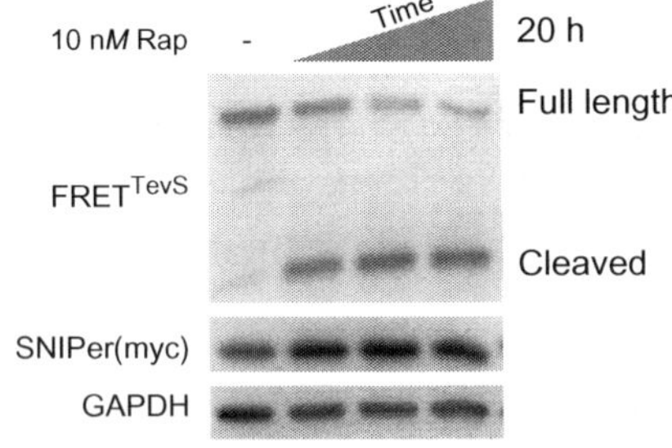

Figure 8.2 Cell engineering overview. (A) Summary of the protocol used for virus production and stable cell line engineering of SNIPer expressing cells. (B) A representative Western immunoblot of SNIPer engineered cells, transfected with $FRET^{TevS}$ reporter, shows selective proteolysis of the reporter over time.

4. The viral supernatant is clarified of cellular debris by either centrifugation at $900 \times g$ for 10 min or passage through a SFCA 0.4 μm filter. Optional: virus concentration by Clontech Lenti-X concentrator.
5. Cleared supernatant is added to previously plated adherent target cell lines (at 60–70% confluency) and incubated overnight. Polybrene

increases effective viral titers by neutralizing the charge on the cell surface. Appropriate concentrations of polybrene should be screened on each new cell types. Typically, we use 8 μg/mL for HEK293 and HeLa cells. Standard spinoculation protocols may be employed when targeting suspension cell lines.

6. We start selection 24 h post target cell transduction. At this point, puromycin and hygromycin are simultaneously added to select for double-positive cells, those that have integrated both halves of the SNIPer. Each half of the SNIPer also coexpresses via internal ribosomal entry site (IRES)-dependent translation of an antibiotic resistance marker (Fig. 8.1E). Dual selection usually takes ~14 days and cells are replated as necessary.
7. Tips on successful selections
   a. Adjusting the growing density of the cells aids in the successful recovery of cells difficult to transduce. Drug selection is also very dependent upon cell density and rate of cell division. We always use a nontransduced cell line treated with selection antibiotics to ensure that all nontransduced cells are eliminated.
   b. In most cases, heterogeneous pools of SNIPer positive cells are sufficient for further study and transfection with caspase^{TevS} isoforms. The SNIPer positive cell line can be used as a parental cell line for further transduction with caspase^{TevS} isoforms.
   c. Typical ranges for selections when using puromycin and hygromycin are 1–10 μg/mL and 100–800 μg/mL, respectively. The effective concentration is cell-type specific. To ensure proper selection, we perform serial selections, first with puromycin since it is fast acting and then with the slower acting hygromycin. When working with a new cell line, we utilize fluorescent proteins with puromycin and hygromycin resistance to get immediate and observable feedback on transduction efficiency and progression of selection.

1.4.2 Assessing SNIPer-mediated proteolysis and rapamycin side effects

In addition to binding the SNIPer, rapamycin also binds to endogenous FK506-binding protein (FKBP) forming a complex that binds the FRB domain of mTOR and results in mTOR inhibition. We, and others, have shown that this problem is reduced by minimizing the dose of rapamycin (typically 5–10 n*M* needed to activate the SNIPer) and in combination the overexpression of the FKBP and FRB from the SNIPer, acts as a buffer to compete with endogenous FK506 binding. Nonetheless, we recommend measuring the level of mTOR inhibition by monitoring the phosphorylation state of the mTOR substrate S6 kinase via Western immunoblotting. We typically monitor this using antibodies for p70 S6 kinase and phoso-p70 S6 kinase (Cell Signaling Technology, #9202 and #9205). One should expect levels of mTOR inhibition in the presence of SNIPer to vary depending on the particular cell line used and the SNIPer expression level.

As a first screen of SNIPer function on a TevS target substrate, it is convenient to test it in transiently transfected cells. We typically use a molar ratio of 6:1 SNIPer halves to substrate^{TevS}. This ratio yields the most consistent results for substrate cleavage. Once substrate cleavage is observed to be rapamycin dependent, then one should move to stably transfected cells to ensure more homogeneity in the cell population.

Western immunoblotting is our standard technique for tracking cleavage of engineered TEV cleavage site containing-substrates and expression of the SNIPer (each half is N-terminally myc tagged). It should be noted that FRB fusions are regularly observed to be less stable than FKBP fusions (Edwards & Wandless, 2007). The kinetics of cleavage by the SNIPer varies depending on the substrate being studied and the readout. We initially test for substrate cleavage following 8 h of rapamycin treatment, but we observe rapamycin-dependent cleavage in as little as 1 h after its addition with certain substrates.

1.4.3 Considerations when employing SNIPer-mediated proteolysis

When designing a SNIPer-mediated research project, it is important to be aware of some of its limitations. As is common with any protein-engineering strategy that introduces conditional control of an enzyme, background activity, for example, TEV protease activity in the absence of rapamycin, presents a challenge and needs to be thoroughly considered when looking at the overall signaling pathway. An additional consideration is the partial loss of catalytic activity of engineered variants. It has been shown that split-TEV constructs suffer from both constitutive background and low maximal activity level (Wehr et al., 2006; Williams, Puhl, & Ikeda, 2009). The SNIPer was designed to minimize background proteolysis as this was essential for selective caspase^{TevS} activation. For other substrates, where complete cleavage of the substrate is necessary, there are additional challenges with the SNIPer. First, the endogenous substrate will compete with the function of the engineered TevS allele, thereby masking the functional consequence of substrate^{TevS} cleavage. Second, the SNIPer has only about ~30% the native activity of intact full-length TEV. In some cases, the transcriptional expression rate of the full-length substrate may exceed the catalytic rate of the SNIPer. If complete cleavage of the substrate is required, then other approaches may need to be considered.

1.4.4 Next-generation SNIPer-based stable cell line creation (lentiviral-based engineering and single-vector versions)

The adaptation of the SNIPer to a lentiviral-based backbone enables the creation of a wider range of cell lines with higher titers, including nondividing cells and cell lines more resistant to traditional retroviral delivery of transgenes. A single-vector version with bicistronic expression of the SNIPer has been constructed and varying iterations are currently being tested for robust and reliable inducible proteolysis. We will make these available upon request once validation is completed (C. W. Morgan & J. A. Wells unpublished).

1.5. Monitoring cell death phenotypes

Following the successful observation of rapamycin-induced cleavage of the targeted protein in cells using Western immunoblotting, the phenotype of this cleavage event can be characterized. This volume offers many detailed protocols, and in addition, Galluzzi et al. reviewed the latest techniques for monitoring cell death (Galluzzi et al., 2009). Assay selection should be matched with the specific cell line and available equipment for characterization. Often, the most convenient method for apoptosis is to measure cell viability following activation by standard methods such as Cell Titer Glo (Promega), MTT assay, or propidium iodide staining. Depending on the cell type, concentration of rapamycin, expression levels of the split-TEV halves, caspase^{TevS} allele, the time to death, and homogeneity of the cells will need to be screened. Generally, with caspase-3^{TevS} we initially observe a decrease in viability starting at 3 h following activation. Once an initial dose response and corresponding time course have been completed, more detailed characterization of the engineered cell line is suggested using TUNEL and Annexin V staining, DNA laddering, and caspase-3 and PARP cleavage. Together, these phenotypic observations constitute hallmarks of apoptosis.

1.6. Conclusions and future questions

Harnessing CID strategies has expanded our ability to test different activation modes for apoptotic caspases, oligomerization, and processing. Previous research on selective activation on caspases has given us crucial insight into the process that cells use to dismantle themselves during apoptosis. The chemical genetic tools outlined above give researchers an unprecedented tool to selectively induce apoptosis in a controlled circuit that conveniently feeds into the endogenous apoptotic pathway. Further engineering for inducible and selective activation needs to be undertaken for the inflammatory caspases-1, -4, -5, -11, and the remaining apoptotic caspase-2. The ability to specifically control caspase signal transduction through selective activation could play important roles in future cell-based therapies in the form of death switches. The conditional control of caspase activation will certainly be important for basic science research, as we uncover more details about the role of proteolysis in controlling cell fate decisions and the cellular mechanisms that antagonizes caspase activation. The chemical genetic strategies outlined above enable researchers highly selective control to turn on caspase-mediated signaling pathways.

2. USE OF Cre-LoxP AND A SELF-ACTIVATING CASPASE-3^{TevS} FOR CONDITIONAL APOPTOSIS OF NEURONS

2.1. Introduction

Unlike the cells in most other tissues, neurons that are born during development are primarily the same neurons that will make up the nervous system for the remainder of an organism's life. Neurons are postmitotic cells and do not readily regenerate, meaning that any lesions incurred during an organism's life will be permanent. Moreover, neurons are remarkably heterogeneous, anatomically, molecularly, and functionally. In order to meaningfully study the function of such diverse neuronal cell types in anatomically distinct regions, we need a means of performing targeted lesions and other manipulations to assess their function. We have utilized the caspase-3^{TevS} to specifically ablate progesterone receptor (PR)-expressing neurons in the ventrolateral region of the ventromedial hypothalamus (VMHvl) in mice, and showed that this region is critical for sexual behavior in both sexes and aggression in males.

2.2. Viral-mediated neuronal ablation

We have encoded the caspase-3^{TevS} described in Section 1, and a full-length TEV protease, in a single adeno-associated virus (referred to as AAV-flex-taCasp3-TEVp) such that expression of both transgenes and enzymatic activation of caspase-3^{TevS} is dependent on Cre recombinase. In this review, we refer to this virus as AAV-flex-C3-Tp. Cre-loxP is a bipartite genetic system that allows for reproducible, inducible targeting of any molecularly defined cell type. We injected this virus into the VMHvl of PR-IRES-Cre mice and

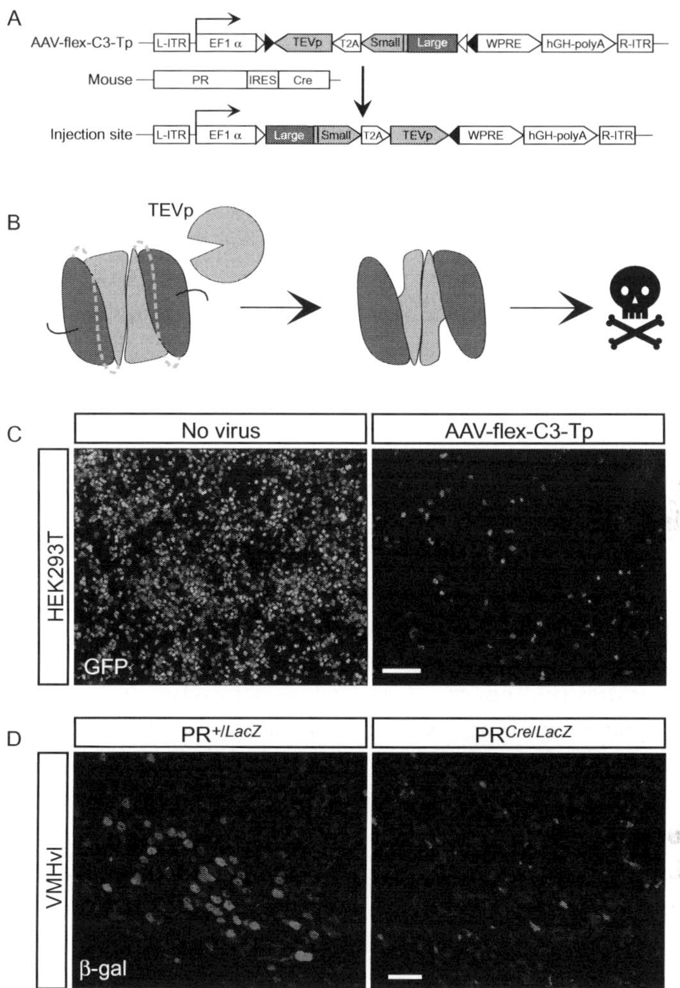

Figure 8.3 Utilizing AAV-flex-C3-Tp. (A) Strategy to ablate PR-expressing cells *in vivo*. (B) Within Cre+ cells, the TEV protease will cleave its consensus recognition sequence (dashed lines) between the large and small subunits of the caspase-3 dimer, activating it and leading to apoptosis. (C) The AAV-flex-C3-Tp encoding plasmid was transfected into HEK293T cells expressing the fusion protein Cre-GFP. Cell death was evaluated after 1 week. (D) AAV-flex-C3-Tp was stereotaxically targeted to the ventrolateral region of the VMH (VMHvl) of adult $PR^{+/LacZ}$ and $PR^{Cre/LacZ}$ mice which harbor the transgenes nuclear LacZ (PR^{LacZ} allele) or Cre recombinase (PR^{Cre} allele) inserted into the PR locus by homologous recombination. Cell death was evaluated after 4 weeks. Scale bars represent 100 μm (C) and 25 μm (D). *(C) and (D) are reproduced from Yang et al. (2013).* (See the color plate.)

were able to ablate, on average, ≥95% of PR neurons while leaving neighboring Cre-nonexpressing cells intact (Fig. 8.3) (Yang et al., 2013). This system is highly modular and applicable not just to PR neurons, but, in principle, any molecularly distinct neuronal population or other Cre-expressing cell type in the body. Indeed, we have successfully used this virus to ablate other distinct neuronal populations engineered to express Cre recombinase *in vivo* (E. K. Unger & N. M. Shah unpublished observations).

2.2.1 Self-activating caspase-3^{TevS} transgene

Caspase-3 was modified to encode the TEV protease consensus recognition sequence, caspase-3^{TevS}, as described above. Our virus coexpresses caspase-3^{TevS} and TEV protease linked by a T2A sequence. T2A is a short, self-cleaving peptide that allows multiple proteins to be expressed from the same transgene (Provost, Rhee, & Leach, 2007). T2A has been shown to be more efficient than an IRES for the expression of multiple products and is effective in neurons (Tang et al., 2009). It is also smaller and thus easier to package within the AAV genome.

2.2.2 Cre-loxP

Our strategy uses the bipartite Cre-loxP system wherein Cre recombinase recognizes and recombines two short palindromic loxP sites eliminating the intervening sequence (Orban, Chui, & Marth, 1992; Sauer, 1987; Sauer & Henderson, 1988). In our strategy, Cre has been knocked-in to the PR locus to generate the PR-IRES-Cre mouse strain, such that expression of Cre mirrors the endogenous expression of PR (Yang et al., 2013). The virally encoded caspase-3^{TevS} and TEV protease are flanked by loxP sites such that in PR-expressing cells, both proteins will be expressed and induce apoptosis. Although AAV can infect cells indiscriminately, only Cre-expressing cells will switch on functional expression of the encoded transgene, thereby enabling genetically targeted ablation of the desired cell type exclusively without bystander toxicity to Cre-nonexpressing cells.

2.2.3 Flex

Caspase-3 is an executioner caspase, and leaky expression of the enzyme even at low levels can elicit apoptosis. We therefore wished to restrict caspase-3 expression exclusively to cells expressing Cre recombinase. Accordingly, we employed a strategy (flex switch) that utilizes flanking heterologous loxP sites to encode the desired transgene, *caspase-3^{TevS}-T2A-TEV protease* in our case, in an orientation reversed to that of the promoter driving expression of the transgene (Atasoy, Aponte, Su, & Sternson, 2008; Schnütgen et al., 2003; Sohal, Zhang, Yizhar, & Deisseroth, 2009). This strategy ensures that the transgene cannot be expressed except in the presence of Cre recombinase, whose activity irreversibly switches the transgene on to the same strand as the promoter. In our case, the flex switch restricts coexpression of caspase-3^{TevS} and TEV protease to Cre-expressing cells.

2.2.4 AAV

We chose to encode the *caspase-3^{TevS}-T2A-TEV protease* transgene in an AAV because it is the best option for our *in vivo* application. First, it is not a retrovirus and thus only requires BSL-1. Second, it has extremely low immunogenicity so there is less worry of an immune response. Third, it has a very broad tropism, which can be made still broader with different serotypes—there are currently 11 in use today with "preferences" for different cell types (Taymans et al., 2007). Fourth, it can be produced at very high titer (10^{12}–10^{13}), and there are several commercial facilities dedicated to producing high-titer AAV for a nominal fee. A potential drawback of using AAV is that its genome is small and it can only accept 4–5 kb DNA inserts. However, this was not an issue for our studies because our transgenic insert is within this range.

An alternative approach to delivering the *caspase-3^{TevS}-T2A-TEV protease* transgene would be to generate a mouse strain harboring this transgene in a Cre-dependent manner. However, promoter sequences that drive transgene expression in a spatiotemporally appropriate manner in the adult VMHvl remain to be defined, and our viral approach circumvents this caveat. Our approach does necessitate, however, that the virus must be injected manually into every experimental animal. Viruses provide the added convenience of being much faster and less resource intensive to generate, produce, and maintain than mouse lines.

2.3. Protocol: Cloning the *caspase-3^{TevS}-T2A-TEVp* construct and AAV plasmid expression

1. Assembling the caspase-3^{TevS}-T2A-TEVp construct

   The *caspase-3^{TevS}* allele described above (see Section 1.3) was stitched to a T2A-TEVp sequence using overlapping PCR. The resulting *caspase-3^{TevS}-T2A-TEVp* transgene was cloned in an orientation reverse to that of the ubiquitous promoter, EF1α, into the pAAV-EF1α-DIO-WPRE-pA plasmid (Sohal et al., 2009). This plasmid is currently available from Addgene.

2. Tips for growing AAV plasmids
   a. Subcloning in AAV vectors can be difficult because the two encoded inverted terminal repeats (ITRs) often render the plasmid unstable. It is therefore desirable to perform as many subcloning steps as possible in a standard plasmid lacking ITRs prior to moving the transgene into the AAV vector.
   b. When transforming, be sure to use high-competency cells that have been specifically designed for unstable sequences. We use MAX Efficiency® Stbl2™ high-competency cells.
   c. Plate these cells on standard LB + ampicillin (amp; 100 μg/mL) plates and grow 16–20 h at 30 °C.
   d. Pick a colony for a starter culture and grow it for 8–10 h at 30 °C in LB with amp (75 μg/mL). It is best to pick several colonies from which to make starter cultures and then expand only the best growing one for the final large-scale maxi-prep culture.
   e. The large-scale culture should be grown in 1:1 LB:2×YT with amp (75 μg/mL) at 30 °C for 16–20 h. Because these cells are grown at 30 °C rather than 37 °C, they will expand much more slowly. These are also low-copy number plasmids meaning that the yield will be quite low. It may be better to split the starter culture into several larger cultures rather than simply growing a single culture for a longer period of time.
   f. We use the Endo-Free Plasmid Maxi Kit from Qiagen to purify the plasmid DNA. In our hands, the yield from these columns is ~300 μg/L culture.
   g. Diagnostic digests of the purified plasmid DNA are essential to exclude recombination events triggered by the ITRs. We digested the AAV-flex-C3-Tp plasmid with eight different enzymes to ensure each element was present and of the expected size (Fig. 8.4A). Some viral cores even require specific digests before they will initiate virus production from the plasmid DNA.
   h. Verifying the sequence of the plasmid is also very important, including all subcloning boundaries and the ITRs. Because of the repeats, sequencing will often fail just inside the ITRs, but it is important to verify that they are present and that the flanking sequences are intact. We sequenced the AAV-flex-C3-Tp plasmid with nine different primers (Table 8.1) that each covered 700–900 bp in an overlapping manner to give the entire sequence of the insert and ITRs.

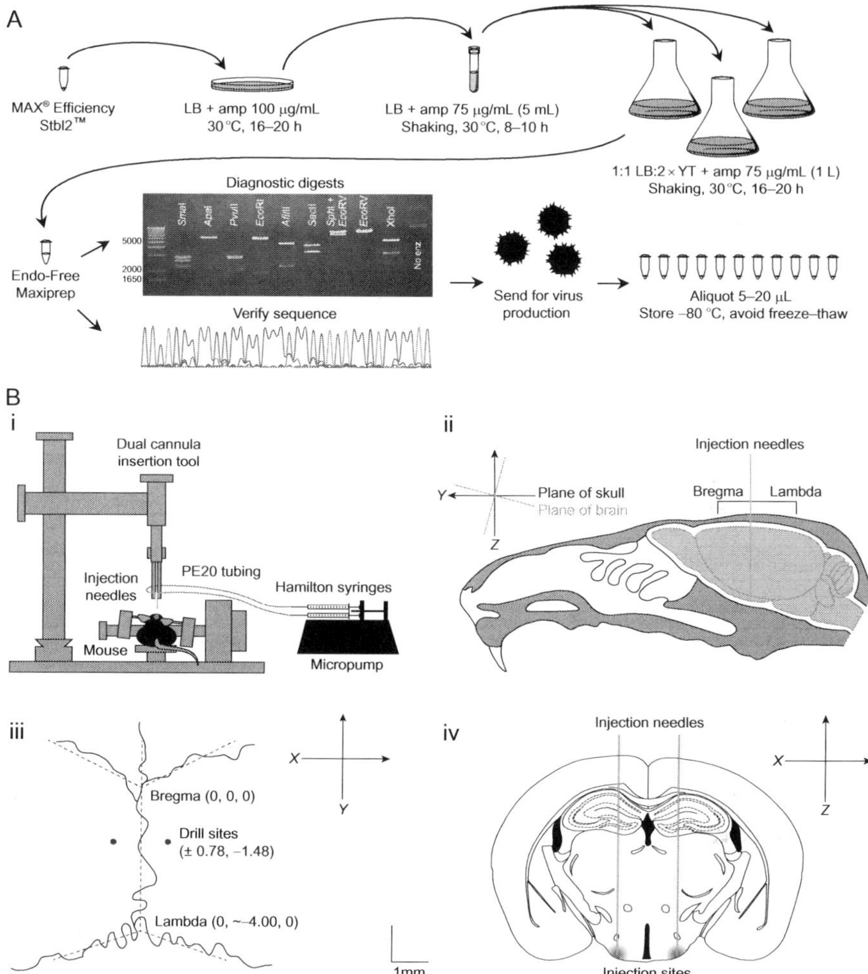

Figure 8.4 Overview of AAV virus production and injection. (A). Summary of the protocol used for AAV plasmid growth and verification for virus production. Maxipreps originating from the same colony were pooled for diagnostics. (B) Schematic of virus injection. (i) Stereotax and micropump setup: injection needles are connected to Hamilton syringes by PE20 tubing. The syringes are depressed steadily and simultaneously by a micropump. (ii) Orientation of the brain within the cranial cavity. Note the 15° slant of the skull relative to the brain which is common in inbred strains. (iii) Cranial sutures with bregma and lambda marked. (iv) Coronal section through the VMHvl where we injected the virus. Note that the *y* coordinate is different from the drilling site in (iii) because of the 15° slant. Scale bar represents 1 mm (iii and iv). *Figure (iv) was modified from Paxinos Brain Atlas (Paxinos & Franklin, 2004).* (See the color plate.)

Table 8.1 List of primers used to sequence the AAV-flex-C3-Tp

Sequence	Orientation	Position	Reads through
CCT CTG ACT TGA GCG TCG AT	Forward	7107	Into left ITR
ACA CGA CAT CAC TTT CCC AG	Reverse	314	Into left ITR
TTC TCA AGC CTC AGA CAG TGG	Forward	1413	lox/lox, TEVpro, T2A, small
AAT CAT GTC CCT GCC GTC GAT C	Forward	2035	T2A, small, TEV, large
AGA GGG GAT CGT TGT AGA AGT C	Reverse	2766	TEV, small, T2A, TEVpro
AAA GCA GCG TAT CCA C	Reverse	3393	lox/lox, large, TEV, small
TAG AAG GAC ACC TAG TCA GA	Reverse	4053	pA, WPRE, lox/lox large
TCA AGC GAT TCT CCT GCC TC	Forward	4282	pA, into right ITR
TAC TAT GGT TGC TTT GAC GT	Reverse	4703	Into right ITR

i. Most AAV's are produced in dedicated cores. Many viral cores have a minimum requirement for the plasmid you provide them, typically 300–500 μg of plasmid DNA.

j. Once the virus is made, it should be aliquoted (typically 5–20 μL) and frozen at −80 °C. Once the aliquot is thawed, it should not be refrozen as freeze–thaw cycles reduce the titer of the virus significantly. Thawed aliquots should remain on ice or at 4 °C, and we typically use each aliquot within 1 week of thawing.

2.4. Protocol: Injecting AAV-flex-C3-Tp into the adult brain

AAV particles, like other viruses and drugs, can be stereotaxically injected into a specific brain region, and the particular characteristics of the virus or chemical determine the subsequent rate and extent of diffusion within the injected area. The spread of AAV particles is limited to 0.5–2 mm from the tip of the injection needle, depending on the volume injected and the size of the needle. The timing of the surgery naturally dictates the onset of the ablation of the desired neuronal subset, offering temporal control, although expression of the virally encoded, Cre-dependent transgene is not immediate, and animals need time to recover from the surgery. Some tissue damage is unavoidable, but this can be reduced by using a smaller needle or pulled glass pipette. There can be significant differences in the tropism of different AAV serotypes for different cell types, including different neuronal populations, and it is important to utilize a serotype that affords maximal infection of the target cells.

The advantages of viral delivery—flexibility in encoding transgenes, temporal and anatomical precision—make it ideal for ablating specific cell types in specific regions.

1. Surgery
   a. Anesthetize the animal. The most common ketamine-based cocktails such as Ketamine/Xylazine/Acepromazine do not induce surgical-plane anesthesia, so addition of an inhalational anesthetic such as 0.5% isoflurane is preferable. If inhalational anesthesia is used as the primary modality of anesthesia, additional analgesics will be required.
   b. Shave and sterilize the scalp.
   c. We use a stereotaxic alignment system from Kopf (model 1900) because it allows the precision in head-angle alignment necessary for our deep brain areas. Small deviations in alignment or rotation will be exaggerated in deeper areas. A deflection of only a 100 μm will miss the target region completely. This model allows for the precise measurement and adjustment of the rotation of the skull in all planes such that variation can be accounted for between individual mice (Fig. 8.4Bi).
   d. Mount the mouse on the bite bar and secure the ear bars as indicated in the instruction manual. Many inbred strains including 129 and B6 have a 15° slant to their skull as compared to wild-caught mice and thus it is advisable to use an angled bite bar. It is also advisable to use nonrupture ear bars as these are more humane.
   e. Make an incision that is roughly 1 cm long along the top of the scalp. It should be as small as possible (to reduce pain and healing time for the animal) while still allowing for good visualization of skull landmarks (bregma and lambda) (Fig. 8.4Bii–iii).
   f. Align the instrument to bregma and position and level the skull according to the instruction manual. Bregma is not always well-defined because of the nonlinear cranial sutures (Fig. 8.4Biii), and visualizing lines that most closely approximate the sutures *in situ* to construct an ideal bregma is recommended (dashed lines in Fig. 8.4Biii). To minimize variability in generating such an ideal bregma, we also recommend that the user maintain a reasonably fixed angle at the intersection of the lines that approximate the sutures. If using the double pressure gauge tool (alignment indicator tool) from Kopf Instruments to level the skull, be sure not to press too hard. The mouse skull is easily deformed, and this will lead to mistargeting.

2. Choosing coordinates

a. Mouse brain atlas-based coordinates are very accurate, with small mouse-to-mouse variability, at least for standard inbred laboratory strains that we use such as C57Bl/6J (Lein et al., 2007; Paxinos & Franklin, 2004). However, because of how the cranial surface is angled relative to the brain, the final coordinates of the needle tip may not match the coordinates where the holes are drilled in the skull (Fig. 8.4Bii–iv). This tilt should be taken into consideration when establishing coordinates. For the VMHvl, we drill at ± 0.78, -1.48 (x,y) and inject at ± 0.78, -1.70, -5.80 $(x,y,z$, bregma$=0,0,0)$. To ensure an accurate set of coordinates, each new region should be tested several times before injecting the experimental virus.

b. We usually establish our initial coordinates by injecting trypan blue, a dye that labels dead cells. For these studies, we euthanize the mouse with a lethal dose of anesthesia just prior to lowering the injection needles. The brain is dissected immediately, cut into slices thin enough to visualize the needle tracks, and placed in a dish of cold PBS to be viewed on a dissecting microscope.

c. These coordinates are subsequently refined by a more precise means involving the injection of a fluorescent-tagged reporter (such as cholera toxin B) or an AAV constitutively expressing a fluorescent reporter protein, and careful histological processing, to ensure accuracy. This is often a reiterative process, with the results of each stereotaxic injection trial leading to further refinement of the injection coordinates.

3. Injecting the virus

a. Assuming coordinates have been chosen, drill holes in the skull above the injection sites (Fig. 8.4Bii).

b. Lower the needles slowly (1–2 mm/min) to the correct z coordinate. Once there, create a space 50–100 μm above and below the target site, by moving the needles up and down a few times slowly to generate a small pocket for the virus solution to fill.

c. We use 33G needles (~200 μm outer diameter) coupled to Hamilton syringes via PE20 tubing. Our syringes are mounted on a micropump to maintain an even flow rate during infusion (Fig. 8.4Bi). If more precision is needed, pulled glass pipettes can be substituted for steel needles; however, glass pipettes are more fragile and are potentially deflected by fiber tracts within the brain.

d. Infuse the virus slowly, 60–100 nL/min. The volume can be anywhere from 50 to 1000 nL depending on the titer and efficacy of the virus. Larger volumes will spread farther. It is not recommended that you exceed 1 μL as this will potentially damage the target site.
e. Allow an additional 5 min for the virus to diffuse away from the needle tips before pulling the needles back up. Retract the needles at the same rate they were lowered.

4. Finishing
   a. Close the wound with skin glue (vetbond, dermabond, or liquid bandaid). You may also use sutures, but in most cases this is not necessary. Wound clips are not recommended for the scalp because the mice will pull these out.
   b. Be sure to use an appropriate analgesic. If an opiate-based analgesic is used (e.g., buprenorphine) monitor the mouse's respiration until it returns to normal before administering the drug, as opiates are a respiratory depressant.
   c. Wait 1–4 weeks before analyzing the mouse to allow adequate recovery time and for expression of the virus to peak. A time course should be performed for each neuronal population to establish a timepoint at which maximal cell loss has occurred. We wait 3–4 weeks after injecting AAV-flex-C3-Tp prior to behavioral testing.

2.5. The future of viral-mediated ablation in the brain

While stroke, cancer, trauma, and neurodegenerative victims have provided invaluable insights into localizing brain function, genetically targeted ablations of specific neuronal populations will enable high-resolution functional mapping of the neural circuits that encode behavior in health and disease. The electrolytic or surgical lesions of the past have offered us insight into regional functionalization of the brain, but now we have the means to determine the function of any group of adult neurons or model neurodegenerative disorders with remarkable accuracy. These genetically targeted studies will provide us with more precise understanding of the function of discrete neuronal populations and a testing ground for therapeutic interventions. As we mentioned earlier, our strategy is modular and can, in principle, be applied to any cell type in the brain or other organ system. We are also developing means of ablating even more specific neuronal populations with enhanced temporal control in the form of light- and ligand-activated caspases.

3. SMALL-MOLECULE ACTIVATORS OF CASPASES

3.1. Introduction

Allosteric modulation of enzymes with unnatural small molecules is an attractive way of modifying enzymatic activities by often offering greater selectivity over targeting conserved active sites. Allosteric control also allows the possibility to turn enzymes on, and there have been important recent discoveries in this area for kinases, deacetylases, dehydrogenases, phosphatases, and nucleases (for review, see Zorn & Wells, 2010). Upregulation of apoptotic signaling is common in precancerous cells and silencing of apoptosis is a hallmark of cancer cells. This apoptotic suppression is usually achieved by making lesions in signaling pathways upstream of the caspases, such as p53 (Lowe & Sherr, 2003). Thus, there has been a significant effort to activate executioner caspase with small molecules in hopes of differentially killing cancer cells.

The caspases are expressed as inactive proenzymes, or zymogens, and their activation is under tight regulation in cells due to their fate-determining activity. For example, the zymogenicity (defined as the mature enzyme k_{cat}/K_m divided by the proenzyme k_{cat}/K_m) of procaspase-3 is greater than 10^7, making it the most inactive protease zymogen yet reported (Pop et al., 2007; Stennicke & Salvesen, 2000; Zorn, Wolan, Agard, & Wells, 2012). Upon proteolysis by upstream proteases, the N-terminal prodomain of the procaspase is removed, and the intersubunit linker is cut, the later being crucial to activation (Fig. 8.1B). This proteolytic activation leads to a dimer of heterodimers consisting of a large and small subunit, as illustrated by the crystal structure of mature caspase-7 (Fig. 8.5A) (Chai et al., 2001; Riedl et al., 2001; Wei et al., 2000). In this active state, the executioner caspases exist in a conformational equilibrium between an "off" and "on" state, but predominantly in the on-state to allow substrate binding in the active site (Fig. 8.5B). Interestingly, mature caspase-1 is predominantly in the off-state until driven to the on-state by binding substrate (Gao, Sidhu, & Wells, 2009). In the "off" state, the active caspase can be trapped in a nonfunctional conformation by allosteric inhibitors that bind to the dimer interface, as shown in previous studies of caspase-1, -5, and -7 (Gao & Wells, 2012; Hardy, Lam, Nguyen, O'Brien, & Wells, 2004; Scheer, Romanowski, & Wells, 2006). These results also raise the question as to whether small molecules can potentially be found to activate the proenzyme (Fig. 8.5C). More recently, our group demonstrated the feasibility

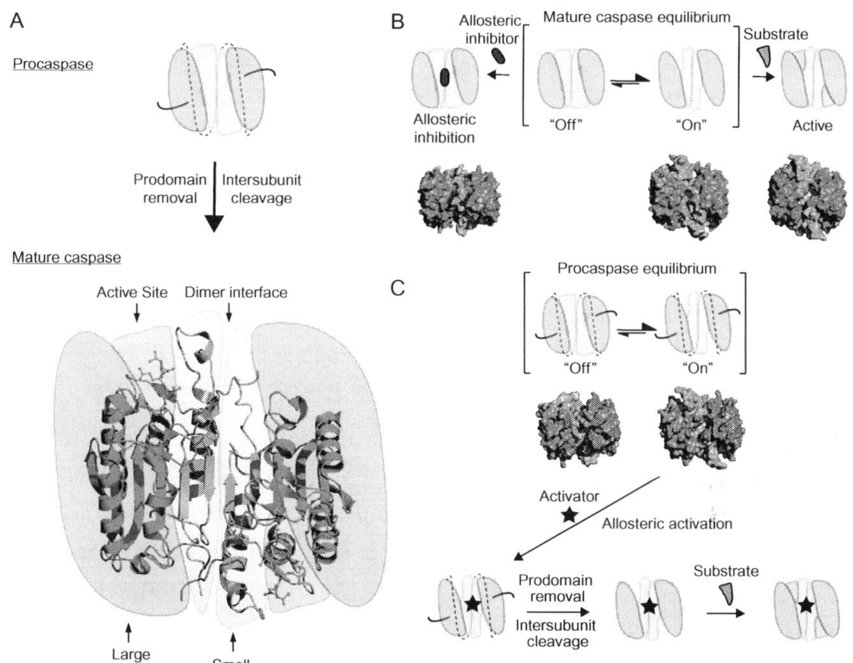

Figure 8.5 Mechanism of caspase regulation. (A) Executioner procaspase maturation requires proteolysis of the N-terminal prodomain (black lines) and cleavage of the intersubunit linker (dashed lines) between the large and small subunits. The crystal structure of mature caspase-7 is shown (PDB ID 1F1J) highlighting the dimer interface, the large and small subunits and the Ac-DEVD-CHO peptide occupying the active site (shown in sticks). (B) Mechanism of allosteric inhibition of caspases, showing the hotspot for allosteric binding located at the dimer interface. The surface representation of experimentally determined X-ray structures of caspase-7 is shown below the cartoons. From left to right, the allosteric-site ligand-bound structure of caspase-7 complexed with DICA, the ligand-free apo caspase-7 structure in an "on" conformation, and the active-site ligand-bound structure of caspase-7 in complex with the Ac-DEVD-CHO peptide (PDB ID 1SHJ, 1K86, 1F1J, respectively). (C) Mechanism of allosteric activation of procaspases. The brackets denote the open and closed active-site equilibrium, with the surface of the X-ray structures of procaspase-3 in the "on" and "off" states (PDB ID 4JR0 and 4JQY, respectively). Upon binding of a hypothetical allosteric activator, the proenzyme is locked into an "on" conformation, allowing possible proteolytic cleavage of the prodomain and intersubunit linkers. (See the color plate.)

of activating procaspase-3 and -7. Antibody fragments were selected for binding the on-state and could stimulate procaspase-3 activity by 1000-fold (Thomsen, Koerber, & Wells, 2013). These results also highlight the enormous barrier of activation of procaspases and suggest that finding robust allosteric activators of procaspase-3 will be challenging.

3.2. Small-molecule activators of caspases

At present, four small-molecule compounds have been reported to activate procaspases *in vitro* and to induce cell death in cell culture. Neither of the binding sites for these small molecules have been structurally defined, nor have their stoichiometry of binding been determined. Interestingly, the mechanisms for the two that have been most extensively characterized are not through traditional allosteric means, that is, stoichiometric binding of a small molecule to the protein. Here, we review the discovery of each of these four compounds and provide practical suggestions for assays that may be useful for future discovery efforts.

3.2.1 PAC-1

PAC-1, or procaspase-activating compound 1, was a small molecule discovered by the Hergenrother group as being capable of activating procaspase-3 *in vitro* and inducing apoptosis in various cell lines (Fig. 8.6A) (Putt et al., 2006). This compound was discovered by performing a high-throughput screen for 20,500 diverse small molecules, and looking for procaspase-3 activation by monitoring cleavage of a peptidic substrate (acetyl Asp-Glu-Val-Asp-*p*-nitroanilide (Ac-DEVD-*p*Na) at 200 μM). The compound PAC-1 activated procaspase-3 up to fourfold over background after 12 h of incubation with an EC_{50} of 0.22 and 4.7 μM for procaspase-3 and -7, respectively (Peterson et al., 2009). This activation is small considering the 10^7-fold change in activity for full activation of the proenzyme (Zorn et al., 2012). Nonetheless, PAC-1 was shown to induce cell death in several cancer cell lines in a manner proportional to the cellular concentration of procaspase-3, and was ineffective in MCF-7 cells that lack procaspase-3, although another group was unable to repeat this result (Denault et al., 2007). PAC-1 was also shown to significantly delay the progression of tumors in mice (Putt et al., 2006). The group later showed that PAC-1 is a zinc chelator agent that leads to activation of procaspase-3 by relieving the zinc inhibition of caspase-3 from their buffers that were missing EDTA (Peterson et al., 2009). These findings are consistent with previous studies showing that zinc inhibits caspase-3 with an IC_{50} of 0.1 μM (Perry et al., 1997). They also reported that the zinc-chelation mechanism was responsible for the cell death observed in cell culture leading to altered intracellular Ca^{2+} concentration, indicative of endoplasmic reticulum stress-induced apoptosis (West et al., 2012). In summary, PAC-1 may not be a direct activator of procaspase, but led to interesting discoveries about the role of metal chelation in caspase

A
PAC-1
(Putt et al., 2006)
1541
(Wolan et al., 2009)
Compound-42
(Schipper et al., 2011)
Compound-2
(Vickers et al., 2013)
B
Propagation
Fibril formation
Initiation
Activation
100 nm
100 nm

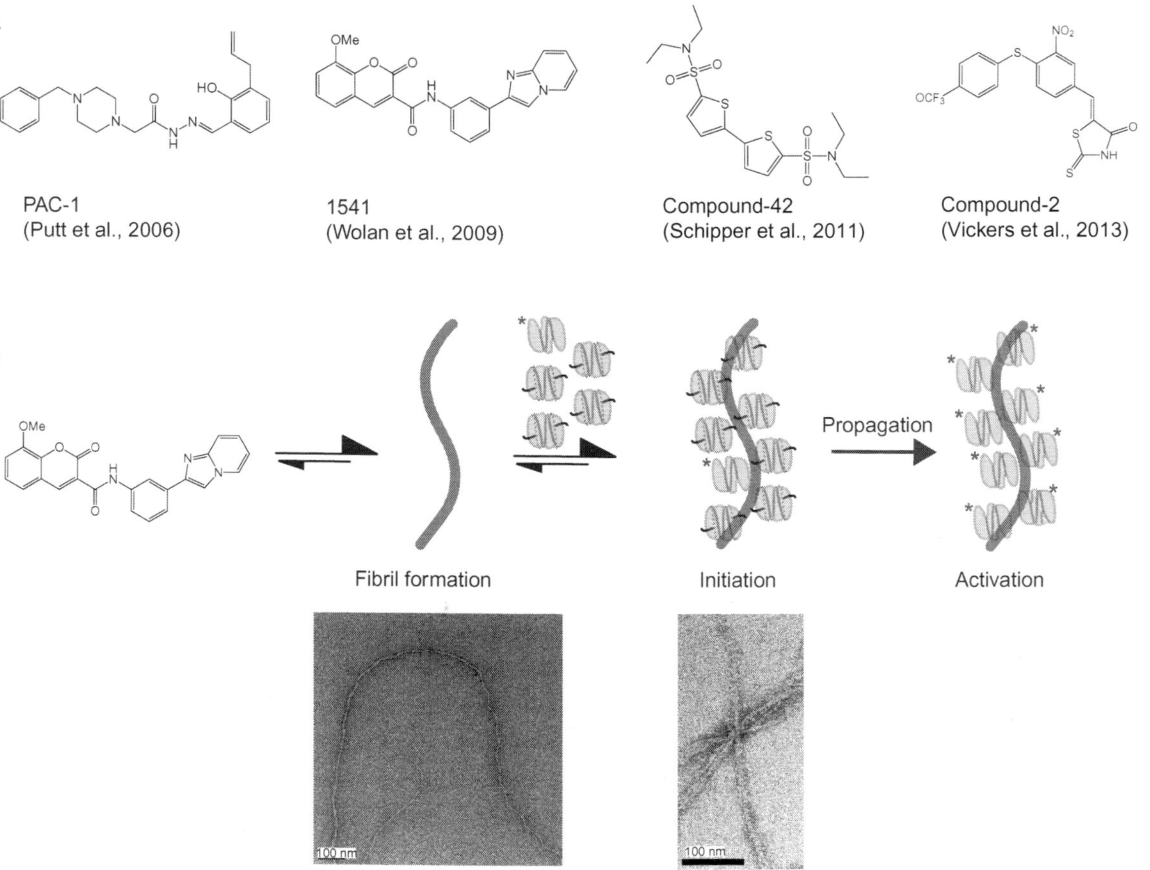

(*Continued*)

activation and initiation of cell death. Whether the cellular activity is a direct effect of chelating zinc from caspase-3 remains challenging to prove, given the many roles of zinc in cells.

3.2.2 Compound-1541

In a separate HTS of over 62,000 compounds, our lab discovered a compound called 1541 (Fig. 8.6A), that lead to specific activation of procaspases-3 and -6 *in vitro* but not procaspase-7 (Wolan et al., 2009). Compound-1541 induces rapid and complete apoptosis with an EC_{50} ~2 μ*M*. The apoptosis is minimally affected by deletion of caspase-8, a driver of the extrinsic pathway, and apoptosis proceeds without significant release of cytochrome *C*, a marker of the intrinsic pathway. These data were consistent with direct activation of caspase-3. Moreover, the rate of cell death is slowed in MCF-7 cells lacking caspase-3. However, we were unable to determine the structure of the compound bound to either the mature or zymogen form of caspase-3. Nor were we able to obtain clear biophysical evidence for stoichiometric binding, for example, by surface plasmon resonance (SPR) or isothermal titration calorimetry (ITC). Further characterization of the compounds by electron microcopy (EM) revealed the surprising finding that 1541 and its active analogs self-assemble into nanofibrils in solution (Fig. 8.5B) (Zorn et al., 2011). Using various biochemical and biophysical methods, we conclusively found that these nanofibrils colocalize caspase-3 with procaspase-3 and promote activation *in vitro* (Zorn et al., 2012).

Globular aggregates of small molecules are known to act as inhibitors and can be diagnosed by testing for detergent sensitivity, β-lactamase inhibition, and sensitivity to bovine serum albumin (BSA) (Coan & Shoichet, 2008; Feng et al., 2007; Seidler, McGovern, Doman, & Shoichet, 2003). However, we obtained mixed results by performing these tests on 1541 (Zorn et al., 2011). For example, detergents such as Triton or CHAPS did not

Figure 8.6—Cont'd Procaspase activation by small molecules. (A) Four small molecules have been reported to act as allosteric activators of procaspases; PAC-1 (Putt et al., 2006), 1541 (Wolan, Zorn, Gray, & Wells, 2009), compound-42 (Schipper, MacKenzie, Sharma, & Clark, 2011), and compound-2 (Vickers et al., 2013). (B) Mechanism of procaspase activation by 1541 nanofibrils. A small amount of active caspase (labeled with asterisk) is already present in *E. coli*-purified procaspase samples. The nanofibrils serve as a scaffold for colocalization of active caspase with the proenzyme, leading to activation. *Figure (B) is adapted with permission from Zorn, Wille, Wolan, and Wells, (2011). Copyright (2011) American Chemical Society.*

disrupt procaspase-3 activation, nor did 1541 inhibit β-lactamase activity. However, the addition of BSA into the assay prevented procaspase-3 activation. A careful characterization of 1541 in various conditions (e.g., buffer, temperature, concentration, etc.) led us to observe higher molecular weight species using dynamic light scattering (DLS). Spin-down assays showed coprecipitation of the nanofibrils with procaspase-3. Moreover, sequestering procaspase-3 from the nanofibrils using a dialysis bag containing the proenzyme prevented activation. We hypothesize that unlike typical amorphous aggregates, that are in free equilibrium between monomer and aggregate, the nanofibrils have fewer ends available for monomer–nanofibrils exchange. This makes it extremely slow for the compound to reform fibrils, making it very unlikely for the compound to enter the dialysis bag and to reform fibrils on the inside of the membrane during the incubation period.

We found that activation *in vitro* depends on the presence of a small amount of active caspase to initiate proenzyme cleavage. For example, adding as little as 0.01 equivalent of the covalent caspase-3 inhibitor (Ac-DEVD-cmk) to procaspase-3 in our assays prevented procaspase activation by 1541. We believe this contaminant comes from tiny amounts of mature enzyme that is processed from the proenzyme, perhaps by endogenous *E. coli* proteases or by the procaspase itself during expression in bacteria. There is also a pronounced lag in the activation process *in vitro,* as is typical for the activation of proenzymes, in general. This lag is not due to assembly of the fibrils; we observe them forming instantly upon dilution into aqueous solution, as monitored by EM or DLS. Moreover, the inclusion of even a 1% stoichiometric amount of active caspase-3 shortens the lag period by 2.5-fold (Zorn et al., 2012). Taken together, our experiments show that 1541 nanofibrils colocalize caspase-3 with procaspase-3 *in vitro*, leading to activation of the proenzyme in a similar manner as a firecracker fuse. The spark of activation is provided by a small amount of active caspase that is colocalized to the fibril. Additional structural characterization of the nanofibrils with procaspase-3 will be important to understand the assembly and activation mechanism further.

Many questions remain to be answered regarding the cellular mechanism for 1541. How does 1541 induce apoptosis in cell culture with such potency? Our preliminary studies indicate that the fibrils are responsible for inducing apoptosis in cell culture (O. Julien & J. A. Wells, unpublished results), but the exact molecular mechanism of cell death is currently under investigation. Are the nanofibrils activating procaspases in cells in the same way they do *in vitro*? There are some intriguing similarities, both at the functional and structural levels, between 1541 nanofibrils and some amyloid fibrils that induce apoptosis such as Aβ (Zorn et al., 2011). The ease of making 1541 and the simplicity of forming these fibrils make it a potentially important model system for understanding cell death induced by such fibril-forming molecules.

3.2.3 Compound-42

Another small molecule, named compound-42 (Fig. 8.6A), was recently discovered by the Clark group that was capable of enhancing procaspase-3 activation by 27-fold (Schipper et al., 2011). Again, this is a small change considering the zymogenicity of the procaspase-3. The compound was discovered using computational docking screen of 62 compounds that targeted the allosteric inhibitory site located at the dimer interface of procaspase-3 (Hardy et al., 2004). They hypothesized that stabilizing the same interface on procaspase-3 could lead to allosteric activation of the zymogen. Based on their docking score, 13 compounds were tested in a procaspase-3 activation assay *in vitro*. One compound, compound-42, was found to increase procaspase-3 activity by 27-fold at 400 μ*M* after 1 h incubation with the proenzyme. The authors hypothesized that compound-42 activates procaspase-3 by releasing the intersubunit linker from the dimer interface allowing the enzyme to auto-process this linker in *trans*. The group measured a K_D of 4 μ*M* for compound-42 to bind mature caspase-3 by ITC, but could not determine the binding affinity of compound-42 to procaspase-3. No cellular activity was reported.

3.2.4 Compound-2

Earlier this year, the Wolan group discovered a new compound that was capable of promoting the maturation of executioner procaspase-3 and -7 (Vickers et al., 2013). They used a clever fluorescence-polarization-based screen. This utilized a newly designed canonical DEVD-aldehyde peptide recognition motif, and a 5(6)-carboxyfluorescein at the N-terminus linked to a 3 × 6-aminohexanoic acid. The main advantage of this new probe is that it binds and inhibits any contaminating mature caspase-3 and -7, in a reversible covalent manner that may be present or produced during the activation assay, and does not get consumed like standard caspase probes that are based on the release of a fluorophore. Using HTS, they identified a small molecule, compound-2 (Fig. 8.6A), that the activated executioner procaspases *in vitro* with an EC_{50} of ~1–5 μ*M*. Importantly, compound-2 and its analogs are capable of inducing apoptosis in cell culture with the same low μ*M* EC_{50}.

- **Detergent**
  - HTS and general assays: Triton X-100
  - Cell culture: Tween-80
- **Dynamic light scattering** (DLS)
  - The gold standard test
- **Enzyme inhibition**
  - β-lactamase assay
  - Effect of BSA on drug activity
- **Transmission electron microscopy** (TEM)
  - Negative stain TEM
- **Stoichiometry of binding**
  - Fluorescence polarization
  - Isothermal titration calorimetry (ITC)
  - Surface plasmon resonance (SPR)
- **Structural data**
  - X-ray crystallography
  - NMR spectroscopy

Figure 8.7 Useful tests to ensure compounds are acting through a stoichiometric binding mechanism.

These values are very similar to 1541. Further studies will be needed to confirm the stoichiometry and site of binding, as well as the mechanism of inducing apoptosis in cells.

3.3. A practical guide to avoid aggregating small molecules: The case of procaspase activators

3.3.1 *Detergent* in vitro

The use of small amounts of detergent during the high-throughput screen is the first step toward avoiding false positive results caused by small molecules (Fig. 8.7). This did not preclude the discovery of 1541 as the screen included 0.1% CHAPS to avoid standard aggregators. Vickers et al. also included some detergent in their assay buffer to discourage small-molecule aggregation. Once hits are found, it is useful to test the effect of increasing the detergent concentration on the enzymatic activity of the target of interest. If a decrease in compound potency is observed, aggregation is a strong possibility.

3.3.2 *Dynamic light scattering*

In our hands, the gold standard test for detecting small-molecule aggregators is DLS. If performed cautiously, this test is capable of detecting all forms of aggregation (e.g., colloid particles, micelles, fibrils, etc.). We recommend all buffer solutions to be filtered at 0.22 μm before adding the compound of interest. Measurements should be made in the same time frame used for the enzyme and cellular assays. Low and high concentration of compounds should also be tested at various temperatures. Measurements should be performed in triplicate. Finally, it is useful to test a known aggregator as positive control (e.g., like tetra-iodophenolphthalein) and a known well-behaved small molecule as negative control (e.g., coumarins).

3.3.3 Nonspecific enzyme inhibition

Various assays can be used to detect the formation of small-molecule aggregates *in vitro*. For example, Shoichet and coworkers developed a β-lactamase inhibition assay based on the high sensitivity of this enzyme to be inhibited by colloidal particles (McGovern, Caselli, Grigorieff, & Shoichet, 2002). While useful for colloidal particles, 1541 nanofibrils did not inhibit β-lactamase. Another simple assay is to test if BSA perturbs the enzymatic activity caused by the compound of interest (Coan & Shoichet, 2007).

3.3.4 Transmission electron microscopy

Transmission electron microscopy is a powerful and relatively easy method to verify the presence of aggregators. However, it is important to remember that a negative result here does not rule out the presence of aggregation, since not all aggregators can be effectively detected using this method. This method was especially useful for observing 1541 nanofibrils.

3.3.5 Stoichiometry of binding

Various techniques allow the determination of the stoichiometry of binding of a compound of interest to its target. One can use ITC, X-ray crystallography, SPR, or nuclear magnetic resonance (NMR) spectroscopy. In addition, X-ray crystallography and NMR spectroscopy are powerful tools to determine the binding location of the small molecule on its target. As with EM, a single negative result does not reveal much, but one should be worried if all fails.

3.3.6 Detergent in cell culture

An easy way to test the presence of small-molecule aggregation is to use small amounts of detergent in cell culture (Owen, Doak, Wassam, Shoichet, & Shoichet, 2012). Specifically, Tween-80 is able to prevent small-molecule aggregation and has been shown by Owen and coworkers to have negligible toxicity in cell culture at low concentration (i.e., less than 0.1%). Recently, we have had success using this assay to block 1541 activity (O. Julien and J. A. Wells, data to be published).

3.3.7 Use of active-site titrants to rule out the involvement of mature caspases in proenzyme activation

Activation by 1541 requires a small amount of mature caspase, which can be tested for and blocked by addition of substoichiometric amounts of DEVD-chloromethyl ketone. We found mature caspase contaminants are unavoidable when using *E. coli* to express the procaspase. Such contaminants are even present if one uses the "noncleavable" D_3A mutant. It is possible that alternate cleavage sites by host proteases are the cause of this mature enzyme contamination. Indeed, proteolysis at D169 has been shown to occur when the normal cleavage site D175 is unavailable (MacKenzie et al., 2013). In our hands, such spurious activation in *E. coli* expression systems can be reduced by using shorter expression times, but it is impossible to eliminate completely. Thus, the safest technique is to inactivate trace levels of mature caspase molecules in proenzyme preparations by adding as little as 0.01–0.05 equivalent of a covalent and irreversible caspase inhibitor (e.g., Ac-DEVD-chloromethyl ketone).

3.4. Conclusions

The most thoroughly characterized procaspase activators, PAC-1 and 1541, were shown to activate the proenzyme through unexpected mechanisms. PAC-1 is believed to remove zinc inhibition, and 1541 through nanofibril formation and activation by colocalization of mature and inactive caspase. These are still very useful tools and have provided important new insights about mechanisms of activation of caspases. Direct activation of executioner caspases remains a big challenge due to the high zymogenicity barrier of the proenzyme (Zorn et al., 2012). It may be easier to target procaspase-7 instead of procaspase-3 because of its lower zymogen activation barrier (Thomsen et al., 2013). It may even be more achievable to target the initiator caspases since they have even lower barriers and compounds that induce dimerization could potentially be found. Finding new drugs here would have a big impact, but many challenges remain.

ACKNOWLEDGMENTS

We are grateful to Dan Gray, Julie Zorn, and Dennis Wolan for their pioneering work on Sections 1 and 3, and Cindy Yang and Dan Gray for their work on Section 2. We also thank

Allison Doak, Jason Porter, Justin Rettenmaier, Cheryl Tajon, Nathan Thomsen, for their thoughtful discussions and careful reading of the chapter. C. W. M. is supported by a NSF Graduate Fellowship. O. J. is supported by a Banting Fellowship from the Government of Canada and the Canadian Institutes of Health Research. E. K. U. is supported by NIH grant F31NS078959. N. M. S. is supported by The Ellison Medical Foundation and the NIH (grants DP1MH099900, R01NS049488, and R01NS083872). J. A. W. is supported by NIH (grants R01 CA136779 and R01 GM081051).

REFERENCES

Amara, J. F., Clackson, T., Rivera, V. M., Guo, T., Keenan, T., Natesan, S., et al. (1997). A versatile synthetic dimerizer for the regulation of protein-protein interactions. *Proceedings of the National Academy of Sciences of the United States of America*, *94*, 10618–10623.

Atasoy, D., Aponte, Y., Su, H. H., & Sternson, S. M. (2008). A FLEX switch targets Channelrhodopsin-2 to multiple cell types for imaging and long-range circuit mapping. *The Journal of Neuroscience*, *28*, 7025–7030.

Belshaw, P. J., Spencer, D. M., Crabtree, G. R., & Schreiber, S. L. (1996). Controlling programmed cell death with a cyclophilin-cyclosporin-based chemical inducer of dimerization. *Chemistry & Biology*, *3*, 731–738.

Chai, J., Wu, Q., Shiozaki, E., Srinivasula, S. M., Alnemri, E. S., & Shi, Y. (2001). Crystal structure of a procaspase-7 zymogen: Mechanisms of activation and substrate binding. *Cell*, *107*, 399–407.

Chang, D. W., Xing, Z., Capacio, V. L., Peter, M. E., & Yang, X. (2003). Interdimer processing mechanism of procaspase-8 activation. *The EMBO Journal*, *22*, 4132–4142.

Chelur, D. S., & Chalfie, M. (2007). Targeted cell killing by reconstituted caspases. *Proceedings of the National Academy of Sciences of the United States of America*, *104*, 2283–2288.

Chen, M., Orozco, A., Spencer, D. M., & Wang, J. (2002). Activation of initiator caspases through a stable dimeric intermediate. *The Journal of Biological Chemistry*, *277*, 50761–50767.

Coan, K. E., & Shoichet, B. K. (2007). Stability and equilibria of promiscuous aggregates in high protein milieus. *Molecular Biosystems*, *3*, 208–213.

Coan, K. E., & Shoichet, B. K. (2008). Stoichiometry and physical chemistry of promiscuous aggregate-based inhibitors. *Journal of the American Chemical Society*, *130*, 9606–9612.

Cooray, S., Howe, S. J., & Thrasher, A. J. (2012). Retrovirus and lentivirus vector design and methods of cell conditioning. *Methods in Enzymology*, *507*, 29–57.

Denault, J. B., Drag, M., Salvesen, G. S., Alves, J., Heidt, A. B., Deveraux, Q., et al. (2007). Small molecules not direct activators of caspases. *Nature Chemical Biology*, *3*, 519, author reply 520.

Di Stasi, A., Tey, S. K., Dotti, G., Fujita, Y., Kennedy-Nasser, A., Martinez, C., et al. (2011). Inducible apoptosis as a safety switch for adoptive cell therapy. *The New England Journal of Medicine*, *365*, 1673–1683.

Edwards, S. R., & Wandless, T. J. (2007). The rapamycin-binding domain of the protein kinase mammalian target of rapamycin is a destabilizing domain. *The Journal of Biological Chemistry*, *282*, 13395–13401.

Fegan, A., White, B., Carlson, J. C., & Wagner, C. R. (2010). Chemically controlled protein assembly: Techniques and applications. *Chemical Reviews*, *110*, 3315–3336.

Feng, B. Y., Simeonov, A., Jadhav, A., Babaoglu, K., Inglese, J., Shoichet, B. K., et al. (2007). A high-throughput screen for aggregation-based inhibition in a large compound library. *Journal of Medicinal Chemistry*, *50*, 2385–2390.

Galluzzi, L., Aaronson, S. A., Abrams, J., Alnemri, E. S., Andrews, D. W., Baehrecke, E. H., et al. (2009). Guidelines for the use and interpretation of assays for monitoring cell death in higher eukaryotes. *Cell Death and Differentiation*, *16*, 1093–1107.

Gao, J., Sidhu, S. S., & Wells, J. A. (2009). Two-state selection of conformation-specific antibodies. *Proceedings of the National Academy of Sciences of the United States of America, 106*, 3071–3076.

Gao, J., & Wells, J. A. (2012). Identification of specific tethered inhibitors for caspase-5. *Chemical Biology & Drug Design, 79*, 209–215.

Gray, D. C., Mahrus, S., & Wells, J. A. (2010). Activation of specific apoptotic caspases with an engineered small-molecule-activated protease. *Cell, 142*, 637–646.

Hardy, J. A., Lam, J., Nguyen, J. T., O'Brien, T., & Wells, J. A. (2004). Discovery of an allosteric site in the caspases. *Proceedings of the National Academy of Sciences of the United States of America, 101*, 12461–12466.

Hughes, M. A., Harper, N., Butterworth, M., Cain, K., Cohen, G. M., & MacFarlane, M. (2009). Reconstitution of the death-inducing signaling complex reveals a substrate switch that determines CD95-mediated death or survival. *Molecular Cell, 35*, 265–279.

Kang, T. B., Oh, G. S., Scandella, E., Bolinger, B., Ludewig, B., Kovalenko, A., et al. (2008). Mutation of a self-processing site in caspase-8 compromises its apoptotic but not its nonapoptotic functions in bacterial artificial chromosome-transgenic mice. *Journal of Immunology, 181*, 2522–2532.

Lein, E. S., Hawrylycz, M. J., Ao, N., Ayres, M., Bensinger, A., Bernard, A., et al. (2007). Genome-wide atlas of gene expression in the adult mouse brain. *Nature, 445*, 168–176.

Lowe, S. W., & Sherr, C. J. (2003). Tumor suppression by Ink4a-Arf: Progress and puzzles. *Current Opinion in Genetics & Development, 13*, 77–83.

MacKenzie, S. H., Schipper, J. L., England, E. J., Thomas, M. E., 3rd., Blackburn, K., Swartz, P., et al. (2013). Lengthening the intersubunit linker of procaspase 3 leads to constitutive activation. *Biochemistry, 52*, 6219–6231.

McGovern, S. L., Caselli, E., Grigorieff, N., & Shoichet, B. K. (2002). A common mechanism underlying promiscuous inhibitors from virtual and high-throughput screening. *Journal of Medicinal Chemistry, 45*, 1712–1722.

Murphy, B. M., Creagh, E. M., & Martin, S. J. (2004). Interchain proteolysis, in the absence of a dimerization stimulus, can initiate apoptosis-associated caspase-8 activation. *The Journal of Biological Chemistry, 279*, 36916–36922.

Muzio, M., Stockwell, B. R., Stennicke, H. R., Salvesen, G. S., & Dixit, V. M. (1998). An induced proximity model for caspase-8 activation. *The Journal of Biological Chemistry, 273*, 2926–2930.

Oberst, A., Pop, C., Tremblay, A. G., Blais, V., Denault, J. B., Salvesen, G. S., et al. (2010). Inducible dimerization and inducible cleavage reveal a requirement for both processes in caspase-8 activation. *The Journal of Biological Chemistry, 285*, 16632–16642.

Orban, P. C., Chui, D., & Marth, J. D. (1992). Tissue- and site-specific DNA recombination in transgenic mice. *Proceedings of the National Academy of Sciences of the United States of America, 89*, 6861–6865.

Ory, D. S., Neugeboren, B. A., & Mulligan, R. C. (1996). A stable human-derived packaging cell line for production of high titer retrovirus/vesicular stomatitis virus G pseudotypes. *Proceedings of the National Academy of Sciences of the United States of America, 93*, 11400–11406.

Owen, S. C., Doak, A. K., Wassam, P., Shoichet, M. S., & Shoichet, B. K. (2012). Colloidal aggregation affects the efficacy of anticancer drugs in cell culture. *ACS Chemical Biology, 7*, 1429–1435.

Pajvani, U. B., Trujillo, M. E., Combs, T. P., Iyengar, P., Jelicks, L., Roth, K. A., et al. (2005). Fat apoptosis through targeted activation of caspase 8: A new mouse model of inducible and reversible lipoatrophy. *Nature Medicine, 11*, 797–803.

Paxinos, G., & Franklin, K. B. J. (2004). *The mouse brain in stereotaxic coordinates*. San Diego, CA: Elsevier Academic Press.

Perry, D. K., Smyth, M. J., Stennicke, H. R., Salvesen, G. S., Duriez, P., Poirier, G. G., et al. (1997). Zinc is a potent inhibitor of the apoptotic protease, caspase-3. A novel target for zinc in the inhibition of apoptosis. *The Journal of Biological Chemistry, 272*, 18530–18533.

Peterson, Q. P., Goode, D. R., West, D. C., Ramsey, K. N., Lee, J. J., & Hergenrother, P. J. (2009). PAC-1 activates procaspase-3 in vitro through relief of zinc-mediated inhibition. *Journal of Molecular Biology, 388*, 144–158.

Pop, C., Fitzgerald, P., Green, D. R., & Salvesen, G. S. (2007). Role of proteolysis in caspase-8 activation and stabilization. *Biochemistry, 46*, 4398–4407.

Provost, E., Rhee, J., & Leach, S. D. (2007). Viral 2A peptides allow expression of multiple proteins from a single ORF in transgenic zebrafish embryos. *Genesis, 45*, 625–629.

Putt, K. S., Chen, G. W., Pearson, J. M., Sandhorst, J. S., Hoagland, M. S., Kwon, J. T., et al. (2006). Small-molecule activation of procaspase-3 to caspase-3 as a personalized anticancer strategy. *Nature Chemical Biology, 2*, 543–550.

Remy, I., & Michnick, S. W. (2007). Application of protein-fragment complementation assays in cell biology. *BioTechniques, 42*, 137, 139, 141 passim.

Riedl, S. J., Fuentes-Prior, P., Renatus, M., Kairies, N., Krapp, S., Huber, R., et al. (2001). Structural basis for the activation of human procaspase-7. *Proceedings of the National Academy of Sciences of the United States of America, 98*, 14790–14795.

Salvesen, G. S., & Dixit, V. M. (1999). Caspase activation: The induced-proximity model. *Proceedings of the National Academy of Sciences of the United States of America, 96*, 10964–10967.

Sauer, B. (1987). Functional expression of the cre-lox site-specific recombination system in the yeast Saccharomyces cerevisiae. *Molecular and Cellular Biology*, 7, 2087–2096.

Sauer, B., & Henderson, N. (1988). Site-specific DNA recombination in mammalian cells by the Cre recombinase of bacteriophage P1. *Proceedings of the National Academy of Sciences of the United States of America, 85*, 5166–5170.

Scheer, J. M., Romanowski, M. J., & Wells, J. A. (2006). A common allosteric site and mechanism in caspases. *Proceedings of the National Academy of Sciences of the United States of America, 103*, 7595–7600.

Schipper, J. L., MacKenzie, S. H., Sharma, A., & Clark, A. C. (2011). A bifunctional allosteric site in the dimer interface of procaspase-3. *Biophysical Chemistry, 159*, 100–109.

Schnütgen, F., Doerflinger, N., Calléja, C., Wendling, O., Chambon, P., & Ghyselinck, N. B. (2003). A directional strategy for monitoring Cre-mediated recombination at the cellular level in the mouse. *Nature Biotechnology, 21*, 562–565.

Seidler, J., McGovern, S. L., Doman, T. N., & Shoichet, B. K. (2003). Identification and prediction of promiscuous aggregating inhibitors among known drugs. *Journal of Medicinal Chemistry, 46*, 4477–4486.

Sohal, V. S., Zhang, F., Yizhar, O., & Deisseroth, K. (2009). Parvalbumin neurons and gamma rhythms enhance cortical circuit performance. *Nature, 459*, 698–702.

Sohn, D., Schulze-Osthoff, K., & Jänicke, R. U. (2005). Caspase-8 can be activated by interchain proteolysis without receptor-triggered dimerization during drug-induced apoptosis. *The Journal of Biological Chemistry, 280*, 5267–5273.

Spencer, D. M., Belshaw, P. J., Chen, L., Ho, S. N., Randazzo, F., Crabtree, G. R., et al. (1996). Functional analysis of Fas signaling in vivo using synthetic inducers of dimerization. *Current Biology, 6*, 839–847.

Spencer, D. M., Wandless, T. J., Schreiber, S. L., & Crabtree, G. R. (1993). Controlling signal transduction with synthetic ligands. *Science, 262*, 1019–1024.

Steller, H. (1998). Artificial death switches: Induction of apoptosis by chemically induced caspase multimerization. *Proceedings of the National Academy of Sciences of the United States of America, 95*, 5421–5422.

Stennicke, H. R., & Salvesen, G. S. (2000). Caspases—Controlling intracellular signals by protease zymogen activation. *Biochimica and Biophysica Acta, 1477*, 299–306.

Straathof, K. C., Pulè, M. A., Yotnda, P., Dotti, G., Vanin, E. F., Brenner, M. K., et al. (2005). An inducible caspase 9 safety switch for T-cell therapy. *Blood*, *105*, 4247–4254.

Tang, W., Ehrlich, I., Wolff, S. B., Michalski, A. M., Wölfl, S., Hasan, M. T., et al. (2009). Faithful expression of multiple proteins via 2A-peptide self-processing: A versatile and reliable method for manipulating brain circuits. *The Journal of Neuroscience*, *29*, 8621–8629.

Taymans, J. M., Vandenberghe, L. H., Haute, C. V., Thiry, I., Deroose, C. M., Mortelmans, L., et al. (2007). Comparative analysis of adeno-associated viral vector serotypes 1, 2, 5, 7, and 8 in mouse brain. *Human Gene Therapy*, *18*, 195–206.

Thomsen, N. D., Koerber, J. T., & Wells, J. A. (2013). Structural snapshots reveal distinct mechanisms of procaspase-3 and -7 activation. *Proceedings of the National Academy of Sciences of the United States of America*, *110*, 8477–8482.

Tonikian, R., Zhang, Y., Boone, C., & Sidhu, S. S. (2007). Identifying specificity profiles for peptide recognition modules from phage-displayed peptide libraries. *Nature Protocols*, *2*, 1368–1386.

Vickers, C. J., González-Páez, G. E., Umotoy, J. C., Cayanan-Garrett, C., Brown, S. J., & Wolan, D. W. (2013). Small-molecule procaspase activators identified using fluorescence polarization. *Chembiochem*, *14*, 1419–1422.

Wehr, M. C., Laage, R., Bolz, U., Fischer, T. M., Grünewald, S., Scheek, S., et al. (2006). Monitoring regulated protein-protein interactions using split TEV. *Nature Methods*, *3*, 985–993.

Wei, Y., Fox, T., Chambers, S. P., Sintchak, J., Coll, J. T., Golec, J. M., et al. (2000). The structures of caspases-1, -3, -7 and -8 reveal the basis for substrate and inhibitor selectivity. *Chemistry & Biology*, *7*, 423–432.

West, D. C., Qin, Y., Peterson, Q. P., Thomas, D. L., Palchaudhuri, R., Morrison, K. C., et al. (2012). Differential effects of procaspase-3 activating compounds in the induction of cancer cell death. *Molecular Pharmaceutics*, *9*, 1425–1434.

Williams, D. J., Puhl, H. L., 3rd., & Ikeda, S. R. (2009). Rapid modification of proteins using a rapamycin-inducible tobacco etch virus protease system. *PLoS One*, *4*, e7474.

Wolan, D. W., Zorn, J. A., Gray, D. C., & Wells, J. A. (2009). Small-molecule activators of a proenzyme. *Science*, *326*, 853–858.

Yang, X., Chang, H. Y., & Baltimore, D. (1998). Autoproteolytic activation of pro-caspases by oligomerization. *Molecular Cell*, *1*, 319–325.

Yang, C. F., Chiang, M. C., Gray, D. C., Prabhakaran, M., Alvarado, M., Juntti, S. A., et al. (2013). Sexually dimorphic neurons in the ventromedial hypothalamus govern mating in both sexes and aggression in males. *Cell*, *153*, 896–909.

Yuan, J., & Kroemer, G. (2010). Alternative cell death mechanisms in development and beyond. *Genes & Development*, *24*, 2592–2602.

Zorn, J. A., & Wells, J. A. (2010). Turning enzymes ON with small molecules. *Nature Chemical Biology*, *6*, 179–188.

Zorn, J. A., Wille, H., Wolan, D. W., & Wells, J. A. (2011). Self-assembling small molecules form nanofibrils that bind procaspase-3 to promote activation. *Journal of the American Chemical Society*, *133*, 19630–19633.

Zorn, J. A., Wolan, D. W., Agard, N. J., & Wells, J. A. (2012). Fibrils colocalize caspase-3 with procaspase-3 to foster maturation. *The Journal of Biological Chemistry*, *287*, 33781–33795.

CHAPTER NINE

A Multipronged Approach for Compiling a Global Map of Allosteric Regulation in the Apoptotic Caspases

Kevin Dagbay[1], Scott J. Eron[1], Banyuhay P. Serrano[1], Elih M. Velázquez-Delgado[1,2], Yunlong Zhao[1], Di Lin[3], Sravanti Vaidya[4], Jeanne A. Hardy[5]

Department of Chemistry, University of Massachusetts, Amherst, Massachusetts, USA

[1]These authors contributed equally to the work of producing this paper.

[5]Corresponding author: e-mail address: hardy@chem.umass.edu

Contents

[2] Present address: Department of Structural Biology, St. Jude Children's Research Hospital, Memphis, TN, USA.

[3] Present address: College of Pharmacy, Purdue University, West Lafayette, IN, USA.

[4] Present address: Biotechnology Department, MS Ramaiah Institute of Technology, Bangalore, India.

Methods in Enzymology, Volume 544
ISSN 0076-6879
http://dx.doi.org/10.1016/B978-0-12-417158-9.00009-1

Abstract

One of the most promising and as yet underutilized means of regulating protein function is exploitation of allosteric sites. All caspases catalyze the same overall reaction, but they perform different biological roles and are differentially regulated. It is our hypothesis that many allosteric sites exist on various caspases and that understanding both the distinct and overlapping mechanisms by which each caspase can be allosterically controlled should ultimately enable caspase-specific inhibition. Here we describe the ongoing work and methods for compiling a comprehensive map of apoptotic caspase allostery. Central to this approach are the use of (i) the embedded record of naturally evolved allosteric sites that are sensitive to zinc-mediated inhibition, phosphorylation, and other posttranslational modifications, (ii) structural and mutagenic approaches, and (iii) novel binding sites identified by both rationally-designed and screening-derived small-molecule inhibitors.

1. INTRODUCTION

As has been so skillfully outlined in Chapter 7, the apoptotic caspases are central players in both the initiation and execution stages of apoptotic cell death and also function in a number of other biological processes. Due to these roles, control of caspase function is of great interest for applications in medicine, genetic engineering, and developmental biology of multicellular organisms. To date, no caspase-directed therapeutics have successfully emerged from clinical trials. One likely contributor of these failures is that the vast majority of caspase-directed drug candidates to date have targeted the most conserved region of caspases, the active (orthosteric), substrate-binding site. Although the substrate-binding sites of caspases are strongly conserved, other regions of caspases are far less conserved. From a biological perspective, it is quite clear that each caspase is regulated independently from other family members and performs unique roles under various homeostatic conditions due to structural and functional differences that originate away from the active site. It is our view that a more comprehensive and nuanced understanding of the constellations of allosteric[1] regulation of each caspase would not only further our appreciation of the function of each

[1] In this work, we define allosteric as non-active-site regions of the protein.

caspase but would also allow us to select the most effective and selective inhibitory or activating site for uniquely targeting caspases individually.

The central hypothesis of this work is that inhibitors or activators that function at different allosteric sites may yield critical functional differences in cell-based activity, which may ultimately be exploited therapeutically. Evidence for this continues to emerge in caspases, as an exosite required for unique substrate recognition in caspase-7 relative to caspase-3 has recently been discovered (Boucher, Blais, & Denault, 2012). We envision that inhibitors that bind to that site could block a particular subset of cellular caspase-7 substrates. Similarly, the caspase-6 S257D phosphomimetic is catalytically inactive against self-processing and peptide substrates, but can cleave its own prodomain. This suggests that different substrates interact uniquely with caspase-6. Two active-site tetrafluorophenoxymethyl inhibitors with extremely similar biochemical inhibition of caspases-3 and -6 also show very different inhibition of cleavage of mutant huntingtin protein intracellularly (Leyva et al., 2010), further underscoring this idea. Together these observations and the data that are described below suggest that inhibitors that function at different caspase allosteric sites should yield different cellular outcomes, which should prove to be therapeutically useful.

Allosteric drugs, in general, are proving to be an effective means for achieving selectivity within families of highly related proteins. To date, the vast majority of allosteric sites used were discovered by chance. In this chapter, we describe a multipronged approach that we have developed and are using in an ongoing way to identify the full spectrum of allosteric regulation that is applicable in the apoptotic caspases. We envision that the same approach that we describe may also be applicable in other protein families, which is an additional motivation for this chapter. Here we describe how we combine the information encrypted in the record of naturally evolved allosteric sites that can be accessed by interrogating posttranslationally modified caspases, zinc-inhibited forms of caspases, mutagenic studies, and the interactions of both novel and known chemical ligands to produce a comprehensive map of caspase allostery. The current map of caspase allostery (Fig. 9.1) combines data from our work with data from many other labs and has identified at least seven nonoverlapping allosteric sites on various caspases. The emerging themes from this work are (1) various caspases are differentially regulated by the same chemical signals (e.g. zinc, phosphorylation), (2) many different sites within the caspase catalytic core domain are sensitive to regulation, and (3) indirect allosteric control of the loops that compose the active site by many inputs modulates caspase function.

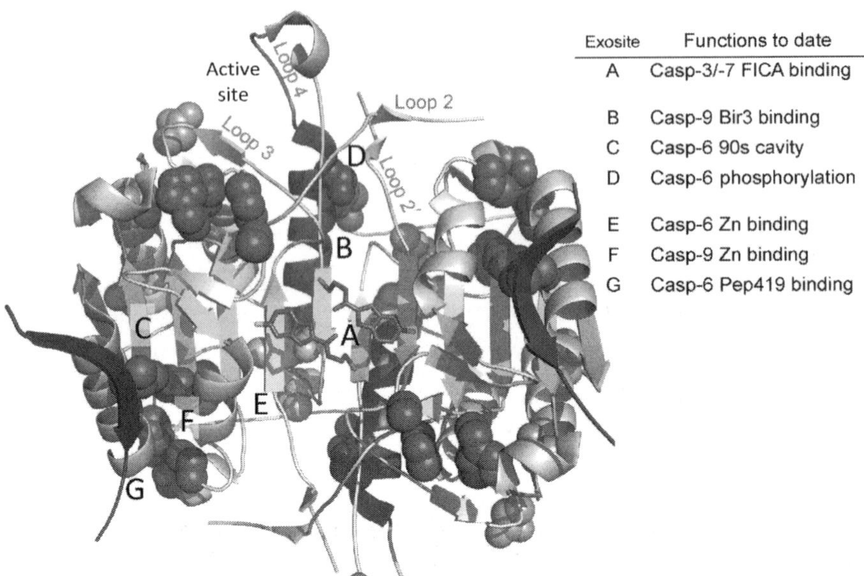

Figure 9.1 Current map of global apoptotic caspase allostery. Exosites that have been identified by the methods included in this chapter are annotated on a canonical caspase structure (gray cartoon). Red, phosphorylation; blue, zinc binding; green, chemical ligands. Various shades of red are used to denote sites of phosphorylation in different caspases. The green caspase helix that forms exosite B is highlighted in the chemical ligands to indicate that this is the region that interacts with the BIR3 region of XIAP- and BIR3-derived peptide inhibitors. *Adapted from Velazquez-Delgado (2012) with permission.* (See the color plate.)

2. ZINC-MEDIATED ALLOSTERIC INHIBITION OF CASPASES

Inhibition of caspases by zinc was reported soon after the discovery of caspases (Chimienti, Seve, Richard, Mathieu, & Favier, 2001; Schrantz et al., 2001; Stennicke & Salvesen, 1997). Until recently, the molecular details and sites of inhibition were not known. We have shown that each apoptotic caspase has unique metal-binding and inhibition profiles (Huber & Hardy, 2012; Velazquez-Delgado & Hardy, 2012b). Caspases-6 and -7 each bind just one zinc per monomer, whereas caspase-9 binds two zincs (Huber & Hardy, 2012) and caspase-3 binds three zincs (Velazquez-Delgado & Hardy, 2012b). Caspase-6 binds zinc only at exosite E (Fig. 9.2A), but not at the active site (Velazquez-Delgado & Hardy, 2012b). Thus, understanding metal-binding preferences provides insights

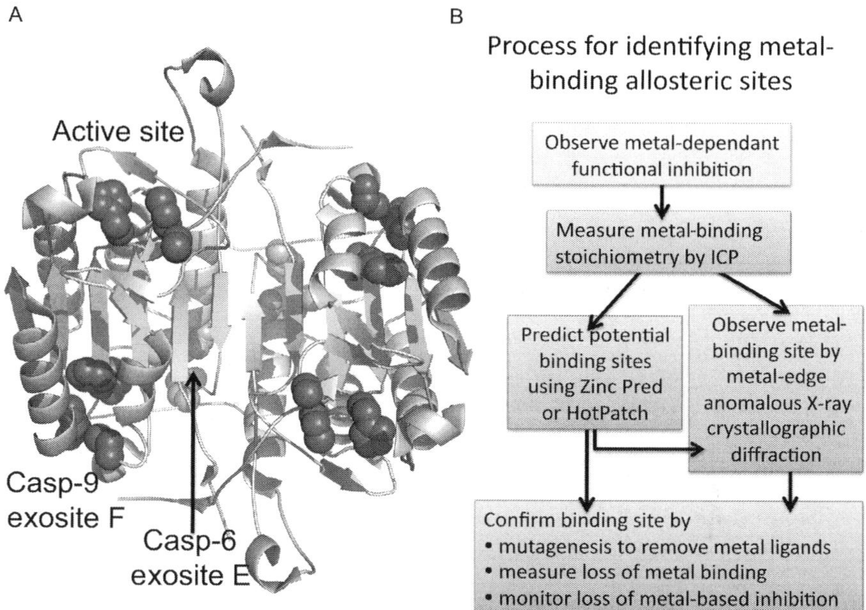

Figure 9.2 Discovery of zinc-binding sites in caspases. (A) Exosites for zinc binding that have been identified to date are shown as spheres on a canonical caspase structure (cartoon). The caspase-9 active site is capable of binding zinc; however, the active site of caspase-6 is not the dominant zinc-binding site in the unliganded protein. Caspase-6 is inhibited by zinc at exosite E. (B) Flowchart enumerating the methods used to identify metal-binding exosites and their mechanisms of allosteric inhibition in caspases. *(A) Adapted from Velazquez-Delgado (2012) with permission.*

into mechanisms that can be exploited to achieve caspase-specific inhibition, so long as the preferences are understood for all related caspases. The process of utilizing zinc binding to define allosteric sites (Fig. 9.2B), described in Sections 2.1 and 2.2, utilizes widely available approaches that are applicable for all caspases as well as for virtually any other protein in which the inhibition by zinc or other metals has been observed.

To date, no clear biological role for zinc-mediated regulation of caspases has been reported; nevertheless, the currently available data suggest that zinc may play a role in regulation of apoptosis. The life–death balance in the cell is tightly coupled to zinc levels. It is becoming increasingly clear that zinc is a critical cellular regulator that some suggest may play roles as significant for cellular homeostasis as calcium is to signal transduction (for review, see Fukada & Kambe, 2011). Even the smallest fluctuation in cellular zinc concentration appears to tip a cell toward survival or apoptotic cell death

(Zalewski, Forbes, & Betts, 1993), which could be in part due to the role of zinc-mediated inhibition of caspases (Perry et al., 1997; Stennicke & Salvesen, 1997). Moreover, controlling zinc levels has been harnessed in several biological contexts. The pathogenic bacterium *Helicobacter pylori* sequesters and then releases zinc into infected cells, inhibiting caspase activity and avoiding apoptotic cell death (Kohler et al., 2010, 2009). These data may be consistent with a model in which at low zinc levels, zinc-mediated inhibition of caspases is released, allowing apoptosis to be induced. Patients with asthma and chronic bronchitis are prone to zinc deficiency leading to increased levels of apoptosis in airway epithelium (Carter et al., 2002; Truong-Tran, Grosser, Ruffin, Murgia, & Zalewski, 2003). PAC-1 is a serendipitous small-molecule procaspase-3 activator which does not directly activate procaspase-3 (Denault, Drag, et al., 2007) but works by relieving zinc-mediated inhibition (Peterson et al., 2009), showing that zinc-mediated inhibition of caspases may be therapeutically exploited. Although physiologically "free" or unliganded zinc concentrations are reported to be in the femto- to picomolar range (Bozym, Thompson, Stoddard, & Fierke, 2006; Krezel & Maret, 2006), the "available" zinc pool appears to be much higher. Eukaryotic cells contain ~200 μM zinc (Krezel & Maret, 2006) where small shifts in glutathione concentration or oxidative stress release zinc from the metallothioneins (Krezel, Hao, & Maret, 2007) or secretory vesicles. Although it is too early to conclude that zinc plays a significant physiological role directly in caspase regulation, it is possible that dissecting the relationship of zinc and caspases may itself be therapeutically relevant, in addition to its utility in defining the allosteric map for caspases described here.

2.1. Allosteric site identification in caspase-9 using metal-binding site prediction algorithms

Caspase-9 is inhibited in the core domain by zinc but not by other metals (Fig. 9.3A; Huber & Hardy, 2012) and binds two zincs per monomer (Huber & Hardy, 2012). We predicted eight putative zinc-binding sites by combining HotPatch and PREDZINC analysis with visual inspection of the caspase-9 structure. Mutagenesis of the true zinc ligands prevented zinc binding (as measured by inductively coupled plasma–optical emission spectroscopy (ICP–OES)), whereas mutagenesis of the other sites had no effect on zinc binding. Zinc binds to the caspase-9 active site as well as to exosite F comprising C230, H224, and C272 (Huber & Hardy, 2012) (Fig. 9.3B). Zinc binding to the active site is the primary site of inhibition,

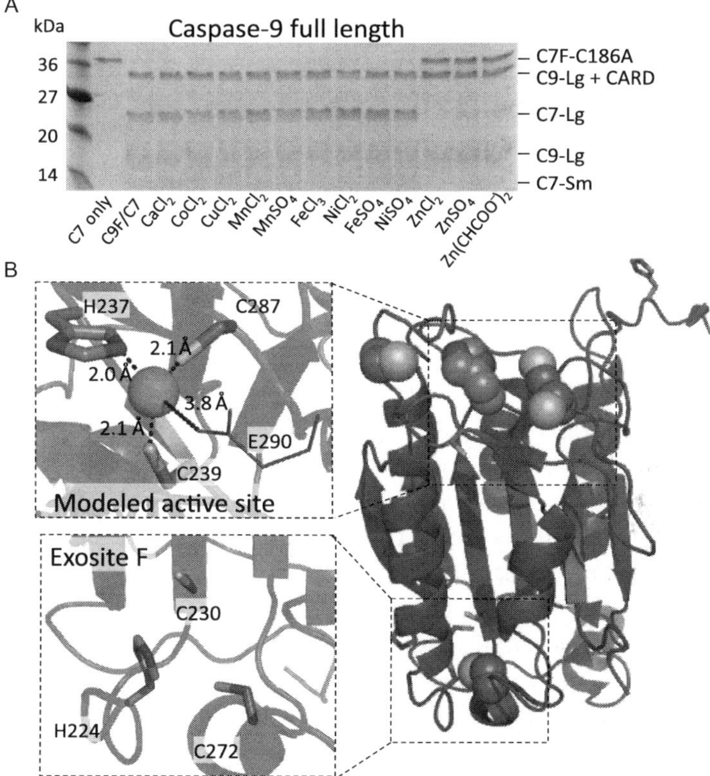

Figure 9.3 Zinc binds and inhibits both active and allosteric sites in caspase-9. (A) Zinc is the predominant metal cation to inhibit full-length caspase-9 (C9 FL) as monitored by cleavage of a natural caspase-9 substrate, the caspase-7 zymogen (C7 C186A) to the caspase-7 large (C7-Lg) and small (C7-Sm) subunits. (B) Location of the conserved active-site and exosite-ligand clusters on caspase-9 (PDB ID 1JXQ). A model of caspase-9 active-site ligand interactions with a modeled zinc ion was obtained by altering the H237, C239, and C287 rotamers in PyMol. *Adapted from Huber and Hardy (2012) with permission.* (See the color plate.)

since the K_i for zinc when exosite F is ablated (5.0 ± 2.8 μ*M*) is similar to that of WT caspase-9 (1.5 ± 0.3 μ*M*). Detailed kinetic analysis shows a mixed mode of inhibition, which suggests that exosite zinc binding is also involved in inhibition and indicates that exosite F should be categorized as a functional allosteric site. Given the ligand sphere identified, we generated a model of zinc binding to caspase-9 (Fig. 9.3B). Three of the four zinc-liganding residues in exosite F are conserved across the caspase family, but we have shown that caspase-6 does not bind zinc at exosite F (Velazquez-Delgado & Hardy, 2012b). This suggests that exosite F may require all four zinc ligands to robustly bind zinc and may therefore not be functional in other caspases, making it unique to caspase-9. This approach using zinc-binding site prediction coupled with mutagenesis can also feed into the methods outlined in Section 2.2 (Fig. 9.2), provided the zinc-anomalous diffraction experiment is technically feasible.

2.2. Allosteric site identification in caspase-6 using X-ray crystallography with anomalous diffraction

Caspase-6 is inhibited by zinc but not by other transition metals tested (Fig. 9.4A) (Velazquez-Delgado & Hardy, 2012b). Using ICP–OES to quantify zinc, we observed that each inhibited caspase-6 monomer bound just one zinc (Fig. 9.4B). To identify the ligands composing the zinc-binding site, we soaked crystals of caspase-6 with zinc and then performed an anomalous diffraction experiment with X-rays tuned at the zinc edge. Anomalous X-ray diffraction is a robust method for unambiguously locating metal-binding sites as only those atoms that diffract anomalously are observed at high intensity. One peak (5σ) per monomer was observed in the zinc-anomalous map (Fig. 9.4C). This site is composed of three liganding residues, K36, E244, and H287, and one water molecule. Removal of any of these ligands prevents zinc binding as assessed by ICP–OES (Fig. 9.4B), further validating the location observed in the anomalous X-ray diffraction experiment. E244 makes a bidentate interaction with the zinc, which exists in a distorted tetrahedral geometry. Lysine residues are uncommon but not unheard of ligands for zinc. In fact, 3% of zincs present in structures in the protein data bank use lysine as a ligand. Given that this preferred site in caspase-6 utilizes lysine as a ligand, we can envision that a site such as this could be overlooked by prediction algorithms, making anomalous X-ray diffraction combined with ligand removal by mutagenesis the preferred method for identifying zinc-sensitive allosteric sites. A structure-based sequence alignment indicates that exosite E is not present in any of the other caspases. Inhibition for the exosite E mutants (K36A, E244A, H287A) by zinc is much weaker than the WT suggesting that although zinc may be able to inhibit caspase-6 at the active site, exosite E is the predominant inhibitory site. Most importantly, no zinc is observed bound to caspase-6 at low or neutral pH when exosite E is knocked out. This result demonstrates that caspase-6 exists in the helical conformation, wherein the 120's and 60's regions are converted from strand or loop to helix. This helical conformation is observed for apo caspase-6 (Baumgartner et al., 2009; Vaidya, Velazquez-Delgado, Abbruzzese, & Hardy, 2011). In this helical

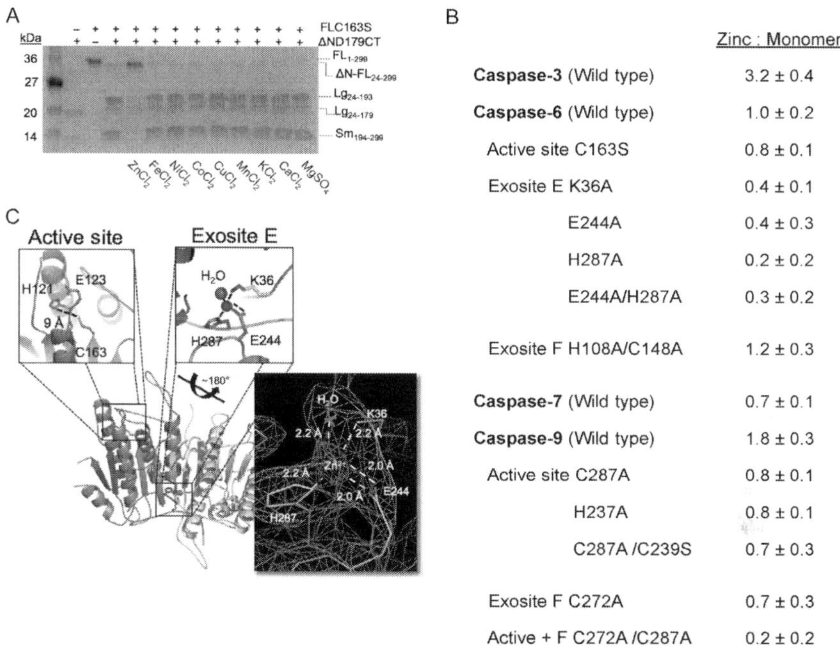

	Zinc : Monomer
Caspase-3 (Wild type)	3.2 ± 0.4
Caspase-6 (Wild type)	1.0 ± 0.2
Active site C163S	0.8 ± 0.1
Exosite E K36A	0.4 ± 0.1
E244A	0.4 ± 0.3
H287A	0.2 ± 0.2
E244A/H287A	0.3 ± 0.2
Exosite F H108A/C148A	1.2 ± 0.3
Caspase-7 (Wild type)	0.7 ± 0.1
Caspase-9 (Wild type)	1.8 ± 0.3
Active site C287A	0.8 ± 0.1
H237A	0.8 ± 0.1
C287A /C239S	0.7 ± 0.3
Exosite F C272A	0.7 ± 0.3
Active + F C272A /C287A	0.2 ± 0.2

Figure 9.4 Zinc binds an allosteric site to inhibit caspase-6. (A) Metal inhibition was tested in an electrophoretic mobility gel-based assay as fragments produced from active caspase-6-(ΔND179CT) mediated cleavage of the full-length C163S zymogen (substrate). Fragments from cleavage include the following: full length (FL), FL lacking the N-terminal prodomain (ΔN-FL), large (Lg), and small subunits (Sm) with the amino acids present in those bands (subscripts) labeled. Zinc is the only metal cation that inhibits caspase-6 activity. (B) The zinc-binding stoichiometry for various caspases, as well as mutants designed to ablate zinc binding are shown. Zinc binding was measured by inductively coupled plasma–optical emission spectroscopy. (C) The structure of zinc-bound caspase-6 (ribbons, PDB ID 4FXO). An anomalous difference map calculated from data collected above the zinc absorbance edge is contoured at 5σ (inset, purple mesh), clearly indicating the location of zinc at the allosteric site (black inset). The side chains and water serving as ligands are drawn as sticks. The 2Fo-Fc electron density map (blue) into which the structure was build is contoured at 1σ. The boxed regions show active site, which contains residues appropriate for metal binding including H121, E126, and C163. In this structure, these residues are not properly positioned to coordinate zinc. The caspase-6 zinc-binding exosite is shown with the side chain ligands for zinc in sticks (K36, E244, H287), zinc (blue), and the water molecule (red). *Adapted from Velazquez-Delgado and Hardy (2012b) with permission.* (See the color plate.)

conformation, the active-site liganding residues H121 and C163 are too far apart to robustly bind zinc (Fig. 9.4C) at both low and neutral pHs. Zn inhibits caspase-6 by locking the enzyme into the inactive helical conformation (discussed at greater length in Section 5.2). We envision that similar zinc-binding allosteric sites may exist on a wide number of proteins, given that up to 10% of proteins are functionally sensitive to zinc (Fukada & Kambe, 2011). The methods outlined here (inhibition and ICP studies combined with mutagenesis of zinc-liganding side chains and structure determination with anomalous diffraction tuned at the zinc edge) should be applicable to the study any of these zinc-sensitive sites.

3. USING THE EMBEDDED RECORD OF FUNCTIONALLY IMPORTANT POSTTRANSLATIONAL MODIFICATIONS TO IDENTIFY ALLOSTERICALLY SENSITIVE SITES

The majority of posttranslational modifications (PTMs) occur because they play a functional role in the modified protein (Minguez et al., 2012; Seet, Dikic, Zhou, & Pawson, 2006). Many examples of PTMs inducing protein–protein interactions or altering substrate recognition which mediate many other crucial signaling pathways exist. Like many proteins, caspases are posttranslationally modified. In addition to proteolytic cleavage (zymogen activation), caspases are also phosphorylated, nitrosylated, ubiquitinated and glutathionylated in a manner that impacts function. We have seen that probing the way that nature has exploited functionally sensitive sites by PTM is a useful means of uncovering new allosteric sites.

3.1. Phosphorylation in apoptotic caspases

Caspases and kinases work in concert to regulate one another in a complex, intertwining web. Many caspases, including the apoptotic caspases-2, -3, -6, -7, -8, and -9, have been reported to be extensively phosphorylated, typically leading to inactivation and avoidance of apoptosis. In the active state, caspases often cleave and inactivate the very kinase that phosphorylates them. Caspases are highly phosphorylated and the sites and impact of phosphorylation continue to be reported (for review, see Kurokawa & Kornbluth, 2009; Lopez-Otin & Hunter, 2010). There are a large number of ongoing studies to map the caspase–kinase interplay. For example, the Cravatt lab recently completed a comprehensive mapping of phosphorylation and caspase cleavage (Dix et al., 2012). Some sites of phosphorylation directly block sites of autoprocessing or cleavage by upstream caspases (Duncan et al., 2011), but most others appear to lead to caspase inactivation. Thus, many of the sites of caspase phosphorylation are an embedded historical record of sensitive sites from which caspase function can be activated or inhibited.

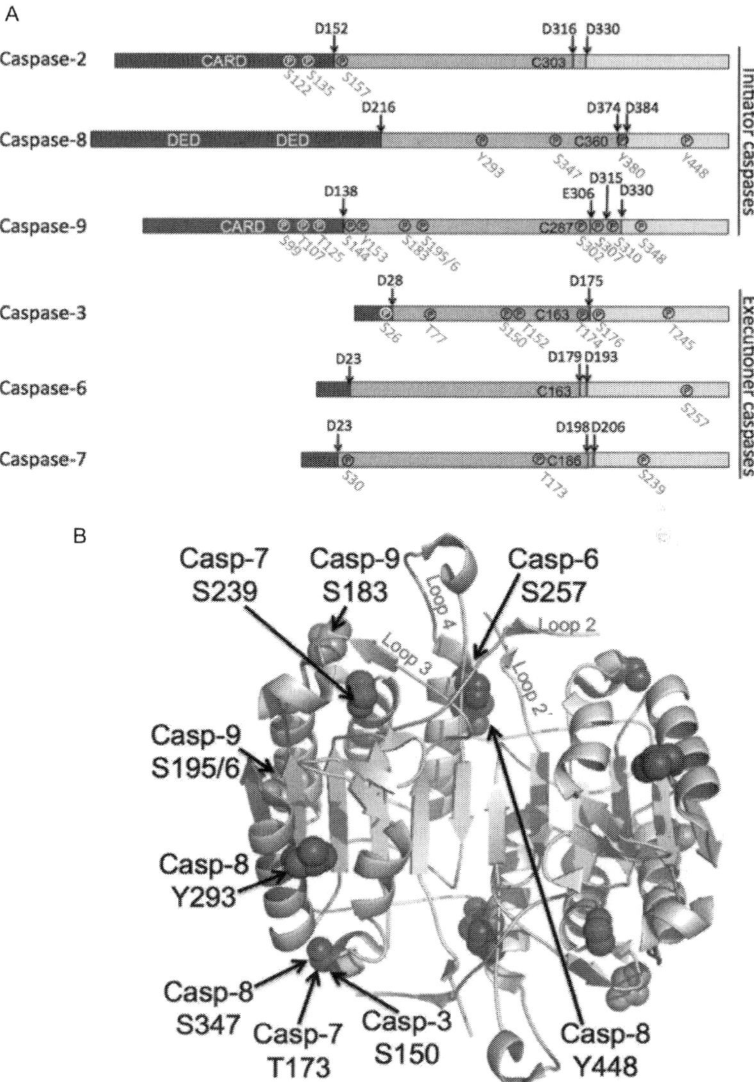

Figure 9.5 Confirmed sites of phosphorylation in apoptotic caspases. (A) This map represents the major, confirmed sites of phosphorylation in the apoptotic caspases. We recognize that some other sites of phosphorylation have been reported in various databases. (B) Structural distribution of functionally critical sites of phosphorylation. *(A) Adapted from Velazquez-Delgado and Hardy (2012a) with permission. (B) Adapted from Velazquez-Delgado (2012) with permission.* (See the color plate.)

A large number of sites of phosphorylation have been identified to date (Fig. 9.5A and B); however, in many cases, the exact biological conditions that lead to phosphorylation and the extent to which particular sites are phosphorylated have not yet been extensively studied. So far, three

PAK-2 sites: T77, T152, and T245 in caspase-3 and three sites: S30, T173, and S239 in caspase-7 were identified via peptide-based mapping (Li et al., 2011). In neutrophils, caspases-3 and -8 are phosphorylated by p38 MAPK at S150 and S347, respectively (Alvarado-Kristensson et al., 2004). S257 was the confirmed phosphorylation site of caspase-6 (Suzuki et al., 2004). Three more phosphorylation sites in caspase-3, S26, T174, and S176, are adjacent to the cleavage sites in the caspase-3 prodomain and intersubunit linker (ISL) (Dix et al., 2012; Duncan et al., 2011). In addition to the single serine phosphorylation on caspase-8, S347, there are three tyrosine phosphorylation sites at Y293, Y380, and Y448 (Jia, Parodo, Kapus, Rotstein, & Marshall, 2008). Research in neutrophils indicates that an intense competition occurs between these sites. The winner decides the fate of the cell by either maintaining the inhibited state or activating apoptosis. Both Y380 and Y448 are phosphorylated by the Src tyrosine kinase Lyn, which results in resistance to activational cleavage and thus a dearth of caspase-8-mediated apoptosis. However, phosphorylation on Y293 is reported to stimulate an interaction with the phosphatase SHP-1, which then dephosphorylates caspase-8 and allows for apoptosis to proceed.

To date, caspase-9 appears to be the most heavily phosphorylated caspase, with 12 known phosphorylation sites variously recognized by 9 different kinases. This complex regulatory system spans the entirety of the protein, including phosphorylation sites on the caspase activation and recruitment domain (CARD), the large and small subunits, as well as the ISL. The most studied phosphorylation event occurs at T125, which is modified by four different kinases: ERK1, ERK2, DYRK1A, and CDK1. Phosphorylation at T125 results in the inability of caspase-9 to be activated (Allan & Clarke, 2007; Allan et al., 2003; Seifert, Allan, & Clarke, 2008) and thus a suppression of apoptosis. This has been linked to cancer and tumorigenesis, with specific implications of this phosphothreonine as a hallmark for gastric carcinomas (Yoo, Lee, Jeong, & Lee, 2007). In addition, three other kinases, Akt, PKA, and PKCζ, phosphorylate caspase-9 and lead to inactivation, stifling cell death. In contrast to these inhibitory effects, it is reported that the kinase c-Abl phosphorylates caspase-9 and promotes autoprocessing (Raina et al., 2005). It was observed that c-Abl phosphorylates caspase-9 at Y153 after exposing the cell to DNA-damaging agents. The results indicated an enhancement of caspase-9 autoprocessing and acceleration of apoptosis after genotoxic stress. Given that the vast majority of phosphorylation sites in caspases lead to a change in function, by examining sites that are phosphorylated (Fig. 9.5), we can come to understand divergent mechanisms of allosteric regulation that can be exploited within this family.

3.2. Mechanism of phosphorylation-based allosteric caspase-6 inhibition

Only one site of phosphorylation, S257, has been reported to date for caspase-6. ARK5 kinase phosphorylates caspase-6 S257 (Fig. 9.5A), leading to inactivation (Suzuki et al., 2004). Given that this residue is not in the active site or substrate-binding groove, we hypothesized that this phosphorylation constituted a natural mechanism of allosteric regulation. Our work (Velazquez-Delgado & Hardy, 2012a) was reported nearly concurrently with another study of S257 phosphorylation (Cao et al., 2012). These two papers were the first to provide any structural detail of how phosphorylation leads to inactivation of any caspase. To understand the mechanistic details of allosteric inactivation at this site, both groups utilized a phosphomimetic approach, in which S257 was replaced by Asp (S257D) (Velazquez-Delgado & Hardy, 2012a) or Glu (S257E) (Cao et al., 2012). Both Glu and Asp have been successful phosphomimetics, as the length and charge are similar to phosphoserine. S257D is a robust phosphomimetic that is inactive, similar to caspase-6 phosphorylated by purified ARK5 kinase (Fig. 9.6A). By subjecting the phosphomimetic to limited proteolysis by caspases-3, -6, or -9, we observed that the active-site loops were ordered differently from the wild-type (WT) enzyme (Fig. 9.6B). To visualize the newly ordered active-site loop bundle, we determined the crystal structure of caspase-6 S257D. The crystal structure of caspase-6 S257D suggests that phosphorylation of S257 inactivates caspase-6 through a steric clash with a single residue, P201, in the L2′ loop (Velazquez-Delgado & Hardy, 2012a) (Fig. 9.6C and E). This steric clash causes the L2′ loop to be dramatically reordered into the inactive down position (Fig. 9.6C). The phosphomimetic S257D is dramatically inactivated with a k_{cat}/K_M that is three orders of magnitude lower than WT (Fig. 9.6D). S257E is also dramatically inactivated relative to WT (Cao et al., 2012). Substitution of S257 with uncharged but large residues also leads to inactivation, suggesting that charge is not critical to the mechanism of inhibition. Removal of the clashing proline side chain by the P201G substitution relieves the inhibition (Fig. 9.6D and E), demonstrating that steric clash is responsible for the phosphorylation-mediated inhibition of caspase-6. Thus, it is clear that phosphomimetic versions of caspases are a powerful method for

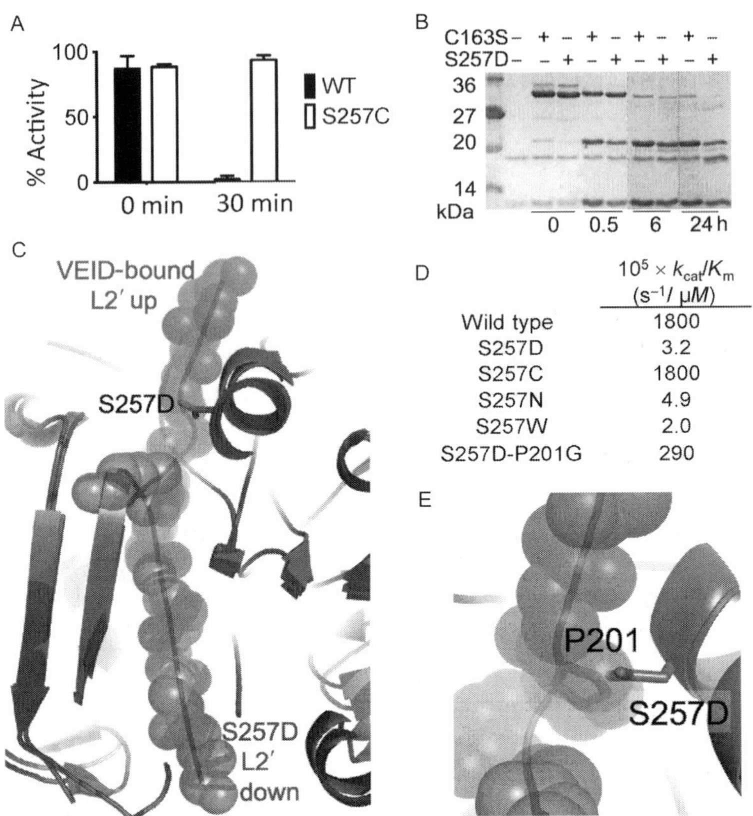

Figure 9.6 S257D is a robust phosphomimetic that elucidated the mechanism of allosteric inhibition by phosphorylation. (A) WT caspase-6 is inhibited by 30 min incubation with ARK5 kinase, whereas the S257C unphosphorylatable mutant is unaffected by ARK5. The S257D phosphomimetic is inactive similar to phosphorylated WT caspase-6. (B) The S257D active-site loops are ordered differently from the C163S zymogen, leading to different rates of cleavage of S257D from C163S. (C) In the structure of S257D (PDB ID 3S8E), the L2′ loop is in a down position (inactive, blue) in contrast to the VEID (substrate)-bound structure in which the L2′ loop is in the up position (active, red). (D) Mutations at S257 show that size is more important than charge in S257 inhibition and that any residue larger than cysteine or serine results in inactivation. Removal of P201 by mutation to glycine restores activity to S257D indicating that a steric clash between P201 and S257D leads to loss of activity. (E) The steric clash between P201 and S257D mirrors that of phosphoserine 257. *Adapted from Velazquez-Delgado and Hardy (2012a) with permission.* (See the color plate.)

understanding how phosphorylation leads to structural and functional changes allosterically. Because this site has not been reported to be phosphorylated in other caspases, this particular site seems to be unique to caspase-6. Close inspection of sites of phosphorylation in other caspases lead us to predict that this region may be ripe for allosteric regulation by phosphorylation by other kinases. For example, in caspase-8, the Y448 site in the same region (Fig. 9.5B) is phosphorylated by Lyn kinase leading to inactivation (Jia et al., 2008).

3.3. Other PTMs that allosterically regulate caspases

While phosphorylation of caspases has been extensively reported, other PTMs, which mediate many other crucial signaling pathways, also occur on caspases to regulate structure, function, translocation, or interactions with other proteins, sometimes in an allosteric manner.

To date, no robust proteomics analysis to identify the global PTMs in caspases has been reported. Although the current list of caspase PTMs has been compiled in a case-by-case manner, a significant number of PTMs that are related to the oxidation of cysteine have been identified. As cysteine proteases, caspases are sensitive to redox signaling, so cellular oxidative stress provides an abundant source for functional PTMs at thiol groups in caspases. Nitric oxide was first reported to be able to directly inhibit caspases-1, -2, -3, -4, -6, -7, and -8 through S-nitrosylation and thereby inhibit apoptosis (Dimmeler, Haendeler, Nehls, & Zeiher, 1997; Kim, Talanian, & Billiar, 1997; Li, Billiar, Talanian, & Kim, 1997). Recently, the S-nitrosylation site in caspase-3 was identified to be the catalytic cysteine C163 in nitrite-treated endothelial cells (Lai et al., 2011). GSSG is another oxidative agent with dose-dependent behavior that inhibits caspase-3 activity via protein glutathionylation at the catalytic cysteine and C220 in helix 4 (Huang, Pinto, Deng, & Richie, 2008). It is of note that oxidizing PTMs are usually reversible and sensitive to reducing agents. Both denitrosylation and deglutathionylation were observed in DTT-treated samples with a resulting restoration of caspase-3-like activity (Huang et al., 2008; Li et al., 1997).

Although the primary mode of inhibitory redox-sensitive PTMs is to directly ablate the nucleophilicity of the catalytic cysteine (e.g., caspase-3 C163), other cysteines in caspases are modified. Caspase-3 C220 is a known secondary glutathionylation site (Huang et al., 2008), but there are other solvent-accessible cysteines that could be alternate PTM sites. At least three cysteines in caspase-3, for instance, are located in regions that are significant to caspase function. C264 (caspase-7 C290, see Section 4.1) is located at the dimer interface where modification can lead to inhibition or activation. Other possible mechanisms of inhibition we may expect to observe include

PTM influencing the orientation of the active-site loops at a distance as we have seen for small-molecule caspase modulators (see Section 4.1). Both C148 which is at the bottom of a helix and C184 in the L2 loop are located near a region that could make electrostatic interactions with the active-site loop bundle. Whether any of these sites could become potential targets for redox-related PTMs and contribute to an allosteric inhibition of caspases is still awaiting further elucidation. As an example of allosteric inhibition using redox PTM, the engineered "handcuff-like" disulfide bond locks caspase-7 in to the inactive apo-conformation under oxidative conditions when a reactive cysteine pair that was introduced across the dimer interface of caspase-7 is oxidized (Witkowski & Hardy, 2011).

In addition to the redox-sensitive sites, one functionally relevant nonredox PTM has been reported. In contrast to other caspases, caspase-8 activity is enhanced by polyubiquitination because it forms an aggregate with p62, a polyubiquitin-binding protein, and increases the stability of cleaved caspase-8 which enhances proteolytic activity (Jin et al., 2009). In addition, sumoylation of caspase-7 has been reported, which appears to impact nuclear localization of caspase-7 (Hayashi, Shirakura, Uehara, & Nomura, 2006).

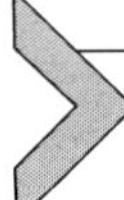

4. SYNTHETIC SMALL-MOLECULE INHIBITORS TO IDENTIFY ALLOSTERIC SITES

Synthetic chemical ligands have proven to be one excellent source of new information about novel allosteric sites in proteins (for review, see Hardy & Wells, 2004), conformational flexibility within proteins, and biological roles of enzymes. In caspases, the use of disulfide tethering, a method of site-directed ligand discovery (Erlanson et al., 2000; Erlanson, Wells, & Braisted, 2004), led Jim Well's group at Sunesis Pharmaceuticals to identify the first allosteric site in caspases-3 and -7 (Hardy, Lam, Nguyen, O'Brien, & Wells, 2004a) (exosite A, Fig. 9.7). Additional ligands that inhibited the inflammatory caspase-1 at the same site were discovered soon after (Datta, Scheer, Romanowski, & Wells, 2008; Scheer, Romanowski, & Wells, 2006), and this site continues to be a target of new small-molecule pan-caspase inhibitors (Feldman et al., 2012). Inhibition of procaspase-3 by zinc was defined upon exploration of the mechanism of activation by the chemical ligand PAC-1 (Peterson et al., 2009) which directly competes with the caspase-3 active site for zinc. We have shown that synthetic peptide ligands can be used to modulate dimerization of caspase-9 (Huber, Ghosh, & Hardy, 2012) (exosite B, Fig. 9.7) and a small peptide, Pep419, has been

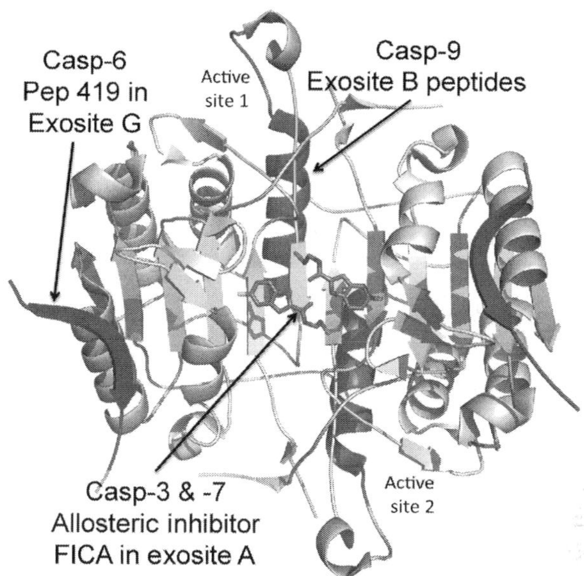

Figure 9.7 Sites for binding of chemical ligands that affect caspase function. The green caspase helix that forms exosite B is highlighted since it is the region that interacts with the BIR3 region of XIAP- and BIR3-derived peptide inhibitors. *Adapted from Velazquez-Delgado (2012) with permission.* (See the color plate.)

developed at Genentech, which inhibits caspase-6 allosterically (Stanger et al., 2012) (exosite G, Fig. 9.7). Thus, the use of chemical ligands has played a central role in uncovering new mechanisms of allosteric regulation in the caspases as well. Here we describe the methods that have allowed identification and characterization of these allosteric sites.

4.1. Allosteric inhibitors acting at a cavity at the apoptotic caspase dimer interface

The first synthetic allosteric inhibitors of caspases were discovered using a disulfide-tethering approach developed at Sunesis Pharmaceuticals (Erlanson et al., 2000, 2004). Tethering makes use of libraries of monophores (pharmacophores) attached to thiol-containing moieties. The presence of a thiol in each library compound allows formation of a disulfide bond with cysteine residues adjacent to the monophore-binding site. The affinity of the monophore binding to the protein modulates the residence time of the compound on the protein, increasing the likelihood of disulfide bond formation and allowing covalent selection of strongly interacting compounds (Erlanson et al., 2000, 2004). DICA (Fig. 9.8A) was identified using

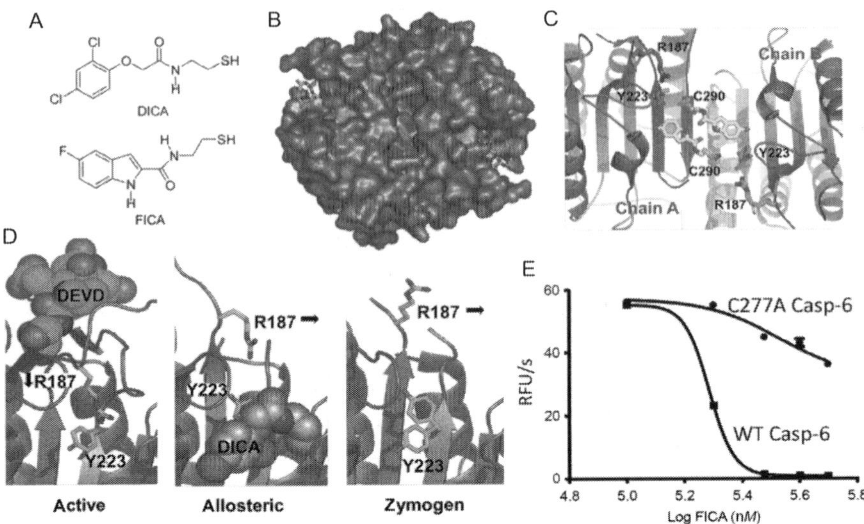

Figure 9.8 Allosteric inhibition at exosite A. (A) The structures of DICA and FICA. (B) The allosteric pocket at the caspase-7 (PDB ID 1F1J) dimer interface exosite A (pink). The FICA/DICA-binding residues C290 on each half of the dimer are shown as yellow patches at the bottom of the pocket. Active sites are denoted by DEVD active-site inhibitor (green sticks) bound to the catalytic cysteine (green patch). (C) The crystal structure of the FICA (yellow) bound caspase-7 exosite A at C290. (D) Conformations of Y233 and R187 shown for active, allosteric inhibitor (DICA, salmon spheres) bound and zymogen caspase-7. DICA bound at the allosteric site prevents R187 and Y233 from pointing down to generate a properly ordered substrate-binding pocket. (E) FICA inhibits wild-type (WT) caspase-6 much more efficiently than C277A caspase-6, indicating that FICA binds and inhibits caspase-6 at the same site and likely by the same mechanism as observed for caspase-7. *Panels A–D are adapted from Hardy et al. (2004a) with permission. Panel E adapted from Velazquez-Delgado (2012) with permission.* (See the color plate.)

a mass spectrometry-based screen for caspase-3-binding compounds (Hardy et al., 2004a). FICA (Fig. 9.8A) was identified using an apoptosis assay in whole-cell lysates and was found to functionally block cleavage and activation of caspase-3 (Hardy et al., 2004a). Both FICA and DICA control the activity of these caspases by binding specifically in a central cavity at the dimer interface (Fig. 9.8B). The allosteric inhibitors (DICA and FICA) bind covalently to a cysteine (C290 in caspase-7) in the allosteric site (Fig. 9.8C), trapping tyrosine 223 (Y223) in an up position. This conformation of Y223 prevents arginine (R187) from assuming the down conformation, which is required for substrate binding and activity (Fig. 9.8D). The substrate-binding groove in caspases is formed of three loops, L2, L3, and L4 from one half of the dimer, and L2′ from the other half of the dimer. In the allosterically inhibited forms of caspase-7, L2′ is bound over the allosteric site and thus unable to assist in the assembly of the substrate-binding groove. The relevant conformational changes that lead to allosteric inhibition can be readily visualized in movie form (Hardy, Lam, Nguyen, O'Brien, & Wells, 2004b). Therefore, disruption of the substrate-binding loops is sufficient to lead to inhibition (Hardy & Wells, 2009). Based on sequence and structural homology, we predicted that FICA and DICA would likewise inhibit caspase-6. C290 in caspase-7 is homologous to C277 in caspase-6. FICA inhibits WT caspase-6 but the inhibition is dramatically reduced when the allosteric site thiol (C277) is removed by mutation to alanine (Fig. 9.8E). This demonstrates that the dominant site of inhibition for caspase-6 is C277. Thus, we see that the three prominent apoptotic executioner caspases, caspases-3, -6, and -7, all possess the same allosteric site at the dimer interface.

Two other classes of small molecules that target the same cavity on the dimerization interface of caspases have been identified. Multiring, pyridinyl, copper-containing compounds (Comp-A, -B, -C, and -D), identified by high-throughput screening, act as pan-caspase inhibitors which inactivate both initiators (caspases-2, -8, and -9) and executioners (caspases-3 and -7) with submicromolar IC_{50} values (Feldman et al., 2012). Using Cu anomalous X-ray diffraction similar to that used for caspase-6 inhibition by zinc (Section 2.2) to locate the compounds in caspase-7 crystals, Comp-A was shown to bind at the same location where FICA and DICA bind (exosite A) and cause the active-site loops to be disordered. However, unlike FICA and DICA, Comp-A does not form a covalent bond with C290, thus functioning as a reversible inhibitor. In addition, this exosite A cavity in caspase-8 was shown to noncovalently bind a beta-strand peptidomimetic inhibitor developed at Pfizer (Wang, Watt, et al., 2010). Binding of the inhibitor induced conformational changes in this pocket. Interestingly, because the residues within this pocket are not conserved across caspases, the authors suggested that this site may be amenable to specific targeting of caspase-8. Given our experience with this site (exosite A, Fig. 9.7) which allows inhibition by the same ligands (FICA and DICA) in caspases-3, -6, and -7, we think this site is functionally conserved across caspases. The level of homology in this pocket between caspase-8 and other caspases will likely dictate whether this cavity will indeed be effective for use in a caspase-8-specific manner.

4.2. Rationally designed synthetic peptide allosteric inhibitors mimic XIAP BIR3 dimerization site blocking in caspase-9

Regulation of caspase-9 is distinct from other caspases. Caspase-9 is active only as a dimer (Fig. 9.9A), but can be held in an inactive state by the BIR3 domain of the XIAP protein which blocks dimerization allosterically (Fig. 9.9B). This precise mechanism has not been observed for any other caspase, although it shares some similarities with the interactions of caspase-8 with FLIP (Boatright et al., 2003). Knowing how nature has exploited this allosteric site, we made a set of allosteric inhibitors that mimic

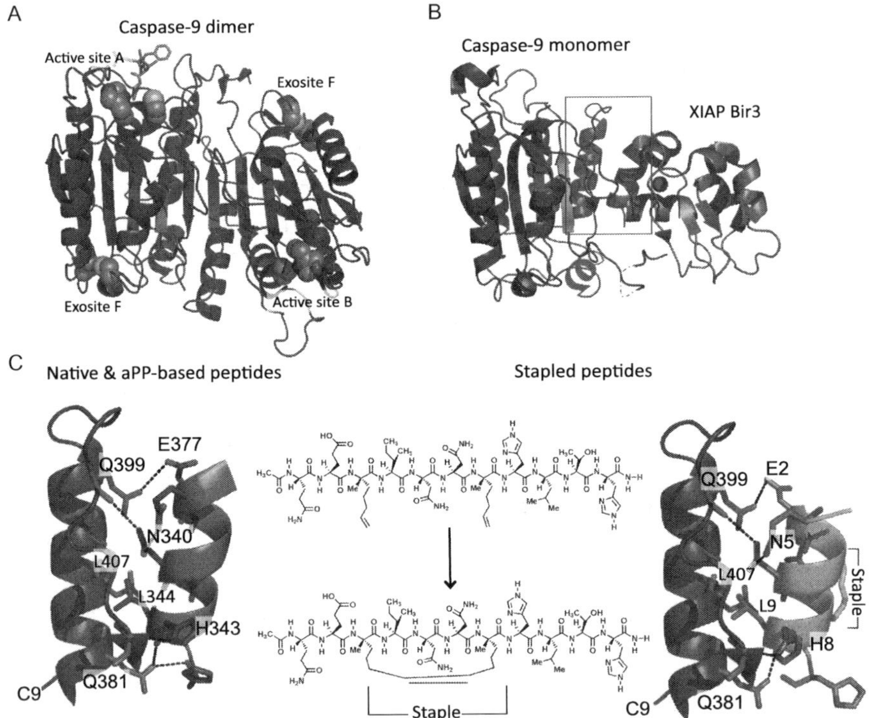

Figure 9.9 Allosteric inhibition of caspase-9 at the dimer interface. (A) Caspase-9 is a dimer composed of A and B chains, each of which has the same amino acid sequence and each possessing the residues required to form one active site. In the structure of mature caspase-9 (PDB ID 1JXQ), only one active site (active site A) is observed in a catalytically competent conformation. Green sticks mark peptide inhibitor in the A-chain active site. For reference, the active site and a zinc-binding exosite F are drawn as orange spheres. (B) XIAP BIR3 domain prevents dimerization and activation of caspase-9 (PDB ID 1NW9). (C) Three classes of BIR3-mimicking peptides produced by novel synthetic schemes developed are shown. *Adapted from Huber et al. (2012) with permission.* (See the color plate.)

the interaction of XIAP BIR3 with caspase-9 in an effort to develop caspase-9-specific inhibitors. The BIR3 domain has been reported to be conformationally unstable (Shin et al., 2005) so we predicted that it would be necessary to stabilize the α5 helix, which interacts directly with caspase-9, in order to develop potent caspase-9 inhibitors. We engineered three types of stabilized peptides, each designed to remain helical and block caspase-9 dimerization (Fig. 9.9C) (Huber, Ghosh, & Hardy, 2012). We envision that both BIR3 and these peptides compete with caspase-9 monomers for binding the dimerization interface, such that caspase-9 monomers are positive allosteric modulators and BIR3-based peptides are negative allosteric modulators. We built native peptides derived directly from BIR3 and peptides in which caspase-9-interacting residues were grafted on the stable avian pancreatic polypeptide scaffold. For peptides made of native amino acids, we developed an expression and purification method for producing long peptides more quickly and with higher fidelity than chemical synthesis (Huber, Olson, & Hardy, 2009). We also incorporated the nonnative amino acid aminoisobutyric acid, which is α-methylated and enforces peptide helicity in some of the designed peptides. In addition, we synthesized a number of peptides stabilized by an aliphatic "staple" (Fig. 9.9C). We introduced staples 8 and 11 atoms long using nonnative alkenyl amino acids and a ring closing metathesis reaction. All three types of stabilized peptides are helical in solution, in contrast to native peptides, which were not helical. The best inhibitor to date shows a K_i of 16 μM. This level of inhibition by properly folded, helical peptides not only suggests that the entire BIR3 domain is essential for optimal caspase-9 inhibition but also indicates that the dimerization interface of caspase-9 is exploitable as an allosteric site.

Other studies have also targeted this region of the initiator caspases to prevent dimerization. Caspase-9 activated by the apoptosome and an engineered constitutively dimeric caspase-9 were inhibited by small molecules (Comp-A, -B, -C, -D) binding to exosite A at the dimerization interface (Feldman et al., 2012). Active caspase-8 is dimeric, but also exists in equilibrium with its monomeric form, which can be resolved by size-exclusion chromatography. Binding of Comp-A shifted the dimeric caspase-8 to its monomeric form with a concurrent decrease in activity. This is in contrast to caspase-7 which remained dimeric upon binding of Comp-A, although in an inactive, incompetent conformation (Feldman et al., 2012). Thus, it appears that allosteric regulation of dimerization may be uniquely applicable to the initiator caspases which have a low intrinsic dimerization affinity relative to the executioner caspases.

4.3. Synthetic inhibitors of caspase-6

A small peptide, Pep419, developed at Genentech, inhibits caspase-6 allosterically (Stanger et al., 2012) and led to the discovery of exosite G (Fig. 9.7). Pep419 was identified using a phage display screen against caspase-6 zymogen and selectively inhibits caspase-6 function *in vitro* and *in vivo*. Pep419 binds at a novel tetramer interface that centrally involves residue H126 in the binding interface, which is not conserved among executioner caspases. The tetrameric state of caspase-6 zymogen they observed by multiangle light scattering is novel relative to the many reported structures of executioner caspase zymogens as dimers (Chai et al., 2001; Riedl et al., 2001; Wang, Watt, et al., 2010). Pep419 is an allosteric regulator of both active and zymogen caspase-6, which follows a novel noncompetitive mechanism that involves a seemingly global effect on protein dynamics affecting the overall oligomeric state of the enzyme rather than on a specific conformational change. The authors suggest that Pep419 binding stabilizes the tetrameric caspase-6 zymogen form and traps active caspase-6 in that tetrameric, zymogen-like form.

While most reported caspase allosteric inhibitors bind to exosites (Fig. 9.1), a unique uncompetitive, 11 n*M* inhibitor of caspase-6 binds selectively to the enzyme–substrate complex that impedes substrate turnover (Heise et al., 2012). Generated from a chemical optimization screen, compound 3 is an optimized analog of an *N*-furoyl-phenylalanine designed to contain a meta-cyano-substituted D-phenylalanine and a demethylated furan ring. The binding of compound 3 to caspase-6 is dependent on the identity of the peptide substrate and, surprisingly, on the identity of the attached fluorophore. The ternary crystal structure of caspase-6/z-VEID/compound 3 reveals a novel platform for uncompetitive inhibition for any members of the caspase family. This type of inhibition could occur only in the presence of a second small molecule like the fluorophore or in a larger compound composed of both the compound 3 pharmacophore and a fluorophore mimic. Nevertheless, compound 3 and Pep419 provide novel mechanisms for caspase inhibitors that work allosterically or unconventionally.

5. EXPLOITING STRUCTURAL DIFFERENCES AMONG CASPASES

A wealth of structural information is available for all of the apoptotic caspases. These data suggest that the core domain of caspases is highly conserved at a structural level. On the other hand, the prodomains are highly divergent and very poorly structurally characterized. In addition, at least one apoptotic caspase, caspase-6, can attain a unique helical structure that is unattainable by any other caspase. Structurally unique features are prime for exploitation by allosteric regulators.

5.1. Structural and organizational differences between the apoptotic caspases

Structurally, all caspases contain the highly homologous protease or core domain that is further subdivided into a large (17–20 kDa) and a small (10–12 kDa) subunit that are held together in the zymogen precursor state by an intersubunit linker (ISL). The N-terminus of caspases is a stretch of residues that comprise the prodomain, with the initiator caspases having longer prodomains than the executioners (Fig. 9.10). This difference in prodomain size and structure reflects their respective function in caspase activation. Initiator caspases utilize their prodomains to serve as platforms for protein–protein interactions that are critical during their full activation. Caspase-9 has a caspase activation and recruitment domain (CARD), while caspase-8 contains the death effector domain (DED) to facilitate initial activation. Thus, the CARD and DED are potential targets to inhibit

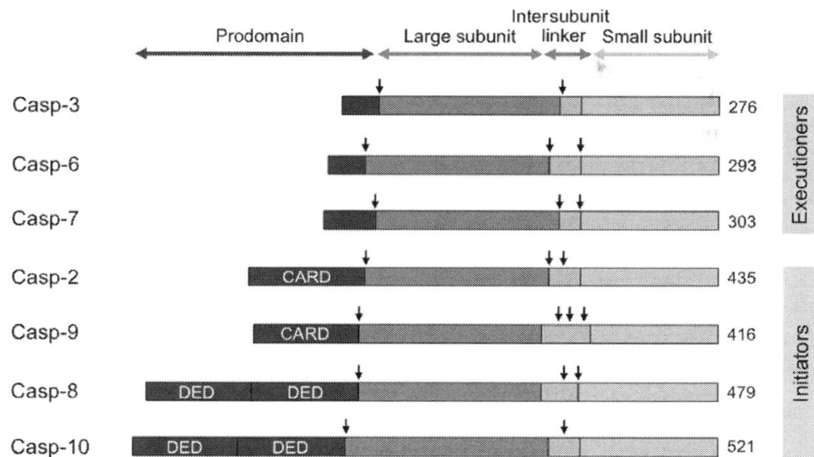

Figure 9.10 Domain organization among the apoptotic caspases. Caspases consist of a prodomain, large subunit, intersubunit linker, and small subunit. Executioner caspases have a short prodomain involved in substrate recognition which has also been proposed to function as a folding chaperone, whereas initiator caspases have a longer prodomain involved in protein:protein interactions. The cleavage sites that lead to production of the mature caspases are indicated by vertical arrows.

initiator caspases. To date, no impact on catalytic function has been attributed to the prodomains of executioner caspases, as prodomain-deleted versions of executioner caspases have been observed to be as active as the full-length protein.

Caspases are synthesized and stored in the zymogen form (procaspase). Cleavage or processing of the ISL as well as dimerization confers maximum activity for caspases (Chai et al., 2001; Pop, Timmer, Sperandio, & Salvesen, 2006; Riedl et al., 2001; Vaidya et al., 2011; Watt et al., 1999; Wilson et al., 1994). Mature, active caspases exist as homodimers. Initiator caspases such as caspases-2, -8, -9, and -10 are monomeric when synthesized (Boatright et al., 2003; Butt, Harvey, Parasivam, & Kumar, 1998; Renatus, Stennicke, Scott, Liddington, & Salvesen, 2001; Wachmann et al., 2010) while executioners dimerize immediately upon ribosome release (Kang, Ko, Kwon, & Choi, 2002; Pop et al., 2001; Talanian et al., 1996). Due to low sequence homology of residues comprising the dimerization interface, the size and nature of the cavity formed at this interface also varies, allowing small molecules to differentially act on this region, whether by preventing the formation of a properly formed active site (e.g., caspase-7; Hardy et al., 2004a) or by inhibiting dimerization (caspases-9; Huber, Ghosh, & Hardy, 2012; Feldman et al., 2012 and -8 Feldman et al., 2012) (see Section 4).

Cleavage of the ISL switches caspases from their inactive to highly active forms. In the case of caspase-9, however, having a longer ISL allows it to facilely support a properly formed active-site loop bundle even prior to cleavage. Wells and colleagues have shown that procaspases-3 and -7, containing short ISLs, can also bind peptide inhibitors and adopt the active conformation even prior to ISL cleavage (Thomsen, Koerber, & Wells, 2013). This suggests that linker cleavage may not be solely a steric requirement but also an allosteric requirement due to the enhanced ability cleaved linker forms the loop bundle which, in turn, stabilizes the active state. In this regard, perhaps the loop bundles themselves could also be considered allosteric sites, although they are proximal to the active site. Almost all caspases have more than one cleavage site in their ISL and the order and pattern of cleavage influences the activity and regulation of the enzyme. In caspase-6, for example, cleavage at D193 occurs before cleavage at D179, as it has been observed that D193A variants are unable to cleave D179 (Vaidya & Hardy, 2011). For caspase-9, the initial cleavage event at D315 generates an epitope that is recognized by the BIR3 domain of XIAP leading to inhibition, while cleavage at D330 by caspase-3 relieves this inhibition (Denault, Eckelman, Shin, Pop, & Salvesen, 2007; Shiozaki et al., 2003; Srinivasula et al., 2001;

Zou et al., 2003). ISL cleavage also plays a role in ubiquitination of caspases, as only cleaved (mature) caspases-9 (Morizane, Honda, Fukami, & Yasuda, 2005) and -8 (Jin et al., 2009), but not their uncleaved zymogen forms are ubiquitinated. We think it is likely that allosteric regulation specific to each caspase will involve exploiting differences in the domain organization of the caspases as discussed here (Fig. 9.10).

5.2. Unique conformation of caspase-6 relative to other caspases

A group at Novartis Pharmaceuticals reported the structure of cleaved (mature) unliganded (apo) caspase-6 (Baumgartner et al., 2009) simultaneous with our determination of the same structures in the presence and absence of the ISL (Vaidya et al., 2011) (Fig. 9.11). All of these structures showed caspase-6 in a conformation that is different than any other caspase structure before or since. Significantly, the 60's and 130's helices were extended, occluding the active site, and the 90's helix is rotated by 21° outward away from the core of the protein (Fig. 9.12). In the past several years, additional structures of caspase-6 in the uncleaved (immature) zymogen (Wang, Cao, et al., 2010) and bound to peptide-based, substrate-like active-site inhibitors have also emerged (Liu, Zhang, Wang, Li, & Su, 2011; Muller, Lamers, Ritchie, Dominguez, et al., 2011; Wang, Cao, et al., 2010) (Fig. 9.11). All of these structures have unique features, particularly in the loops (L1, L2, L3, and L4) which comprise the substrate-binding groove (detailed distinctions noted in Fig. 9.11). An additional structure of apo, mature caspase-6 showed the 130's region in the canonical conformation (Muller, Lamers, Ritchie, Park, et al., 2011) similar to that observed in the zymogen, substrate-bound conformations. This canonical apo mature structure crystallizes at neutral pH, whereas the helical conformation crystallizes at low pH. We have shown that caspase-6 exhibits a helical CD signal at neutral or low pH in the absence of active-site substrate-like ligands and loses ~20% of the helical signal when substrate-like ligands bind, which causes 20% of the helical residues convert to strand or loops (Vaidya et al., 2011). In addition, the fact that the active site of caspase-6 does not bind zinc in solution at neutral or low pH strongly suggests that caspase-6 exists in an equilibrium that favors predominantly the helical conformation at all pHs until substrate binds (Velazquez-Delgado & Hardy, 2012b). Prior to substrate binding, the active-site residues are positioned too far apart to facilitate zinc binding. Together these data lead us to conclude that caspase-6 is capable of interconverting between the helical conformation

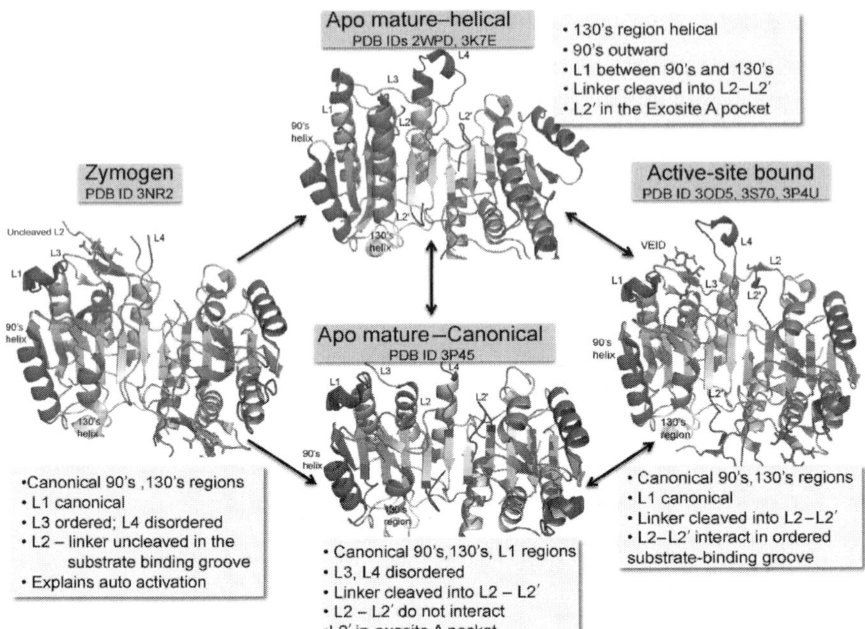

Figure 9.11 Conformational changes during caspase-6 lifecycle. Crystal structures of the caspase-6 are available for zymogen, mature and active-site bound forms. The zymogen of caspase-6 exists in a conformation that explains the structural basis of self-processing. The mature form exists in a conformational equilibrium between a helical and a canonical conformation, though in the canonical conformation far more regions of the protein are disordered. The equilibrium seems to favor the helical form at all pHs. The active-site bound form attains a structure that is extremely similar to all other active-site bound caspases. The colored regions highlight conformational changes that occur during the transitions between these states in the 90's helix (red), 130's region (orange), L1 loop (brown), L2 (green), L2′ (blue), L3 pink, and L4 (purple). Significant structural characteristics are denoted for each conformation (yellow boxes). (See the color plate.)

and the canonical conformation, but that caspase-6 exists predominantly in the helical conformation in solution.

The helical conformation (Fig. 9.12, orange) is of interest because it cannot bind substrate and is thus naturally inactive. This helical conformation cannot be adopted by any other caspase (Vaidya & Hardy, 2011). This is because caspase-6 is the only caspase lacking helix-breaking glycine and proline residues in this region, so it is the only caspase that can form an extended 130's helix (Vaidya & Hardy, 2011). Thus, the helical conformation appears to be unique to caspase-6. Any allosteric sites that exist in the helical

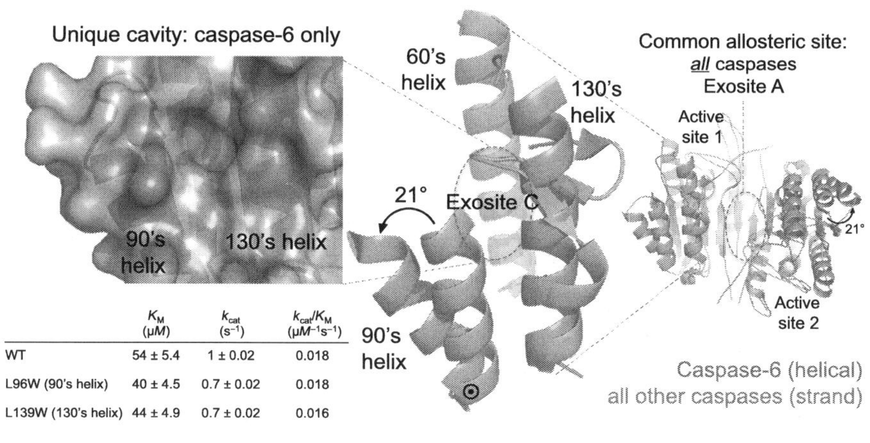

	K_M (μM)	k_{cat} (s^{-1})	k_{cat}/K_M ($\mu M^{-1}s^{-1}$)
WT	54 ± 5.4	1 ± 0.02	0.018
L96W (90's helix)	40 ± 4.5	0.7 ± 0.02	0.018
L139W (130's helix)	44 ± 4.9	0.7 ± 0.02	0.016
L96W/L139W	47 ± 2.0	$4.6 \pm 4.2 \times 10^{-3}$	9.8×10^{-5}

Figure 9.12 Caspase-6 (orange, PDB ID 3NKF) exists in an extended helical conformation that cannot be attained by any other caspase, which all have the same canonical strand structure as caspase-7 (green, PDB ID 1F1J) in the 90's helix. A cavity formed when the 90's helix rotates away from the 130's helix can function as a unique allosteric site at exosite C, as shown by insertion of tryptophan residues designed to hold the 90's helix in an open conformation. *Adapted from Vaidya et al. (2011) with permission.* (See the color plate.)

conformation but not in the canonical conformation should be extremely specific for caspase-6. We have identified one such potentially exploitable cavity (exosite C, Fig. 9.12). When residues in the 90's and 130's helices were replaced by tryptophan (L96W/L139W), the enzyme was locked in the helical conformation and was allosterically inactivated (Fig. 9.12, table). We have noted a marked circular dichroism signal in this helical state, which also serves as a robust spectroscopic signature for this unique, inactivated state (Vaidya et al., 2011), enabling identification of inhibitors that bind to the helical conformation.

5.3. Caspase allostery identified by mutagenesis

Savvy use of mutagenesis has also pinpointed allosterically sensitive sites in caspases. Clark and his group have shown that variants at dimer interface residue V266 exploit caspase-3 allostery. V266 is extremely sensitive in maintaining the dimeric state, and thus the conformation and catalytic function of caspase-3. The glutamate variant V266E in mature caspase-3 shows a lower activity and efficiency of apoptotic death than WT. Intriguingly, the V266E variant was also observed to have intrinsic activity even in an uncleavable version of procaspase-3 (D9A/D28A/D175A) (Walters et al., 2009), which shows an extremely low catalytic efficiency for the WT (V266) version (Bose, Pop, Feeney, & Clark, 2003). This suggests that V266E is capable of activating caspase-3 even before caspase-3 is processed. The authors propose that the negative charge of E266 disrupts the original hydrophobic cluster across the dimer interface and that a specified series of conformational changes in the loops allow the correct orientation of the catalytic and substrate-binding residues (Walters et al., 2009). In particular, the V266E mutation forces the ISL to an active position forming the substrate-binding pocket and loop bundle. This mechanistic suggestion is supported by the structure of procaspase-7 (Riedl et al., 2001) and by a recently published structure of WT procaspase-3 (Thomsen et al., 2013). In this structure, although the ISL is partially disordered, the residues (K186–V189) substantially shift to the dimer interface, confirming that the central cavity is blocked by the uncleaved linker. In addition, another crucial loop LH, which is parallel with loop 3, appears to be expelled out of the central cavity in the inactive WT procaspase-3. During the procaspase-3 activation by V266E, LH might also require an insertion into the relieved central cavity, similar to the motion of the L3 loop, for the active orientation of catalytic residue H121. Interestingly, a different variant at the same position, V266H, inhibits mature caspase-3 activity because it could prevent R164 and the L2 loop from inserting into the central cavity and thus increase the population of the inactive ensemble (Walters et al., 2009; Walters, Schipper, Swartz, Mattos, & Clark, 2012). Thus, it is clear based on these mutational and structural analyses that this region of the protein, overlapping with exosite A, is an allosterically modulatable region. This is significant because the same site has been identified by both mutagenesis and a number of small-molecule ligands, suggesting that mutagenesis alone can be very predictive of allosteric sites.

5.4. Exosites for substrate recognition in caspases

Proteases require precise cleavage instructions to suppress potentially devastating and irreversible effects. One effective method for cleavage control is through extremely selective sequence recognition, where only specific combinations of residues are recognized at the active site and cleaved by the protease. In addition to this, proteases have also evolved exosites that allow for further exclusivity of substrates as well as increasing the rate of catalysis. Thrombin is a prominent example of a protease that utilizes exosites to perform its many duties in the coagulation cascade (for review, see Huntington, 2005; Lane, Philippou, & Huntington, 2005). Exosites on the surface of thrombin are recognized by various substrates and cofactors that decide its ultimate fate (Stubbs & Bode, 1995; Verhamme, Olson, Tollefsen, & Bock, 2002).

It now appears clear that caspases also use exosites to increase efficiency of catalysis and improve selectivity for substrates. As the cell undergoes apoptosis, it is imperative that the active caspases carry out their duties swiftly and accurately to ensure a clean death. Two substrates, poly(ADP ribose)polymerase 1 (PARP) and p23, are critical apoptotic caspase targets. Both of these substrates are primarily cleaved by caspase-7, yet their cleavage site shares a recognition sequence for caspase-3, which has a much higher intrinsic activity than caspase-7. The Denault group discovered that caspase-7 predominately cleaves these substrates due to a charged exosite near the N-terminus of the protein. Four consecutive lysine residues create a basic patch that allows for caspase-7, the slower executioner, to selectively cleave PARP much more efficiently than caspase-3. Removal of the caspase-7 prodomain is essential for engaging this exosite, providing a first glimpse of the role of prodomain removal in the executioner caspases.

Discovery of this exosite indicates that caspase-7 requires more than just a consensus cleavage motif in the substrate to attain full specificity. Due to the sensitive nature of apoptosis and similarities across the caspase family, we hypothesize that other caspases likewise contain exosites that aid in achieving full proteolytic specificity. If these sites resemble the caspase-7 exosite, then they too may be distant from the active site. Discovery of these allosteric regions offers tremendous potential for precise control of individual caspases in both apoptotic and nonapoptotic functions.

6. FUTURE OF ALLOSTERIC SITE DISCOVERY

Work from a large number of labs has contributed and will continue to contribute to a global map of apoptotic caspase allostery. In the current map of caspase allostery (Fig. 9.1), important trends are already emerging. For example, the fact that chemical ligands, phosphorylation, and zinc all exploit the F/G region suggests that this region is important in controlling the function of multiple caspases. On the other hand, we have shown that exosite C is only present in caspase-6 (Fig. 9.1). Thus, the A or F/G exosite should be exploited to control caspases globally and the C exosite would be appropriate for inhibiting caspase-6 selectively. As we expand and improve our map of caspase allostery, we will identify regions that play critical roles in caspase function, improve our understanding of natural methods of regulation: phosphorylation and other PTMs as well as binding of zinc and other metals. This continued understanding will thus increase our chances for strategic control of apoptosis.

The methods outlined in this chapter are of interest because they can be undertaken by a wide number of groups. In a similar manner to the approach described here for the apoptotic caspases, any other family of proteins could likewise be studied. Several obvious approaches remain to be undertaken in the study of the apoptotic caspases. In addition to zinc, any other metals capable of inhibiting caspases could be used with anomalous diffraction to unambiguously identify the binding sites. Coupling this with a mutagenesis approach is a rapid way to confirm the binding site and determine whether the site is an allosteric inhibitory site. Whereas this chapter has focused on work in the apoptotic caspases, a full understanding of this class of enzymes requires comparative work with the inflammatory caspases and less functionally well-characterized caspases including caspase-2.

The most important outcome of this type of work in the future will be the ability to assess which type of inhibition (orthosteric vs. various allosteric sites) is most effective and specific. These results will also address the important hypothesis that engaging different allosteric regulatory mechanism will impact only certain subsets of caspase function. Confirming that certain kinds of inhibition (e.g. zinc vs. phosphorylation) result in unique cellular outcomes will allow development of new caspase-directed drugs that will likely avoid the problems with toxicity that have plagued caspase inhibitors in the clinic to date.

ACKNOWLEDGMENTS

S. J. E. was supported by NSF Grant IGERT DGE-0654128. B. P. S. was supported in part by National Research Service Award T32 GM08515 from the National Institute of Health. This work was supported by the National Institutes of Health GM80532.

REFERENCES

Allan, L. A., & Clarke, P. R. (2007). Phosphorylation of caspase-9 by CDK1/cyclin B1 protects mitotic cells against apoptosis. *Molecular Cell*, *26*(2), 301–310.

Allan, L. A., Morrice, N., Brady, S., Magee, G., Pathak, S., & Clarke, P. R. (2003). Inhibition of caspase-9 through phosphorylation at Thr 125 by ERK MAPK. *Nature Cell Biology*, *5*(7), 647–654.

Alvarado-Kristensson, M., Melander, F., Leandersson, K., Ronnstrand, L., Wernstedt, C., & Andersson, T. (2004). p38-MAPK signals survival by phosphorylation of caspase-8 and caspase-3 in human neutrophils. *Journal of Experimental Medicine*, *199*(4), 449–458.

Baumgartner, R., Meder, G., Briand, C., Decock, A., D'arcy, A., Hassiepen, U., et al. (2009). The crystal structure of caspase-6, a selective effector of axonal degeneration. *Biochemistry Journal, 423*, 429–439.

Boatright, K. M., Renatus, M., Scott, F. L., Sperandio, S., Shin, H., Pedersen, I. M., et al. (2003). A unified model for apical caspase activation. *Molecular Cell, 11*(2), 529–541.

Bose, K., Pop, C., Feeney, B., & Clark, A. C. (2003). An uncleavable procaspase-3 mutant has a lower catalytic efficiency but an active site similar to that of mature caspase-3. *Biochemistry, 42*(42), 12298–12310.

Boucher, D., Blais, V., & Denault, J. B. (2012). Caspase-7 uses an exosite to promote poly(ADP ribose) polymerase 1 proteolysis. *Proceedings of the National Academy of Sciences of the United States of America, 109*(15), 5669–5674.

Bozym, R. A., Thompson, R. B., Stoddard, A. K., & Fierke, C. A. (2006). Measuring picomolar intracellular exchangeable zinc in PC-12 cells using a ratiometric fluorescence biosensor. *ACS Chemical Biology, 1*(2), 103–111.

Butt, A. J., Harvey, N. L., Parasivam, G., & Kumar, S. (1998). Dimerization and autoprocessing of the Nedd2 (caspase-2) precursor requires both the prodomain and the carboxyl-terminal regions. *The Journal of Biological Chemistry, 273*(12), 6763–6768.

Cao, Q., Wang, X. J., Liu, C. W., Liu, D. F., Li, L. F., Gao, Y. Q., et al. (2012). Inhibitory mechanism of caspase-6 phosphorylation revealed by crystal structures, molecular dynamics simulations, and biochemical assays. *The Journal of Biological Chemistry, 287*(19), 15371–15379.

Carter, J. E., Truong-Tran, A. Q., Grosser, D., Ho, L., Ruffin, R. E., & Zalewski, P. D. (2002). Involvement of redox events in caspase activation in zinc-depleted airway epithelial cells. *Biochemical and Biophysical Research Communications, 297*(4), 1062–1070.

Chai, J., Wu, Q., Shiozaki, E., Srinivasula, S. M., Alnemri, E. S., & Shi, Y. (2001). Crystal structure of a procaspase-7 zymogen: Mechanisms of activation and substrate binding. *Cell, 107*(3), 399–407.

Chimienti, F., Seve, M., Richard, S., Mathieu, J., & Favier, A. (2001). Role of cellular zinc in programmed cell death: Temporal relationship between zinc depletion, activation of caspases, and cleavage of Sp family transcription factors. *Biochemical Pharmacology, 62*(1), 51–62.

Datta, D., Scheer, J. M., Romanowski, M. J., & Wells, J. A. (2008). An allosteric circuit in caspase-1. *Journal of Molecular Biology, 381*(5), 1157–1167.

Denault, J. B., Drag, M., Salvesen, G. S., Alves, J., Heidt, A. B., Deveraux, Q., et al. (2007). Small molecules not direct activators of caspases. *Nature Chemical Biology, 3*(9), 519, author reply 520.

Denault, J. B., Eckelman, B. P., Shin, H., Pop, C., & Salvesen, G. S. (2007). Caspase 3 attenuates XIAP (X-linked inhibitor of apoptosis protein)-mediated inhibition of caspase 9. *Biochemistry Journal, 405*(1), 11–19.

Dimmeler, S., Haendeler, J., Nehls, M., & Zeiher, A. M. (1997). Suppression of apoptosis by nitric oxide via inhibition of interleukin-1beta-converting enzyme (ICE)-like and cysteine protease protein (CPP)-32-like proteases. *Journal of Experimental Medicine, 185*(4), 601–607.

Dix, M. M., Simon, G. M., Wang, C., Okerberg, E., Patricelli, M. P., & Cravatt, B. F. (2012). Functional interplay between caspase cleavage and phosphorylation sculpts the apoptotic proteome. *Cell, 150*(2), 426–440.

Duncan, J. S., Turowec, J. P., Duncan, K. E., Vilk, G., Wu, C., Luscher, B., et al. (2011). A peptide-based target screen implicates the protein kinase CK2 in the global regulation of caspase signaling. *Science Signaling, 4*(172), ra30.

Erlanson, D. A., Braisted, A. C., Raphael, D. R., Randal, M., Stroud, R. M., Gordon, E. M., et al. (2000). Site-directed ligand discovery. *Proceedings of the National Academy of Sciences of the United States of America, 97*(17), 9367–9372.

Erlanson, D. A., Wells, J. A., & Braisted, A. C. (2004). Tethering: Fragment-based drug discovery. *Annual Review of Biophysics and Biomolecular Structure*, *33*, 199–223.

Feldman, T., Kabaleeswaran, V., Jang, S. B., Antczak, C., Djaballah, H., Wu, H., et al. (2012). A class of allosteric caspase inhibitors identified by high-throughput screening. *Molecular Cell*, *47*(4), 585–595.

Fukada, T., & Kambe, T. (2011). Molecular and genetic features of zinc transporters in physiology and pathogenesis. *Metallomics*, *3*(7), 662–674.

Hardy, J. A., Lam, J., Nguyen, J. T., O'Brien, T., & Wells, J. A. (2004). Discovery of an allosteric site in the caspases. *Proceedings of the National Academy of Sciences of the United States of America*, *101*(34), 12461.

Hardy, J. A., Lam, J., Nguyen, J. T., O'Brien, T., & Wells, J. A. (2004b). http://www.pnas.org/content/suppl/2004/08/09/0404781101.DC1/04781Movie2.mov.

Hardy, J. A., & Wells, J. A. (2004). Searching for new allosteric sites in enzymes. *Current Opinion in Structural Biology*, *14*(6), 706–715.

Hardy, J. A., & Wells, J. A. (2009). Dissecting an allosteric switch in caspase-7 using chemical and mutational probes. *The Journal of Biological Chemistry*, *284*(38), 26063.

Hayashi, N., Shirakura, H., Uehara, T., & Nomura, Y. (2006). Relationship between SUMO-1 modification of caspase-7 and its nuclear localization in human neuronal cells. *Neuroscience Letters*, *397*(1–2), 5–9.

Heise, C. E., Murray, J., Augustyn, K. E., Bravo, B., Chugha, P., Cohen, F., et al. (2012). Mechanistic and structural understanding of uncompetitive inhibitors of caspase-6. *PLoS One*, 7(12), e50864.

Huang, Z., Pinto, J. T., Deng, H., & Richie, J. P., Jr. (2008). Inhibition of caspase-3 activity and activation by protein glutathionylation. *Biochemical Pharmacology*, *75*(11), 2234–2244 [Research Support, N.I.H., Extramural].

Huber, K. L., Ghosh, S., & Hardy, J. A. (2012). Inhibition of caspase-9 by stabilized peptides targeting the dimerization interface. *Biopolymers Peptide Science*, *98*(5), 451–465.

Huber, K. L., & Hardy, J. A. (2012). Mechanism of zinc-mediated inhibition of caspase-9. *Protein Sciences*, *20*(7), 1056–1065.

Huber, K. L., Olson, K. D., & Hardy, J. A. (2009). Robust production of a peptide library using methodological synchronization. *Protein Expression and Purification*, *67*(2), 139–147.

Huntington, J. A. (2005). Molecular recognition mechanisms of thrombin. *Journal of Thrombosis and Haemostasis*, *3*(8), 1861–1872.

Jia, S. H., Parodo, J., Kapus, A., Rotstein, O. D., & Marshall, J. C. (2008). Dynamic regulation of neutrophil survival through tyrosine phosphorylation or dephosphorylation of caspase-8. *The Journal of Biological Chemistry*, *283*(9), 5402–5413.

Jin, Z., Li, Y., Pitti, R., Lawrence, D., Pham, V. C., Lill, J. R., et al. (2009). Cullin3-based polyubiquitination and p62-dependent aggregation of caspase-8 mediate extrinsic apoptosis signaling. *Cell*, *137*(4), 721–735.

Kang, B. H., Ko, E., Kwon, O. K., & Choi, K. Y. (2002). The structure of procaspase 6 is similar to that of active mature caspase 6. *Biochemistry Journal*, *364*(Pt. 3), 629–634.

Kim, Y. M., Talanian, R. V., & Billiar, T. R. (1997). Nitric oxide inhibits apoptosis by preventing increases in caspase-3-like activity via two distinct mechanisms. *The Journal of Biological Chemistry*, *272*(49), 31138–31148.

Kohler, J. E., Dubach, J. M., Naik, H. B., Tai, K., Blass, A. L., & Soybel, D. I. (2010). Monochloramine-induced toxicity and dysregulation of intracellular Zn2+ in parietal cells of rabbit gastric glands. *American Journal of Physiology—Gastrointestinal and Liver Physiology*, *299*(1), G170–G178.

Kohler, J. E., Mathew, J., Tai, K., Blass, A. L., Kelly, E., & Soybel, D. I. (2009). Monochloramine impairs caspase-3 through thiol oxidation and Zn2+ release. *Journal of Surgical Research*, *153*(1), 121–127.

Krezel, A., Hao, Q., & Maret, W. (2007). The zinc/thiolate redox biochemistry of metallothionein and the control of zinc ion fluctuations in cell signaling. *Archives of Biochemistry and Biophysics, 463*(2), 188–200.

Krezel, A., & Maret, W. (2006). Zinc-buffering capacity of a eukaryotic cell at physiological pZn. *Journal of Biological Inorganic Chemistry, 11*(8), 1049–1062.

Kurokawa, M., & Kornbluth, S. (2009). Caspases and kinases in a death grip. *Cell, 138*(5), 838–854.

Lai, Y. C., Pan, K. T., Chang, G. F., Hsu, C. H., Khoo, K. H., Hung, C. H., et al. (2011). Nitrite-mediated S-nitrosylation of caspase-3 prevents hypoxia-induced endothelial barrier dysfunction. *Circulation Research, 109*(12), 1375–1386.

Lane, D. A., Philippou, H., & Huntington, J. A. (2005). Directing thrombin. *Blood, 106*(8), 2605–2612.

Leyva, M. J., Degiacomo, F., Kaltenbach, L. S., Holcomb, J., Zhang, N., Gafni, J., et al. (2010). Identification and evaluation of small molecule pan-caspase inhibitors in Huntington's disease models. *Chemistry & Biology, 17*(11), 1189–1200.

Li, J., Billiar, T. R., Talanian, R. V., & Kim, Y. M. (1997). Nitric oxide reversibly inhibits seven members of the caspase family via S-nitrosylation. *Biochemical and Biophysical Research Communications, 240*(2), 419–424.

Li, X., Wen, W., Liu, K., Zhu, F., Malakhova, M., Peng, C., et al. (2011). Phosphorylation of caspase-7 by p21-activated protein kinase (PAK) 2 inhibits chemotherapeutic drug-induced apoptosis of breast cancer cell lines. *The Journal of Biological Chemistry, 286*(25), 22291–22299.

Liu, X., Zhang, H., Wang, X. J., Li, L. F., & Su, X. D. (2011). Get phases from arsenic anomalous scattering: De novo SAD phasing of two protein structures crystallized in cacodylate buffer. *PLoS One, 6*(9), e24227.

Lopez-Otin, C., & Hunter, T. (2010). The regulatory crosstalk between kinases and proteases in cancer. *Nature Reviews Cancer, 10*(4), 278–292.

Minguez, P., Parca, L., Diella, F., Mende, D. R., Kumar, R., Helmer-Citterich, M., et al. (2012). Deciphering a global network of functionally associated post-translational modifications. *Molecular Systems Biology, 8*, 599.

Morizane, Y., Honda, R., Fukami, K., & Yasuda, H. (2005). X-linked inhibitor of apoptosis functions as ubiquitin ligase toward mature caspase-9 and cytosolic Smac/DIABLO. *The Journal of Biochemistry, 137*(2), 125–132.

Muller, I., Lamers, M. B., Ritchie, A. J., Dominguez, C., Munoz-Sanjuan, I., & Kiselyov, A. (2011). Structure of human caspase-6 in complex with Z-VAD-FMK: New peptide binding mode observed for the non-canonical caspase conformation. *Bioorganic and Medicinal Chemistry Letters, 21*(18), 5244–5247.

Muller, I., Lamers, M. B., Ritchie, A. J., Park, H., Dominguez, C., Munoz-Sanjuan, I., et al. (2011). A new apo-caspase-6 crystal form reveals the active conformation of the apoenzyme. *Journal of Molecular Biology, 410*(2), 307–315.

Perry, D. K., Smyth, M. J., Stennicke, H. R., Salvesen, G. S., Duriez, P., Poirier, G. G., et al. (1997). Zinc is a potent inhibitor of the apoptotic protease, caspase-3. A novel target for zinc in the inhibition of apoptosis. *The Journal of Biological Chemistry, 272*(30), 18530–18533.

Peterson, Q. P., Goode, D. R., West, D. C., Ramsey, K. N., Lee, J. J., & Hergenrother, P. J. (2009). PAC-1 activates procaspase-3 in vitro through relief of zinc-mediated inhibition. *Journal of Molecular Biology, 388*(1), 144–158.

Pop, C., Chen, Y. R., Smith, B., Bose, K., Bobay, B., Tripathy, A., et al. (2001). Removal of the pro-domain does not affect the conformation of the procaspase-3 dimer. *Biochemistry, 40*(47), 14224–14235.

Pop, C., Timmer, J., Sperandio, S., & Salvesen, G. S. (2006). The apoptosome activates caspase-9 by dimerization. *Molecular Cell, 22*(2), 269–275.

Raina, D., Pandey, P., Ahmad, R., Bharti, A., Ren, J., Kharbanda, S., et al. (2005). c-Abl tyrosine kinase regulates caspase-9 autocleavage in the apoptotic response to DNA damage. *The Journal of Biological Chemistry*, *280*(12), 11147–11151.

Renatus, M., Stennicke, H. R., Scott, F. L., Liddington, R. C., & Salvesen, G. S. (2001). Dimer formation drives the activation of the cell death protease caspase 9. *Proceedings of the National Academy of Sciences of the United States of America*, *98*(25), 14250.

Riedl, S. J., Fuentes-Prior, P., Renatus, M., Kairies, N., Krapp, S., Huber, R., et al. (2001). Structural basis for the activation of human procaspase-7. *Proceedings of the National Academy of Sciences of the United States of America*, *98*(26), 14790.

Scheer, J. M., Romanowski, M. J., & Wells, J. A. (2006). A common allosteric site and mechanism in caspases. *Proceedings of the National Academy of Sciences of the United States of America*, *103*(20), 7595–7600.

Schrantz, N., Auffredou, M. T., Bourgeade, M. F., Besnault, L., Leca, G., & Vazquez, A. (2001). Zinc-mediated regulation of caspases activity: Dose-dependent inhibition or activation of caspase-3 in the human Burkitt lymphoma B cells (Ramos). *Cell Death and Differentiation*, *8*(2), 152–161.

Seet, B. T., Dikic, I., Zhou, M. M., & Pawson, T. (2006). Reading protein modifications with interaction domains. *Nature Reviews. Molecular Cell Biology*, 7(7), 473–483.

Seifert, A., Allan, L. A., & Clarke, P. R. (2008). DYRK1A phosphorylates caspase 9 at an inhibitory site and is potently inhibited in human cells by harmine. *The FEBS Journal*, *275*(24), 6268–6280.

Shin, H., Renatus, M., Eckelman, B. P., Nunes, V. A., Sampaio, C. A., & Salvesen, G. S. (2005). The BIR domain of IAP-like protein 2 is conformationally unstable: Implications for caspase inhibition. *Biochemistry Journal*, *385*(Pt. 1), 1–10.

Shiozaki, E. N., Chai, J., Rigotti, D. J., Riedl, S. J., Li, P., Srinivasula, S. M., et al. (2003). Mechanism of XIAP-mediated inhibition of caspase-9. *Molecular Cell*, *11*(2), 519–527.

Srinivasula, S. M., Hegde, R., Saleh, A., Datta, P., Shiozaki, E., Chai, J., et al. (2001). A conserved XIAP-interaction motif in caspase-9 and Smac/DIABLO regulates caspase activity and apoptosis. *Nature*, *410*(6824), 112–116.

Stanger, K., Steffek, M., Zhou, L., Pozniak, C. D., Quan, C., Franke, Y., et al. (2012). Allosteric peptides bind a caspase zymogen and mediate caspase tetramerization. *Nature Chemical Biology*, *8*(7), 655–660.

Stennicke, H. R., & Salvesen, G. S. (1997). Biochemical characteristics of caspases-3, -6, -7, and -8. *The Journal of Biological Chemistry*, *272*(41), 25719.

Stubbs, M. T., & Bode, W. (1995). The clot thickens: Cues provided by thrombin structure. *Trends in Biochemical Sciences*, *20*(1), 23–28.

Suzuki, A., Kusakai, G., Kishimoto, A., Shimojo, Y., Miyamoto, S., Ogura, T., et al. (2004). Regulation of caspase-6 and FLIP by the AMPK family member ARK5. *Oncogene*, *23*(42), 7067–7075.

Talanian, R. V., Dang, L. C., Ferenz, C. R., Hackett, M. C., Mankovich, J. A., Welch, J. P., et al. (1996). Stability and oligomeric equilibria of refolded interleukin-1beta converting enzyme. *The Journal of Biological Chemistry*, *271*(36), 21853–21858.

Thomsen, N. D., Koerber, J. T., & Wells, J. A. (2013). Structural snapshots reveal distinct mechanisms of procaspase-3 and -7 activation. *Proceedings of the National Academy of Sciences of the United States of America*, *110*(21), 8477–8482.

Truong-Tran, A. Q., Grosser, D., Ruffin, R. E., Murgia, C., & Zalewski, P. D. (2003). Apoptosis in the normal and inflamed airway epithelium: Role of zinc in epithelial protection and procaspase-3 regulation. *Biochemical Pharmacology*, *66*(8), 1459–1468.

Vaidya, S., & Hardy, J. A. (2011). Caspase-6 latent state stability relies on helical propensity. *Biochemistry*, *50*(16), 3282–3287.

Vaidya, S., Velazquez-Delgado, E. M., Abbruzzese, G., & Hardy, J. A. (2011). Substrate-induced conformational changes occur in all cleaved forms of caspase-6. *Journal of Molecular Biology*, *406*(1), 75–91.

Velazquez-Delgado, E. M. (2012). *Allosteric regulation of caspase-6 proteolytic activity*. Amherst: University of Massachusetts, unpublished dissertation.

Velazquez-Delgado, E. M., & Hardy, J. A. (2012a). Phosphorylation regulates assembly of the caspase-6 substrate-binding groove. *Structure*, *20*(4), 742–751.

Velazquez-Delgado, E. M., & Hardy, J. A. (2012b). Zinc-mediated allosteric inhibition of caspase-6. *The Journal of Biological Chemistry*, *287*(43), 36000–36011.

Verhamme, I. M., Olson, S. T., Tollefsen, D. M., & Bock, P. E. (2002). Binding of exosite ligands to human thrombin. Re-evaluation of allosteric linkage between thrombin exosites I and II. *TheJournal of Biological Chemistry*, *277*(9), 6788–6798.

Wachmann, K., Pop, C., van Raam, B. J., Drag, M., Mace, P. D., Snipas, S. J., et al. (2010). Activation and specificity of human caspase-10. *Biochemistry*, *49*(38), 8307–8315.

Walters, J., Pop, C., Scott, F. L., Drag, M., Swartz, P., Mattos, C., et al. (2009). A constitutively active and uninhibitable caspase-3 zymogen efficiently induces apoptosis. *Biochemistry Journal*, *424*(3), 335–345.

Walters, J., Schipper, J. L., Swartz, P., Mattos, C., & Clark, A. C. (2012). Allosteric modulation of caspase 3 through mutagenesis. *Bioscience Reports*, *32*(4), 401–411.

Wang, X. J., Cao, Q., Liu, X., Wang, K. T., Mi, W., Zhang, Y., et al. (2010). Crystal structures of human caspase 6 reveal a new mechanism for intramolecular cleavage self-activation. *EMBO Reports*, *11*(11), 841–847.

Wang, Z., Watt, W., Brooks, N. A., Harris, M. S., Urban, J., Boatman, D., et al. (2010). Kinetic and structural characterization of caspase-3 and caspase-8 inhibition by a novel class of irreversible inhibitors. *Biochimica et Biophysica Acta*, *1804*(9), 1817–1831.

Watt, W., Koeplinger, K. A., Mildner, A. M., Heinrikson, R. L., Tomasselli, A. G., & Watenpaugh, K. D. (1999). The atomic-resolution structure of human caspase-8, a key activator of apoptosis. *Structure*, 7(9), 1135–1143.

Wilson, K. P., Black, J. A., Thomson, J. A., Kim, E. E., Griffith, J. P., Navia, M. A., et al. (1994). Structure and mechanism of interleukin-1 beta converting enzyme. *Nature*, *370*(6487), 270–275.

Witkowski, W. A., & Hardy, J. A. (2011). A designed redox-controlled caspase. *Protein Sciences*, *20*(8), 1421–1431.

Yoo, N. J., Lee, S. H., Jeong, E. G., & Lee, S. H. (2007). Expression of phosphorylated caspase-9 in gastric carcinomas. *APMIS*, *115*(4), 354–359 [Research Support, Non-U.S. Gov't].

Zalewski, P. D., Forbes, I. J., & Betts, W. H. (1993). Correlation of apoptosis with change in intracellular labile Zn(II) using zinquin [(2-methyl-8-p-toluenesulphonamido-6-quinolyloxy)acetic acid], a new specific fluorescent probe for Zn(II). *Biochemistry Journal*, *296*(Pt. 2), 403–408.

Zou, H., Yang, R., Hao, J., Wang, J., Sun, C., Fesik, S. W., et al. (2003). Regulation of the Apaf-1/caspase-9 apoptosome by caspase-3 and XIAP. *The Journal of Biological Chemistry*, *278*(10), 8091–8098.

CHAPTER TEN

Measuring Caspase Activity *In Vivo*

Samantha B. Nicholls, Bradley T. Hyman[1]
MassGeneral Institute for Neurodegenerative Disease, Department of Neurology, Alzheimer's Disease Research Laboratory, Massachusetts General Hospital, Harvard Medical School, Charlestown, Massachusetts, USA
[1]Corresponding author: e-mail address: bhyman@partners.org

Contents

Abstract

Caspases are a family of integral proteases playing a role in apoptosis. The importance of apoptosis in disease has made these proteases not only an attractive drug target but also a focal point for measuring apoptosis *in vivo*. The critical role caspases play in determining cell death has led to the development of a wide array of technologies to measure caspase activity *in vivo*, ranging from small molecule PET imaging reagents to fluorescent and luminescent protein-based reporters used in whole animal and cell-based applications. This chapter reviews this wide range of technologies available as well as the most appropriate applications for each reagent and the mechanism of how it measures caspase activity *in vivo*.

1. INTRODUCTION

The ability to selectively activate or inhibit cell death at the caspase level is an attractive but elusive target in diseases from cancer to

Methods in Enzymology, Volume 544
ISSN 0076-6879
http://dx.doi.org/10.1016/B978-0-12-417158-9.00010-8

neurodegeneration. The major difficulty in this approach lies in the increasingly appreciated but complex roles of the caspase family for normal cellular function and cell turnover. Moreover, the task is complicated by the difficulty in distinguishing the roles of the 14 individual members of the family in both health and disease, especially given the fact that caspases tend to activate one another, initiating and carrying out a complicated series of cascades that involve many (overlapping) substrates each of which has the potential to feedback on the cascade. In order to better understand the roles of the individual caspase members, several technologies have been developed over the past decade to measure the activity of different caspases both *in vitro* and *in vivo*. Because of the high level of structural and functional homology among the caspases observed *in vitro* as well as the critical role they play in activation of other caspases, it has become increasingly obvious that to fully understand the activation and role of caspases we need to measure their activity *in vivo*.

There are three main classifications of caspases, initiator, execution, and inflammatory. Most advanced from the perspective of *in vivo* imaging are reagents to detect activation of the initiator (caspases-8 and -9) and executioner caspases (caspases-3 and -7). Caspases (cysteine aspartic acid proteases) are expressed as a proenzyme with either an N-terminal caspase activation and recruitment domain (caspases-1, -2, -4, and -9), a death effector domain (caspases-8 and -10) or a shorter N-terminal prodomain (the executioners, caspases-3, -6, and -7), which is cleaved in the initiation of apoptosis, often by another upstream caspase. While initiator and executioner caspases have different substrate preferences there is some overlap in activity, especially in the case of caspases-3 and -7 which are considered to have nearly identical activity against the recognition sequence DEVD. This is a major limitation of the field of caspase research in general and applies to all the technologies discussed in this chapter. Though there are groups working toward understanding the subtle differences, such as allosteric or exo-sites, between seemingly similar caspases in an effort to specifically differentiate them *in vivo*, the current challenge of designing specific substrates for enzymes with overlapping activity should be kept in mind throughout the discussion of the following technologies. Thus, the focus of this review is on the techniques and approaches to detect caspase activation in living cells and organisms appreciating that reagents are sometimes limited in terms of detecting individual-specific caspase activity.

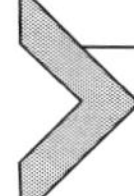

2. SMALL MOLECULE REAGENTS

2.1. Fluorescence

Peptide substrates, designed to provide some specificity for preferred recognition sequences for various caspases, frequently report a signal by dequenching of a fluorophore if the peptide is cleaved, thereby physically separating a fluorescent moiety from a quencher. Though popular for the measurement of caspase activity *in vitro*, peptides and small molecules have been challenging reagents for *in vivo* use due to their poor cell penetrance. However, several groups have developed a variety of modifications to the small peptide substrates used *in vitro* to improve their ability to enter the cell thus making them increasingly popular options for *in vivo* measurements (Table 10.1).

One popular fluorescent indicator is fluorochrome-labeled inhibitor of caspases (FLICA) (Amstad et al., 2001; Bedner et al., 2000). FLICA contains a fluoromethyl ketone moiety that irreversibly binds and inhibits the active site of caspases, which are only accessible when in the active form. It also contains a fluorescein derivative allowing for fluorescent monitoring of binding. It was originally designed and intended for high-throughput applications such as flow cytometry and microplate analysis of apoptosis in cells although, like many of the reagents discussed, specificity issues must be kept in mind (Kuzelova, Grebenova, & Hrkal, 2007; Pozarowski et al., 2003). The inhibition of caspases by FLICA also makes it impossible to simply monitor the caspase activity in a cell without affecting its function, although its relatively short half-life mitigates this liability. One recent application of FLICA reagents was to image caspase activation in an Alzheimer model of neurofibrillary tangles. Using multiphoton microscopy of a living mouse, neurons were monitored to determine the temporal sequence of neurofibrillary tangle formation and caspase activation (Calignon et al., 2012). In the experiments, neurons became FLICA positive but did not die over the ensuing days of observation. Although the FLICA reagents inhibit caspases, this effect was short-lived; reapplication of a FLICA reagent 20 min after initial application revealed continued caspase activity. Interestingly, in this model, neurofibrillary tangles occurred after caspase activation with the development of caspase-induced neoepitopes, supporting the conclusion that the FLICA reagent reported caspase activation.

Also in the family of small molecule fluorescent reporters is AB50-Cy5 (Fig. 10.1; Edgington et al., 2009). AB50 is an optimized acyloxymethyl

Table 10.1 Small molecule reagents bound to DEVD caspase cleavable peptide

Reagent	Method of measurement	Reporter molecule	Applications	References
FLICA	Fluorescence	Fluorescein	High-throughput FACS Microplate	Amstad et al. (2001) and Bedner, Smolewski, Amstad, and Darzynkiewicz (2000)
AB50	Fluorescence	Cy5	Cells Microscopy	Edgington et al. (2009)
ICE-NIRF	Fluorescence	Cy5.5	Cells Whole animal models Microscopy	Messerli et al. (2004)
PhiPhiLuxG1D2	Fluorescence	G1D2	High-throughput FACS	Packard, Toptygin, Komoriya, and Brand (1997) and Packard, Toptygin, Komoriya, and Brand (1996)
DEVD-NucView	Fluorescence	NucView488	Cells Whole animal models Microscopy	Cen, Mao, Aronchik, Fuentes, and Firestone (2008)
CapQ peptides	Fluorescence	Alexa Flour 647	Cells Whole animal models Microscopy	Barnett, Zhang, Maxwell, Chang, and Piwnica-Worms (2009), Bullok and Piwnica-Worms (2005), and Bullok et al. (2007)
CP18	PET	[^{18}F]	Cells Whole animal models	Xia et al. (2013)

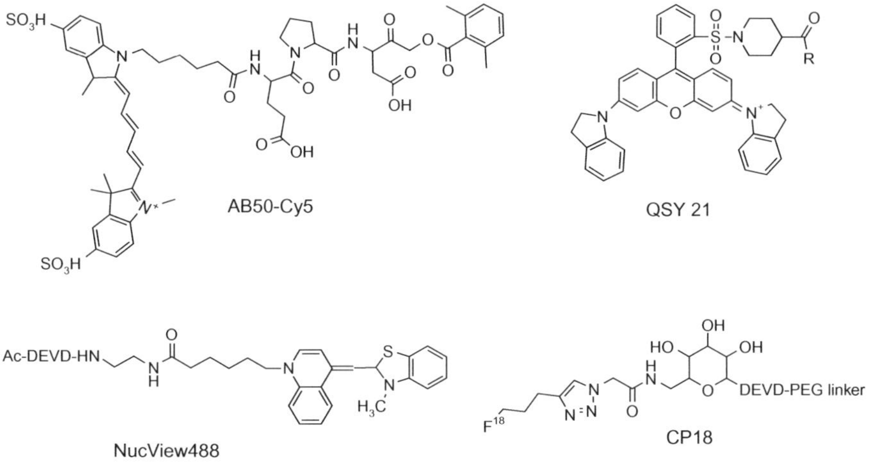

AB50-Cy5

QSY 21

NucView488

CP18

Figure 10.1 Small molecule caspase activity reagents. The molecule on the top left is AB50-Cy5 a reporter molecule fused to a TAT peptide to assist in cell penetrance with a Cy5 fluorescent moiety (Edgington et al., 2009). The top right is the quencher portion (QSY 21) of the K and TCapQ small molecules, which are fused to a peptide linker containing the amino acid sequence DEVD to the fluorophores Alexa Fluor 647 as well as one of the different (K or T) cell-penetrating peptide sequences (Barnett et al., 2009; Maxwell, Chang, Zhang, Barnett, & Piwnica-Worms, 2009). The fluorophore for the NucView488 small molecule reporter is shown on the bottom left (Cen et al., 2008). The PET reagent [^{18}F] CP18 is shown on the bottom right (Xia et al., 2013).

ketone caspase inhibitor engineered to specifically label caspase with no reactivity to cathepsins, which was a challenge with previous versions (Rozman-Pungercar et al., 2003). This molecule, fused with a cell-penetrating TAT peptide and a Cy5 fluorophore, was effectively used in a model of tumor-bearing mice to monitor apoptosis after treatment with dexamethasone or with an apoptosis-inducing antibody (Apomab).

One of the leading challenges in *in vivo* imaging is the ability to use these reagents in live animal models. As one of the primary potential applications of an apoptosis-indicating reagent is to report on tumor response to treatment, there is often an issue of depth of tissue penetrance of the wavelength of the reagent. As a result, reagents that respond in the far red or near-IR range are often more desirable. One reagent designed with this in mind is the caspase-1 near-infrared fluorescent probe (ICE-NIRF) (Messerli et al., 2004). ICE-NIRF is a GWEHDGK cleavable peptide fused to a FITCY-cys-NH2 cy5.5. This peptide is then attached through a poly-L-lys delivery vehicle for delivery into the cell. ICE-NIRF has the advantages of a high-signal-to-noise ratio (78-fold *in vitro*) and is specific to caspase-1 activation. One of the major disadvantages to ICE-NIRF is the variability in cell delivery. Due to the nature of the cell delivery molecule, there is not necessarily uniform labeling of the vehicle, leaving a high degree of variability in signal and making quantification of signal difficult.

A similar reagent is the reporter PhiPhiLuxG1D2 (which stands for fluorescent–fluorescent light; PhiPhiLux; with the specific fluorophores used; G1D2) (Packard et al., 1996). This reagent is a cell permeable substrate GDEVDGI fused to the fluorophores G1D2 (505/530 nm). Upon cleavage of the substrate by active caspase-3, the fluorophore's signal increases significantly over the background fluorescence of the uncleaved reporter. Similar to FLICA, the ability of this substrate to enter the cell has made it an attractive substrate for FACS applications to monitor apoptosis in high-throughput applications (Telford, Komoriya, & Packard, 2002). A family of CaspaLux reporters has been developed based on this technology to respond to caspases-1, -6, -8, and -9 with various fluorophores in both green and red wavelengths. Although this family of reagents is useful in cell suspensions and high-throughput screening applications, it is not intended for whole/live animal measurements of caspase activity.

Building on the principles of ICE-NIRF, PhiPhiLuxG1D1, and DEVD-NucView (Cen et al., 2008) was further developed as a cell permeable fluorescence indicator similar in design to the caspase-1 indicator ICE-NIRF (Messerli et al., 2004). However, DEVD-NucView488 was designed to be a more uniform reporter than the cellular delivery mode of ICE-NIRF. NucView488, the fluorophore, is not fluorescent until bound to DNA, but is nonfluorescent in the cytoplasm. When the N-terminally fused DEVD is cleaved by active caspase, the NucView488 binds to DNA in the nuclei allowing for real-time monitoring of apoptosis in the cell, which is detected both by the presence of the fluorophores in the nucleus and the classical morphological changes in the nucleus that occur. As with all localization-based reporters, the inherent limitation of relying on cellular transport mechanisms cellular imaging is required for interpretation of results.

As in the case of ICE-NIRF, there is an advantage to moving the fluorescence of these *in vivo* probes toward the far red or near-IR range due to a higher tissue penetrance of the excitation light as well as the emitted photons at the longer wavelength. An activatable red probe, TCapQ (Bullok et al., 2007; Bullok & Piwnica-Worms, 2005), was developed which uses a TAT (cell-penetrating peptide) to enter the cell. TCapQ has been shown to detect apoptosis on the single cell level in rat models of glaucoma *in vivo* (Barnett et al., 2009). A quenching portion of the molecule (QSY 21) is cleaved by active caspases-3, -6, and -7 via a DEVD linker, allowing visualization of the fluorescent indicator (Alexa Fluor 647). KCapQ is the second generation of the CapQ series using the same fluorophores and quencher, with an improved cell-penetrating peptide sequence (KKKRKV) (Maxwell et al., 2009). The CapQ series is not only recognized primarily by caspase-3 but also showed activity to caspases-6 and -7. The revised peptide of KCapQ showed not only higher cell penetration but also lower toxicity to the cell as well. KCapQ was used in cell culture and in a similar live animal application to detect apoptosis in rat retinal ganglion cells after NMDA treatment in a rat model of retinal neuronal excitotoxicity.

2.2. PET imaging

The prospect of the ability to monitor caspase activity in human patients is an increasingly tantalizing one since it holds out the promise of monitoring response to therapies, especially in cancer, as well as to detect aberrant activation of caspase or apoptotic pathways in a wide variety of human disease conditions. Although no PET ligands have been approved yet, substantial progress has been made to move this idea closer to the clinic. Prior to the development of the extensive range of reporters discussed in this chapter, the most common way to monitor apoptosis in cells was the labeling of phosphatidylserine, which is translocated to the plasma membrane during the process of apoptosis, using the phospholipid binding protein Annexin V. Annexin V could then be labeled with a variety of tags including fluorophores or radiolabels for PET. However, this method of monitoring apoptosis is indirect and does not distinguish different forms of cell death.

A direct monitor of apoptosis, as opposed to necrosis which is not caspase dependent, is much more desirable. A promising reagent for monitoring apoptosis via caspase activity is the PET reagent [^{18}F]-CP18 (Xia et al., 2013). It is designed as a DEVD peptide substrate fused to polyethylene glycol and galactose moieties designed to allow CP18 to cross-intact plasma membranes. When the cell-permeating moiety is cleaved by active caspase-3, the ^{18}F-radiolabeled DEVD accumulates in the cytoplasm. Though CP18 was effective in identifying caspase-3 activity in different tumor types, it has only been tested in xenograft tumors thus far. Because it is based on a small peptide substrate, CP18 also potentially faces similar challenges as other caspase indicators in being only relatively selective for caspases-3 and -7. While a diverse panel of proteases was tested for their activity against CP18, only caspases-1 and -3 were included in the screen. It is possible that other caspases, such as caspase-8 is also active against CP18, although the extent to which this would limit clinical utility is not clear. One can also imagine indicators that are designed with relative preference for other caspases such as ICE (caspase-1), presumably activated in NLRP3 and other innate immunity conditions and potentially allowing direct monitoring of the state of inflammatory response in some autoimmune or inflammatory conditions.

3. PROTEIN INDICATORS

Protein-based indicators are an important tool because they are genetically encoded. This allows for applications both in cells *in vitro* and *in vivo*, potentially allowing monitoring in cells expressing the reporter without disrupting the cell prior to reading, in many instances provides a tool for longitudinal studies. It also allows for the monitoring of caspase activity without the necessity of binding or inhibiting the enzymatic activity as in the case of some of the small molecule reagents described above.

3.1. Luciferase

One of the most popular protein systems used for reporting caspase activity is a split luciferase reporter. There are several versions of this type of reporter (Coppola, Ross, & Rehemtulla, 2008; Gross & Piwnica-Worms, 2005; Shekhawat, Porter, Sriprasad, & Ghosh, 2009). One of the early versions was a split luciferase construct with strongly interacting peptides, PepA and PepB fused to the termini of each luciferase half, separated by the DEVD caspase recognition site (Coppola et al., 2008). While the protein substrate was easily cleaved by the caspase, there was a high-background level from the luciferase. To get around this challenge, Galban et al. (2013) screened for circularly permuted luciferase variants with a lower background and found that permutation at residue 358V3 with an additional point mutation at residue T151I (507 in nonpermuted protein) was the optimal construct. This variant, which they named "GloSensor 3/7" showed a greater than 50-fold increase in signal-to-noise, overcoming the previous challenges and making GloSensor a useful *in vivo* tool. The *in vivo* testing of GloSensor was done by grafting tumor cells transfected with the reporter into bone and measuring apoptosis throughout tumor treatment. Although GloSensor may present a challenge in terms of delivery for some *in vivo* experiments, it likely will prove to be a useful tool for high-throughput screening *in vitro*.

One group has developed a dual-cleavage construct where luciferase (Luc) is flanked by DEVD linkers fusing estrogen receptor (ER) regulatory domains on either terminus, which act to silence the luciferase as an inactive form (Laxman et al., 2002). When caspase-3 is activated, for example, when the cell undergoes apoptosis, the DEVD sites are cleaved by caspase-3, releasing the luciferase activity, and allowing for bioluminescent imaging of caspase activity. This reporter was shown to be effective in xenografted glioma tumors expressing the ER-DEVD-Luc-DEVD-ER reporter in mice after treatment with the apoptosis inducer TRAIL.

In an interesting approach to luciferase sensors, one group has developed a specialized luciferase substrate; Z-DEVD-Luciferin (Scabini et al., 2011). This reagent can be used in luciferase transfected cells or luciferase transgenic animals (Kadonosono et al., 2011) and is available as a commercial reagent, VivoGlo™ Caspase-3/-7 substrate from Promega. In this system, the DEVD peptide tag on the luciferin prevents it from binding in the active site of luciferase. Once active caspase cleaves the peptide the luciferin becomes a viable substrate and caspase activity is measured indirectly as a measure of released luciferin. This approach to engineering luciferase substrate has become increasingly popular recently with a "split-luciferin" substrate being developed to offer a more targeted approach to bioluminescence (Godinat et al., 2013). In this substrate, the DEVD peptide is fused to a D-cysteine, which when released by caspase cleavage reacts with cyanobenzothiazole to form a version of the luciferin substrate. This substrate is more efficient at binding and brighter than luciferin, giving it an advantage over the traditional substrate.

One disadvantage to the split luciferase design is background signal from intermolecular reactions vs. the desired intramolecular signal. This can be overcome in the design of the reporter. Shekhawat et al. specifically designed split reporters with an autoinhibitory coiled-coil design included in the reporter. This prevents intermolecular interactions via steric hindrance prior to linker cleavage (Shekhawat et al., 2009). This design is intended for use in both split luciferase and split fluorescent protein reporters.

Another disadvantage to using luciferase is that, although it is highly sensitive, there are the additional challenges of intercellular delivery of the enzyme, moderate access *in vivo* of the substrate, and the short half-life of signal after substrate delivery. For this reason, many groups have avoided luciferase reporters and rely on the extensive family of fluorescent proteins as indicators.

3.2. Fluorescent proteins

Fluorescent protein reporters have two distinct advantages over other classes. The first is that it is genetically encoded and do not require any additional substrate for fluorescence in contrast to luciferase reporters. They therefore cause less disruption to the cellular environment they are monitoring. There is no disruption to the cell membrane for transport of the probes and though some specific fluorescent protein variants have a level of toxicity when overexpressed (Liu, Jan, Chou, Chen, & Ke, 1999), in general, they do not overtly perturb the cell. The second advantage is direct measurement of caspase activity; many of the fluorescent protein-based reporters are initiated by direct cleavage of a linker region within the protein either triggering an Förster resonance energy transfer (FRET) response or a refolding, as in the case of split luciferase or fluorescent protein-based reporters, as well as localization-based reporters. Because of this variation in design, the most appropriate reporter depends on the experimental question being asked. When trying to determine the specific kinetics of caspase activation within the cell or in relation to other cellular events, FRET reporters are more appropriately applied. For longitudinal measurement of caspase activation and cellular apoptosis, dark-to-bright reporters or localization-based probes may be more appropriate as they often have improved signal-to-noise and longevity of signal.

A reporter that takes advantage of genetically encoded fluorophores is the caspase activatable-green fluorescent protein (CA-GFP) (Nicholls, Chu, Abbruzzese, Tremblay, & Hardy, 2011; Nicholls & Hardy, 2013). CA-GFP is based on the green variant S65T, which is held in a semi-folded state until active caspase cleaves a peptide at the C-terminus allowing the barrel of the GFP to properly fold for fluorescence. This dark-to-bright activation gives CA-GFP an advantage of a high-signal-to-noise, up to 45-fold *in vitro* and three-fold *in vivo* (Nicholls et al., 2011). Co-transfection or use of an IRES or similar construct using a bright fluorophore of another color gives a denominator to control for efficiency of delivery of the reagent. As an example the response of CA-GFP in a vector containing an IRES site followed by a constitutively active red fluorescent protein, mLumin was transfected in H4 human neuronal gliomal cells. After treatment with the apoptosis-inducing drug staurosporine, there is a clear increase in green fluorescence (Fig. 10.2). As the cells undergo apoptosis, the green signal increases relative to the red. CA-GFP also has the potential to be modified to screen for a variety of other proteases making it an attractive reporter system which is adaptable to many applications (Wu, Nicholls, & Hardy, 2013). However, the specific kinetics of cleavage and refolding are not well-understood making this reporter more useful in longitudinal monitoring applications than to assess rapid induction of caspase activation.

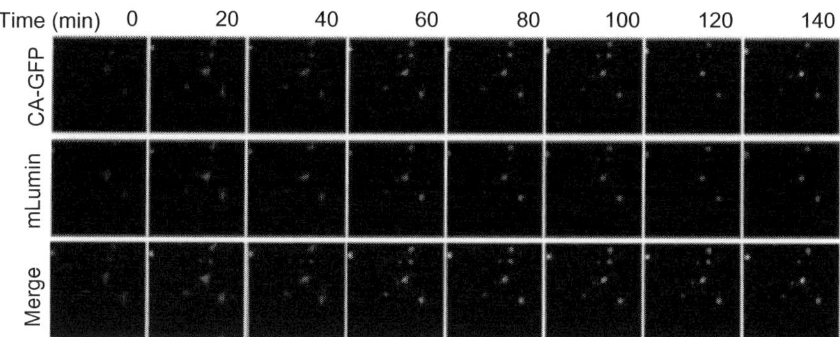

Figure 10.2 Dark-to-bright response of CA-GFP in H4 human neuronal gliomal cell line. CA-GFP was transiently transfected in a vector containing an IRES followed by a constitutively bright red fluorescent protein, mLumin. There is a green response which increases in intensity and colocalizes with the red in the H4 cells treated with 1 mM staurosprine, a known apoptosis inducer, as the cell begins to bleb and undergo apoptosis. (See the color plate.)

Some designs seek to incorporate multiple imaging facets, as with the multimodality reporter (Ray, De, Patel, & Gambhir, 2008). This reporter is designed by fusing the red fluorescent protein mRFP C-terminally through a DEVD containing linker to thymidine kinase, a PET reporter, which is fused through a second DEVD linker to firefly luciferase. The intrinsic fluorescence or luminescence of each protein is partially quenched in the fusion and as a result the signal of all three reporters increases after cleavage of the linkers with active caspase. This allows for one reporter to be useful in multimodal applications from noninvasive imaging to cell sorting applications. The major challenge with this reporter, however, is that although the fusion protein causes some level of quenching there is still a significant amount of background, especially in the case of the mRFP which makes the signal-to-noise on the order of two- to threefold for the three reporters. In addition, the design of the probe does not necessarily allow for a "loading control" type of denominator in order to take into account the efficiency of transfection or expression levels in individual cells.

3.3. Localization indicators

Most fluorescent proteins are constitutively bright, making signal-to-noise a difficult challenge to overcome with many fluorescent protein-based reporters. One way around this obstacle is to design a molecule that undergoes intracellular translocation or unique localization after interaction with caspases or apoptosis-related cellular events. One of these reporters,

Apoliner (Bardet et al., 2008) is designed so that the fluorescent protein moieties are targeted to the nucleus following cleavage by active caspases-3/-7. Apoliner is a fusion of mRFP and eGFP fused through a DEVD, caspases-3 and -7 cleavable linker. Upon cleavage by active caspase the eGFP, which is tagged with a nuclear localization ligand, relocates to the nucleus, leaving the mRFP at the membrane. This reporter is primarily useful in microscopy applications as both fluorescent proteins are constitutively bright making bulk or high-throughput assessments difficult, although high-content imaging can be used to convert this to a medium- to high-throughput assay.

3.4. Degradation-based tags

Another way around the challenge of the constitutively fluorescent protein is a creative design by Lee et al. They designed eGFP with an N-terminal arginine residue, which is rapidly degraded in the cell. In the presence of active caspase, an N-terminal cleavage site removes the N-terminal arginine leaving a methionine at the terminus. This is stable in the cell and the eGFP is stably fluorescent (Lee, Beem, & Segal, 2002). Though there are other degradation tags used to mark cellular processes, this is specific for caspase activity making it unique. It is also sensitive enough to indicate intrinsic caspase activity in nonapoptotic cells because the signal is cumulative.

3.5. Förster resonance energy transfer

The protein reporters discussed thus far are in general most useful in the setting where one is trying to observe apoptosis "turning on" and provide a way to monitor activation but not necessarily to provide a quantitative read out. One of the most popular methodologies to provide quantitation is to use either ratiometric or FRET-based approaches. FRET is a biophysical assay that takes advantage of a transfer of energy between two matched fluorophores if they are in very close proximity. FRET-based caspase reporters are designed with a fluorescent donor and acceptor protein fused through a caspase cleavable linker (Fig. 10.3). When caspases are inactive, the linker remains intact, keeping the donor and acceptor proteins in close proximity allowing for high-FRET efficiency. After cleavage by active caspase the two proteins diffuse apart and the FRET efficiency decreases. The different FRET pairs and applications of these reporters are summarized in Table 10.2. Cyan and yellow fluorescent proteins are efficient FRET donor and acceptors, and fusion of the two across a DEVD linker recognized by the executioner caspases allows measurement of both intact and cleaved

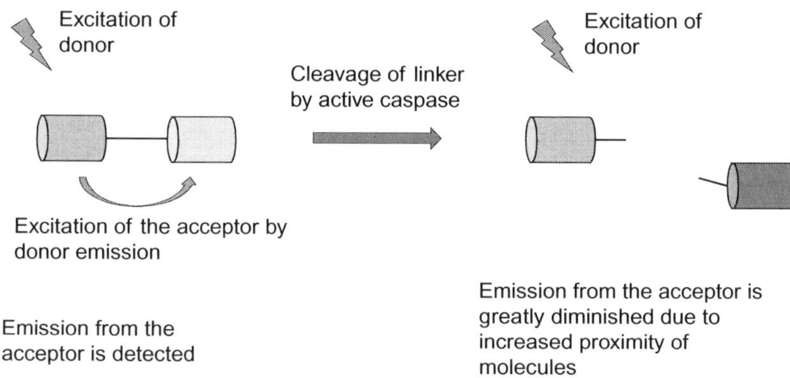

Figure 10.3 FRET response in caspase activatable fluorescent proteins. Regardless of which fluorescent proteins used, most of the FRET-based reporters described have a similar mechanism by which the donor and acceptor fluorescent proteins are held in close physical proximity by a caspase cleavable linker allowing for excitation of the acceptor. After cleavage of the linker by active caspase, the two fluorescent proteins diffuse further from each other and the signal from the acceptor is greatly diminished. (See the color plate.)

molecules, as FRET efficiency (as measured by fluorescent emission ratios or by fluorescent lifetime) changes due to cleavage of the linker. Several versions of this type of reporter, which have mainly been designed for high-throughput screening of drug compounds, have been successfully used in FACS analysis and screening of drug libraries (He et al., 2004; Jones et al., 2000; Savitsky et al., 2012; Wu et al., 2006). A SCAT3 reporter, which has been moved into a transgenic mouse line, has been developed by Yamaguchi et al. using an eCFP/Venus FRET pair (since CFP/YFP transgenic lines proved difficult to sustain) (Yamaguchi et al., 2011). The availability of a transgenic line expressing a FRET caspase reporter, combined with newer approaches of *in vivo* microscopy including multiphoton imaging and fluorescent endoscopy, promises to be a powerful approach to monitoring caspase biology *in vivo*.

All of these reporters share a similar design with fluorescence proteins representing a FRET pair fused through a linker containing a caspase cleavage recognition site. Once cleaved the FRET efficiency decreases. The overall FRET efficiency varies depending on the fluorescent proteins used in the fusion. Nguyen specifically developed the FRET pair CyPet/YPet through directed evolution to have an optimal pair for FACS sorting and showed the enhanced independent fluorescence of these fluorescent proteins as well as optimal FRET efficiency using a caspase-3 cleavage site as a linker between the two fluorescent proteins. They observed a high level

Table 10.2 Fluorescent protein based, caspase cleavable FRET pairs

Reporter	FRET pair	Applications	Cleavage site	References
FRET caspase reporter	CFP/YFP	Cells High-throughput FACS	DEVD	He et al. (2004), Jones, Heim, Hare, Stack, and Pollok (2000), and Savitsky et al. (2012)
SCAT3	eCFP/mVenus	Cells Transgenic mouse model	DEVD	Yamaguchi et al. (2011)
CyPet/YPet	CyPet/YPet	FACS	DEVD	Nguyen and Daugherty (2005)
CRY83	CFP/Venus Venus/mRFP1	Cells	IETD DEVD	Kominami et al. (2013)
Microtubule-anchored FRET sensors	eCFP/eYFPTau	Cells Microscopy	DEVD VEID LEHD LEVA	Figueroa et al. (2011)
Simultaneous FRET	CFP/DsRed YFP/DsRed	Cells FACS	DEVD IETD	Kawai et al. (2005)
FLIM FRET	TagRFP/KFP	Cells Microscopy	DEVD	Savitsky et al. (2012)

of sensitivity to detect apoptotic cells (Nguyen & Daugherty, 2005). One version that was designed to specifically be used in neuronal cells is anchored to microtubules by a fusion of eYFP to Tau microtubule associated protein (Figueroa et al., 2011). This reporter was used to monitor caspase cleavage in relation to neurodegeneration.

In addition to the popularity of measuring response at the executioner level of the caspase cascade, the initiator family has also been of interest. In one if the first attempts to distinguish between the initiator and effector caspases, Kawai et al. developed two FRET reporters to be used simultaneously, one fused YFP to DsRed with an IETD (caspases-8/-9 cleavage site) and the second fused CFP to DsRed with a DEVD (caspases-3/-7) linker. They were able to simultaneously monitor both cleavage events in

HeLa cells after induction of apoptosis with TNF-α (Kawai et al., 2005). From this design several double FRET reporters, which incorporate both cleavage sites into a single fusion protein, have been developed to look at the dynamics of initiator versus executioner caspase activation. One of these is the dual-FRET-based reporter CYR83 (Kominami et al., 2013). This reporter takes advantage of the distinction between the substrate preferences of the initiator (caspases-8 and -9) and executioner (caspases-3 and -7) caspases with a fusion protein seCFP-IETD-Venus-DEVD-mRFP1 where IETD is recognized by the initiator and DEVD is recognized by the executioner caspases. The loss of the CFP/Venus FRET indicates the activation of the caspase-8, while loss of the Venus/mRFP FRET signals activation of the executioner caspases.

These dual-FRET reporters have also been developed to distinguish which initiator caspases are activated in response to specific stimuli. This is a more feasible approach in distinguishing caspases since the initiators in particular have the lowest homology to one another in terms of substrate specificity. An adaptation of the dual-FRET reporter reported by Kawai et al. (2005) was used to distinguish the response of caspases-2 and -9 in the case of anticancer agent cisplatin-induced cell death (Chu, Wang, Luo, & Zhang, 2008). They found that in fact both pathways are induced by the drug in HeLa cells stably expressing the reporter.

3.6. Quantum dots

Another way to enhance signal-to-noise is to increase the signal and the efficiency of quenching in the noncaspase activated circumstance. Quantum dots (QD) are intensely bright and long-lived fluorophores that can nonetheless be efficiently quenched with an appropriate reagent. QD-fused fluorescent protein FRET reporters have been generated, although the ability to deliver these widely *in vitro* and *in vivo* remains difficult in some experimental circumstances (Shi, Paoli, Rosenzweig, & Rosenzweig, 2006). An intriguing alternative: much in the same way that dual-fluorescent protein reporters use FRET, it is possible to fuse a fluorescent protein, such as mCherry through a linker containing a caspase cleavage site to the QD as the FRET partner (Boeneman et al., 2009). This gives a certain measure of flexibility to the system as the QD can be tuned much more easily than a fluorescent protein in terms of wavelength, allowing an optimal fluorescent protein to be chosen for its independent fluorescent properties without concern for the availability of a FRET partner.

4. FUTURE DIRECTIONS

One of the main challenges remaining in caspase imaging is the ability to distinguish the activities of each of the individual caspases. This challenge is daunting due to the high similarity of many of the caspase family members in terms of their cleavage preference and relative ability to cleave similar if not identical substrates. Some strides toward this have been successful in distinguishing initiator caspases from executioners as with the dual-FRET reporters (Galban et al., 2013). As new protein imaging substrates are developed, it will be interesting to note if specificity can be increased as there has been some recent evidence that caspases that are recognized as having virtually identical peptide substrate preference (e.g., caspases-3 and -7) may show differences in preference when presented with full protein substrate (Boucher, Blais, & Denault, 2012).

The second major challenge is the quantification of caspase activity within a cell. Many of the reporters described above are very sensitive qualitative indicators of cell death. However, the variability in caspase expression and activation from cell to cell as well as the variability of the expression or delivery of the reporter molecules in the same cells leads to a high level of error in determining the specific activities of a caspase within the live cell. Some FRET reporters take advantage of the ratiometric readout to quantify activity; however, these reporters are more often used to look at comparative reactivity of the caspases. This issue is of particular importance in the examination of nonapoptotic roles of caspases during normal cellular activities. Essentially by definition, activation of caspases in these circumstances occurs to a far weaker level than occurs during cell death processes, so that better anatomical, temporal, and signal-to-noise resolution will be necessary to examine these more subtle events. For example, recent data suggest that caspases are activated in the dendritic spine during memory formation (D'Amelio et al., 2011), but the very small size of the dendritic spine compartment and the very low levels of activation will make imaging this potentially important neuronal phenomenon difficult.

Despite these challenges, dramatic advances across a wide spectrum of modalities has already led to the ability to monitor caspase activation in cells and in living animals, with further refinements promising even more detailed temporal, spatial, and enzyme specificity advances likely to come.

REFERENCES

Amstad, P., Yu, G., Johnson, G. L., Lee, B. W., Dhawan, S., & Phelps, D. J. (2001). Detection of caspase activation in situ by fluorochrome-labeled caspase inhibitors. *Biotechniques*, *31*(3), 608–616.

Bardet, P.-L., Kolahgar, G., Mynett, A., Miguel-Aliaga, I., Briscoe, J., Meier, P., et al. (2008). A fluorescent reporter of caspase activity for live imaging. *PNAS*, *105*(37), 13901–13905.

Barnett, E. M., Zhang, X., Maxwell, D., Chang, Q., & Piwnica-Worms, D. (2009). Single-cell imaging of retinal ganglion cell apoptosis with a cell-penetrating, activatable peptide probe in an in vivo glaucoma model. *PNAS*, *106*(23), 9391–9396.

Bedner, E., Smolewski, P., Amstad, P., & Darzynkiewicz, Z. (2000). Activation of caspases measured in situ by binding of fluorochrome-labeled inhibitors of caspases (FLICA): Correlation with DNA fragmentation. *Experimental Cell Research*, *259*(1), 308–313.

Boeneman, K., Mei, B. C., Dennis, A. M., Bao, G., Deschamps, J. R., Mattoussi, H., et al. (2009). Sensing caspase 3 activity with quantum dot-fluorescent protein assemblies. *Journal of the American Chemical Society*, *131*, 3828–3829.

Boucher, D., Blais, V., & Denault, J.-B. (2012). Caspase-7 uses an exosite to promote poly(ADP ribose) polymerase 1 proteolysis. *PNAS*, *109*(15), 5669–5674.

Bullok, K. E., Maxwell, D., Kesarwala, A. H., Gammon, S., Prior, J. L., Snow, M., et al. (2007). Biochemical and in vivo characterization of a small membrane-permeant, caspase-activatable far-red fluorescent peptide for imaging apoptosis. *Biochemistry*, *46*, 4055–4065.

Bullok, K. E., & Piwnica-Worms, D. (2005). Synthesis and characterization of a small, membrane-permeant, caspase-activatable far-red fluorescent peptide for imaging apoptosis. *Journal of Medicinal Chemistry*, *48*, 5404–5407.

Calignon, A. d., Polydoro, M., Suarez-Calvet, M., William, C., Adamowicz, D. H., Kopeikina, K. J., et al. (2012). Propagation of tau pathology in a model of early Alzheimer's disease. *Neuron*, *73*, 685–697.

Cen, H., Mao, F., Aronchik, I., Fuentes, R. J., & Firestone, G. L. (2008). DEVD-NucView488: A novel class of enzyme substrates for real-time detection of caspase-3 activity in live cells. *The FASEB Journal*, *22*, 2243–2252.

Chu, J., Wang, L., Luo, Q., & Zhang, Z. (2008). Simultaneous imaging of two initiator caspases during cisplatin-induced HeLa apoptosis. *Proceedings of SPIE*, *6857*, 68570R-1–68570R-7.

Coppola, J. M., Ross, B. D., & Rehemtulla, A. (2008). Noninvasive imaging of apoptosis and its application in cancer therapeutics. *Clinical Cancer Research*, *14*, 2492.

D'Amelio, M., Cavallucci, V., Meddei, S., Marchetti, C., Pacioni, S., Ferri, A., et al. (2011). Caspase-3 triggers early synaptic dysfunction in a mouse model of Alzheimer's disease. *Nature Neuroscience*, *14*(1), 69–76.

Edgington, L. E., Berger, A. B., Blum, G., Albrow, V. E., Paulick, M. G., Lineberry, N., et al. (2009). Noninvasive optical imaging of apoptosis by caspase-targeted activity-based probes. *Nature Medicine*, *15*(8), 967–973.

Figueroa, R. A., Ramberg, V., Gatsinzi, T., Samuelsson, M., Zhang, M., Iverfeldt, K., et al. (2011). Anchored FRET sensors detect local caspase activation prior to neuronal degeneration. *Molecular Neurodegeneration*, *6*, 35.

Galban, S., Jeon, Y. H., Bowman, B. M., Stevenson, J., Sebolt, K. A., Sharkey, L. M., et al. (2013). Imaging proteolytic activity in live cells and animal models. *PLoS One*, *8*(6), e66248.

Godinat, A., Park, H., Miller, S., Cheng, K., Hanahan, D., Sanman, L., et al. (2013). A biocompatible in vivo ligation reaction and its applications for noninvasive bioluminescent imaging of protease activity in living mice. *ACS Chemical Biology*, *8*(5), 987–999.

Gross, S., & Piwnica-Worms, D. (2005). Spying on cancer: Molecular imaging in vivo with genetically encoded reporters. *Cancer Cell*, *7*, 5–15.

He, L., Wu, X., Meylan, F., Olson, D. P., Simone, J., Hewgill, D., et al. (2004). Monitoring caspase activity in living cells using fluorescent proteins and flow cytometry. *The American Journal of Pathology*, *164*(6), 1901–1913.

Jones, J., Heim, R., Hare, E., Stack, J., & Pollok, B. A. (2000). Development and application of a GFP-FRET intracellular caspase assay for drug screening. *Journal of Biomolecular Screening*, *5*, 307–317.

Kadonosono, T., Kuchimaru, T., Yamada, S., Takahashi, Y., Murakami, A., Tani, T., et al. (2011). Detection of the onset of ischemia and carcinogenesis by hypoxia-inducible transcription factor-based *in vivo* bioluminescence imaging. *PLoS One*, *6*(11), e26640.

Kawai, H., Suzuki, T., Kobayashi, T., Sakurai, H., Ohata, H., Honda, K., et al. (2005). Simultaneous real-time detection of initiator- and effector-caspase activation by double fluorescence resonance energy transfer analysis. *Journal of Pharmacological Sciences*, *97*(3), 361–368.

Kominami, K., Nagai, T., Sawasaki, T., Tsujimura, Y., Yashima, K., Sunaga, Y., et al. (2013). In vivo imaging of hierarchical spatiotemporal activation of caspase-8 during apoptosis. *PLoS One*, *7*(11), e50218.

Kuzelova, K., Grebenova, D., & Hrkal, Z. (2007). Labeling of apoptotic JURL-MK1 cells by fluorescent caspase-3 inhibitor FAM-DEVD-fmk occurs mainly at site(s) different from caspase-3 active site. *Cytometry. Part A*, *71*(8), 605–611.

Laxman, B., Hall, D. E., Bhojani, M. S., Hamstra, D. A., Chenevert, T. L., Ross, B. D., et al. (2002). Noninvasive real-time imaging of apoptosis. *PNAS*, *99*(26), 16551–16555.

Lee, P., Beem, E., & Segal, M. S. (2002). Marker for real-time analysis of caspase activity in intact cells. *Biotechniques*, *33*(6), 1284–1291.

Liu, H.-S., Jan, M.-S., Chou, C.-K., Chen, P.-H., & Ke, N. J. (1999). Is green fluorescent protein toxic to the living cells. *Biochemical and Biophysical Research Communications*, *260*(3), 712–717.

Maxwell, D., Chang, Q., Zhang, X., Barnett, E. M., & Piwnica-Worms, D. (2009). An improved cell-penetrating, caspase-activatable, near-infrared fluorescent peptide for apoptosis imaging. *Bioconjugate Chemistry*, *4*, 702–709.

Messerli, S. M., Prabhakar, S., Tang, Y., Shah, K., Cortes, M. L., Murthy, V., et al. (2004). A novel method for imaging apoptosis using a caspase-1 near-infrared fluorescent probe. *Neoplasia*, *6*(2), 95–105.

Nguyen, A. W., & Daugherty, P. S. (2005). Evolutionary optimization of fluorescent proteins for intracellular FRET. *Nature Biotechnology*, *23*, 355–360.

Nicholls, S. B., Chu, J., Abbruzzese, G., Tremblay, K. D., & Hardy, J. A. (2011). Mechanism of a genetically encoded dark-to-bright reporter of caspase activity. *Journal of Biological Chemistry*, *286*(28), 24977–24986.

Nicholls, S. B., & Hardy, J. A. (2013). Structural basis of fluorescence quenching in caspase activatable-GFP. *Protein Science*, *22*(3), 247–257.

Packard, B. Z., Toptygin, D. D., Komoriya, A., & Brand, L. (1996). Profluorescent protease substrates: Intramolecular dimers described by the exciton model. *PNAS*, *93*, 11640–11645.

Packard, B., Toptygin, D., Komoriya, A., & Brand, L. (1997). Design of a profluorescent protease substrates guided by exciton theory. *Methods Enzymology*, *278*, 15–23.

Pozarowski, P., Huang, X., Halicka, D. H., Lee, B., Johnson, G., & Darzynkiewicz, Z. (2003). Interactions of fluorochrome-leveled caspase inhibitors with apoptotic cells: A caution in data interpretation. *Cytometry. Part A*, *55*(1), 50–60.

Ray, P., De, A., Patel, M., & Gambhir, S. S. (2008). Monitoring caspase-3 activation with a multimodality imaging sensor in living subjects. *Clinical Cancer Research*, *14*(18), 5801–5809.

Rozman-Pungercar, J., Kopitar-Jerala, N., Bogyo, M., Turk, D., Vasiljeva, O., Stefe, I., et al. (2003). Inhibition of papain-like cysteine proteases and legumain by caspase-specific inhibitors: When reaction mechanism is more important than specificity. *Cell Death and Differentiation*, *10*, 881–888.

Savitsky, A. P., Rusanov, A. L., Zherdeva, V. V., Gorodnicheva, T. V., Khrenova, M. G., & Nemukhin, A. V. (2012). FLIM-FRET imaging of caspase-3 activity in live cells using pair of red fluorescent proteins. *Theranostics*, *2*(2), 215–226.

Scabini, M., Stellari, F., Cappella, P., Rizzitano, S., Texido, G., & Pesenti, E. (2011). In vivo imaging of early stage apoptosis by measuring real-time caspase-3/7 activation. *Apoptosis*, *16*, 198–207.

Shekhawat, S. S., Porter, J. R., Sriprasad, A., & Ghosh, I. (2009). An autoinhibited coiled-coil design strategy for split-protein protease sensors. *Journal of the American Chemical Society*, *131*(42), 15284–15290.

Shi, L., Paoli, V. D., Rosenzweig, N., & Rosenzweig, Z. (2006). Synthesis and application of quantum dots FRET-based protease sensors. *Journal of the American Chemical Society*, *128*, 10378–10379.

Telford, W. G., Komoriya, A., & Packard, B. Z. (2002). Detection of localized caspase activity in early apoptotic cells by laser scanning cytometry. *Cytometry*, *47*, 81–88.

Wu, P., Nicholls, S. B., & Hardy, J. A. (2013). A tunable, modular approach to fluorescent protease-activated reporters. *Biophysical Journal*, *104*(7), 1605–1614.

Wu, X., Simone, J., Hewgill, D., Siegel, R., Lipski, P. E., & He, L. (2006). Measurement of two caspase activities simultaneously in living cells by a novel dual FRET fluorescent indicator probe. *Cytometry. Part A*, *69*(6), 477–486.

Xia, C.-F., Chen, G., Gangadharmath, U., Gomez, L. F., Liang, Q., Mu, F., et al. (2013). In vitro and in vivo evaluation of the caspase-3 substrate-based radiotracer [18F]-CP18 for PET imaging of apoptosis in tumors. *Molecular Imaging and Biology*, *15*(6), 748–757.

Yamaguchi, Y., Shinotsuka, N., Nonomura, K., Takemoto, K., Kuida, K., Yosida, H., et al. (2011). Live imaging of apoptosis in a novel transgenic mouse highlights its role in neural tube closure. *The Journal of Cell Biology*, *195*, 1047–1060.

CHAPTER ELEVEN

Single-Molecule Sensing of Caspase Activation in Live Cells via Plasmon Coupling Nanotechnology

Cheryl Tajon[*,†], **Young-Wook Jun**[†,‡], **Charles S. Craik**[*,†,1]
[*]Department of Pharmaceutical Chemistry, University of California, San Francisco, California, USA
[†]Graduate Program in Chemistry and Chemical Biology, University of California, San Francisco, California, USA
[‡]Department of Otolaryngology, University of California, San Francisco, California, USA
[1]Corresponding author: e-mail address: charles.craik@ucsf.edu

Contents

Abstract

Apoptotic caspases execute programmed cell death, where low levels of caspase activity are linked to cancer (Kasibhatla & Tseng, 2003). Chemotherapies utilize induction of

Methods in Enzymology, Volume 544
ISSN 0076-6879
http://dx.doi.org/10.1016/B978-0-12-417158-9.00011-X

apoptosis as a key mechanism for cancer treatment, where caspase-3 is a major player involved in dismantling these aberrant cells. The ability to sensitively measure the initial caspase-3 cleavage events during apoptosis is important for understanding the initiation of this complex cellular process; however, current ensemble methods are not sensitive enough to measure single cleavage events in cells. To overcome this, we describe a procedure to develop peptide-linked gold nanoparticles that have unique optical properties and can serve as beacons to visualize the apoptotic drug response in cancer cells at the single-molecule level. By thorough analyses of their trajectories, one can reveal early-stage caspase-3 activation in live cells continuously and with no ambiguity.

1. IMAGING OF CASPASE ACTIVATION

1.1. Caspases can be targeted to understand cell death mechanisms

Proteases catalyze the permanent hydrolysis of amide bonds and have evolved to participate in processes that are fundamentally irreversible: apoptosis, the degradation of intracellular proteins, fibrin clot deposition, and digestion (Chapman, Riese, & Shi, 1997). Among these, understanding the onset of apoptotic processes in the context of cancer and its response to drug treatment has been of considerable importance. Chemotherapies utilize induction of apoptosis as a key mechanism for cancer treatment, where cell death is tightly regulated by the dynamic actions of multiple caspases (*c*ysteine–*asp*artic acid prote*ases*) that mediate the consecutive and irreversible proteolytic events leading to cellular destruction. Two pathways can be engaged that both occur as proteolytically driven processes: (1) the extrinsic or death receptor-induced pathway via caspase-8 or (2) the intrinsic or Apaf-1 (apoptotic protease-activating factor 1) apoptosome pathway via caspase-9 (Logue & Martin, 2008). Upon commencement of the caspase activation cascades, hundreds of different protein substrates are cleaved and signs of apoptosis (nuclear condensation, cell shrinkage, membrane blebbing, and DNA fragmentation) begin to appear. It has been postulated that the trajectory of cell death occurs on the order of hours or days, where once a certain internal signaling threshold has been met, the cell is doomed to die within 10 minutes (Green, 2005). This so-called "10 minutes to dead" begs the emergence of a technology that can directly test and visualize this hypothesis. This challenge calls for the construction of a sensitive imaging tool to continuously monitor caspase activity that may transpire along a time course unanticipated by the observer.

1.2. Fluorescent probes can detect caspase activation

Since proteases produce a permanent insult to their targets, their surveillance is achieved by probes that can mimic this vulnerability to hydrolysis. Many current probes used to monitor caspase-3/7 activation utilize fluorescence. They come in the form of conventional organic dyes, fluorescent proteins, or quantum dots (QDs) that are paired with quencher moieties that operate via FRET. For example, commercialized assay kits (PhiPhiLux, Caspase Glo 3/7) can measure ensemble behaviors in live cells against drug treatments. Near-infrared fluorescent probes were used to interrogate the activities of the executioner caspases (3, 6, and 7) (Bullok et al., 2007; Bullok & Piwnica-Worms, 2005; Maxwell, Chang, Zhang, Barnett, & Piwnica-Worms, 2009), while QD–fluorescent protein conjugates have been used to detect *in vitro* caspase-3 activity at concentrations as low as 20 p*M* (Boeneman et al., 2009).

At the single-cell level, FRET studies have reported heterogeneity among a cell population. By TIRF microscopy, Angres, Steuer, Weber, Wagner, and Schneckenburgerr (2009) constructed an ECFP-DEVD-EYFP fusion protein bound to the inner leaflet of the plasma membrane by a palmitoyl anchor. HeLa cells transiently transfected with this sensor were challenged with staurosporine and ECFP-DEVD-EYFP cleavage became apparent from the redistribution of EYFP from the plasma membrane to the cytoplasm. This was detected with a simultaneous rise in fluorescence intensity of ECFP at the plasma membrane. Spatial rearrangement of EYFP along the time course in several cells of the same population indicated temporally different caspase-3/7 activation. Moreover, Sorger and colleagues (Albeck et al., 2008) prepared three separate fluorescent fusion proteins to monitor initiator and executioner caspases and mitochondrial outer membrane permeabilization. They showed that activation can take several minutes to hours and that a population of cells treated with the same apoptotic stimulus display heterogeneity.

Dickson and coworkers (Zheng, Zhang, & Dickson, 2004) were the first to develop subnanometer fluorescent gold nanoclusters, called gold QDs (Au-QDs), for their anticipated use as biological labels. These Au-QDs have narrower excitation and emission spectra and are smaller in size (5–31 atoms) compared to the larger semiconductor QDs that contain hundreds to thousands of atoms. Lin et al. (2010) utilized Au-QDs to prepare protease-mediated nucleus shuttles for real-time monitoring of apoptosis. Au-QDs were attached to peptides containing a nuclear export signal (NES), DEVD peptide sequence, and a nuclear localization signal. Initially, the probes were homogeneously distributed throughout HeLa cells. Upon induction with staurosporine, the localization of the cleaved Au-QD moiety bearing the NES showed an accumulation in the cytoplasmic compartment as early as 1 h by confocal microscopy. In this single-cell study, these constructs imitated the action of nuclear shuttle proteins that are activated during cell death.

Conventional dyes have been attached to particles that are at least tens of nanometers in diameter as in Au nanoparticle–dye (AuNP–dye) conjugates (Sun et al., 2010), where the AuNPs act as a quencher, or hyaluronic acid-based polymeric nanoparticles attached to linkers containing a Cy 5.5/black hole quencher 3 (BHQ-3) pair (Lee et al., 2011) to monitor caspase-3/7 activity. To image multiple proteases simultaneously, multistep synthesis with distinguishing dye/quencher pair selections would be necessary. To overcome this, Huang et al. (2012) prepared a multiplex imaging tool using a nonfluorescent, broad-spectrum, organic nanoquencher readily doped with BHQs onto the surface capable of quenching a variety of common dyes in the visible to NIR range. A mixture of linkers containing the optimal cleavage sequence of caspase-3/7, -8, or -9 with a terminal Cy 5.5, FPG 456, or FPR 560 dye, respectively, were distributed on the particle surface. This modality provided imaging of the progression of the apoptotic signaling cascade from the activation of initiator to executioner caspases within minutes by single-cell fluorescence microscopy.

Though *in vivo* single-cell fluorescence imaging has displayed features such as heterogeneity, allowed visualization of intracellular traffic upon cell death, and multiplex imaging of the caspase signaling cascade, limitations with this technique for understanding caspase activation at the single-molecule level exist. Though TIRF microscopy has permitted single-molecule imaging of components residing on the cell surface, this technique is not applicable for detection of intracellular species such as caspases in live cells. While QDs provide a reasonable signal-to-noise ratio for single-molecule imaging under confocal microscopy, their blinking characteristics prevent unambiguous detection of these events in live cells (Resch-Genger, Grabolle, Cavaliere-Jaricot, Nitschke, & Nann, 2008). In contrast, gold nanoparticles are free of these characteristics. They offer an enhanced set of properties, with their large signal-to-noise ratio, unlimited observation time (temporal resolution), and applicable dynamic range.

In Section 2, we describe the basic principles of gold nanoparticle optical sensors, and how their use can benefit sensitive caspase detection in live cells at the single-molecule level.

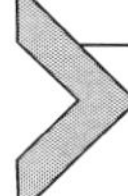

2. GOLD NANOPARTICLES AND THEIR POTENTIAL AS PLASMON RULERS

2.1. Gold nanoparticles have unique optical properties

The fourth century Lycurgus Cup[1] and medieval stained glass windows display the array of colors produced by the presence of noble metal nanoparticles, like gold, that are held in localized space within the glass material. These early works that contributed to the culture and esthetics of their day would undoubtedly stand the test of time in their beauty and elegance. From a reductionist standpoint, they represent evidence of light interacting with particles whose size is overshadowed by that of the wavelength of light and the subsequent curiosity of their optical properties that would be further uncovered and characterized.

Among the numerous contributions of Michael Faraday (1791–1867), his general curiosity surrounding the appearance of gold particles spanning any one of the colors of the electromagnetic spectrum provoked his hypothesis that such a distribution is due to a variation in particle size, as proposed in his Bakerian Lecture (Faraday, 1857; Kerker, 1991). In 1902, Richard Zsigmondy (1865–1929) and Henry Siedentopf (1872–1940) at Carl Zeiss AG developed the ultramicroscope, so aptly named for visualizing particles below microscale resolution as airy disks displaying their inherent scattering color and intensity. It delivered light scattering against a dark background as in darkfield illumination that is utilized today for noble metal nanoparticles. Zsigmondy would later be awarded the Nobel Prize for his contributions to both colloidal chemistry and the ultramicroscope in 1925. This instrument provided the workhorse for demonstrating Faraday's hypothesis that nanoparticle color was indeed contingent upon its size. This was achieved when physicist and mathematician Gustav Mie (1868–1957) provided ample mathematical treatment to describe the scattering and absorption of spherical gold nanoparticles (<200 nm) interacting with incident light (Helmuth, 2009; Mie, 1908). He leveraged observations on the ultramicroscope and infinite expansions of Maxwell's equations (Maxwell, 1861) which were

[1] Ancient Roman glass cup with gold and silver nanoparticles, giving it unusual optical properties. See https://www.britishmuseum.org/explore/highlights/highlight_objects/pe_mla/t/the_lycurgus_cup.aspx

entirely calculated by hand. Mie's work established the mathematical framework for gold nanoparticle optical properties and has found its influence in other fields.

The tunability of nanoparticle color is due to a number of factors not only limited to size. Both maximal light scattering color and intensity are dependent on other parameters including the nanoparticle's shape, composition, the dielectric constant of the dispersion medium, and close proximity of nanoparticles in space (Kelly, Coronado, Zhao, & Schatz, 2003). Upon illumination, photons of the incident light and the conduction electrons (plasmons) on the surface of metal nanoparticles oscillate cooperatively. This phenomenon is called plasmon resonance, and at nanometer scale, confinement of the oscillating surface electron motion causes an enhancement of the electromagnetic field strength by many orders of magnitude (Willets & Van Duyne, 2007). This provides light scattering at well-defined wavelengths in the visible range, and additionally, no limit to the duration of signal stability (Kreibig & Vollmer, 1995). Second, the confinement effects of plasmonic nanoparticles can lead to the concentration of electromagnetic fields of absorbed photons at the nanoparticle surface. The electromagnetic field concentrated by the nanoparticle can couple with other molecules and particles (Eustis & El-Sayed, 2006; Loo, Lowery, Halas, West, & Drezek, 2005; Lyandres, Shah, Zhao, & Van Duyne, 2008; Reinhard, Yassif, Vach, & Liphardt, 2010). Sections 2.2 and 2.3 describe plasmon coupling involving the interaction of two metal particles for single biomolecule imaging.

2.2. Plasmon coupling of a nanoparticle dimer and its scaling law

When particles in close proximity form a pair, their plasmon dipoles strongly couple when brought within a distance of the particle diameter. This coupling has been described in terms of molecular orbital theory, and a hybridization model illustrates the occupancy of the relative energies based on favorable or nonfavorable interactions of plasmons within higher order nanostructures (Fig. 11.1A; Jain, Eustis, & El-Sayed, 2006; Prodan, Radloff, Halas, & Nordlander, 2003; Wang, Brandl, Nordlander, & Halas, 2007). The plasmon dipole modes (Ψ_1 and Ψ_2) of a pair of two identical metal nanoparticles in close proximity can hybridize either in-phase ($\Psi_1 + \Psi_2$) or out-of-phase ($\Psi_1 - \Psi_2$). Interaction of longitudinally polarized light (i.e., along the interparticle axis) with the in-phase or out-of-phase combination produces energy splitting into bonding and antibonding configurations, respectively, denoted as σ and σ^* (Fig. 11.1A). On the other

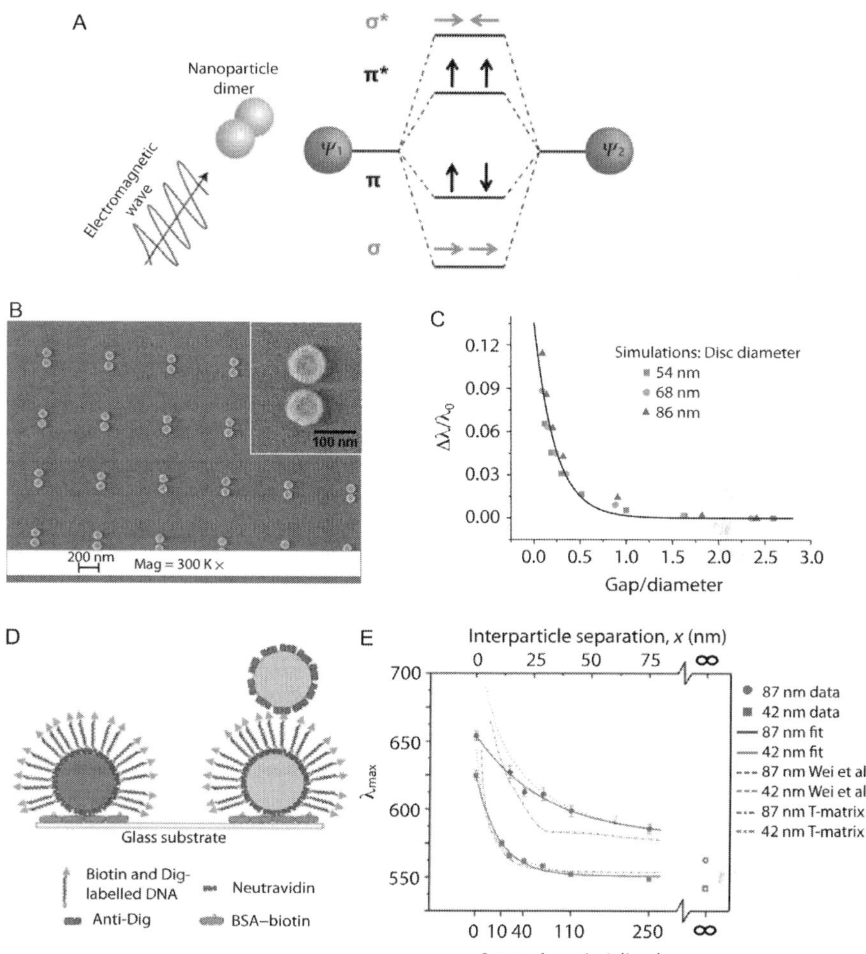

Figure 11.1 Plasmon rulers to measure single-molecule dynamic structural changes. (A) A hybridization model of plasmon coupling in a nanocrystal dimer. (B) SEM image of nanodisc pairs with an interparticle edge-to-edge distance of 12 nm. (C) Calibration of lithography patterned nanodisc dimers showing fractional plasmon shift as a function of interparticle gap to nanodisc diameter. (D) Plasmon ruler is assembled using a neutravidin-coated nanoparticle immobilized on a glass substrate. It is functionalized with biotin–dsDNA–digoxigenin linkers that allow direct conjugation onto an anti-digoxigenin-coated nanoparticle. (E) Calibration of DNA-linked colloidal gold nanosphere dimers. *(A) Reprinted with permission from Sheikholeslami, Jun, Jain, and Alivisatos (2010) Copyright (2010) American Chemical Society. (C) Reprinted with permission from Jain, Huang, and El-Sayed (2007) Copyright (2007) American Chemical Society. (E) Reprinted with permission from Reinhard, Siu, Agarwal, Alivisatos, and Liphardt (2005) Copyright (2005) American Chemical Society.*

hand, transversely polarized light (i.e., perpendicular to the interparticle axis) interacts with these modes yielding an antibonding mode (π^*) and a bonding mode (π). Since the degree of coupling for transverse polarization is weaker, the plasmon energy splitting is smaller compared to longitudinal coupling. It should be noted that the σ^* and π modes are spectrally dark due to the cancellation of the equal, but oppositely oriented dipoles on the two particles and the degree of π modes are relatively weak. Hence, the optical characteristics of a metal nanoparticle pair are strongly influenced by the σ bonding mode with lowered energy and red-shifted frequency.

Since plasmon coupling can simply be interpreted by dipole–dipole interactions, one can easily find that the plasmons of two metal nanoparticles can interact in a distance-dependent manner, where stronger coupling strength is expected as the interparticle separation decreases. As a result, the longitudinal plasmon resonance frequency red shifts, while the scattering intensity increases exponentially. Such a unique distance-dependent optical signature of nanoparticle dimers makes them a useful tool to readily explore the spatial conditions relevant to biology. When a biomolecule links two nanoparticles, the linker can undergo dynamic structural changes upon interaction with a biological entity that can perturb the interparticle distance. This produces a change in the dielectric surrounding the nanoparticle dimer that can be sensed optically at nanometer scale resolution and is analogous to the principles behind surface plasmon resonance devices.

Plasmon coupling has been exploited in top-down fabrication approaches to prepare plasmon rulers as tools to measure distances on the nanometer scale (Jain et al., 2007). In order to provide a universal scaling behavior for different nanoparticle parameters (size, shape, metal type, or medium dielectric constant), the El-Sayed group derived the distance decay equation of plasmon coupling in lithographically patterned metal nanodisc dimers as follows:

$$\frac{\Delta\lambda}{\lambda} = 0.18\exp\left(\frac{-s}{0.23D}\right)$$

where $\frac{\Delta\lambda}{\lambda}$ is the fractional plasmon shift, s is the interparticle edge-to-edge separation, and D is the particle diameter (Fig. 11.1B and C; Jain et al., 2007). Though plasmon coupling of lithographically patterned nanodisc dimers has provided fundamental principles to the field, the realization of coupled nanoparticles for single biomolecule imaging requires a method to fabricate nanoparticle dimers in a freestanding form using bottom-up techniques. For biological use, a valid calibration curve should be generated under typical experimental conditions, where the functionalized nanoparticle dimers are dynamically translating and randomly oriented. One such advance demonstrated a DNA conjugation method to allow precise tuning of the interparticle distance of the nanoparticle dimer by modulating the DNA length, enabling the calibration of colloidal plasmon rulers (Reinhard et al., 2005). The Alivisatos and Liphardt team immobilized DNA-linked colloidal gold nanoparticle dimers on a glass substrate and examined a wavelength versus distance relationship (Fig. 11.1D and E). In fact, double-stranded DNA was an ideal variable spacer because of its stiffness under 100 nm and synthetic feasibility with fine length tenability (Mastroianni, Sivak, Geissler, & Alivisatos, 2009).

2.3. Plasmon rulers for imaging applications

The plasmon ruler uses the unique distance-dependent scattering properties of a metal nanoparticle pair, where the conformational changes of a biomolecular linker can be translated to optical signal changes. There have been many other rulers that measure biological distances via a similar mechanism (Deniz et al., 2001; Roy, Hohng, & Ha, 2008). For example, the single-molecule fluorescence resonance energy transfer technique uses a relationship of intermolecular distance versus energy transfer efficiency that is determined by the dipole–dipole interaction between transition dipoles of a donor and an acceptor. This technique has been extremely useful in studying conformational changes of biomolecules in many systems, including enzyme activity, transcription, protein dynamics, identification of rare intermediates, and kinetic heterogeneity during RNA folding and RNA–protein interactions (Ha & Tinnefeld, 2012). This smFRET technique includes many features: (1) The FRET dyes allow facile conjugation with target biomolecules via either well-established bioconjugation chemistry or fluorescence tagging at the genetic level. (2) The dipole–dipole interaction is inversely proportional to the power of six of the intermolecular distance, allowing high-precision subnanometer spatial resolution in a short distance range (1–10 nm). (3) Theoretical and experimental background has provided very precise calibration curves for each set of FRET pairs. However, this technique has inherent limitations in lifetime and applicable distance. Plasmon rulers can resolve the problems of FRET techniques because of the following unique properties:

1. *Long applicable distance*: Plasmon coupling shows near-exponential distance decay in the range of 1–70 nm (see Section 2.2). Considering that some biomacromolecules exceed several tens of nanometers, the extended applicable distance of plasmon rulers is advantageous for macromolecular imaging.
2. *No photobleaching and no blinking*: Fluorescence dyes and proteins photobleach within several minutes and QDs blink. The plasmon ruler scattering signal is invariant over time; any timescale of biological events can be monitored continuously and indefinitely.
3. *Continuous colorimetric sensing*: FRET pairs typically use two-color switching and sometimes allow multicolor switching. The plasmon rulers exhibit a gradual spectral shift as a function of interparticle distance.
4. *Simple equipment required*: Unlike single-molecule fluorescence imaging, plasmon ruler imaging does not require high-end microscope systems, such as laser-TIRF, because the scattering signal from a single plasmon ruler is extremely high even with conventional halogen lamps.

Together, these superior properties of plasmon rulers have led to the development of a single-molecule nanosensor for ultrasensitive caspase detection in live cells, which is described in Section 3.

3. LIVE CELL IMAGING USING CROWN NANOPARTICLES

3.1. Development of a nanosensor by using plasmon coupling

Despite remarkable progress in understanding the molecular mechanisms of cancer cell death, conventional imaging techniques only allow static assessment of cell status. Instead, we monitor the dynamic activity of caspase-3 in real-time, high-resolution, and with ultrasensitive (single-molecule level) capabilities using gold nanoparticles.

Additional determinants of a nanotechnology tool for properly sensing caspase-3 activity at single-molecule resolution in live cells include: (1) introducing caspase-3 selectivity and (2) utilizing optical properties that can deliver quantitative measurements at a heightened signal-to-noise ratio. A systematic evaluation of substrate selectivity (at nonprime P4–P1 subsites) has revealed that caspase substrates can be cleaved by other members of the family (McStay, Salvesen, & Green, 2008). Most notably, both caspase-3 and -7 share a preference to cleave after the P1 Asp of the sequence, DEVD (P4–P1). Tailoring them to selectively measure caspase-3 activity requires incorporation of known amino acid preferences at both nonprime and prime side subsites of the targeting moiety. This can be achieved by using a peptide-based library to screen for the prime side residues that could impart selectivity (Demon et al., 2009; O'Donoghue et al., 2012; Schilling & Overall, 2008; Stennicke, Renatus, Meldal, & Salvesen, 2002). By exploiting plasmon coupling with our methodology (Jun et al., 2009), gold nanoparticles linked together by a caspase-3/7 peptide cleavage sequence provided a sensor that directly correlates scattering intensity changes to every caspase-3/7 cleavage event (Fig. 11.2A). To ensure ample signal-to-noise above that of the high- and heterogenous-background scattering of cells,

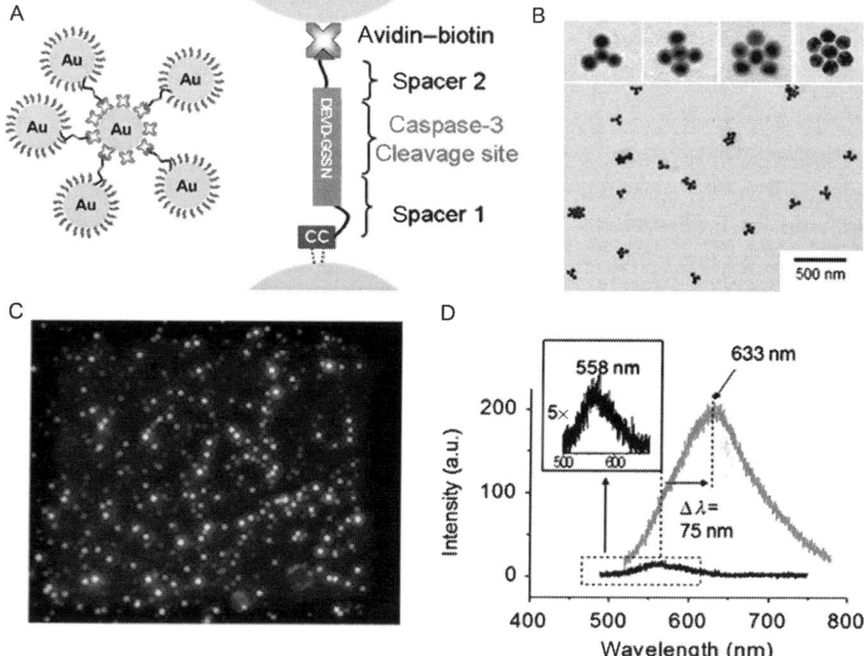

Figure 11.2 Crown nanoparticle sensors. (A) Illustration of a crown nanoparticle containing an avidin-coated gold core nanoparticle with multiple biotinylated gold satellite nanoparticles. Linkers bearing the caspase-3 cleavage sequence DEVD tether the core and satellite nanoparticles together. (B) Transmission electron microscopy (TEM) shows different configurations of the crown nanoparticles. They contain either three to six satellite nanoparticles linked to the core nanoparticle. (C) A representative scattering image of crown nanoparticles by darkfield microscopy. Each red (white in print) spot corresponds to a single crown nanoparticle. (D) Representative scattering spectra of crown (red, gray in print) and monomeric gold nanoparticles (black). The crown nanoparticles exhibited an increased scattering intensity ($\sim$44 ×) and a red shift ($\Delta\lambda \sim 75$ nm), compared to monomeric gold nanoparticles. *Reprinted from Jun et al. (2009).* (See the color plate.)

multiple nanoparticle assemblies were used to provide a sufficient signal during imaging.

Nanoparticle assemblies were prepared with a central core nanoparticle and peripheral satellite nanoparticles (Fig. 11.2A) held together by linkers that take advantage of two key interactions: (1) biotin–avidin ($K_d = 10^{-15}$ *M*) and (2) Au-thiol (bond strength ~45 kcal mol^{-1}; Dubois & Nuzzo, 1992). Linkers were designed with a biotinylated N-terminus and tandem cysteine residues at the C-terminus. These regions flank either a caspase-3 DEVD-containing peptide sequence or an extended PEG spacer. The peptide-containing linker was conjugated to the satellite nanoparticles, while the biotinylated extended PEG spacer populated the core nanoparticles. Both sets were brought together by avidin to ultimately yield the crown nanoparticles, named for their resemblance to crown ethers. They are comprised of multiple (three to six) satellite gold nanoparticles surrounding the core nanoparticle (Fig. 11.2B). Comparing its scattering intensity to that of monomeric gold nanoparticles, we found that crown nanoparticles exhibited strong red-colored optical signals (Fig. 11.2C). A hexameric crown nanoparticle scattered light approximately 44 times more intensely than a single particle (Fig. 11.2D). The protease sensors provided stable signals during the observation time that lasted for hours with no noticeable change (data not shown).

3.2. Crown nanoparticles directly measured caspase-3 activity *in vitro*

Since caspase-3 promotes proteolytic cleavage following the aspartic acid in the P1 position of the peptide sequence DEVD, loss of a satellite nanoparticle mediated by caspase-3 resulted in changes to both scattering color and intensity. Nanosensors were immobilized on a glass flow chamber and treated with recombinant caspase-3 (Fig. 11.3A). Under darkfield microscopy, initial red spots gradually turned into yellow and then green spots as time elapsed (Fig. 11.3B–D). Consistent with these observations, stepwise spectral blue shifts were detected in the scattering spectrum of a single crown nanoparticle as a result of successive loss of satellite nanoparticles by caspase-3 proteolysis (Fig. 11.3E). Single particle trajectories of the scattering intensity showed a stepwise decrease in the signal, corresponding to each individual proteolytic event (Fig. 11.3F).

Protease sensors do not perturb caspase-3 activity as cleavage kinetics fit a Michaelis–Menten first-order kinetic model. For this type of quantitation, each proteolytic event was enumerated and displayed as a progress curve (Fig. 11.3G) showing the cumulative probability as a function of time. Cumulative probability is defined here as the number of cutting

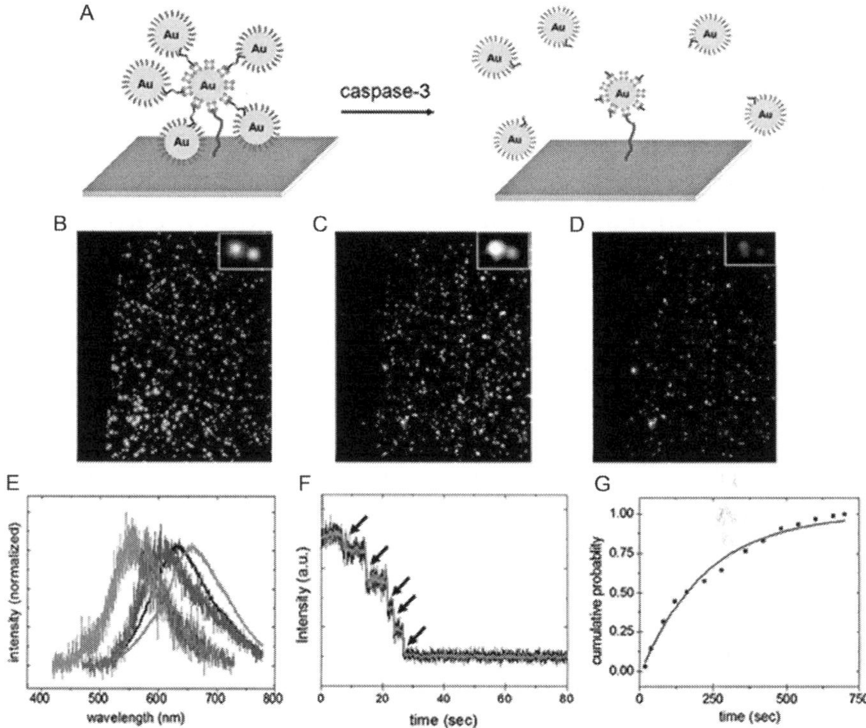

Figure 11.3 *In vitro* caspase-3 activity reported by crown nanoparticles. (A) Illustration of immobilized crown nanoparticles on a glass flow chamber by an avidin–biotin linkage. After equilibration, recombinant caspase-3 was presented to the crown nanoparticles. Color and intensity changes were visualized by darkfield microscopy. (B–D) Scattering color and intensity changes were observed upon treatment with caspase-3, from red to yellow to green spots (bright white to dim white in print) over time. (E) A representative spectral shift of a single crown nanoparticle upon exposure to caspase-3. At 654 nm, crown nanoparticles exhibited their scattering peak maximum. As the crown nanoparticles responded to caspase-3 proteolysis, blue shifts were observed over time. (F) A single crown nanoparticle intensity trace versus time was recorded at 100 Hz. Each arrow corresponds to the time at which caspase-3 mediated a cutting event of the crown nanoparticle substrate. (G) Progress curve showing cumulative probability of cutting events as a function of time that enabled extraction of a k_{cat} value ($6.2\ s^{-1}$). *Reprinted from Jun et al. (2009).* (See the color plate.)

events at a given time point per total number of cutting events observed during the experiment. Since $[S] < \ll [E_T]$, we applied the relationship $[k_{cat}/K_M] = [k_{obs}/E_T]$ where K_M is taken from the literature and k_{obs} is derived from the initial slope of the progress curve as obtained by fitting to a nonlinear regression model. We ultimately obtained a k_{cat} value ($6.2\ s^{-1}$) that falls within the range of previously reported values

(2.4–8.2 s^{-1}) obtained from small molecule substrates (Talanian et al., 1997). Therefore, we verified that geometric and steric effects do not alter caspase-3 kinetics, since the peptide sequences are surrounded by nanometer-sized particles.

3.3. Monitoring real-time caspase-3 signaling in a colon cancer cell line

To monitor intracellular caspase-3 activity, we visualized the response of nanoparticle assemblies of human colon carcinoma (SW620) cells upon stimulation of the extrinsic pathway with tumor necrosis factor-α (TNF-α) and cycloheximide (CHX). As demonstrated in the literature (Ndozangue-Touriguine et al., 2008), the SW620 cell line is resistant to death receptor-induced apoptosis due to high expression of an endogenous inhibitor, XIAP. Additionally, these cells do not express caspase-7 at the mRNA level (http://biogps.gnf.org/#goto=welcome). Therefore, this unique cell line was chosen to demonstrate the sensitivity of the crown nanoparticle technology to exclusively detect caspase-3 activation.

We conjugated Au crown nanoparticles with biotinylated TAT and found that intracellular delivery was successful in the SW620 colon cancer cell line (Fig. 11.4A). The red spots that represent each crown nanoparticle inside the cell boundary suggest efficient delivery of the nanosensors. The scattering signals from the nanosensors are highly intense compared to the background scattering from the cell. Since the crown nanoparticles are relatively large, they are held within the cytoskeletal network of the cells and afforded straightforward particle tracking during data analysis.

Caspase-3 activation at early time points was not observed with conventional ensemble analyses (Fig. 11.4B–C) upon induction of SW620 cells with TNF-α and CHX. Detection of caspase-3 activity in a lysate, with a luminescent substrate (Caspase Glo 3/7®), showed minimal activity at early time points. Robust caspase-3 activity was evident only at 23 h. However, use of a cell permeable fluorescent substrate allowed single-cell analysis in a flow cytometer and showed subtle caspase-3 activity at early time points. In contrast, reduction of the crown nanoparticles' scattering signals were observed as early as 30 min, which corresponds to the proteolytic cleavage of the DEVD peptide sequence (Fig. 11.4D, left). Vehicle- and inhibitor-treated samples showed no change in scattering color or intensity during the time course (Fig. 11.4D, middle, right). Unambiguous detection of early caspase-3 activity was made possible only by continuous imaging of the crown nanoparticles.

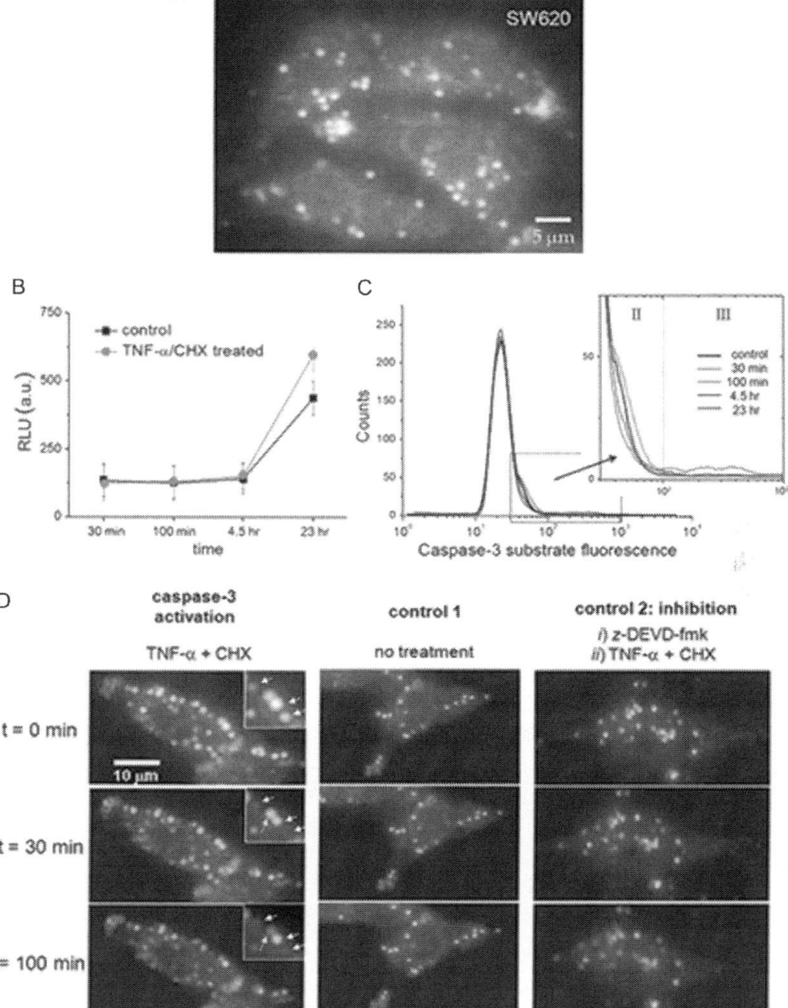

Figure 11.4 Caspase-3 activation in live cells reported by Caspase Glo 3/7® assay, flow cytometry, and crown nanoparticles. (A) Crown nanoparticles were modified with the cell-penetration peptide, TAT, and delivered into SW620 colon cancer cells. Red-colored spots (white in print) show the location of each individual crown nanoparticle. (B and C) Ensemble caspase-3 activity was monitored by either a luminescence assay (Caspase Glo 3/7®) or flow cytometry after treatment of cells with TNF-α/CHX. (B) With the luminescence assay, minimal caspase-3 activation was observed at early time points and was apparent only at 23 h after induction of cells. (C) Flow cytometry spectra show similar results to luminescence assay. Shoulder peaks may indicate caspase-3 activity at earlier time points. (D) (Left) After treatment of cells with TNF-α/CHX, caspase-3 activation was evident as early as 30 min. This is indicated by the change in scattering color and intensity across the time points. (Middle) Crown nanoparticles show no response in vehicle-treated cells. (Right) Pretreatment with inhibitor, z-DEVD-fmk, followed by induction with TNF-α/CHX, showed no response from crown nanoparticles. *Reprinted from Jun et al. (2009).* (See the color plate.)

In contrast to the ensemble measurements, the single particle trajectory displayed in Fig. 11.5A shows that caspase-3 proteolysis was not evident until 23 min after the addition of TNFα/CHX. All caspase-3 cleavage events within a single cell were counted, and the statistics from 10 individual cells were plotted in Fig. 11.5B and C. We observed significant variations in the induction times between different cells in the same population, ranging from 10 to 40 min. Ensemble studies have shown that cells in a group can each exhibit varying levels of caspase-3 when exposed to the same apoptotic stimulus and that this variation governs their viability after a certain period of time (Albeck et al., 2008; Angres et al., 2009; Morgan & Thorburn, 2001). The single-molecule results obtained here suggest that not only do active caspase-3 levels vary across cells, but that

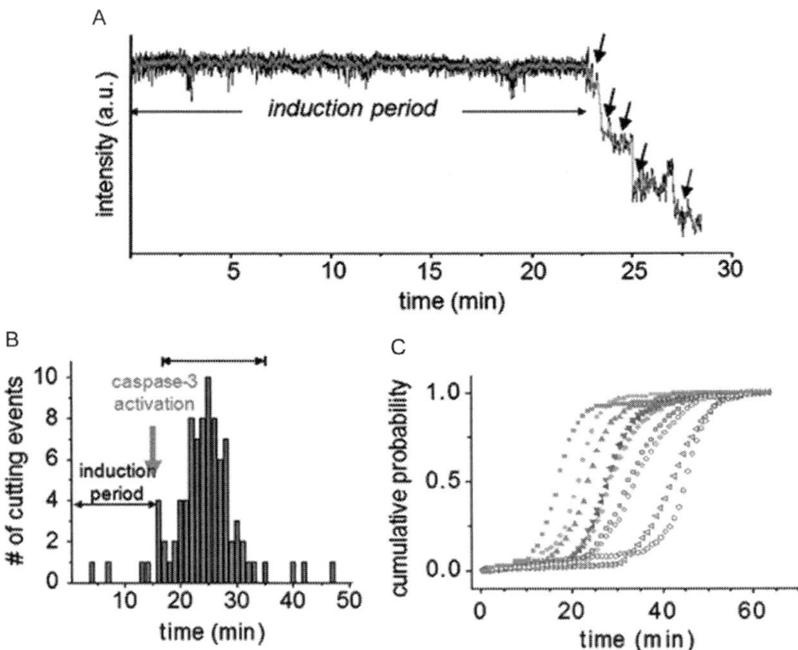

Figure 11.5 Statistical analyses acquired from crown nanoparticle trajectories after induction of SW620 cells. (A) Trajectory of a single crown nanoparticle displaying scattering intensity as a function of time. Successive decreases in intensity report cleavage of satellite nanoparticles by caspase-3. (B) Histogram of cutting events as a function of time from a single cell. The induction period (0–16 min) shows nominal response by crown nanoparticles. This is followed by rapid cutting events from 16 to 35 min indicating early-stage caspase-3 activation. (C) Caspase-3 activation across 10 identically treated cells show cell-by-cell heterogeneity. *Reprinted from Jun et al. (2009).*

the induction time required for caspase-3 activation also varies. This variability is likely a result of each cell's specific resistance to death receptor-induced apoptosis and is a clear representation of the "individuality" of cells and their response to a drug.

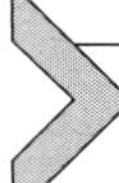

4. METHODS

4.1. Synthesis of satellite and core nanoparticles (Day 1)

- React citrate protected 40 nm gold particles with bis(*p*-sulfonatophenyl) phenylphosphine dihydrate dipotassium (BSPP) salt at 1 mg ml^{-1} colloids, as reported previously (Reinhard, Sheikholeslami, Mastroianni, Alivisatos, & Liphardt, 2007). Stir particles with BSPP for at least 8 h. Wash 2 × via centrifugation (5000 rpm, 15 min) with resuspension in 10 m*M* Tris, 40 m*M* NaCl, pH 8 (T40).
- Obtain absorbance of AuNPs (40 nm) at 520 nm using a UV/vis spectrophotometer and calculate yield ($\varepsilon = 9.3$ ml $pmol^{-1}$). Adjust nanoparticle concentration to 3 n*M*.
- Prepare stocks of biotinylated peptide linker (biotin—GSEGGS ESEDEVDGGSNSGGRLCC, biotin—$CH_2CH_2(OCH_2CH_2)_6$DEVDG —$GCH_2CH_2(OCH_2CH_2)_{12}CC$, or other linker with an optimal sequence), biotin—$CH_2(OCH_2CH_2)_6S$—$S(OCH_2CH_2)_6$—CH_2—biotin, and $HOOCCH_2(OCH_2CH_2)_6S$—$S(OCH_2CH_2)_6$—CH_2COOH in deionized water.
- Treat biotin—$CH_2(OCH_2CH_2)_6S$—$S(OCH_2CH_2)_6$—CH_2—biotin and $HOOCCH_2(OCH_2CH_2)_6S$—$S(OCH_2CH_2)_6$—CH_2COOH solutions with BSPP (1000 × molar excess) and shake for 10 min.
- *Peptide-conjugated AuNPs (satellites)*: Add 6 pmol biotinylated peptide to 3 n*M* gold nanoparticles. React overnight.
- *Peg-conjugated AuNPs (cores)*: Add 25 *M* excess of reduced biotin —$CH_2(OCH_2CH_2)_6S$—$S(OCH_2CH_2)_6$—CH_2—biotin solution and 10^5 *M* excess reduced $HOOCCH_2(OCH_2CH_2)_6S$—S $(OCH_2CH_2)_6$—CH_2COOH solution to 3 n*M* gold nanoparticles. React for 2 h.
- Separate cores via electrophoresis in a 0.7% agarose gel with 0.5 × TBE buffer. Run at 100 V for 1 h. Isolate by cutting band out of gel, placing it into dialysis bag containing T40 buffer, and running dialysis bag in the electrophoresis chamber containing 0.5 × TBE buffer. Run until particles move out of gel band and into buffer.

- Prepare neutravidin (Ntv) in deionized water and filter sterilize using a 0.2-μm syringe filter. Add 10,000 equivalents of Ntv to cores and react overnight at 4 °C.

4.2. Synthesis of crown nanoparticles (Day 2)

- Reduce $HOOCCH_2(OCH_2CH_2)_6S—S(OCH_2CH_2)_6—CH_2COOH$ with BSPP (1000 × *M* excess) and shake for 10 min.
- *Satellites*: Add 10^5 *M* excess of reduced $HOOCCH_2(OCH_2CH_2)_6S—S(OCH_2CH_2)_6—CH_2COOH$ solution to satellite nanoparticles and react for 2 h. Separate satellites by gel electrophoresis and isolate. Centrifuge at 1800 RCF for 10 min and disperse in 10 m*M* Tris, 100 m*M* NaCl, pH 8 (T100). Determine satellite nanoparticle concentration by UV/vis spectrometry and adjust concentration to 5 n*M*.
- *Ntv cores*: Centrifuge Ntv cores at 1800 RCF for 10 min and disperse in T100. Remove excess Ntv using a Centricon (Amicon) by washing 3 × with T100. Determine concentration of Ntv cores by UV/vis spectrometry and adjust nanoparticle concentration to 1 n*M*.
- *Crowns*: Add 15 μl 1 n*M* Ntv core nanoparticles to 300 μl 5 n*M* satellite nanoparticles. React overnight at 4 °C. Separate crowns across a sucrose density gradient (15–35%) at 4000 rpm for 30 min, with UV/vis detection. Use second sucrose density gradient containing control nanoparticles for comparison. Isolate crowns. Buffer exchange crown nanoparticles with T100 using a Centricon (Amicon) at 1500 rpm for 15 min until particles are dispersed entirely in T100. Characterize crown nanoparticles by TEM.

4.3. Preparing flow chambers

The procedures mentioned below have been adapted from Joo and Haa (2012) (Scheme 11.1).

- With a permanent marker, draw markings at 1 cm from the short edge and 0.5 cm from the long edge of a standard glass slide. Cover the entire slide with invisible tape.
- Lay new standard glass slide directly above the marked glass slide. Introduce a layer of water between both slides and on the top layer of glass to prevent fracturing while drilling. Drill four holes through the glass slide at each of the markings using a 0.75-mm diamond drill bit. Repeat for the number of chambers to be prepared (Scheme 11.1A).

* *Exercise caution when operating a drill and wear personal protective equipment.*

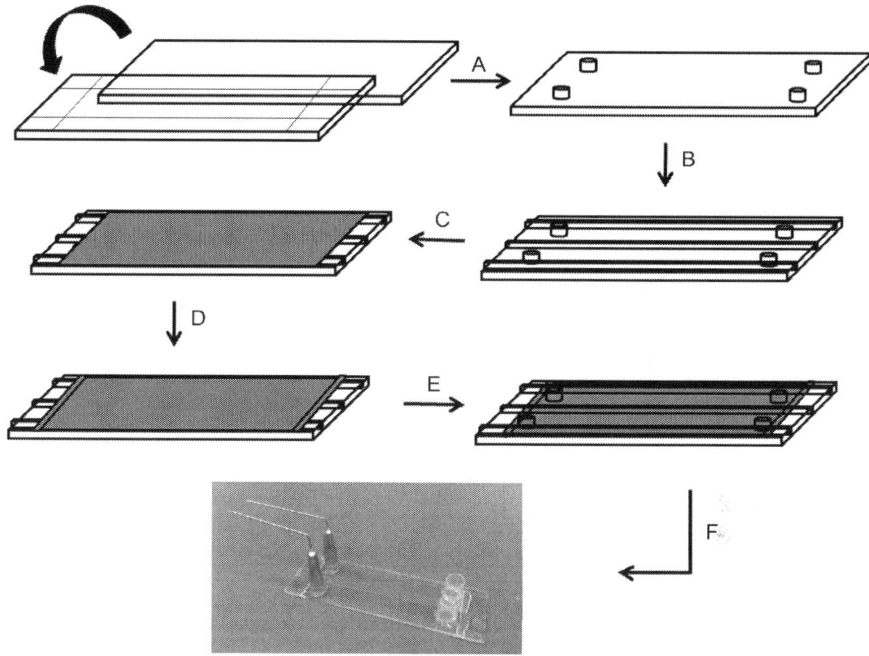

Scheme 11.1 Flow chamber assembly. Place standard slide directly above marked slide. (A) Add H_2O between and above glass layers. Drill holes using markings as guides. (B) Wash and dry glass before adding double-sided tape along the top, middle, and bottom areas of the slide. (C) Adhere a coverslip (blue) (dark gray in print) to the taped areas, while covering the holes. (D) Secure with epoxy glue (gray). (E) Once glue is dry, flip the slide over. (F) Modify the chambers with nanosensors and finish assembly by gluing needles and trimmed pipette tips. *Adapted from Joo and Haa (2012).*

- Place glass slides and cover slips (24 × 60) in an enclosable container. Fill container with 5% alconox and sonicate for 20 min in a water bath.
- Discard 5% alconox solution and wash glass 5 × with filtered water. Fill container with filtered water and sonicate for 20 min in a water bath.
- Discard water from container and wash glass 5 × with filtered water. Fill container with 1 *M* KOH and sonicate for 20 min in a water bath.
- Discard 1 *M* KOH and wash glass 5 × with filtered water. Fill container with filtered water and sonicate for 20 min in a water bath.
- Discard water and wash glass 5 × with filtered water. Fill container with filtered water.
- Dry glass using high purity Ar or N_2 gas sprayed at 40 psi. Immediately, store glass separately in 50 ml conical tubes to protect from dust.

- Examine drilled holes on each face of the glass slide and orient the slide such that the side with smaller holes is facing down.
- With a razor blade, cut 3 in. long strip of double-sided tape into smaller strips that are 3 mm in width. Carefully place a strip of tape along the top, middle, and bottom of the glass slide along its length (Scheme 11.1B).
- Lay coverslip on top of tape, while covering holes. Carefully press coverslip along taped areas using a pipette tip (Scheme 11.1C).
- Trim excess tape hanging beyond the edges of the glass slide.
- Use epoxy glue to seal the edges only along the width of the coverslip. Allow glue to dry for at least 15 min (Scheme 11.1D).
- Once glue is completely dry, flip chamber over (Scheme 11.1E). Wash inner chambers 3 × with 200 μl filtered 10 m*M* Tris, 30 m*M* NaCl, pH 8 (T30).
- Wash inner chambers with 150 μl filtered 1 mg ml^{-1} BSA–biotin and incubate for 10 min. To prevent formation of BSA crystals in chambers, do not incubate BSA–biotin for longer periods of time.
- Wash 3 × with 200 μl filtered T30.
- Dilute crown nanoparticle solution four-fold using T30 and apply 50 μl per chamber. Incubate for 10 min.
- Wash 3 × with 200 μl filtered T30.
- Cut off wider end of two P200 pipette tips and use epoxy glue to adhere to flow chamber. This will serve as the inlet.
- Glue needles to outlet of chambers. Once glue is dry, use pliers to bend needles in an "L" shape (Scheme 11.1F). Attach appropriately sized tubing to outlet and connect a syringe with a needle to serve as a waste reservoir. A syringe pump or peristaltic pump can be used with some added modifications.
- Apply T30 to inlet and pull back on syringe to ensure that flow chambers are completely sealed and working properly. Wash 3 × with T30 per chamber.
- Prepare reagents to be introduced to the flow chamber.
- When finished with experiments, chambers can be left in acetone overnight. This will separate the glass slide from the coverslip. The glass slide can then be reused after undergoing the washing steps above.

4.4. Darkfield microscopy

- Setup flow chamber at microscope and perform alignment of darkfield condenser and objective lens. Find field of view to image.

- Prepare settings on EMCCD detector.
- Evacuate inlet of buffer using disposable pipette. Add fresh buffer to inlet and pull back on syringe to introduce buffer to sensors for equilibration. Discard waste and attach syringe back onto needle. Wash chamber with buffer 3 × and ensure that chamber does not go dry.
- Introduce recombinant caspase-3 to the flow chamber.
- Begin acquisition (10–120 Hz) using an EMCCD camera to record changes in scattering intensity. Record interval time. To detect changes in maximum resonance wavelength, use a color camera attached to the port of a spectrometer.

4.5. Data analysis of *in vitro* caspase-3 activity

Each crown sensor provides quantitative information where each stepwise decrease in scattering intensity acts as a direct measure of caspase-3 activity.

- Use ImageJ, MatLab, or other software to obtain scattering intensity trajectories of crown nanoparticles in the field of view.
- *ImageJ*: Open folder containing images (as .tiff files) in ImageJ. Go to Analyze, then Tools, and open ROI Manager. Check "Show All." During manual particle picking, ImageJ will mark each selected particle when "Show All" is checked. Draw box around particle in the field of view. Press Control+t or click "Add [t]" in ROI Manager. The ROI manager will then record the coordinates. Continue until all particles in the field of view are selected. Click "More" and then "Save" to save the coordinates. Go to Analyze, then Set Measurements, and check "Integrated Density" as the parameter to be measured. Select "More" and then "Multi Measure" to analyze all the trajectories in the stack of images. When completed, a spreadsheet will open containing all the scattering intensities for each selected particle for each frame.
- For each set of scattering data for a given crown nanoparticle, plot scattering intensity as a function of time. Record time at which a decrease in scattering intensity is observed.
- Plot cumulative probability as a function of time and fit to a Michaelis–Menten first-order kinetic model. To extract a k_{cat} value, use a K_M parameter reported in literature for caspase-3. If the kinetic parameters of the peptide sequence are known, extract a k_{cat}/K_M value from the nanoparticle experiment for comparison to the parameters of the characterized peptide substrate.

4.6. Live cell imaging

- Before live cell imaging, ensure that the culture media being used is compatible with the conditions of the microscope setup. If a CO_2 control setup is absent, utilize CO_2-independent media, such as L-15, to prevent compromising cell viability.
- Seed flow chamber (Ibidi μ-slide I Luer) with adherent cell line based on manufacturer's instructions. It will be optimal to use a low seeding density to prevent high confluency and premature apoptosis, though this should be empirically determined for a given cell line. *Note: Caution should be taken when inducing cells that readily respond to apoptotic inducers. Immediate collapse of the cell body halts the experiment and prevents further acquisition of data from the nanosensors.*
- *Modification of crown nanoparticles with biotinylated TAT*: Add 3 pmol biotinylated TAT to 200 μl crown nanoparticle solution having an OD 4.3. React at RT for 1 h. Dilute nanoparticle solution with 1.5 ml culture media and filter through a 0.4 μm syringe filter. Add filtered solution to the seeded flow chamber. Allow overnight reaction (at least 12 h) for crown nanoparticle delivery to cells. *Note: Although cytosolic delivery of crown nanoparticles in adherent SW620 cells was reasonably successful, some sensors may be trapped in endosomes, so delivery of nanoparticles to different cell lines may lead to some endosomal trapping. To ensure cytosolic delivery of nanoparticles, others have explored methods such as modifying the TAT cell-penetration peptide* (Wadia, Stan, & Dowdy, 2004), *electroporation (Amaxa), microinjection, nanostraws* (VanDersarl, Xu, & Melosh, 2012), *and endosome-disrupting polymers* (Bayles et al., 2010).
- The next day, wash chamber 5 × with fresh culture media.
- Prepare microscope by performing alignment of darkfield condenser and objective lens. If temperature control is available, set to 37 °C. Find cell(s) to image.
- Wash chamber 3 × with fresh culture media for equilibration.
- Prepare settings on EMCCD detector.
- Introduce apoptotic inducer-containing culture media to the flow chamber.
- Begin acquisition (up to 250 Hz) using an EMCCD camera to record changes in scattering intensity. Record interval time. To detect changes in maximum resonance wavelength, use a color camera attached to the port of a spectrometer.
- Repeat above steps for vehicle or other drug treatment conditions, as necessary.

4.7. Data analysis of live cell imaging

- Analyze single particle trajectories and statistics, as in Section 4.5.
- Plot histogram displaying number of cutting events as a function of time to show the onset of robust proteolytic activity. Bin data along the *x*-axis using an appropriate time interval. Next, plot cumulative probability as a function of time to display any heterogeneity among a population of cells.

5. CONCLUSIONS

Before the twentieth century, noble metal nanoparticles were new and unusual materials whose reputation evolved to command an indelible visual presence in both art and science. Their study was exceedingly valued in the beginning of the twentieth century given the number of Nobel Prizes awarded and seminal papers published in that field. Utilization of these nanoparticles has since trickled down into the discrete space of the cellular environment to understand a myriad of biological processes. In this chapter, we focused on protease signaling, in particular, monitoring the enzymatic activity of caspase-3 whose activation is the desired outcome of certain drug interventions in cancer.

At single-molecule resolution, the outlook for live cell imaging using nanoparticle assays is promising given the ability to manipulate their optical properties to report the exquisite details of enzymatic behavior and concomitant cellular function at the single-cell level. In our single-molecule studies, unlike ensemble measurements that reported cell death 23 h after exposure to inducer, the crown nanoparticles allowed full display of the colon cancer (SW620) cell line's vulnerability to drug within minutes and indicates that the cell's internal signaling threshold was met at an early stage. With the continuous, unambiguous, and sensitive observation afforded by gold nanosensors, visualization of caspase-3 in other systems or interrogation of different proteases can be explored by use of a specific cell line or modification of the targeting moiety, respectively.

Since the crown nanoparticles offered a clear indication of when cells in a population were responding to drug at such high resolution, their use holds great promise for translational applications. With specific interest in monitoring caspases to evaluate patient response to chemotherapy, nanogold sensors can act as a screening tool to assess the effectiveness of preclinical and clinically approved drugs, and/or drug combinations on both cell lines- and patient-derived cells. Furthermore, diagnostic intervention with regard to a blood cancer, like leukemia, can be a powerful application where sample acquisition is minimally invasive, unlike that of solid tumors. Observation of elevated caspase-3 levels in circulating leukemia cells and/or blood could be used to evaluate patient response to drug treatment and identify variant cells displaying heterogeneity in terms of their low death response to apoptotic stimuli.

Finally, the study of noble metal nanoparticles has shaped a diverse set of fields like biophotonics, medicine, aerosol optics, atmospheric science, defense, and security. The same can be said for its emerging influence on provocative inquiries into biology and only time will tell how nanotechnologists of today will move the trajectory of colloidal chemistry.

HIGHLIGHTS

- Peptide-linked gold nanoparticles were synthesized as single molecule imaging tools to follow protease activity in live cells.
- These tools allowed continuous, unambiguous, and sensitive observation of caspase-3 activity in a colon cancer (SW620) cell line that is resistant to death receptor-induced apoptosis.
- They permitted reporting of early-stage caspase-3 activation that was undetectable by ensemble measurements.

ACKNOWLEDGMENTS

C. T. was supported as an appointee on the Microbial Pathogenesis and Host Defense Training Grant (T32-GM64337) and received the UCSF Quantitative Biosciences Consortium (QBC) Fellowship and NIGMS-IMSD award (R25-GM56847). C. S. C. received support from the National Institutes of Health (NIH R01CA128765). Y. J. was supported from the National Institute of Biomedical Imaging and Bioengineering and National Institutes of Health (NIH 1R21EB015088).

REFERENCES

Albeck, J. G., Burke, J. M., Aldridge, B. B., Zhang, M., Lauffenburger, D. A., & Sorger, P. K. (2008). Quantitative analysis of pathways controlling extrinsic apoptosis in single cells. *Molecular Cell*, *30*, 11–25.

Ancient Roman glass cup with gold and silver nanoparticles, giving it unusual optical properties. See https://www.britishmuseum.org/explore/highlights/highlight_objects/pe_mla/t/the_lycurgus_cup.aspx.

Angres, B., Steuer, H., Weber, P., Wagner, M., & Schneckenburger, H. (2009). A membrane-bound FRET-based caspase sensor for detection of apoptosis using fluorescence lifetime and total internal reflection microscopy. *Cytometry. Part A*, *75*, 420–427.

Bayles, A. R., Chahal, H. S., Chahal, D. S., Goldbeck, C. P., Cohen, B. E., & Helms, B. A. (2010). Rapid cytosolic delivery of luminescent nanocrystals in live cells with endosome-disrupting polymer colloids. *Nano Letters*, *10*, 4086–4092.

Boeneman, K., Mei, B. C., Dennis, A. M., Bao, G., Deschamps, J. R., Mattoussi, H., et al. (2009). Sensing caspase 3 activity with quantum dot-fluorescent protein assemblies. *Journal of the American Chemical Society, 131*, 3828–3829.

Bullok, K. E., Maxwell, D., Kesarwala, A. H., Gammon, S., Prior, J. L., Snow, M., et al. (2007). Biochemical and in vivo characterization of a small, membrane-permeant, caspase-activatable far-red fluorescent peptide for imaging apoptosis. *Biochemistry, 46*, 4055–4065.

Bullok, K., & Piwnica-Worms, D. (2005). Synthesis and characterization of a small, membrane-permeant, caspase-activatable far-red fluorescent peptide for imaging apoptosis. *Journal of Medicinal Chemistry, 48*, 5404–5407.

Chapman, H. A., Riese, R. J., & Shi, G. P. (1997). Emerging roles for cysteine proteases in human biology. *Annual Review of Physiology, 59*, 63–88.

Demon, D., Van Damme, P., Vanden Berghe, T., Deceuninck, A., Van Durme, J., Verspurten, J., et al. (2009). Proteome-wide substrate analysis indicates substrate exclusion as a mechanism to generate caspase-7 versus caspase-3 specificity. *Molecular and Cellular Proteomics, 8*, 2700–2714.

Deniz, A. A., Laurence, T. A., Dahan, M., Chemla, D. S., Schultz, P. G., & Weiss, S. (2001). Ratiometric single-molecule studies of freely diffusing biomolecules. *Annual Review of Physical Chemistry, 52*, 233–253.

Dubois, L. H., & Nuzzo, R. G. (1992). Synthesis, structure, and properties of model organic surfaces. *Annual Review of Physical Chemistry, 43*, 437.

Eustis, S., & El-Sayed, M. A. (2006). Why gold nanoparticles are more precious than pretty gold: Noble metal surface plasmon resonance and its enhancement of the radiative and nonradiative properties of nanocrystals of different shapes. *Chemical Society Reviews, 35*, 209–217.

Faraday, M. (1857). Experimental relations of gold (and other metals) to light. *Philosophical Transactions of the Royal Society of London, 147*, 145–181.

Green, D. R. (2005). Apoptotic pathways: Ten minutes to dead. *Cell, 121*, 671–674.

Ha, T., & Tinnefeld, P. (2012). Photophysics of fluorescent probes for single molecule biophysics and super-resolution imaging. *Annual Review of Physical Chemistry, 63*(2012), 595–617.

Helmuth, H. (2009). Gustav Mie and the scattering and absorption of light by particles: Historic developments and basics. *Journal of Quantitative Spectroscopy & Radiative Transfer, 110*, 787–799.

Huang, X., Swierczewska, M., Choi, K. Y., Zhu, L., Bhirde, A., Park, J., et al. (2012). Multiplex imaging of an intracellular proteolytic cascade by using a broad-spectrum nanoquencher. *Angewandte Chemie, International Edition in English, 51*, 1625–1630.

Jain, P. K., Eustis, S., & El-Sayed, M. A. (2006). Plasmon coupling in nanorod assemblies: Optical absorption, discrete dipole approximation simulation, and exciton-coupling model. *Journal of Physical Chemistry B, 110*, 18243–18253.

Jain, P. K., Huang, W., & El-Sayed, M. A. (2007). On the universal scaling behavior of the distance decay of plasmon coupling in metal nanoparticle pairs: A plasmon ruler equation. *Nano Letters, 7*, 2080–2088.

Joo, C., & Ha, T. (2012). Single-molecule FRET with total internal reflection microscopy. *Cold Spring Harbor Protocols, 12*, 22–25.

Jun, Y., Sheikholeslami, S., Hostetter, D. R., Tajon, C., Craik, C. S., & Alivisatos, A. P. (2009). Continuous imaging of plasmon rulers in live cells reveals early-stage caspase-3 activation at the single-molecule level. *Proceedings of the National Academy of Sciences of the United States of America, 106*, 17735–17740.

Kasibhatla, S., & Tseng, B. (2003). Why target apoptosis in cancer treatment? *Molecular Cancer Therapeutics, 2*, 573–580.

Kelly, K. L., Coronado, E., Zhao, L. L., & Schatz, G. C. (2003). The optical properties of metal nanoparticles: The influence of size, shape, and dielectric environment. *Journal of Physical Chemistry B*, *107*, 668–677.

Kerker, M. (1991). Founding fathers of light scattering and surface-enhanced Raman scattering. *Applied Optics*, *30*, 4699–4705.

Kreibig, U., & Vollmer, M. (1995). *Optical properties of metal clusters*. Berlin: Springer-Verlag, pp. 13–193.

Lee, S., Choi, K. Y., Chung, H., Ryu, J. H., Lee, A., Koo, H., et al. (2011). Real time, high resolution video imaging of apoptosis in single cells with a polymeric nanoprobe. *Bioconjugate Chemistry*, *22*, 125–131.

Lin, S. Y., Chen, N. T., Sun, S. P., Chang, J. C., Wang, Y. C., Yang, C. S., et al. (2010). The protease-mediated nucleus shuttles of subnanometer gold quantum dots for real-time monitoring of apoptotic cell death. *Journal of the American Chemical Society*, *132*, 8309–8315.

Logue, S. E., & Martin, S. J. (2008). Caspase activation cascades in apoptosis. *Biochemical Society Transactions*, *36*(Pt 1), 1–9.

Loo, C., Lowery, A., Halas, N., West, J., & Drezek, R. (2005). Immunotargeted nanoshells for integrated cancer imaging and therapy. *Nano Letters*, *5*, 709–711.

Lyandres, O., Shah, N. C., Zhao, J., & Van Duyne, R. P. (2008). Biosensing with plasmonic nanosensors. *Nature Materials*, 7, 442–453.

Mastroianni, A. J., Sivak, D. A., Geissler, P. L., & Alivisatos, A. P. (2009). Probing the conformational distributions of subpersistence length DNA. *Biophysical Journal*, *97*, 1408–1417.

Maxwell, J. C. (1861). On physical lines of force. *Philosophical Magazine*, *21*, 161–175 , 281–291, 338–348; 23, 12–24, 85–95.

Maxwell, D., Chang, Q., Zhang, X., Barnett, E. M., & Piwnica-Worms, D. (2009). An improved cell-penetrating, caspase-activatable, near-infrared fluorescent peptide for apoptosis imaging. *Bioconjugate Chemistry*, *4*, 702–709.

McStay, G. P., Salvesen, G. S., & Green, D. R. (2008). Overlapping cleavage motif selectivity of caspases: Implications for analysis of apoptotic pathways. *Cell Death and Differentiation*, *15*, 322–331.

Mie, G. (1908). Beitrage zur Optik Truber medien, special kolloidaler Metalldsungen. *Annals of Physics*, *26*, 329–445.

Morgan, M. J., & Thorburn, A. (2001). Measurement of caspase activity in individual cells reveals differences in the kinetics of caspase activation between cells. *Cell Death and Differentiation*, *8*, 38–43.

Ndozangue-Touriguine, O., Sebbagh, M., Merino, D., Micheau, O., Bertoglio, J., & Breard, J. (2008). A mitochondrial block and expression of XIAP lead to resistance to TRAIL-induced apoptosis during progression to metastasis of a colon carcinoma. *Oncogene*, *27*, 6012–6022.

O'Donoghue, A. J., Eroy-Reveles, A. A., Knudsen, G. M., Ingram, J., Zhou, M., Statnekov, J. B., et al. (2012). Global identification of peptidase specificity by multiplex substrate profiling. *Nature Methods*, *9*, 1095–1100.

Prodan, E., Radloff, C., Halas, N. J., & Nordlander, P. (2003). A hybridization model for the plasmon response of complex nanostructures. *Science*, *302*, 419–422.

Reinhard, B. M., Sheikholeslami, S., Mastroianni, A., Alivisatos, A. P., & Liphardt, J. (2007). Use of plasmon coupling to reveal the dynamics of DNA bending and cleavage by single EcoRV restriction enzymes. *Proceedings of the National Academy of Sciences of the United States of America*, *104*, 2667–2672.

Reinhard, B. M., Siu, M., Agarwal, H., Alivisatos, A. P., & Liphardt, J. (2005). Calibration of dynamic molecular rulers based on plasmon coupling between gold nanoparticles. *Nano Letters*, *5*, 2246–2252.

Reinhard, B. M., Yassif, J. M., Vach, P., & Liphardt, J. (2010). Plasmon rulers as dynamic molecular rulers in enzymology. *Methods in Enzymology*, *475*, 175–198.

Resch-Genger, U., Grabolle, M., Cavaliere-Jaricot, S., Nitschke, R., & Nann, T. (2008). Quantum dots versus organic dyes as fluorescent labels. *Nature Methods*, *5*, 763–775.

Roy, R., Hohng, S., & Ha, T. (2008). A practical guide to single-molecule FRET. *Nature Methods*, *5*, 507–516.

Schilling, O., & Overall, C. M. (2008). Proteome-derived, database-searchable peptide libraries for identifying protease cleavage sites. *Nature Biotechnology*, *26*, 685–694.

Sheikholeslami, S., Jun, Y., Jain, P. K., & Alivisatos, A. P. (2010). Coupling of optical resonances in a compositionally asymmetric plasmonic nanoparticle dimer. *Nano Letters*, *10*, 2655–2660.

Stennicke, H. R., Renatus, M., Meldal, M., & Salvesen, G. S. (2002). Internally quenched fluorescent peptide substrates disclose the subsite preferences of human caspases 1, 3, 6, 7 and 8. *Biochemistry Journal*, *350*(Pt 2), 563–568.

Sun, I. C., Lee, S., Koo, H., Kwon, I. C., Choi, K., Ahn, C. H., et al. (2010). Caspase sensitive gold nanoparticle for apoptosis imaging in live cells. *Bioconjugate Chemistry*, *21*, 1939–1942.

Talanian, R. V., Quinlan, C., Trautz, S., Hackett, M. C., Mankovich, J. A., Banach, D., et al. (1997). Substrate specificities of caspase family proteases. *Journal of Biological Chemistry*, *272*, 9677–9682.

VanDersarl, J. J., Xu, A. M., & Melosh, N. A. (2012). Nanostraws for direct fluidic intracellular access. *Nano Letters*, *12*, 3881–3886.

Wadia, J. S., Stan, R. V., & Dowdy, S. F. (2004). Transducible TAT-HA fusogenic peptide enhances escape of TAT-fusion proteins after lipid raft macropinocytosis. *Nature Medicine*, *10*, 310–315.

Wang, H., Brandl, D. W., Nordlander, P., & Halas, N. J. (2007). Plasmonic nanostructures: Artificial molecules. *Accounts of Chemical Research*, *40*, 53–62.

Willets, K. A., & Van Duyne, R. P. (2007). Localized surface plasmon resonance spectroscopy and sensing. *Annual Review of Physical Chemistry*, *58*, 267–297.

Zheng, J., Zhang, C., & Dickson, R. M. (2004). Highly fluorescent, water-soluble, size-tunable gold quantum dots. *Physical Review Letters*, *93*(7), 077402.

CHAPTER TWELVE

In Vivo Monitoring of Caspase Activation Using a Fluorescence Resonance Energy Transfer-Based Fluorescent Probe

Yoshifumi Yamaguchi[*,†,1], **Erina Kuranaga**[‡], **Yu-ichiro Nakajima**[§], **Akiko Koto**[*], **Kiwamu Takemoto**[†,¶], **Masayuki Miura**[*,||,1]

[*]Department of Genetics, Graduate School of Pharmaceutical Sciences, The University of Tokyo, Tokyo, Japan
[†]PRESTO, JST, Tokyo, Japan
[‡]Laboratory for Histogenetic Dynamics, RIKEN CDB, Kobe, Japan
[§]Stowers Institute for Medical Research, Kansas, Missouri, USA
[¶]Department of Physiology, Graduate School of Medicine, Yokohama City University, Yokohama, Japan
[||]CREST, JST, Tokyo, Japan
[1]Corresponding author: e-mail address: bunbun@mol.f.u-tokyo.ac.jp or miura@mol.f.u-tokyo.ac.jp

Contents

Methods in Enzymology, Volume 544
ISSN 0076-6879
http://dx.doi.org/10.1016/B978-0-12-417158-9.00012-1

Abstract

Caspases, which constitute a family of cysteine proteases, are highly conserved in multicellular organisms and function as a central player in apoptosis. The detection of apoptosis is intrinsically difficult because dying cells are rapidly removed from tissues by phagocytosis. Thus, the development of a method for detecting caspase activation is critical for the *in vivo* study of apoptosis. In this chapter, we describe a genetically encoded fluorescent probe for live imaging of caspase activation.

1. INTRODUCTION

Cell death is commonly observed during development in animals. However, except for regions such as the anterior necrotic zone or interdigital region of developing vertebrate limbs in which massive cell death occurs, cell death is intrinsically difficult to detect. This is because apoptotic cells are rapidly removed by macrophages or the neighboring cells. To increase the sensitivity and reliability of the methods used to detect apoptosis, several genetically encoded probes have been developed for the detection of caspase activation by different mechanisms such as immunogenic detection, detection by subcellular relocalization, and fluorescence activation or fluorescence resonance energy transfer (FRET).

In fixed tissues, the CD8::poly-ADP-ribose polymerase-1 (PARP)::Venus probe can sensitively detect caspase activation in *Drosophila* (Schoenmann et al., 2010; Takeishi et al., 2013; Williams, Kondo, Krzyzanowska, Hiromi, & Truman, 2006). The typical caspase-3-cleavage site "DEVD" is found in human PARP and is efficiently cleaved by caspase-3-like effector caspases. A specific antibody that recognizes the caspase-cleaved neoepitope of PARP (anti-cPARP antibody) is commercially available. Overexpression of CD8::PARP::Venus, a caspase substrate, greatly enhances the sensitivity of immunological detection of the cPARP product relative to that achieved using conventional immunohistochemical techniques. Because of the high sensitivity of this approach, localized caspase activation was detected in the pruning dendrites of sensory neurons in *Drosophila* at the pupal stage (Williams et al., 2006).

The "Apoliner" probe, which was also designed in *Drosophila*, utilizes subcellular relocalization triggered by caspase cleavage. Apoliner allows for the real-time detection of caspase activation at a single-cell level. Starting from the N-terminus, the probe consists of the transmembrane domain of mouse CD8, a monomeric red fluorescent protein, the caspase-cleavage site of the *Drosophila* inhibitor of apoptosis 1, and enhanced green fluorescent protein (EGFP) with a nuclear localization signal (NLS–EGFP). The activated caspase cleaves Apoliner and separates mCD8-monomeric red fluorescent protein and NLS–EGFP. Thus, caspase activation can be monitored by changes in the localization of the fluorescent protein (Bardet et al., 2008). Subcellular relocalization in combination with a ratiometric approach has also been used in mammals (Ohsawa, Hamada, Yoshida, & Miura, 2008).

Then, various approaches have been used to develop a fluorescent protein reporter through the either cleavage of a destabilizing peptide, to increase the abundance of a fluorescent reporter, or the cleavage of a linker sequence that might prevent the proper folding of the fluorescent reporter. The proteasome-based caspase-3 reporter was developed in mammals. A protein-degradation sequence (polyubiquitin domain) followed by the caspase-3-cleavage site was fused with the chimeric luciferase/EGFP protein. Caspase activation removes the protein-degradation sequence; this stabilizes luciferase/EGFP, which can be used as a reporter for caspase-activated cells (Huang et al., 2011). Cyclic luciferase (pcFluc–DEVD) was developed for monitoring caspase activity *in vivo*. In this probe, DnaE inteins are fused to the N- and C-terminal ends of luciferase connected with the caspase-3-cleavage site, DEVD. After cleavage of the DEVD sequence by caspase 3, the luciferase activity can be restored (Kanno, Yamanaka, Hirano, Umezawa, & Ozawa, 2007). The switch-on fluorescence-based caspase-3-like protease activity indicator was developed by modifying the circularly permutated Venus protein; it contains a caspase-3-cleavage site and a split *Npu* DnaE intein. This caspase-3 activity indicator named C3AI acquires fluorescent activity after cleavage by DEVDase (Zhang et al., 2013). Another approach for generating a switch-on type (or dark-to-bright) reporter was developed by generating a GFP construct in which the caspase-3-cleavage site (DEVD) was followed by a hydrophobic quenching peptide at the C-terminus. The quenching peptide tetramerizes GFP and prevents maturation of the GFP chromophore. Catalytic removal of the quenching peptide by DEVDase restores the fluorescence (Nicholls, Chu, Abbruzzese, Tremblay, & Hardy, 2011).

In 2000, a FRET-based ratiometric fluorescent caspase probe was reported (Tyas, Brophy, Pope, Rivett, & Tavare, 2000) and further developed as SCAT (a sensor for activated caspases based on FRET) for *in vivo* use (Fig. 12.1; Takemoto, Nagai, Miyawaki, & Miura, 2003). Compared to other methods described earlier, a ratiometric FRET probe detects the activation of caspases with a higher temporal resolution.

In this chapter, we describe a method for monitoring caspase activation using a FRET-based caspase probe in mammalian cells and in two *in vivo* systems, that is, the fly and mouse.

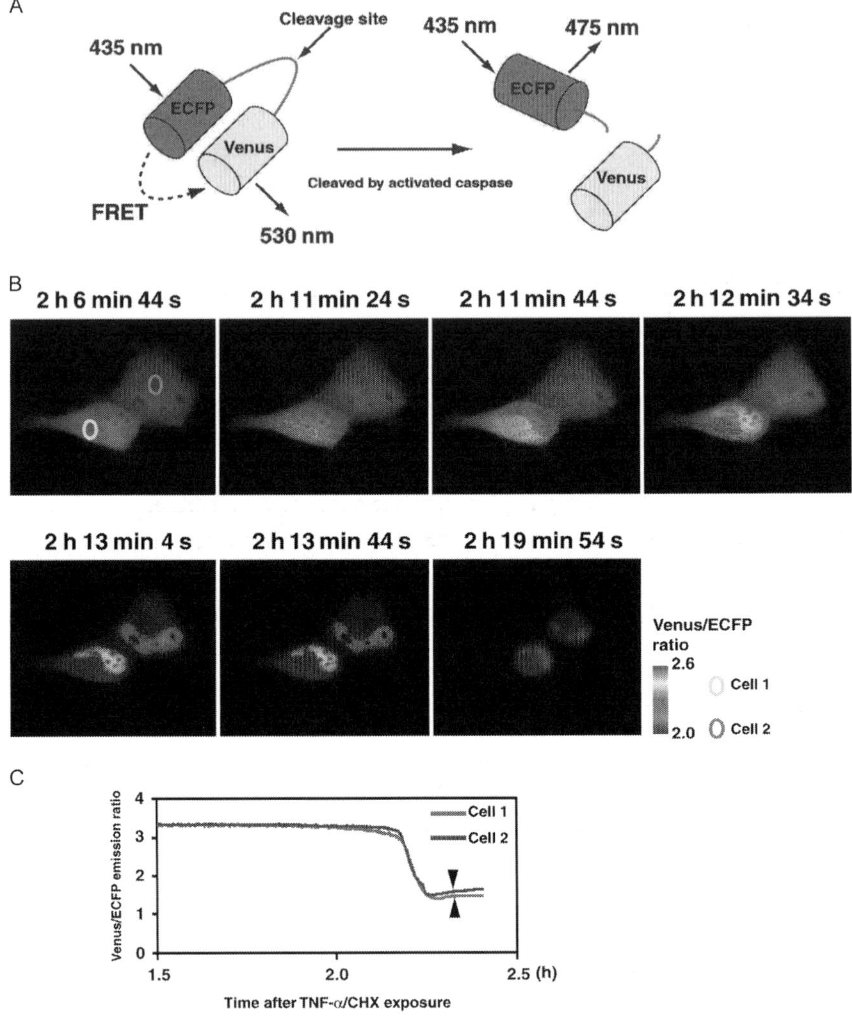

Figure 12.1 FRET imaging analysis of caspase activation with the SCAT3 probe. Cells were treated with TNF/CHX to induce caspase activation and cell death. A Venus/ECFP ratio image is displayed with intensity-modified pseudocolors. (A) Schematic representation of SCAT. (B) Before cell shrinkage, caspase activation was observed from the cytosol to the nucleus. (C) Caspase activation was monitored by observing the "all-or-none" pattern. *Copyright © Takemoto et al. (2003), originally published in* Journal of Cell Biology. *http://dx.doi.org/10.1083/jcb.200207111.* (See the color plate.)

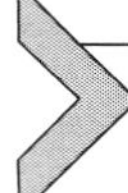

2. MONITORING CASPASE ACTIVITIES IN LIVING CELLS: PRINCIPLE AND TIPS FOR USING A SCAT3 PROBE

2.1. Development of an FRET probe for monitoring caspase activation

FRET technology can provide valuable information about the dynamics and pattern of enzyme activity. Early FRET probes for caspase activation used a hybrid protein, that is, an enhanced cyan fluorescence protein (ECFP) as the FRET donor, and an enhanced yellow fluorescence protein (EYFP) as the FRET acceptor, linked by peptides containing the caspase-3 cleavage sequence, DEVD (Luo, Yu, Pu, & Chang, 2001; Rehm et al., 2002; Tyas et al., 2000). The most important feature of an FRET probe is the specificity for the target molecule. The molar extinction coefficient of the acceptor protein is an important factor for determining the efficiency of FRET; however, the molar extinction coefficient of EYFP, a common fluorescent protein in FRET acceptors, shows high sensitivities for protons and chloride ions, especially under physiological conditions (Jayaraman, Haggie, Wachter, Remington, & Verkman, 2000; Llopis, McCaffery, Miyawaki, Farquhar, & Tsien, 1998). Several reports have suggested that acidification occurs during apoptosis (Matsuyama, Llopis, Deveraux, Tsien, & Reed, 2000; Perez-Sala, Collado-Escobar, & Mollinedo, 1995; Vincent, TenBroeke, & Maiese, 1999) and that a chloride/bicarbonate exchanger is involved in apoptotic events (Araki et al., 2002; Maeno, Ishizaki, Kanaseki, Hazama, & Okada, 2000). As a change in the chloride concentration also occurs in neurons, it would be crucial to use a chloride-insensitive indicator for monitoring caspase activation in neurons. Because of these features, researchers should be prudent in the use of EYFP as the FRET acceptor, especially for *in vivo* apoptosis imaging.

To overcome these limitations, Venus was utilized as the FRET acceptor in the SCAT3 probe (Fig. 12.1; Takemoto et al., 2003) because it shows low sensitivity for protons and chloride ions (Nagai et al., 2002). Indeed, SCAT3 showed high stability *in vitro* and in living cells compared with the previously reported ECFP–EYFP probe (Takemoto et al., 2003). The development of SCAT3 indicators has enabled researchers to monitor caspase activation in living cells and *in vivo* in real time (Campbell & Okamoto, 2013; Koto, Kuranaga, & Miura, 2009; Kuranaga et al., 2011; Nakajima, Kuranaga, Sugimura, Miyawaki, & Miura, 2011; Ohgushi et al., 2010; Takemoto et al., 2007; Yamaguchi et al., 2011).

2.2. Microscope systems for conducting time-lapse imaging with SCAT3

2.2.1 Microscope

Both wide-field and confocal microscopes can be used for SCAT3 imaging, depending on the purpose of the experiments. Any upright or inverted microscope can be practically retrofitted for SCAT3 imaging. In our laboratory, we use inverted wide-field and confocal microscopes because they offer a wide range of applications, from imaging cell cultures in dishes to *in vivo* organ imaging in flies or *ex utero* imaging of mouse embryos. In the following Sections 2.2, 3.2–3.6, and 4, we describe the microscope systems that were used for each imaging experiment in our laboratory. We emphasize that other microscope systems can be retrofitted for SCAT3 imaging, depending on the purpose of the experiment.

2.2.2 Light source, filter sets, and scanning and detection systems

The light source for SCAT3 imaging varies with the microscope settings. It is important to excite ECFP selectively, but not Venus. For laser scanning confocal imaging, the best laser is a 442 nm (or 446 nm) diode laser because it excites only ECFP, but not Venus. However, the major drawback of using this laser is its high costs, apart from the limited applicability to other experiments. The second, practical choice is to use an argon multiline laser (458 nm) in order to excite ECFP. However, this laser also excites Venus at low efficiency, thereby increasing the background signal from Venus and interfering with the FRET signals. Nevertheless, we confirmed that excitation with 458 nm was sufficient to detect caspase activation in apoptotic larval epithelial cells (LECs, see Section 3.4.) during metamorphosis. Thus, a practical solution to conduct SCAT3 imaging would be to use a standard laser confocal microscope equipped with an argon multiline laser (458 nm). For wide-field fluorescent imaging, conventional light sources such as a light-emitting diode or mercury arc lamp were used with appropriate filter sets and dichroic mirrors.

In general, the microscope should be equipped with an excitation filter that corresponds to the fluorescent spectrum of ECFP, two emission filters for ECFP and Venus (EYFP), and an appropriate dichroic mirror. The best way to collect the images per channel is through coregistration, because the ratiometric approach utilized for assessing FRET requires a pixel-by-pixel analysis in image processing. Therefore, a microscope with a detection path that enables simultaneous two-channel detection is preferred, in order to avoid movement artifacts that can influence radiometric measures. This is observed in a laser confocal microscope system equipped with more than two photodetectors such as the photomultiplier tube. However, when using one photodetector system (e.g., a confocal microscope with one CCD camera or a wide-field fluorescent microscope), it is also practical to use a single dichromatic mirror coupled to excitation and an emission filter wheel or slider to acquire images in order to minimize or eliminate image shifts.

A possible setting for SCAT3 imaging using a wide-field microscope (or confocal microscope with a 440- to 446-nm laser) could be as follows:

- Excitation filter: A440AF21 for ECFP
- Two emission filters: 480AF30 for ECFP and 535AF26 for Venus
- Dichroic mirror: 455DRLP (or RSP455, Leica)

This filter/dichroic mirror set is available as a CFP/YFP FRET set (XF88-2) (Omega Optical Inc.).

For a confocal microscope with a 458-nm laser, the dichroic beam splitter could be DD458/514 (Leica) or ZT458/514rpc (Chroma).

In addition, there is a unique detection system equipped with a three-color cooled intensified CCD camera compatible with FRET imaging (Aquacosmos/Ashura system, Hamamatsu Photonics, K.K., Japan). This system has an appropriate prism, which enables simultaneous detection of ECFP and Venus even in using a wide-field fluorescent microscope.

2.2.3 Data collection

In general, high sensitivity is a crucial factor for FRET probe imaging because it would shorten the image acquisition (exposure) time, thereby reducing photobleaching of the probes and phototoxicity to cells. This is absolutely true for SCAT3 imaging.

Achieving a short scanning time requires the improvement of three parameters, that is, intensity of fluorescence, high scanning speed, and high sensitivity of the detection system. The intensity of the SCAT3 fluorescence solely depends on its expression level. Hence, it is important to choose cells or strains that express appropriate amounts of SCAT3 (see also Sections 2.4 and 3.1) for SCAT3 imaging. An effective solution for fast scanning with a wide-field microscope would be to use external filter wheels that enable short switchover times without causing vibration of the microscope. There are two very effective systems for fast scanning with laser confocal microscopes, that is, the Nipkow disk scanner (e.g., CSU10 or CSU-X, Yokogawa Electric Corporation, Japan) and the resonant scanner (Leica Microsystems, Germany; Nikon, Japan). For sensitive detection and high resolution, cooled and intensified CCD camera systems or photomultiplier detectors are available, depending on the microscope configuration.

2.2.4 Data analysis

An image processing software should be used for analyzing datasets obtained with time-lapse imaging. Please see "Section 2.4."

2.3. The procedure

This is a protocol for detecting activation of caspase-3 in HeLa cells undergoing apoptosis (Fig. 12.1; Takemoto et al., 2003).

1. Seed 1–2 × 10^5 HeLa cells in a 35-mm glass-bottom dish (Matsunami glass; NO1S thickness) coated with polyethyleneimine on the day before transfection is carried out.
2. Transfect the cells with a SCAT3-expressing vector (pSCAT3; 1 μg) using transfection reagents such as the Superfect reagent (Qiagen) and culture for at least 18 h.
3. Add tumor necrosis factor alpha (Sigma) with cycloheximide (50 ng/mL and 10 μg/mL, respectively) to cells in Hank's balanced salt solution or phenol red-free medium.
4. Monitor FRET using an inverted microscope (IX70, Olympus) equipped with a CCD camera (CoolSNAP HQ, Roper Scientific) and an illumination pillar (ILL100LH, Olympus). A440AF21 excitation filter, two emission filters (480AF30 for ECFP and 535AF25 for Venus) alternated by a filter changer (Lambda10-2, Sutter Instruments), was used for SCAT3 imaging. All these devices were controlled by the MetaFluor software (Universal Imaging).
5. Process the ratio images (Venus intensity/ECFP intensity) using the MetaMorph software (Universal Imaging). We recommended the "intensity-modified display" (IMD) (Tsien & Harootunian, 1990) for evaluating reliable signals for caspase activation (see also Section 2.4).

2.4. Tips for image acquisition and processing

Here are some tips for obtaining correct results from SCAT3 imaging. Most of them are commonly applicable to imaging experiments using other FRET probes to measure signaling activities. It is also important to keep all conditions constant through an experiment.

Excitation laser power

- Use excitation wavelengths minimally to reduce both phototoxicity and photobleaching. The use of strong light or laser power causes extensive photobleaching during imaging, hampering the proper calculation of the Venus/ECFP ratio, and is often accompanied by strong phototoxicity. Even without extensive photobleaching, phototoxicity can still occur. The intensity of light or laser power used for excitation should depend on the microscope settings and the level of SCAT3 expression observed in cells or tissues.

Signal intensity

- Avoid saturation of signals when acquiring time-lapse images. Signals from ECFP and Venus must not be saturated for measuring the Venus/ECFP ratio in cells of interest throughout the imaging time because saturation of Venus and ECFP signals results in a falsely low and falsely high Venus/ECFP ratio, respectively. Thus, saturation leads to misinterpretation of caspase activation.
- Likewise, avoid a very low signal-to-noise ratio because a low signal-to-noise ratio affects the Venus/ECFP ratio.

Background noise

- If possible, it is recommended to reduce the background noises when acquiring images. Identifying the cause of the background noise would be necessary for this purpose. Note that some cell culture media contain materials that cause background noise (e.g., phenol red) and should not be used for SCAT3 imaging.

2.4.1 Ratio image processing

To create Venus/ECFP ratio images that could be understood easily, we use image processing software packages such as Metafluor, MetaMorph, or Volocity. These programs provide algorithms that simultaneously represent the ratio of each pixel as pseudocolor and the intensity as a compressed intensity gradient from bright to dark. Owing to these techniques, areas of interest are ratioed, background areas are blank, and thresholding of the original images can be avoided. Because of this property, the morphological structure of cells or tissues can be visualized with the Venus/ECFP ratio. Hereafter, we provide tips for generating ratio images with MetaMorph using IMD (Tsien & Harootunian, 1990).

- If possible, subtract the background signals (noise) before creating Venus/ECFP ratio images. This is especially important when the background noise is higher than the signals from the samples.
- Set a proper range for the Venus/ECFP ratio for pseudocolor representation. For determining the range, it is useful to measure the distribution of the ratio in the images.

– Use an appropriate combination of hues and intensities. For example, one can choose 8 hues of 32 intensities or 16 hues of 16 intensities each, and so on. The choice of numbers of hues and intensities solely depends on the dynamic range of the ratio and what has to be emphasized.

2.5. Interpretation of the result

Because of the high stability of the SCAT3 probe in the presence of protons and chloride ions, reliable signals can be detected in living cells. In spite of these advantages, artifacts may be seen, especially in FRET experiments. It should be noted that the concentration of the probe is sometimes critical for the evaluation of FRET signals. Because intermolecular FRET can occur if the FRET probe is used at high concentrations, FRET signals in SCAT3 imaging could be misinterpreted. For example, fluorescence signals are usually weak at the cell edge compared with the cell center. Thus, it is possible that this result is misinterpreted as caspase activation at the cell edge.

To avoid these artifacts, we first recommend creating an IMD image (Tsien & Harootunian, 1990) (described earlier). Because an IMD image is a ratio image modified by the intensity of the probe (value of the Venus/ECFP ratio with the intensity of ECFP or Venus), these artifacts can be identified. Furthermore, we also propose control experiments using SCAT-DEVG, which cannot be cleaved by caspases. Compared with SCAT3 and SCAT-DEVG, you can easily discriminate real signals and artifacts. Another important control experiment for checking caspase activation is to observe the Venus/ECFP ratio under conditions in which caspase activation is prevented, for example, with pharmaceutical inhibitors of caspases or by genetic deletion of caspases. In addition to these SCAT experiments, immunohistochemistry with anti-activated caspase-3 (for instance, 5E1 antibody, Cell Signaling) could be used to detect caspase-3 activation (Yamaguchi et al., 2011).

3. MONITORING CASPASE ACTIVATION WITH SCAT3 IN *DROSOPHILA*

3.1. Detecting caspase activation in various developmental processes in *Drosophila*

Detailed genetic analyses of caspase-deficient *Caenorhabditis elegans*, *Drosophila*, and mice have shown that caspases are essential, not only in controlling the number of cells involved in sculpting or deleting structures in developing animals, but also in dynamic cellular behaviors in a nonapoptotic context during morphogenesis (Kuranaga, 2011, 2012). The advanced procedure for the detection of spatiotemporal caspase activation *in vivo* is useful for analyzing the function of caspases, that is, when or where caspases are activated and how cells activating caspases affect neighboring cells and contribute to morphogenesis. Our understanding of the spatiotemporal caspase activation mechanisms has advanced, primarily through the study of the developmental processes in *Drosophila*. In this chapter, we describe studies on *Drosophila*, focusing on the dynamics of caspase activation and its regulatory system in salivary gland regression (Section 3.3), epidermal replacement (Section 3.4), selection of sensory organ precursor cells (Section 3.5), and male genitalia rotation (Section 3.6) using "SCAT3-transgenic flies."

3.2. Materials and flies

3.2.1 Materials

- Forceps
- Calligraphy brush (ink brush)
- Silicone (HIVAC-G, Shinetsu)
- Filter paper
- Microscope slide
- Cover glass (0.17-mm thickness)
- ø 35-mm plastic dish
- Double-sided tape

3.2.2 Preparation of SCAT3 fly lines

The selection of driver lines to express SCAT3 is a first important step to acquire reliable data for FRET imaging. Appropriate fly lines ensure high expression of SCAT3 and also enable minimization of the laser power and exposure time to avoid photobleaching and phototoxicity. Useful combinations of SCAT3 and drivers are illustrated in Table 12.1 (Koto et al., 2011; Kuranaga et al., 2011; Nakajima et al., 2011; Ninov et al., 2007; Sekyrova et al., 2010; Takemoto et al., 2007). To precisely measure the kinetics of caspase activation in individual cells, use red fluorescent nuclear markers (e.g., *histone 2A variant-monomeric red fluorescent protein*, Fig. 12.3C) to determine the exact location of cells.

3.3. Salivary gland regression

During *Drosophila* metamorphosis, successive pulses of 20-hydroxyecdysone (ecdysone) trigger many events, including programmed cell death of larval cells, to eliminate obsolete tissues (Thummel, 1996). About 12 h after

Table 12.1 Useful combination of SCAT3 lines and Gal4-driver lines to observe apoptosis in Drosophila

SCAT3 line	Drivers	Expression pattern	Tips and comments	References
UAS-SCAT3/		Cytosolic	Useful for visualization of a whole cell	
	tsh-Gal4	Ubiquitous	Strong expression in LECs (see Section 3.4)	Ninov, Chiarelli, and Martin-Blanco (2007)
	056-Gal4	(LECs + histoblasts)		Nakajima et al. (2011)
	en-Gal4	Posterior compartment	Useful for analysis of LEC apoptosis in the posterior compartment	Nakajima et al. (2011)
	hh-Gal4	(LECs + histoblasts)		
	Eip71CD-Gal4	Ubiquitous (LECs only)	Useful for genetic manipulation LECs	Sekyrova, Bohmann, Jindra, and Uhlirova (2010) and Nakajima et al. (2011)
	Ay-Gal4	Clonal expression with *hs-flp*	Useful for clonal analysis	Ninov et al. (2007) and Nakajima et al. (2011)
	N393	Salivary gland	Useful for imaging of salivary gland degeneration (see Section 3.3)	Takemoto et al. (2007)
UAS-nls-SCAT3/		Nucleus	Useful for identification of an individual cell	
	neu-GAL4	Sensory organ lineage	Useful for imaging of sensory organ development (see Section 3.5)	Koto, Kuranaga, and Miura (2011)
	en-GAL4	Posterior compartment of each segment	Useful for imaging of genitalia rotation (see Section 3.6)	Kuranaga et al. (2011)
lexAop-SCAT3/		Cytosolic	Useful for visualization of a whole cell	
	tubp-LexA::GAD	Ubiquitous	Tissue/region-specific genetic manipulations combining GAL4/UAS system	Nakajima et al. (2011)
		(LECs + histoblasts)	(e.g., histoblast-specific cell cycle perturbations with *esg-Gal4*)	Nakajima et al. (2011)

puparium formation (APF), ecdysone pulses induce programmed cell death of salivary gland cells by inducing the expression of the early genes *E93*, *E74*, and *BR-C* (Lee & Baehrecke, 2001), which in turn induce the caspase family cell death executors. Using the following *in vivo* imaging protocol, we found that caspase activation was locally and symmetrically initiated in the anterior cells and then propagated to the posterior cells in the salivary gland (Takemoto et al., 2007; Fig. 12.2). With later *in vitro* and *in vivo* assays, it was hypothesized that this spatiotemporal pattern could be regulated by ecdysone biosynthesis and its secretion into the anterior part of the salivary gland.

3.3.1 The procedure

1. Select late third-instar larvae from appropriate crosses (*N393/+*; *UAS-SCAT3/+*) and monitor them every 10–15 min for pupal formation.
2. Collect white pupae defined as APF 0 h in a new vial to maintain appropriate humidity conditions. We recommend that particles such as dust and food that emit strong autofluorescence should be removed before transfer to a new vial.
3. Transfer 10–11 h APF pupae to a 35-mm dish and place wet paper around the pupae to avoid desiccation.
4. Observe samples with an inverted microscope (IX70, Olympus) equipped with a CCD camera (CoolSNAP HQ, Roper Scientific), illumination pillar (ILL100LH, Olympus), and UPlanAPO 4 × 0.16 NA objective (Olympus). We used 440AF21 excitation filter, 455DRLP dichroic mirror, and two emission filters (480AF30 for ECFP and 535AF25 for Venus) alternated by a filter changer (Lambda10-2, Sutter Instruments) for SCAT3 imaging. All these devices were controlled by the MetaFluor 4.6.5 software (Universal Imaging).

3.4. Epidermal replacement during metamorphosis

The epidermal replacement of the *Drosophila* abdomen is an emerging model for studying cellular events in tissue remodeling *in vivo*. During metamorphosis, LECs are replaced by abdominal histoblasts, which are adult precursor cells of the abdominal epidermis (Madhavan & Madhavan, 1980). The whole replacement process takes approximately 20 h, beginning at 16 h APF, including histoblast nest expansion by cell proliferation and collective migration, as well as LEC removal by cell death and basal cell delamination (Madhavan & Madhavan, 1980; Ninov et al., 2007). The application of long-term live imaging with confocal microscopes, together with *Drosophila* genetics, makes this system suitable for understanding the cellular and

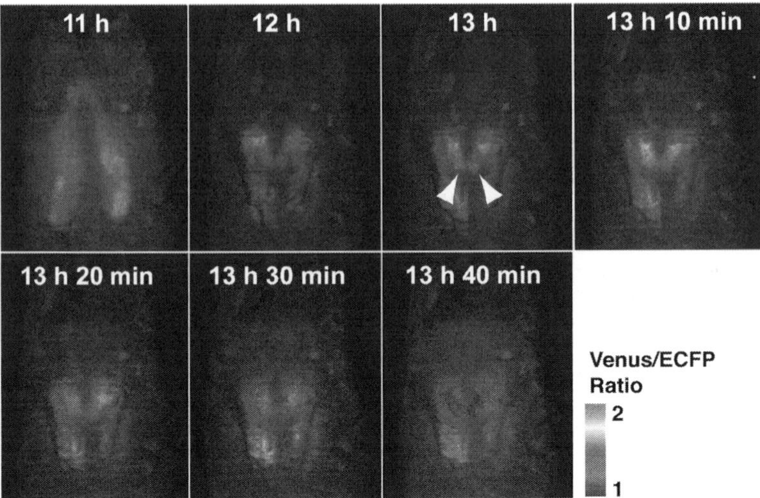

Figure 12.2 Local initiation and propagation of the caspase activities during programmed cell death in the salivary gland *in vivo*. Venus/ECFP ratio images of SCAT3-expressing salivary glands. *In vivo* live imaging analysis of SCAT3 was started from 10 to 11 h APF. Arrowheads indicate the symmetrical initiation of caspase activation. Time indicates APF. *Copyright* © National Academy of Sciences, USA *from Takemoto et al. (2007). http://dx.doi.org/10.1073/pnas.0702733104.* (See the color plate.)

molecular mechanisms underlying the epithelial replacement (Bischoff & Cseresnyes, 2009; Ninov et al., 2007, 2010; Ninov et al., 2010).

During tissue remodeling, caspase-mediated cell death, that is, apoptosis, is closely associated with surrounding proliferating cells. Indeed, one striking feature in epidermal replacement is that LEC apoptosis and histoblast proliferation are coordinately regulated (Ninov et al., 2007). Furthermore, LEC apoptosis is nonautonomously induced at the LEC/histoblast boundary by proliferating histoblasts (Nakajima et al., 2011). By analyzing caspase activation dynamics using SCAT3, spatiotemporal patterns of LEC apoptosis and LEC/histoblast interactions have been studied (Nakajima et al., 2011). We describe a protocol for live imaging of epidermal replacement during metamorphosis, from the selection of fly lines to the microscope setting. Our protocol of pupal mounting is applicable to both inverted and upright microscopes because of the use of simple microscope slides (Fig. 12.3A).

3.4.1 The procedure

1. By using a wet calligraphy brush, collect staged pupae (~15 h APF) from the vial containing fly food. For staging, refer to white pupae (0 h APF)

or the time of head eversion (12 h APF). Keep the pupae at 25 °C until the desired stage unless otherwise needed, according to specific experimental conditions.

2. Select a pupa and clean the pupal case with the calligraphy brush and Kimwipes. Transfer the pupa on a microscope slide where a double-sided tape is attached. Place the pupa on the tape at an appropriate angle.
3. Remove the abdominal part of the pupal case carefully with forceps. To efficiently remove the pupal case, first cut the posterior-most part and then peel the case in the posterior-to-anterior direction.
4. Decide the final position of the pupa. If the pupa is not oriented in the right angle, tilt the pupa with the wet calligraphy brush and forceps. Make sure that the pupa is attached to the tape before proceeding to the next step.
5. Prepare a hand-made chamber as follows (Fig. 12.3A): place a small piece of wet filter paper (~15 × 15 mm) with a hole (use a pierce punch) and apply silicone oil compounds (HIVAC-G, Shinetsu) around the filter paper using a 10-mL syringe. Seal the chamber by placing a cover glass with a drop of water (~5 μL) on the pupa. Make sure that there is no gap between the silicone and the cover glass to prevent the sample from drying and that the drop of water is in the region where the pupal case has been removed.
6. Transfer the slide to a stage of a microscope and start time-lapse imaging with the *z*-series. To monitor caspase activation in LECs, a time-lapse interval between 3 and 5 min using an UPlanApo 20× 0.7 NA dry objective lens (Leica) is adequate. The expected image of the abdominal epidermis surrounding dorsal histoblasts is illustrated in Fig. 12.3B. For FRET imaging, we used an inverted microscope (DMI6000B, Leica) with a spinning disk-type confocal unit (CSU10, Yokogawa) and a CCD camera (CoolSNAP HQ, Roper Scientific), controlled by the MetaMorph software (Molecular Devices). The Venus/ECFP ratio images were created using MetaMorph (Fig. 12.3C; see Section 2.4).

3.5. Sensory organ development

The development of the sensory organ is a suitable model system to study neural cell fate determination, proliferation, differentiation, and death. The whole process of its development takes 20 h, beginning at 16 h APF. Sensory organ precursor cells differentiate from epithelial cells on the notum at around 15 h APF. After asymmetric cell division, four different types of

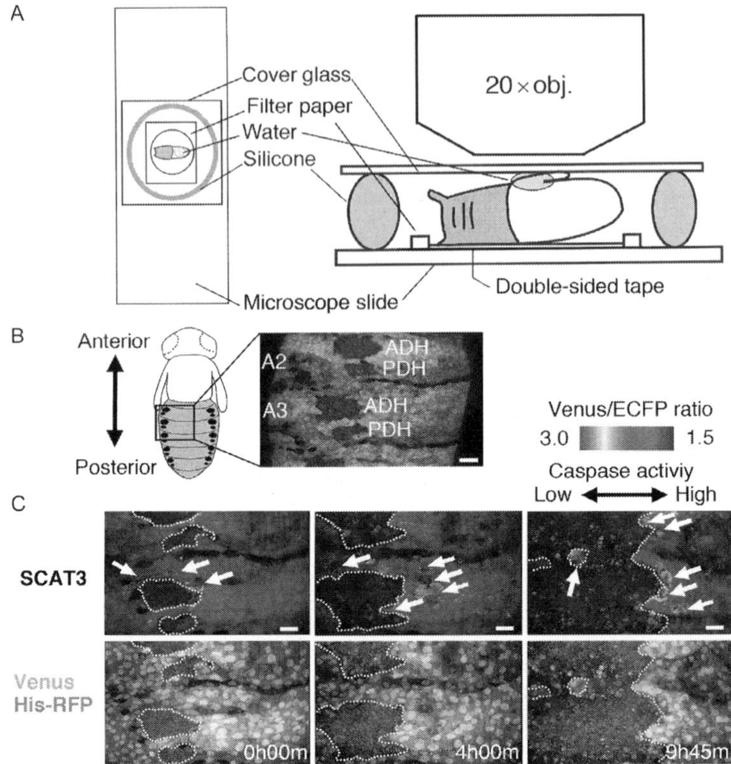

Figure 12.3 (A) Illustration of mounting a pupa for live imaging of the epidermal replacement. The abdominal part of the pupal case is removed, and the pupa is placed on the slide as dorsal histoblasts face the cover glass. (B) Schematic illustration (left) and confocal image (right) of the dorsal abdominal epidermis. Anterior dorsal histoblasts (ADH) and posterior dorsal histoblasts (PDH) exist in each segment (A1–A6). Green (gray in print) color represents LECs. (C) Caspase activation during epidermal replacement, revealed by SCAT3. Arrows indicate caspase-activated LECs. Dashed lines indicate the LEC/histoblast boundary. Pseudocolor ratio images are shown in the upper panels, and nucleus positions are labeled by His-RFP in the lower panels. *UAS-SCAT3* is expressed by *tsh-Gal4*. Scale bars, 50 μm. *Panels (B) and (C) are reprinted and modified from Nakajima et al. (2011). Copyright © American Society for Microbiology,* Molecular and Cellular Biology, *http://dx.doi.org/10.1128/MCB.01046-10.* (See the color plate.)

cells, that is, socket cells, shaft cells, sheath cells, and neurons, are produced and compose the sensory organ in the adult stage (Gho, Bellaiche, & Schweisguth, 1999).

Programmed cell death functions at multiple steps during the development of the sensory organ. During asymmetric cell division, the fate of glial cells is to die by apoptosis soon after birth (Fichelson & Gho, 2003;

Reddy & Rodrigues, 1999). In addition, programmed cell death functions to remove cells that start neural differentiation at aberrant positions and timing to ensure robust sensory organ patterning (Fig. 12.4A and B; Koto & Miura, 2011; Koto et al., 2011). In both processes, caspase activation was observed using a nuclear localization signal-tagged SCAT3 (Fig. 12.4C; Koto et al., 2009, 2011). A neuralized driver was used to express the fluorescent proteins specifically in cells of the sensory organ precursor lineage (Jhaveri, Sen, Reddy, & Rodrigues, 2000). The following is a protocol for time-lapse imaging of sensory organ development.

3.5.1 The procedure

1. Collect white prepupae defined as 0 h APF and raise them in a fresh vial containing food to maintain appropriate humidity conditions.
2. At 15 h APF, wash the staged pupae gently with a wet calligraphy brush to remove dust and food from the surface of the pupal case.
3. Mount the pupae on a glass slide using double-sided tape.
4. Carefully remove the pupal case covering the head and notum region with forceps.
5. Place extremely wet filter paper around the pupae to keep them hydrated (see also Fig. 12.3A).
6. Place silicone (Shinetsu) around the filter paper to seal the chamber.
7. Cover the pupae with a cover glass in a small drop of water to avoid desiccation.
8. Observe the samples with an inverted microscope (DMI6000B, Leica) equipped with a CSU10 confocal unit (Yokogawa) and cooled CCD camera (Coolsnap HQ, Roper Scientific) at 22 °C using an HCX PL 40× 1.25–0.75 NA oil immersion objective (Leica).

3.6. Male genitalia rotation

The *Drosophila* male terminalia is an asymmetric looping organ; the internal genitalia (spermiduct) loops dextrally around the hindgut. During metamorphosis, the male terminalia initiates a 360° clockwise rotation, which starts at 24 h APF and finishes at 36–38 h APF (Gleichauf, 1936; Kuranaga et al., 2011; Fig. 12.5A). An orientation defect in adult male terminalia occurs when this rotation is incomplete (Adam, Perrimon, & Noselli, 2003). It has been thought that caspases somehow contribute to complete genitalia rotation because orientation defects of the adult male terminalia were observed in mutants of caspases or their activators (Abbott & Lengyel, 1991; Krieser et al., 2007; Muro et al., 2006).

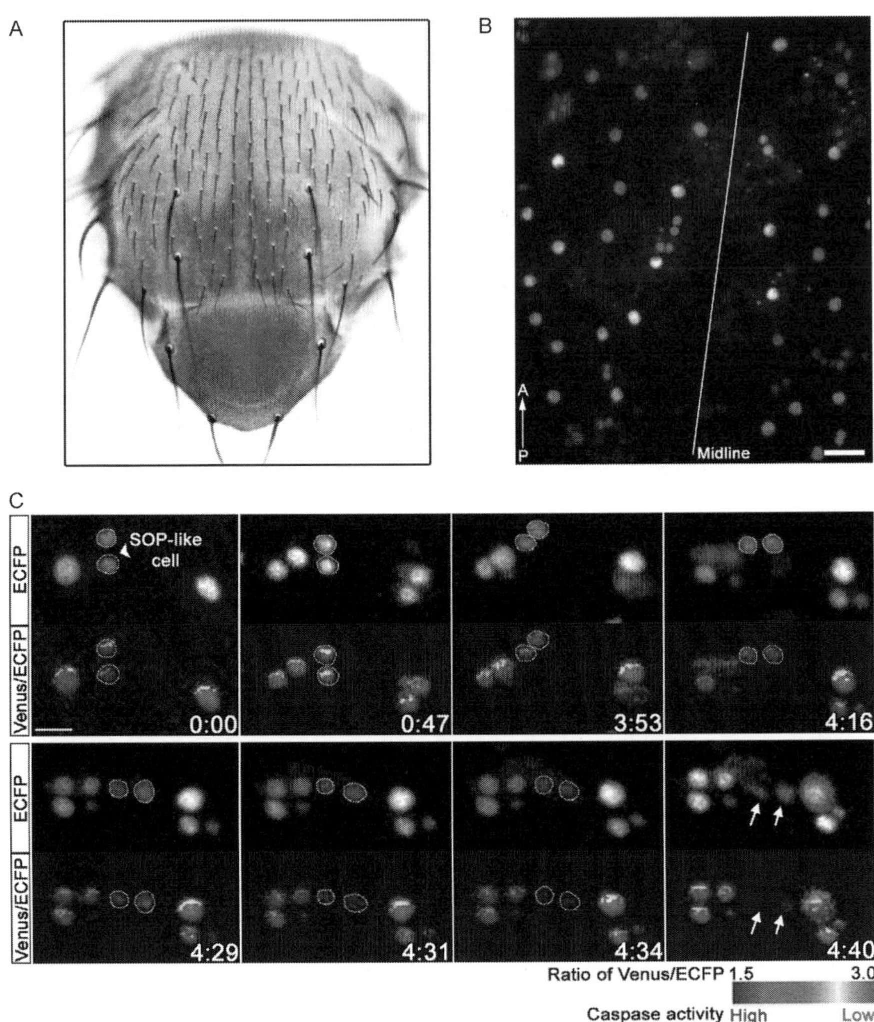

Figure 12.4 (A) Light-microscopy image of sensory organs on the notum of a wild-type fly. (B) Confocal imaging of nls-SCAT3 using a *neu-GAL4* driver. *Neuralized*-positive cells that were removed by apoptosis are marked in red. Scale bar represents 20 μm. (C) Caspase activation was detected prior to nuclear fragmentation (arrows) in *neuralized*-positive cells that appeared in aberrant positions or timing. The lower row of panels shows caspase activity images taken with nls-SCAT3 and shown in pseudocolor. Scale bar represents 10 μm. *Copyright ©Koto et al. (2011), originally published in* Current Biology, *http://dx.doi.org/10.1016/j.cub.2011.01.015.* (See the color plate.)

Time-lapse imaging analyses demonstrated that genitalia rotation accelerated as the development proceeded, completing a full 360° rotation in control flies; however, acceleration was impaired in caspase-inhibiting flies (Kuranaga et al., 2011). Acceleration is produced by two sets of cellular movement in the A8 tergite surrounding the male genitalia: cells in the posterior part of the A8 tergite drive the first 180° of rotation, pulling the genitalia with it, and most of the other cells in the A8 tergite drive the rotation through the remaining 180° (Kuranaga et al., 2011; Fig. 12.5B). In this process, a strong correlation of apoptosis in the A8 tergite with the latter half of rotation was suggested on the basis of the quantification of caspase-activating cells using a nuclear localization signal-tagged SCAT3 (Kuranaga et al., 2011; Fig. 12.5C). The following protocol describes how time-lapse imaging of genitalia rotation can be performed.

3.6.1 The procedure

1. Collect white prepupae of the *en-GAL4 UAS-nls-SCAT3/+* strain defined as 0 h APF and raise them in a fresh vial containing food to maintain appropriate humidity conditions.
2. At 22 h APF, wash the staged pupae gently with a wet calligraphy brush to remove dust and food from the surface of the pupal case.
3. Carefully remove the pupal case covering the caudal part of the abdomen with forceps.
4. Mount the pupae upside down on a glass slide using double-sided tape (see also Fig. 12.3A).
5. Place very wet filter paper around the pupae to keep them hydrated.
6. Place silicone (Shinetsu) around the filter paper to seal the chamber.
7. Cover the pupae with a cover glass in a small drop of water to avoid desiccation.
8. Observe the samples with an inverted microscope (IX71, Olympus) equipped with a CSU10 confocal unit (Yokogawa) and Aquacosmos/Ashura system (Hamamatsu Photonics) at 22 °C using an UPlanApo 20 × 0.7 NA objective (Leica).

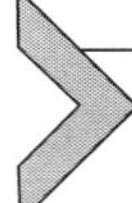

4. MONITORING CASPASE ACTIVATION WITH SCAT3 IN MICE

The generation of transgenic mice expressing SCAT3 has enabled the observation of apoptotic processes in living mouse embryos and tissues (Yamaguchi et al., 2011). While it has long been difficult to produce

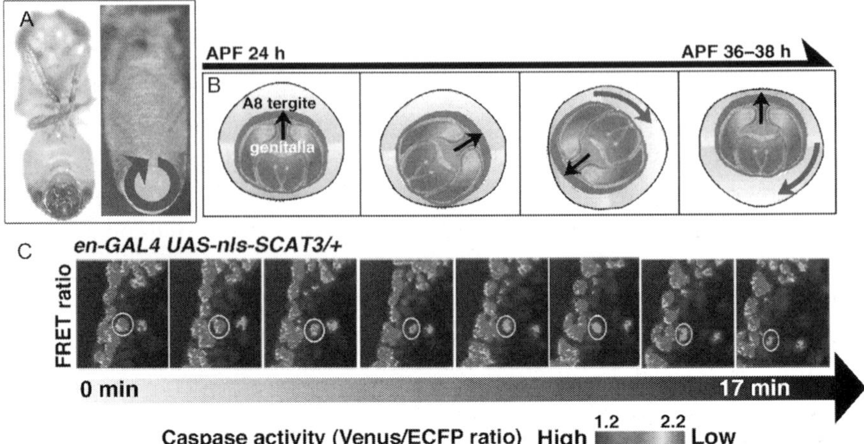

Figure 12.5 (A) Ventral view of an adult male (left) and pupae (right). The 360° clockwise rotation of the genitalia during the pupal stage is indicated by blue arrows (gray in print) (right). (B) Schematic illustration of male genitalia rotation in *Drosophila*. Cells in the posterior part of the A8 tergite (magenta) drive the rotation of the genitalia by 180° (pink–gray); then, the other cells in the A8 tergite (light-green: ventral and blue: dorsal) drive the rotation through the remaining 180° (dark-blue arrow), completing the rotation. The black arrow indicates the anus-to-penis direction on the genital plate. (C) Caspase activity was examined by imaging using a FRET-based probe, nls-SCAT3, and shown in pseudocolor. White circles indicate cells that underwent apoptosis; the pseudocolor gradually changed from red to blue. The genotype was *en-GAL4 UAS-nls-SCAT3/+*. *(C) is reprinted from Kuranaga et al. (2011),* Development, *http://dx.doi.org/10.1242/dev.058958.* (See the color plate.)

transgenic mice that stably express CFP–YFP-based fluorescent reporters, we successfully achieved ubiquitous and stable expression of SCAT3 under CAG promoters in living mouse embryos with the usage of HS4 insulator sequences (Yamaguchi et al., 2011). Another group also reported transgenic mouse in which SCAT3.1, a linker-modified version of SCAT3, could be expressed in a tissue-specific manner by Cre-loxP recombination (Nagai & Miyawaki, 2004; Tomura et al., 2009). SCAT3 and SCAT3.1 transgenic mice are now available from the RIKEN Bioresource Center, Japan. Here, we describe a protocol for conducting time-lapse imaging of mouse neural tube closure, in which numerous apoptotic cells can be observed during normal development (Fig. 12.6; Yamaguchi & Miura, 2013; Yamaguchi et al., 2011).

4.1. Materials and equipment

- Embryos (E8.5) of pregnant mice (ICR mice crossed with SCAT3 transgenic mice; available from RIKEN Bioresource Center, Japan).

- Confocal microscope and lens: We use an inverted confocal microscope (TCS SP5, Leica) equipped with a galvo-stage and a resonant scanner for fast scanning using an HCX PL FLUOTAR 10 × 0.3 NA dry objective (Leica). Fast scanning with a spinning disk-type confocal microscope (e.g., CSU10, CSU-X, Yokogawa) equipped with CCD cameras is also applicable for *ex vivo/in vivo* live imaging analysis.
- Incubation chamber: A humidified cell culture chamber with a continuous supply of 5% CO_2/air at 37 °C (Tokai Hit).
- Dissection medium I: Opti-MEM I (Invitrogen) containing 10% fetal bovine serum.
- Culture medium: Opti-MEM I (Invitrogen) containing 50% immediately centrifuged rat serum (We purchased it from Charles River, Japan.)
- Glass-bottom dish: ø 35-mm with No. 0 glass bottom (thickness of 0.10 mm; Matsunami Glass Ind., Ltd.)
- Low melting point agarose (melting point: 40 °C–60 °C): 2.0g is dissolved in 100 ml of PBS (–), autoclaved and stored in a glass bottle. Redissolve with microwave oven immediately before use.
- For heating before and during dissection:
  - Hot plate for heating medium and dishes during dissection of the embryos
  - 37 °C, 5% CO_2 cell culture incubator
  - 37 °C water bath for heating medium
- Dissection scissors and fine forceps (No. 5)

4.2. The procedure

Live imaging experiments are performed using embryos obtained by intercrossing $SCAT3^{tg/tg}$ or $SCAT3^{tg/+}$ male mice with female ICR mice. For observing neural tube closure, E8.5–9.0 embryos are used.

1. Preparation of the glass-bottom dish and medium: Pour 300 μL of 2% low-melting agarose into a glass-bottom dish and creates holes in which the embryos are placed at least 1 h before filling the dish with medium (Fig. 12.6A).
2. Prewarm all media and dishes to 37 °C in a water bath before the embryos are dissected. Fill No. 0 glass-bottom dishes with culture medium and keep them at 37 °C in a 5% CO_2 incubator.
3. Dissection: Kill the pregnant mice at day 8.5 postcoitum by cervical dislocation. Remove the uterus and place it in Opti-MEM I. Remove the embryos (in their yolk sac) from the uterus and transfer them to dissection medium.

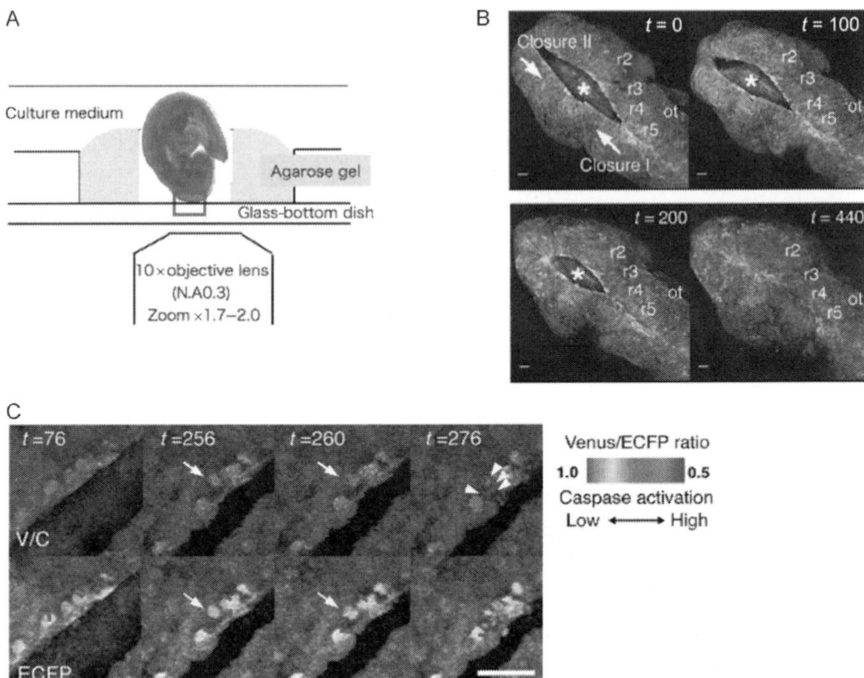

Figure 12.6 (A) System for time-lapse imaging of apoptosis in the cranial NTC. Embryos undergoing NTC (E8.5–9.0) were dissected from the uterus and yolk sac and placed into holes in 2% low-temperature melting agarose in glass-bottom dishes. The dishes were filled with culture medium and placed in a humidified incubator at 37 °C/5% CO_2 in order to visualize the hindbrain region. (B) Time-lapse observation of closure in hindbrain of a SCAT3-transgenic embryo (ECFP images). (C) Live imaging of apoptosis during the hindbrain closure. A cell showing activated caspase (arrows in *V/C* images; $t=260$) shrank ($t=260$) and became fragmented ($t=276$; arrowheads) in the boundary domain before the completion of NTC. The same cell is circled by magenta (gray in print) in the ECFP images in the magnified images of the boxed areas. All scale bar represent 50 μm. *Copyright © Yamaguchi et al. (2011), originally published in* Journal of Cell Biology, *http://dx.doi.org/10.1083/jcb.201104057.* (See the color plate.)

4. Check the fluorescence of the dissected embryos under a fluorescence stereomicroscope and select embryos exhibiting SCAT3 fluorescence.
5. Transfer only fluorescence-positive embryos to a new dish filled with dissection medium. Carefully remove the yolk sac and amnion, while the embryos are kept on a heater.
6. Transfer the dissected embryos to a prewarmed No. 0 glass-bottom dish filled with culture medium. Place the dish in a humidified cell culture incubator with continuous supply of 5% CO_2/air at 37 °C.

7. Live imaging of embryos using a confocal microscope: Place the embryos in the agarose holes created in the glass-bottom dishes and check if they are properly positioned by bright-field observation (Fig. 12.6B).
8. Acquire images by exciting ECFP with a 442-nm diode laser (or 458-nm argon laser) and detect the emission spectra of ECFP and Venus (FRET) simultaneously with two detectors using the resonant scanner (8000 Hz; Leica) (Fig. 12.6C). Use a 10 × lens for observing the entire head region of the embryos. The acquisition interval of the image (512 × 512 pixels) and the thickness of the z slices should be determined depending on the purpose of the experiment. We use an interval of 4–10 min and a slice thickness of 4–8 μm in most experiments (total of 50–100 slices/ time point).

4.3. Tips for image processing

For visualizing neural tube closure movements, 4D (*xyzt*) stacks are projected along the *z*-series with the maximum-intensity projection algorithm to generate surface images of embryos. This algorithm creates an output image in which each pixel contains the maximum value over all images in the stack at the particular pixel location. Because SCAT3 is ubiquitously expressed in embryos, the maximum-intensity projection generates a surface image of the embryo.

- Generate the Venus/ECFP images using the MetaMorph (Molecular Devices) or Volocity (PerkinElmer) software, both of which are suitable for changes in ratio and intensity
- If required, also perform video editing and linear intensity adjustment for visualization with the ImageJ software (National Institutes of Health).
- To eliminate the *XY* drift in the time-lapse series, use the stackreg plug-in for ImageJ.

Venus/ECFP ratio tracking should be manually performed in maximum-projected images using the MetaMorph, the Volocity, or ImageJ software.

To analyze apoptosis in internal morphology of embryos, we generated Venus/ECFP ratio images for each *z*-plane before making a 2D projection of the 3D data. However, this method is not efficient in some cases because a confocal microscope is not sufficient for the visualization of deep inside regions of embryos. Moreover, processing images for the generation of ratio images for all *z*-planes and time-series require computers with high processing requirements. Therefore, we recommend the reduction of data size by cropping or cutting images beyond the scope of analysis. *XZ* transverse images were also processed from raw 4D datasets with the Volocity or ImageJ software. Only for visualization of highly magnified images, a smooth zoom filter can be applied to the Venus/ECFP and ECFP images using the Volocity software. The "Image, Stack, and Time-lapse Arrow Labeling" tool can be used to add arrows to time-lapse videos.

5. CONCLUDING REMARKS

Live imaging of apoptosis with SCAT3 is a relatively easy and powerful tool for observing caspase activation in real time within spatiotemporal contexts. However, several specific pitfalls that may bring about incorrect interpretation of the results may be encountered, as described in Section 2. Researchers should perform appropriate control experiments, and conduct other methods if possible, to determine whether caspases are activated under the applied experimental conditions.

REFERENCES

Abbott, M. K., & Lengyel, J. A. (1991). Embryonic head involution and rotation of male terminalia require the *Drosophila* locus head involution defective. *Genetics*, *129*, 783–789.

Adam, G., Perrimon, N., & Noselli, S. (2003). The retinoic-like juvenile hormone controls the looping of left-right asymmetric organs in *Drosophila*. *Development*, *130*, 2397–2406.

Araki, T., Hayashi, M., Watanabe, N., Kanuka, H., Yoshino, J., Miura, M., et al. (2002). Down-regulation of Mcl-1 by inhibition of the PI3-K/Akt pathway is required for cell shrinkage-dependent cell death. *Biochemical and Biophysical Research Communications*, *290*, 1275–1281.

Bardet, P. L., Kolahgar, G., Mynett, A., Miguel-Aliaga, I., Briscoe, J., Meier, P., et al. (2008). A fluorescent reporter of caspase activity for live imaging. *Proceedings of the National Academy of Sciences of the United States of America*, *105*, 13901–13905.

Bischoff, M., & Cseresnyes, Z. (2009). Cell rearrangements, cell divisions and cell death in a migrating epithelial sheet in the abdomen of *Drosophila*. *Development*, *136*, 2403–2411.

Campbell, D. S., & Okamoto, H. (2013). Local caspase activation interacts with Slit-Robo signaling to restrict axonal arborization. *The Journal of Cell Biology*, *203*, 657–672.

Fichelson, P., & Gho, M. (2003). The glial cell undergoes apoptosis in the microchaete lineage of *Drosophila*. *Development*, *130*, 123–133.

Gho, M., Bellaiche, Y., & Schweisguth, F. (1999). Revisiting the *Drosophila* microchaete lineage: A novel intrinsically asymmetric cell division generates a glial cell. *Development*, *126*, 3573–3584.

Gleichauf, R. (1936). Anatomie und variabilität des geschlechtsapparates von *Drosophila* melanogaster. *Zeitschrift für Wissenschaftliche Zoologie*, *148*, 1–66.

Huang, Q., Li, F., Liu, X., Li, W., Shi, W., Liu, F. F., et al. (2011). Caspase 3-mediated stimulation of tumor cell repopulation during cancer radiotherapy. *Nature Medicine*, *17*, 860–866.

Jayaraman, S., Haggie, P., Wachter, R. M., Remington, S. J., & Verkman, A. S. (2000). Mechanism and cellular applications of a green fluorescent protein-based halide sensor. *The Journal of Biological Chemistry*, *275*, 6047–6050.

Jhaveri, D., Sen, A., Reddy, G. V., & Rodrigues, V. (2000). Sense organ identity in the *Drosophila* antenna is specified by the expression of the proneural gene atonal. *Mechanisms of Development, 99*, 101–111.

Kanno, A., Yamanaka, Y., Hirano, H., Umezawa, Y., & Ozawa, T. (2007). Cyclic luciferase for real-time sensing of caspase-3 activities in living mammals. *Angewandte Chemie (International Ed. in English), 46*, 7595–7599.

Koto, A., Kuranaga, E., & Miura, M. (2009). Temporal regulation of *Drosophila* IAP1 determines caspase functions in sensory organ development. *The Journal of Cell Biology, 187*, 219–231.

Koto, A., Kuranaga, E., & Miura, M. (2011). Apoptosis ensures spacing pattern formation of *Drosophila* sensory organs. *Current Biology, 21*, 278–287.

Koto, A., & Miura, M. (2011). Who lives and who dies: Role of apoptosis in quashing developmental errors. *Communicative & Integrative Biology, 4*, 495–497.

Krieser, R. J., Moore, F. E., Dresnek, D., Pellock, B. J., Patel, R., Huang, A., et al. (2007). The *Drosophila* homolog of the putative phosphatidylserine receptor functions to inhibit apoptosis. *Development, 134*, 2407–2414.

Kuranaga, E. (2011). Caspase signaling in animal development. *Development, Growth & Differentiation, 53*, 137–148.

Kuranaga, E. (2012). Beyond apoptosis: Caspase regulatory mechanisms and functions in vivo. *Genes to Cells, 17*, 83–97.

Kuranaga, E., Matsunuma, T., Kanuka, H., Takemoto, K., Koto, A., Kimura, K., et al. (2011). Apoptosis controls the speed of looping morphogenesis in *Drosophila* male terminalia. *Development, 138*, 1493–1499.

Lee, C. Y., & Baehrecke, E. H. (2001). Steroid regulation of autophagic programmed cell death during development. *Development, 128*, 1443–1455.

Llopis, J., McCaffery, J. M., Miyawaki, A., Farquhar, M. G., & Tsien, R. Y. (1998). Measurement of cytosolic, mitochondrial, and Golgi pH in single living cells with green fluorescent proteins. *Proceedings of the National Academy of Sciences of the United States of America, 95*, 6803–6808.

Luo, K. Q., Yu, V. C., Pu, Y., & Chang, D. C. (2001). Application of the fluorescence resonance energy transfer method for studying the dynamics of caspase-3 activation during UV-induced apoptosis in living HeLa cells. *Biochemical and Biophysical Research Communications, 283*, 1054–1060.

Madhavan, M. M., & Madhavan, K. (1980). Morphogenesis of the epidermis of adult abdomen of *Drosophila. Journal of Embryology and Experimental Morphology, 60*, 1–31.

Maeno, E., Ishizaki, Y., Kanaseki, T., Hazama, A., & Okada, Y. (2000). Normotonic cell shrinkage because of disordered volume regulation is an early prerequisite to apoptosis. *Proceedings of the National Academy of Sciences of the United States of America, 97*, 9487–9492.

Matsuyama, S., Llopis, J., Deveraux, Q. L., Tsien, R. Y., & Reed, J. C. (2000). Changes in intramitochondrial and cytosolic pH: Early events that modulate caspase activation during apoptosis. *Nature Cell Biology, 2*, 318–325.

Muro, I., Berry, D. L., Huh, J. R., Chen, C. H., Huang, H., Yoo, S. J., et al. (2006). The *Drosophila* caspase Ice is important for many apoptotic cell deaths and for spermatid individualization, a nonapoptotic process. *Development, 133*, 3305–3315.

Nagai, T., Ibata, K., Park, E. S., Kubota, M., Mikoshiba, K., & Miyawaki, A. (2002). A variant of yellow fluorescent protein with fast and efficient maturation for cell-biological applications. *Nature Biotechnology, 20*, 87–90.

Nagai, T., & Miyawaki, A. (2004). A high-throughput method for development of FRET-based indicators for proteolysis. *Biochemical and Biophysical Research Communications, 319*, 72–77.

Nakajima, Y., Kuranaga, E., Sugimura, K., Miyawaki, A., & Miura, M. (2011). Nonautonomous apoptosis is triggered by local cell cycle progression during epithelial replacement in *Drosophila. Molecular and Cellular Biology, 31*, 2499–2512.

Nicholls, S. B., Chu, J., Abbruzzese, G., Tremblay, K. D., & Hardy, J. A. (2011). Mechanism of a genetically encoded dark-to-bright reporter for caspase activity. *The Journal of Biological Chemistry, 286*, 24977–24986.

Ninov, N., Chiarelli, D. A., & Martin-Blanco, E. (2007). Extrinsic and intrinsic mechanisms directing epithelial cell sheet replacement during *Drosophila* metamorphosis. *Development, 134*, 367–379.

Ninov, N., Menezes-Cabral, S., Prat-Rojo, C., Manjon, C., Weiss, A., Pyrowolakis, G., et al. (2010). Dpp signaling directs cell motility and invasiveness during epithelial morphogenesis. *Current Biology, 20*, 513–520.

Ohgushi, M., Matsumura, M., Eiraku, M., Murakami, K., Aramaki, T., Nishiyama, A., et al. (2010). Molecular pathway and cell state responsible for dissociation-induced apoptosis in human pluripotent stem cells. *Cell Stem Cell, 7*, 225–239.

Ohsawa, S., Hamada, S., Yoshida, H., & Miura, M. (2008). Caspase-mediated changes in histone H1 in early apoptosis: Prolonged caspase activation in developing olfactory sensory neurons. *Cell Death and Differentiation, 15*, 1429–1439.

Perez-Sala, D., Collado-Escobar, D., & Mollinedo, F. (1995). Intracellular alkalinization suppresses lovastatin-induced apoptosis in HL-60 cells through the inactivation of a pH-dependent endonuclease. *The Journal of Biological Chemistry, 270*, 6235–6242.

Reddy, G. V., & Rodrigues, V. (1999). A glial cell arises from an additional division within the mechanosensory lineage during development of the microchaete on the *Drosophila* notum. *Development, 126*, 4617–4622.

Rehm, M., Dussmann, H., Janicke, R. U., Tavare, J. M., Kogel, D., & Prehn, J. H. (2002). Single-cell fluorescence resonance energy transfer analysis demonstrates that caspase activation during apoptosis is a rapid process. Role of caspase-3. *The Journal of Biological Chemistry, 277*, 24506–24514.

Schoenmann, Z., Assa-Kunik, E., Tiomny, S., Minis, A., Haklai-Topper, L., Arama, E., et al. (2010). Axonal degeneration is regulated by the apoptotic machinery or a NAD^+-sensitive pathway in insects and mammals. *The Journal of Neuroscience, 30*, 6375–6386.

Sekyrova, P., Bohmann, D., Jindra, M., & Uhlirova, M. (2010). Interaction between *Drosophila* bZIP proteins Atf3 and Jun prevents replacement of epithelial cells during metamorphosis. *Development, 137*, 141–150.

Takeishi, A., Kuranaga, E., Tonoki, A., Misaki, K., Yonemura, S., Kanuka, H., et al. (2013). Homeostatic epithelial renewal in the gut is required for dampening a fatal systemic wound response in *Drosophila*. *Cell Reports, 3*, 919–930.

Takemoto, K., Kuranaga, E., Tonoki, A., Nagai, T., Miyawaki, A., & Miura, M. (2007). Local initiation of caspase activation in *Drosophila* salivary gland programmed cell death in vivo. *Proceedings of the National Academy of Sciences of the United States of America, 104*, 13367–13372.

Takemoto, K., Nagai, T., Miyawaki, A., & Miura, M. (2003). Spatio-temporal activation of caspase revealed by indicator that is insensitive to environmental effects. *The Journal of Cell Biology, 160*, 235–243.

Thummel, C. S. (1996). Flies on steroids—*Drosophila* metamorphosis and the mechanisms of steroid hormone action. *Trends in Genetics, 12*, 306–310.

Tomura, M., Mori, Y. S., Watanabe, R., Tanaka, M., Miyawaki, A., & Kanagawa, O. (2009). Time-lapse observation of cellular function with fluorescent probe reveals novel CTL-target cell interactions. *International Immunology, 21*, 1145–1150.

Tsien, R. Y., & Harootunian, A. T. (1990). Practical design criteria for a dynamic ratio imaging system. *Cell Calcium, 11*, 93–109.

Tyas, L., Brophy, V. A., Pope, A., Rivett, A. J., & Tavare, J. M. (2000). Rapid caspase-3 activation during apoptosis revealed using fluorescence-resonance energy transfer. *EMBO Reports, 1*, 266–270.

Vincent, A. M., TenBroeke, M., & Maiese, K. (1999). Neuronal intracellular pH directly mediates nitric oxide-induced programmed cell death. *Journal of Neurobiology*, *40*, 171–184.

Williams, D. W., Kondo, S., Krzyzanowska, A., Hiromi, Y., & Truman, J. W. (2006). Local caspase activity directs engulfment of dendrites during pruning. *Nature Neuroscience*, *9*, 1234–1236.

Yamaguchi, Y., & Miura, M. (2013). How to form and close the brain: Insight into the mechanism of cranial neural tube closure in mammals. *Cellular and Molecular Life Sciences*, *70*, 3171–3186.

Yamaguchi, Y., Shinotsuka, N., Nonomura, K., Takemoto, K., Kuida, K., Yosida, H., et al. (2011). Live imaging of apoptosis in a novel transgenic mouse highlights its role in neural tube closure. *The Journal of Cell Biology*, *195*, 1047–1060.

Zhang, J., Wang, X., Cui, W., Wang, W., Zhang, H., Liu, L., et al. (2013). Visualization of caspase-3-like activity in cells using a genetically encoded fluorescent biosensor activated by protein cleavage. *Nature Communications*, *4*, 2157.

CHAPTER THIRTEEN

Global Analysis of Cellular Proteolysis by Selective Enzymatic Labeling of Protein N-Termini

Arun P. Wiita[*,†,1], **Julia E. Seaman**[*,1], **James A. Wells**[*,‡,2]
[*]Department of Pharmaceutical Chemistry, University of California, San Francisco, California, USA
[†]Department of Laboratory Medicine, University of California, San Francisco, California, USA
[‡]Department of Cellular and Molecular Pharmacology, University of California, San Francisco, California, USA
[1]These authors contributed equally to this work.
[2]Corresponding author: e-mail address: jim.wells@ucsf.edu

Contents

Methods in Enzymology, Volume 544
ISSN 0076-6879
http://dx.doi.org/10.1016/B978-0-12-417158-9.00013-3

Abstract

Proteolysis is a critical modification leading to alteration of protein function with important outcomes in many biological processes. However, for the majority of proteases, we have an incomplete understanding of both cellular substrates and downstream effects. Here, we describe detailed protocols and applications for using the rationally engineered peptide ligase, subtiligase, to specifically label and capture protein N-termini generated by proteases either induced or added to complex biological samples. This method allows identification of the protein targets as well as their precise cleavage locations. This approach has revealed >8000 proteolytic sites in healthy and apoptotic cells including >1700 caspase cleavages. One can further determine substrate preferences through rate analysis with quantitative mass spectrometry, physiological substrate specificities, and even infer the identity of proteases operating in the cell. In this chapter, we also describe how this experimental method can be generalized to investigate proteolysis in any biological sample.

1. INTRODUCTION

1.1. Importance of proteolysis

Proteolysis, the hydrolysis of peptide bonds by proteases, is an essential activity in a wide range of cellular functions. Proteases exist in virtually all forms of life, and are classified into five mechanistic categories: serine, threonine, cysteine, acid, and metallo (Lopez-Otin & Matrisian, 2007). In humans alone there are over 550 identified proteases, but their precise roles and substrates are generally poorly understood. The number of substrates for a given protease ranges widely, from a single sites on a few proteins to cleaving a broad swath of the proteome. Proteases function in digesting and recycling proteins, irreversible posttranslational modification via N-terminal methionine processing, signal or transit peptide removal, cleavage of polypeptide chains into their multiple components, and removal of precursor domains. In addition to their role in protein maturation and function, proteolysis is critical for physiological processes including apoptosis. Endoproteolysis can lead to activation, inhibition, or a change of substrate function, allowing proteases to play important roles in signaling. Dysregulation of proteolysis contributes to many pathological states such as arthritis, inflammation, and cancer. Additionally, proteases are used as tools in the laboratory, industrial manufacturing, and commercial products.

An important step to understanding a protease's role is identification and validation of substrates and cleavage locations. Such information leads naturally to examining the specific functional consequences for individual targets. Thus, there has been a surge in the development of technologies for global and unbiased characterization of proteolysis in complex biological samples (for reviews see Agard & Wells, 2009; Impens et al., 2010; Klingler & Hardt, 2012; Rogers & Overall, 2013). We briefly cover the state-of-the-art in this field and then focus on the detailed implementation and applications of the N-terminomics technology developed in our lab using subtiligase.

1.2. Approaches to substrate cleavages and identification

Historically, the identification of proteases responsible for specific cleavage events has often been driven by knowledge of important substrate proteins that were found cleaved in a biological process. For example, insulin was known to be produced from the precursor pro-insulin leading to the discovery of the protease furin (Smeekens et al., 1992). The processing of pro-IL-1β to IL-1β led to the discovery of the protease caspase-1 (Black, Kronheim, & Sleath, 1989). Until recently, most substrates have been found in a labor intensive, candidate-based approach using a range of focused biochemical approaches.

New proteomic methods have allowed unbiased searches of proteolytic substrates in complex samples (Table 13.1). These global studies often have two goals: to identify all cleavage events in a cell during a particular process and to identify all possible substrates of a given protease. These aims have been greatly aided by the advancements of analytical instruments, specifically liquid chromatography coupled to mass spectrometry (LC–MS). Most methods enrich for proteolytically cleaved peptides by taking advantage of the newly created α-carboxy- or α-amino-termini on either side of the cleavage site. This allows for capture and purification of substrates through specific chemical or enzymatic modification. A single global experiment can generate over a thousand peptide identifications that can be scored and mapped to a specific protein and/or cleavage site. The advantage to capturing the N-terminal side of the cleavage site ("N-terminomics") is that most proteins are acetylated naturally as they are translated from ribosomes, such that virtually all unblocked α-amines are produced by a post-translational proteolytic event.

Table 13.1 A summary of current methods for proteolytic cleavage site and substrate identification

Method	Description	Proteolytic substrates reported	References
Subtiligase	Positive selection of free N-termini α-amines through subtiligase enzymatic labeling with an ester peptide tag.	8090 peptides (1706 caspase) from untreated and apoptotic human cells	Crawford et al. (2013) and Mahrus et al. (2008)
COFRADIC	COmbined FRActional DIagonal Chromatography uses negative selection with a chemical modification at free N-termini (or other modification of interest) to enable separation of modified from unlabeled peptides during chromatography.	68 caspase substrates from recombinant caspases-2, -3, -7; 9729 carboxypeptidase substrates from *in vitro* peptide library	Staes et al. (2008), Tanco et al. (2013), and Wejda et al. (2012)
TAILS	Terminal amine isotopic labeling uses chemical modifications of protein amines and thiols, sample trypsinization, and negative selection to enrich for neo N- or C-termini.	288 MMP-2 cleavage sites; >100 GluC cleavage sites	Kleifeld et al. (2010) and Schilling, Huesgen, Barre, and Overall (2011)
N-CLAP	N-terminalomics by chemical labeling of the α-amines of proteins. Uses Edman degradation chemistry to block lysine amines to label N-terminal amines with a biotinylated tag for positive selection.	278 peptides (23 caspase) in apoptotic Jurkat cells	Xu and Jaffrey (2010) and Xu, Shin, and Jaffrey (2009)

Table 13.1 A summary of current methods for proteolytic cleavage site and substrate identification—cont'd

Method	Description	Proteolytic substrates reported	References
PROTOMAP	PROtein TOpography and Migration Analysis Platform creates visual peptographs from 1D SDS gel migration patterns and sequence coverage from MS of in-gel digestions to identify cleavages from mass shifts.	744 proteins with cleavages in apoptotic Jurkat cells	Dix, Simon, and Cravatt (2008) and Dix et al. (2012)
GASSP + C-terminal immuno-pull down	Global analyzer of SILAC-derived substrates of proteolysis (GASSP) using differential gel analysis combined with pull down of Asp at C-termini using a specific antibody.	360 proteolytic sites in Jurkat cells; 160 known caspase sites	Pham et al. (2012)
2D DiGE + MS	Two-dimensional differential gel electrophoresis (2D-DiGE) separates complex mixtures using orthogonal electrophoresis methods and comparison of induced proteolysis and control sample gels reveal shifted spots due to proteolysis.	21 caspase substrates in Jurkat cells	Tonge et al. (2001)
1D gel + MS	Lysates harvested from *in vivo* induced proteolysis are run on a large gel, separated into 100 slices and	37 peptides in apoptotic Jurkat cells	Thiede, Treumann, Kretschmer, Sohlke, and Rudel (2005)

Continued

Table 13.1 A summary of current methods for proteolytic cleavage site and substrate identification—cont'd

Method	Description	Proteolytic substrates reported	References
	prepared for mass spectrometry with in-gel trypsinization. Substrates are identified as those with less mass than expected values indicating cleavage events.		
2D SDS PAGE	2D SDS PAGE gel electrophoresis with protease addition as an intermediate step to look for spots that differentially migrate compared to control indicating proteolysis.	41 caspase-1 substrates in THP cells	Shao, Yeretssian, Doiron, Hussain, and Saleh (2007)
ProC-TEL	Positive selection through carboxy termini tagging using transpeptidation enzymatic reaction.	76 peptides from *Escherichia coli* lysates	Xu, Shin, and Jaffrey (2011)

2. BASIC FEATURES OF SUBTILIGASE METHOD

2.1. Introduction

Here, we describe a global N-terminomics positive enrichment method using the engineered enzyme subtiligase. This method allows one to specifically tag and identify with LC–MS new N-termini generated by endogenous or exogenous proteases (Fig. 13.1A). With this approach, we have identified over 8000 unique α-amines in healthy and apoptotic cell lysate (publicly available at http://wellslab.ucsf.edu/degrabase) as well as quantitatively monitored kinetics of individual cleavage events (Agard et al., 2012). The subtiligase method is easily applied to many different complex biological samples to identify substrates and sites of proteolysis.

2.2. Types of samples and peptide tags

The method has been successfully used on purified proteins, cell cultures, peripheral blood plasma, and tissue samples from humans, mice, insects, and worms. In general, proteins in complex biological samples can be solubilized into an appropriate buffer for subtiligase labeling. Additionally, one can design the chemical structure of the peptide ester tag to facilitate downstream purification, identification, and quantification for customization in specific applications (Fig. 13.1B).

2.3. Sample setup introduction

2.3.1 Discovery versus targeted protocols

There are two experimental protocols, we currently use that provide the most complete information for global N-terminomics. The initial "discovery" experiments are designed to identify which proteins and specific sites are cleaved. The discovery experiments are qualitative and focus on high confidence identification of tagged N-terminal peptides. These experiments are important to optimize the labeling procedures, determine background, and establish a list of high confidence peptide identifications. Peptides from discovery experiments are then used in "targeted" mass spectrometry experiments. Targeted experiments allow for the specific monitoring of a subset of peptides in a more sensitive and/or quantitative manner across a wider range of samples, such as selective reaction monitoring (SRM). Examples of both types of experiments will be discussed in more detail below.

2.3.2 Forward versus Reverse experiments

There are two main experimental strategies for subtiligase labeling, which we term "Forward" and "Reverse" (Fig. 13.2). Forward experiments use intact biological systems where endogenous proteolysis is induced such as for apoptosis followed by protein isolation and N-terminal labeling. The panel of substrates is compared to those from an uninduced sample. In contrast, Reverse experiments are performed *in vitro* where an exogenous protease is added to the total protein from a sample where endogenous proteases have been inactivated. The Forward experiments allow for the identification of biologically relevant protease cleavage events but may not be able to identify the specific protease responsible. The Reverse setup specifically identifies the activity of the added protease, but occurs in lysates where the intracellular structure is disrupted and thus may be less physiologically relevant.

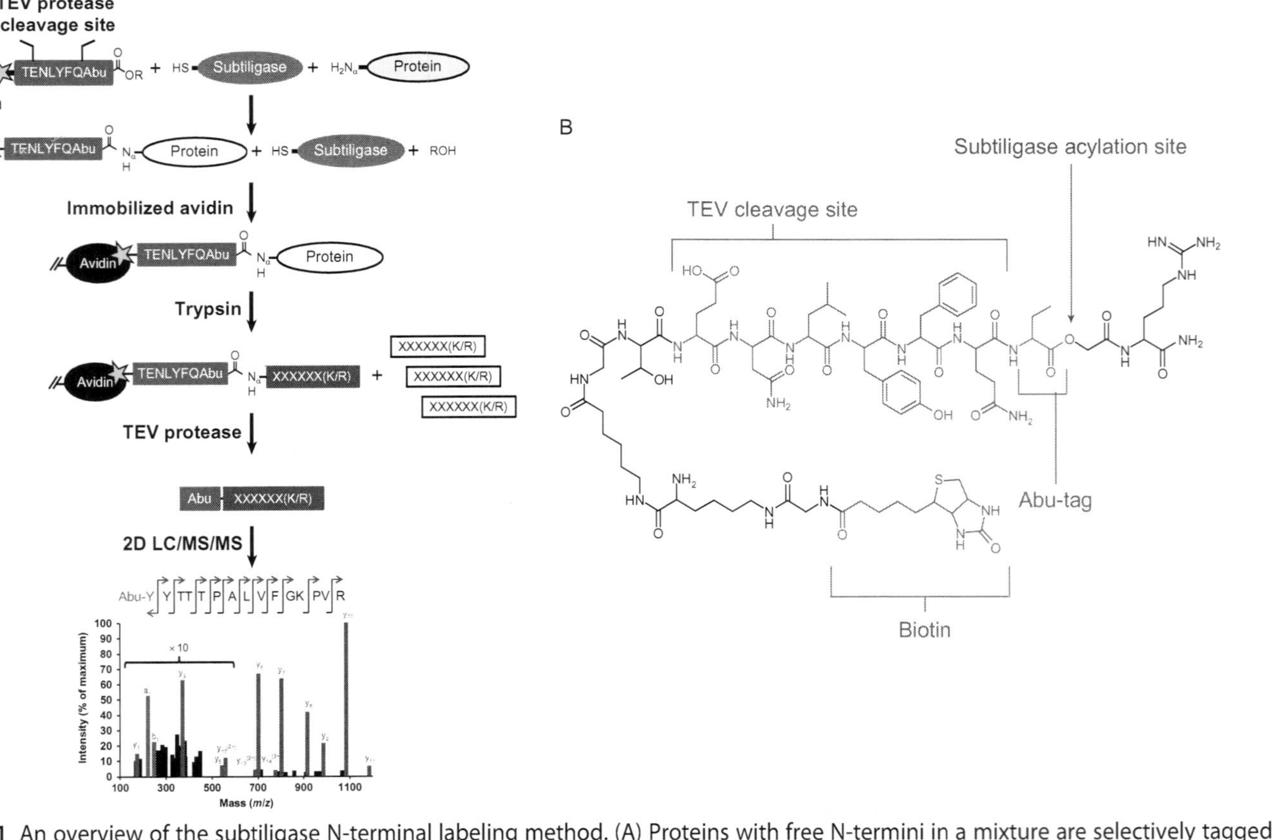

Figure 13.1 An overview of the subtiligase N-terminal labeling method. (A) Proteins with free N-termini in a mixture are selectively tagged using the engineered enzyme, subtiligase. Whole protein samples are incubated with subtiligase and the peptide ester containing a biotin tag. After enzymatic labeling, free N-termini are captured on avidin beads. Proteins are digested by trypsin. The final N-terminal peptide is released from beads via TEV protease cleavage and identified by mass spectrometry. (B) The current peptide ester contains an ester subtiligase acylation site, Abu-tag for positive mass spectrometry identification, a TEV protease site and a biotin label. The peptide ester can be further modified for specific experimental needs. (See the color plate.)

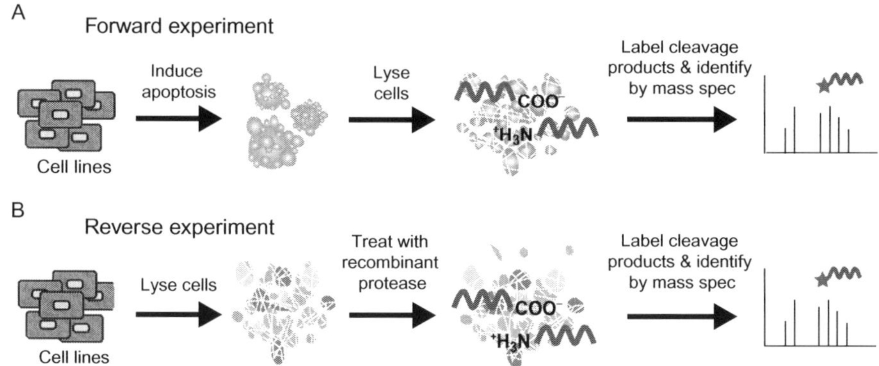

Figure 13.2 A schematic difference between Forward and Reverse experiments. Forward experiments use samples from intact biological systems, either perturbed or unperturbed that is then harvested, lysed, and labeled. Reverse experiments involve exogenous addition of protease to whole cell or tissue lysate of interest followed by labeling.

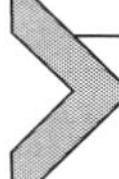

3. SUBTILIGASE-BASED LABELING METHOD

3.1. Overview of method

The subtiligase protocol is designed to positively enrich free protein N-termini. Subtiligase itself is a rationally engineered version of the bacterial serine protease subtilisin BPN′. Two-point mutations simultaneously abolish protease activity and allow ligase activity (Abrahmsen et al., 1991). With these modifications, subtiligase can covalently link free peptide α-amines with an ester-containing synthetic peptide. Importantly, the subtiligase enzyme is exquisitely selective for peptide α-amines over the ε-amines of lysine residues (Braisted, Judice, & Wells, 1997). Furthermore, acetylated N-termini present on 80–90% of native eukaryotic proteins (Polevoda & Sherman, 2003) are ignored by the labeling process, greatly reducing background identifications and focusing on N-termini generated by proteolysis.

The protocol follows a catch-and-release strategy (Fig. 13.1A). In combination with a designed synthetic peptide ester (Fig. 13.1B), subtiligase selectively biotinylates free α-amines in the sample. Following avidin bead-mediated immobilization, proteins are then digested with trypsin and nonbiotinylated protein fragments are washed away. The most N-terminal peptide from each substrate is then released from avidin beads through tobacco etch virus (TEV) protease cleavage. TEV is an extremely specific plant viral protease that can be readily purified (Lucast, Batey, & Doudna, 2001) or purchased. TEV recognizes the amino acid sequence ENLYFQ↓S, which importantly is not found in the mammalian proteome. After TEV cleavage, all labeled peptides have a nonnatural amino acid mass tag (α-aminobutyric acid, or Abu-) remaining on the N-terminus. This tag, which is compatible with both subtiligase and TEV, greatly enhances confidence for identifying subtiligase-labeled peptides over nonspecifically bound background. In our experience, >90% of peptides observed by LC–MS incorporate the Abu-mass tag, providing evidence for the specificity of the labeling procedure and recovery method.

3.2. Specialized reagents for subtiligase labeling and enrichment

3.2.1 Subtiligase enzyme

Plasmid vectors and detailed instructions for subtiligase expression are available on request from the Wells laboratory. Subtiligase is expressed in *Bacillus subtilis* and the enzyme is secreted to the media. The enzyme is purified through ammonium sulfate precipitation, anion exchange, thiopropyl resin capture (for the catalytic cysteine residue in subtiligase), and gel filtration. The enzyme is stored at −80 °C and retains activity for at least 2 years after purification. Activity can be tested and quantified using FRET ester reporters (Shimbo et al., 2012; Yoshihara, Mahrus, & Wells, 2008).

3.2.2 Peptide ester label

The synthetic ester used for labeling is customizable for different experimental goals (Yoshihara et al., 2008). The current version contains four distinct features: (i) an ester linkage for subtiligase acylation and transfer to the free peptide α-amine, (ii) the unique Abu-tag to facilitate MS identification, (iii) the TEV protease cleavage site for elution from avidin beads, and (iv) biotin for initial capture (Fig. 13.1B). The peptide ester is readily synthesized using solid phase fMOC chemistry modified for the more reactive ester bond (Braisted et al., 1997; Jackson et al., 1994). Since each amino acid is added individually, it is possible to change any part of the sequence after the ester bond so long as the first four amino acids can be recognized by subtiligase.

4. EXPERIMENTAL APPLICATIONS OF SUBTILIGASE-BASED N-TERMINOMICS

4.1. Application to cell culture systems undergoing apoptosis (Forward Discovery experiments)

Below, we describe a general protocol using subtiligase-based labeling to identify proteolytic substrates generated during apoptosis in a cell culture system.

4.1.1 Sample size and expected yield

The extent of subtiligase labeling varies but we estimate about 10–15% of the α-amines in a sample are routinely labeled. Hydrolysis of the ester by subtiligase is the biggest impediment to higher labeling efficiency. While the enzyme is very suitable for N-terminomics despite the hydrolysis side-reaction, it requires the use of greater protein sample input. For initial Forward Discovery experiments, we typically use on the scale of $0.5–5 \times 10^9$ cells (~30–300 mg total protein in lysate) to maximize our number of peptides identified by mass spectrometry. The use of more sensitive mass spectrometers or an experimental system which does not require deep coverage allows for smaller amounts of starting sample.

4.1.2 Choice of proteolysis inducer for Forward experiments

The specific proteolysis inducer chosen will depend on the system of interest and research question. Apoptosis can be induced in a cell culture system using a variety of agents but depends on the cell type and organism. For example, we have used both small molecule cytotoxic agents (doxorubicin, bortezomib, staurosporine) to activate the intrinsic pathway of apoptosis and protein-based agents (TRAIL, Fas-ligand) that bind to extracellular surface death receptors and trigger the extrinsic pathway of apoptosis (Agard et al., 2012; Mahrus et al., 2008; Shimbo et al., 2012).

4.1.3 Monitoring proteolysis and sample harvest

Once the desired cell culture system and inducer are chosen, it is recommended to perform validation experiments on a small scale to identify a concentration and time point where proteolysis is most extensive. For apoptosis, we find the maximal number of caspase cleavage events when the extent of apoptosis is >90%. We have primarily used biochemical assays to monitor caspase activity (Caspase-Glo, Promega) and cell viability

(Cell-Titer Glo, Promega), though there are a number of other experimental methods also available (Galluzzi et al., 2009). Figure 13.3A demonstrates a typical time course of cell viability and caspase activation with two different doses of apoptotic inducers in two different human malignancy-derived cell lines.

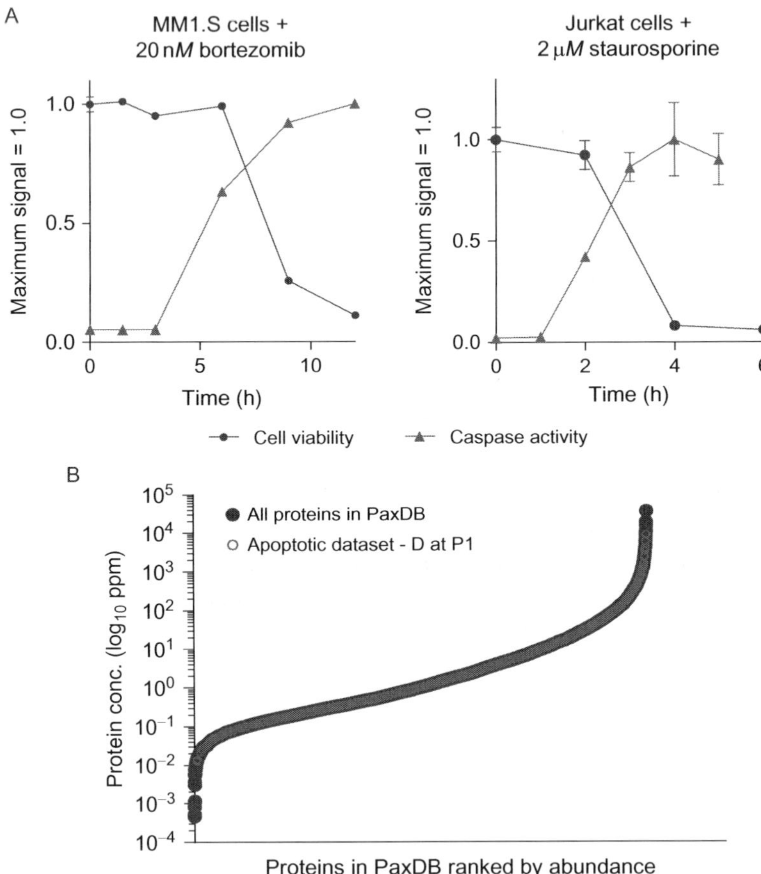

Figure 13.3 Monitoring apoptosis and proteomic distribution of cleavage substrates. (A) Measuring cell viability and caspase activation. It is important to monitor apoptosis versus time after exposure to drug, as the rate of apoptosis can vary substantially depending on the drug. Caspase activity appears before cell viability decreases. (B) Comparison of caspase substrates identified versus broad range of baseline protein abundance. Protein abundance estimated derived from PaxDB. Extensive distribution overlap indicates that subtiligase-based N-terminomics leads to broad coverage across 6-logs of abundance in the proteome. Figure adapted from Crawford et al. (2013) with permission of the authors.

After cells are grown to scale for Forward Discovery and have undergone apoptosis to the desired extent, cell bodies and debris are pelleted by centrifugation and washed once with ice-cold PBS. The washed cell pellet is then lysed directly or flash-frozen, stored at −80 °C, and thawed prior to lysis. For comparison to background cellular proteolysis, one should include a control sample not exposed to proteolysis inducers.

4.1.4 Protocol for forward discovery labeling

1. Cell lysis: Prepare 1 mL per sample of 4 × lysis buffer (ratio 4:4:2 of 10% SDS (w/v):1 *M* bicine pH 8.5:ddH_2O). Also prepare stocks of protease inhibitors (Sigma) to quench ongoing endogenous proteolysis: 10 m*M* z-VAD-fmk (Sigma) caspase inhibitor in DMSO; 10 m*M* E-64 (Sigma) cysteine protease inhibitor in DMSO; 100 m*M* 4-(2-aminoethyl)benzenesulfonyl fluoride (AEBSF, Sigma) serine protease inhibitor in DMSO; 0.5 *M* EDTA pH 8.0 in ddH_2O; 100 m*M* phenylmethylsulfonyl fluoride (PMSF, Sigma) (freshly prepared) in isopropanol. Add 5 μL of each protease inhibitor stock per 1 mL of 4 × lysis buffer. Add 1 mL 4 × lysis buffer with inhibitors to cell pellet and lyse completely by probe ultrasonication. Use clarified lysate sample for protein concentration determination.
2. Cysteine reduction and alkylation: In all proteomic experiments, it is important to first reduce and then irreversibly block-free thiol groups on cysteines to prevent formation of mixed oxidation products that hinder identification by MS. Prepare a fresh stock of 100 m*M* tris(2-carboxyethyl)phosphine (TCEP, Sigma) in ddH_2O. Add TCEP stock to final concentration of 5 m*M* to each lysed sample and mix. Heat at 95 °C for 15 min to ensure free thiols on cysteines. Allow to cool to room temperature (RT). During cooling, prepare a fresh stock of 200 m*M* iodoacetamide (IAM). Add IAM stock to final concentration of 10 m*M* to each sample and mix. Incubate 1 h in the dark at RT to block all cysteines. After incubation, add 1 *M* dithiothreitol (DTT) stock to final concentration of 25 m*M* to quench remaining IAM, as any free IAM will block catalytic cysteines of subtiligase at next step. Vortex briefly. Add Triton-X to a final concentration of 2.5% (v/v) to form micelles with SDS (*note*: subtiligase labeling will not work without removing detergent in this manner).
3. Subtiligase labeling: Centrifuge each sample for 10 min, ~4000 × *g* to pellet out any insoluble debris and transfer clarified supernatant to new tube. Add ddH_2O to final volume of 3.6 mL. Check and adjust

pH to 8.5. Add 400 μL of 10 m*M* peptide ester stock in DMSO to final concentration of 1 m*M*. Vortex briefly. Add 40 μL of 100 μ*M* subtiligase stock to final concentration of 1 μ*M*. Vortex briefly. Incubate for 1 h at RT (note: labeling for >1 h generally does not improve yields as peptide ester is either ligated to N-termini or hydrolyzed by this time). Labeling can be confirmed through a Western blot against biotin using NeutrAvidin-HRP (Pierce) with comparison to a presubtiligase sample.

4. Removal of excess peptide ester and exchange into denaturing conditions by protein precipitation: Biotin moieties on excess and hydrolyzed peptide ester will compete for binding sites on avidin beads and render them unavailable for capturing biotinylated proteins. Therefore, precipitate proteins by adding labeled sample dropwise to 35 mL acetonitrile at RT; short peptides, including excess peptide ester, will remain in solution. Vortex gently. Incubate on ice for at least 15 min up to overnight. Centrifuge 8000 × *g* for 30 min at 4 °C. Carefully decant supernatant to waste. Let precipitated protein pellet air dry for ~15 min. Add 1 mL of 8 *M* Guanidine HCl over pellet and let dissolve at RT for 30 min to 1 h. Swirl gently and pipet up and down to dissolve pellet. Add another 1 mL of 8 *M* Guanidine HCl to fully dissolve; use ultrasonication if necessary to solubilize. Precipitate protein a second time by adding dropwise to 30 mL ice-cold ethanol in a new 50 mL conical vial. Incubate at −80 °C overnight. The next day, centrifuge sample at 30 min, 8000 × *g*, 4 °C to pellet precipitated protein. Decant supernatant and air dry pellet for 15–20 min (*note*: can also freeze pellet at −80 °C for later use). Add 3 mL, 8 *M* Guanidine HCl over pellet and let dissolve at RT for 20 min. Add an additional 2 mL Guanidine HCl to complete dissolution. Transfer dissolved protein to new 15 mL conical vial. Rinse prior 50 mL conical vial with 2.5 mL ddH_2O, and transfer rinsed solution to same 15 mL conical. Take 8 μL aliquot of total dissolved protein sample for dot blot (below) and store at 4 °C.
5. Capture on NeutrAvidin resin: We use NeutrAvidin High Capacity resin (Pierce) to maximize capture of biotinylated proteins. For protein from 5×10^8 cells, we will typically add 1 mL of 50% bead slurry. After adding beads, place overnight at RT with gentle agitation or rotation. A dot blot against NeutrAvidin-HRP (Pierce) is recommended to confirm complete capture of labeled peptides compared to the prebead aliquot, indicated by disappearance of luminescence signal in the postbead sample. If capture is not complete, add additional NeutrAvidin resin and incubate for 2 h up to overnight, dot blot again, and repeat as necessary.

Note that incompletely removed peptide ester will increase the amount of beads necessary for complete capture.

6. On-bead trypsinization: After complete peptide capture, transfer beads to empty polypropylene chromatography column with frit at outlet. Attach the column to vacuum set up and remove the supernatant. Wash beads (add buffer, vortex, flow through) three times with 2 mL biotin wash buffer (10 m*M* bicine pH 8.0, 1 m*M* biotin) to occupy unbound avidin sites. Wash beads 5–10 × with 5 *M* Guanidine HCl to remove nonspecifically bound protein from beads. Wash beads 3 × in trypsin wash buffer (100 m*M* bicine pH 8.0, 200 m*M* NaCl, 20 m*M* $CaCl_2$, 1 *M* Guanidine HCl). Add 10–100 μg sequencing grade modified trypsin (trypsin should be added at 1:50 (w/w) to estimated amount of captured protein) in trypsin wash buffer to each sample. Incubate overnight at 37 °C with gentle agitation.
7. N-terminal peptide elution with TEV protease: Freshly prepare TEV protease buffer (50 m*M* ammonium bicarbonate pH 8.1, 2 m*M* DTT, 1 m*M* EDTA). Remove trypsinization supernatant to waste. Wash beads 5–10 × with 5 *M* Guanidine HCl to remove nonspecifically bound peptides. Wash beads 5 × with TEV protease buffer to completely remove guanidine. For each sample, mix 50 μg TEV protease with 1.5 mL TEV protease buffer and add to beads. Incubate overnight at RT with agitation or rotation. The next day, elute supernatant with TEV-cleaved peptides into 1.5 mL tubes. Evaporate to dryness.
8. Sample cleanup by ZipTip: Resuspend sample in a total of 100 μL 5% TFA to achieve pH ≤ 3. Let stand >10 min at RT. Spin 10 min at 14,000 × *g* at RT to pellet precipitated TEV protease. Transfer supernatant to new tube. For cleanup, we use C18 ZipTips (Millipore) performed with manufacturer protocol with elution into low-protein retention 500 μL tubes. Evaporate to dryness. Peptides may now be stored at −80 °C, resuspended in 0.1% FA to run directly on mass spectrometer, or used for fractionation.
9. Fractionation using reverse-phase high-pH chromatography (optional but recommended): To obtain the greatest depth of peptide coverage and greatest number of substrate identifications in Discovery experiments, it is advisable to perform a separation step prior to MS analysis. We use reverse-phase high-pH chromatography as it has been shown to offer similar separation capabilities to strong cation exchange and does not require additional ZipTip clean up of fractions (Yang, Shen, Camp, & Smith, 2012). If desired, after separation fractions can be

pooled for analysis. Evaporate to dryness. Store at −80 °C or resuspend each fraction in 0.1% FA for evaluation by MS.

4.1.5 Mass spectrometry analysis and bioinformatics

Mass spectrometry analysis of samples is carried out essentially like any other proteomic-based method, incorporating low-pH reverse-phase chromatography in-line with the mass spectrometer, as described in detail by others (for review, see Aebersold & Mann, 2003). Samples are analyzed in data-dependent acquisition mode, with exact parameters dependent on instrument used.

1. General protein database search to identify substrates: To identify N-termini, MS data must be searched against a database of known proteins specific for the organism of interest. Such searches can be performed with a variety of resources (Kapp & Schutz, 2007); we typically use Protein Prospector (http://prospector.ucsf.edu). This search algorithm is able to search a semitryptic peptide space: while the C-terminus has trypsin cleavage (at Arg/Lys), the N-terminus is allowed to be any amino acid in order to capture all potential proteolytic cleavages. In addition to this feature, Protein Prospector also allows a search with a constant N-terminal Abu-modification, which has a mass orthogonal from any naturally occurring amino acid. The completed database search across all analyzed fractions (typically at a false discovery rate of <1% based on a decoy database set) results in a list of N-terminal peptides identified in the sample. Further analysis of the sequence prior to the identified N-terminus (derived from database protein sequence) can reveal proteolytic specificity. In our prior studies, we have found that after apoptosis, there is a large increase in caspase-related Asp residues at the P1 position immediately N-terminal to the identified cleavage site. In nonapoptotic samples, we find a preponderance of Arg or Lys at P1 (Crawford et al., 2013).

   We have found it best for downstream analysis to collect and store all experimental data in a central database. For each experiment, we upload the relevant details (date, cell line, Forward or Reverse, inducer), links to the raw MS files, MS run parameters, protein database search parameters, and the protein database output file. We designed a FileMakerPro database (available at wellslab.ucsf.edu/degrabase) that also contains lookups to UniProtKb (www.uniprot.org) to add further details about the protein. This central database creates an easy and consistent workflow for importing and storing datasets and allows for easy export of data for analysis in another program or publication.

2. Comparison to other substrate databases: The list of potential proteolytic substrates can be compared to existing databases that focus on protease specificity or already identified substrates. Examples include MEROPS (Rawlings, Barrett, & Bateman, 2012), TopFind (Lange, Huesgen, & Overall, 2012), and the DegraBase (Crawford et al., 2013).
3. Analysis of biological function: Additionally, biological function of substrates can be analyzed using tools such as GoMiner (Zeeberg et al., 2005) or Ingenuity Pathway Analysis (www.ingenuity.com). Comparison to a general database of human cellular protein abundance, PaxDB (Wang et al., 2012), demonstrates that subtiligase offers substantial coverage of proteolytic substrates across the concentration range of the cellular proteome (Fig. 13.3B).

4.2. Identifying specific substrates of recombinant proteases in cell lysates (Reverse Discovery experiments)

In a Reverse Discovery experiment, the subtiligase-based method is modified to examine cleavage events catalyzed by the addition of a specific recombinant protease. The Reverse protocol is effective for the study of individual proteases of biological interest and for systems that may not be amenable to cell culture-based experiments.

4.2.1 Choice of sample and preparation

4.2.1.1 Preparation of cell lysate

For Reverse experiments, we typically start with $0.5\text{–}1 \times 10^9$ cells (30–60 mg of total protein in lysate). Our protocol has been optimized for caspase experiments but can be customized for the requirements of other enzymes.

4.2.1.2 Inhibition of endogenous proteases

In order to reduce background protease activity as thoroughly as possible, we include protease inhibitors in our lysis buffer. However, as the lysis buffer is not typically exchanged before protease addition, it is important to not include any inhibitors that would target the protease of interest. Therefore, for Reverse experiments with caspases, we include EDTA, AEBSF, and PMSF but omit the cysteine protease inhibitors z-VAD-fmk and E-64. One also needs to be aware that if the added protease activates an endogenous one of the same class that is not inhibited, then one would get a sum of the protease substrates. This is a distinct possibility when testing individual caspases and will not be eliminated even when comparing with a control sample with no protease added.

4.2.2 Preparation of proteases and characterization in lysate system

The purity and activity of the protease of interest is the most important part of a Reverse experiment. For example, for caspase-3, we express and purify our own enzymes and then perform kinetic activity assays using z-DEVD-AFC fluorescent substrate (CalBioChem) in either a small volume (50 μL–1 mL) of cellular lysate or optimal enzyme buffer. We use a range of caspase-3 concentrations from 1 to 1000 n*M*, near the expected physiological caspase concentration of ~50–200 n*M*. Ideally, activity in both lysate and buffer will be similar. Experiments at this small scale allows for lysis buffer modifications as necessary to enhance activity. These experiments can also determine appropriate enzyme concentration and time points where proteolysis has reached completion for full-scale studies.

4.2.3 Reverse experimental protocol

1. Cell lysis and caspase inactivation: For 1×10^9 cells in a caspase experiment, lyse using 10 mL of cold 100 m*M* HEPES pH 7.4, 0.1% Triton-X-100, 10 m*M* IAM, and 5 m*M* EDTA, 1 m*M* AEBSF, and 1 m*M* PMSF. Use 50 μL of lysed sample for protein concentration determination. The lysate is kept in the dark for 15 min at 4 °C for irreversible alkylation of endogenous caspase catalytic cysteines by IAM. Excess IAM is quenched by the addition of 20 m*M* DTT to prevent inhibition of recombinant caspase added in the next step. The lysate is cleared 2 × by centrifugation at 4000 × *g* for 5 min.
2. Protease addition and quenching: Prepare a 1 mL volume of active protease buffer (for caspases: 10 m*M* DTT, 0.1% CHAPS, 20 m*M* Pipes pH 7.2, 100 m*M* NaCl, 1 m*M* EDTA, and 10% sucrose). Add active protease to the 1 mL buffer, adjusted for the desired final concentration in ~15 mL after addition of enzyme in buffer to lysate. After a desired amount of time (determined in validation experiments), appropriate protease inhibitors are added to quench proteolysis. For caspases, we use 10–1000 n*M* for 10 min to 3 h and quench with 100 μ*M* z-VAD-fmk.
3. Subtiligase labeling: To the lysate, add 10 m*M* ester peptide stock in DMSO to final concentration of 1 m*M*. Vortex briefly. Check and adjust

pH to 8.5. Add 100 μ*M* subtiligase stock to final concentration of 1 μ*M*. Vortex briefly. Incubate for 1 h at RT. Labeling can be confirmed through a Western blot against biotin.

The remaining steps are similar to those in the Forward protocol (Section 4.1.4). Samples are desalted by acetonitrile precipitation. After resuspension of the pellet in 8 *M* Guanidine HCl, thiols are reduced by TCEP and alkylated by IAM. Protein is then precipitated again in EtOH overnight. After resuspension of protein in 8 *M* Guanidine HCl, all steps from NeutrAvidin capture to MS analysis (Section 4.1.4, Steps 5–9) are identical.

4.3. Quantification and kinetics

Thus far, we have described methods by which to identify proteolytic substrates, either in intact cells or as substrates of a specific recombinant enzyme. Next, we describe methods of quantitative proteomics combined with subtiligase-based labeling to determine the rates that substrates are cleaved.

4.3.1 MS quantification methods

There are a number of methods to quantify peptides by mass spectrometry. The relative advantages and disadvantages have been reviewed in detail by others (Bantscheff, Lemeer, Savitski, & Kuster, 2012; Bantscheff, Schirle, Sweetman, Rick, & Kuster, 2007; Liebler & Zimmerman, 2013; Nikolov, Schmidt, & Urlaub, 2012). These quantitative methods can be classified as either labeling or label-free, as well as either untargeted or targeted mode. Notably, essentially all can be implemented in combination with subtiligase-based N-terminomics. Untargeted labeling approaches include metabolic stable isotope labeling (SILAC) (Bushell et al., 2006) or isobaric mass tags coupled to peptides after tryptic digestion (Bantscheff et al., 2012). However, these methods can limit the number of samples that can be compared simultaneously.

Various untargeted, label-free approaches have been implemented too, including the method of "spectral counting" as well as others that rely on the peak area in the MS extracted ion chromatogram (Nahnsen, Bielow, Reinert, & Kohlbacher, 2013). Like all untargeted methods, a significant limitation is that these methods frequently do not identify all peptides of interest across all tested conditions. Targeted, label-free quantification, while limited to a few hundred proteins per run, offers highly sensitive and reproducible quantification as the MS instrument focuses only detecting peptides of interest in a complex sample. The most common method in this category is selected reaction monitoring (SRM; also known as multiple reaction monitoring) (Picotti & Aebersold, 2012). SRM methods can typically be run on unfractionated samples, leading to greatly reduced MS instrument time compared to other methods described. In addition, SRM assays can easily be applied to an unlimited number of samples (Huttenhain et al., 2012; Li et al., 2013). Below, we outline our approach to development of SRM assays to monitor proteolysis during apoptosis.

4.3.2 Design of a kinetic time course

We have primarily used SRM to quantify kinetics of proteolysis across a time course (Agard et al., 2012; Shimbo et al., 2012; Wiita et al., 2013). Similar to the Forward and Reverse Discovery experiments, SRM kinetic analysis can also be performed in intact cells as well as with recombinant proteases. Intact cells may present more variable kinetic data as they will initiate the apoptotic process at different times after inducer addition. Nonetheless, we have found that the kinetics of caspase cleavage in Reverse experiments with recombinant caspase matches quite well with cleavage kinetics in intact cells (Agard et al., 2012). In terms of experimental design, it is important to identify time points which will provide the most information on the proteolytic process of interest. A small scale experiment and Western blotting for cleavage of known substrates or a fluorescent reporter assay can fill this role.

4.3.3 Development and use of SRM assays

The SRM method relies on the use of a triple quadrupole mass spectrometer. In this system, a total peptide sample is injected onto an LC directly in-line with the MS instrument (Fig. 13.4). As peptides are eluted, the first quadrupole is used as a mass filter to only isolate peptides with a targeted m/z. The second quadrupole serves as a collision cell to break the peptide into fragments. The third quadrupole functions as a second mass filter for specified m/z fragments from the initial parent peptide. Each of these parent-fragment ion pairs is termed a "transition," and transition intensity is recorded by the detector. The coelution of multiple fragments from a single parent peptide indicates the specific identification of the peptide of interest. The total peak area reflects the relative abundance of the peptide across conditions (Fig. 13.4B).

1. Development of a spectral library: As SRM is a targeted method, Forward or Reverse Discovery experimental MS identification is required to develop an SRM assay. From this Discovery dataset, a "spectral

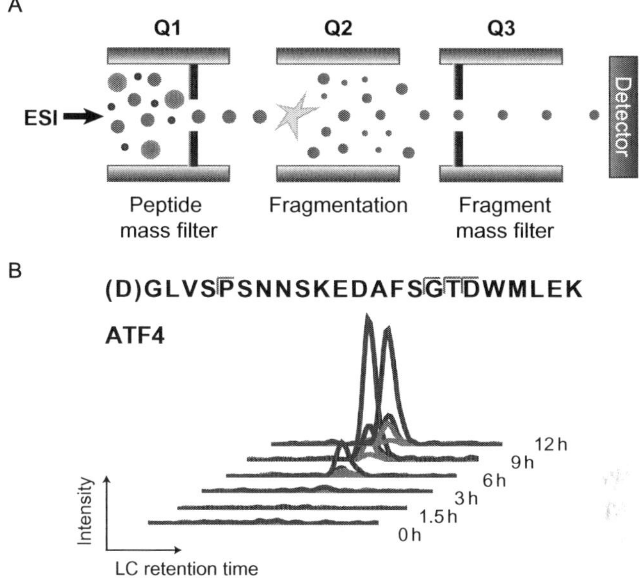

Figure 13.4 Selected reaction monitoring (SRM) for proteolytic substrate quantification. (A) Schematic diagram of triple quadrupole (Q1–Q2–Q3) mass spectrometer used for SRM. ESI, electrospray ionization, representing ionized peptides eluted from liquid chromatography column into the mass spectrometer. (B) Example data from SRM monitoring caspase-cleaved peptide from ATF4 protein during bortezomib-induced apoptosis. Each individual trace represents a parent-fragment ion pair (transition). Coelution of multiple transitions from the same peptide confirms peptide identity. Peak area can be used for quantification, with kinetic parameters derived based on change across the time course.

library" of identified peptides is generated, including the parent peptide mass, fragment ions, and MS signal intensity information. From this spectral library, peptides of further interest are specified for quantitative study. As an alternative source of a peptide data, our laboratory has made mass spectra available for peptides identified in Forward Discovery experiments in apoptotic and nonapoptotic cells (wellslab.ucsf.edu/degrabase).

2. SRM method development: Once a spectral library is obtained and peptides of interest are selected, the next step of method development is selection of the optimal transitions. One of the biggest recent advances in SRM is the implementation of the open-source, freely available Skyline software (MacLean et al., 2010). Skyline quickly generates transition lists from the imported spectral library. For targeted peptides, we first run

unscheduled SRM validation runs (no LC retention time information) monitoring up to seven transitions per protein and ~200 transitions per run, ideally using the same Discovery samples. We typically require a minimum of five of seven coeluting transitions and a retention time consistent with Discovery experiments to proceed with development. Synthetic peptide standards can also aid in confirmation. We then apply collision energy optimization and then create a scheduled method incorporating LC retention time information. Scheduling allows for monitoring significantly more peptides and transitions per run. We have found using the AB SCIEX QTRAP 5500 instrument up to ~250 peptides (~1000 transitions) can be included in a single SRM run. The method development process typically can be completed in less than a week.

3. SRM sample preparation and analysis: Samples are prepared according to the same protocols as in Sections 4.1 or 4.2. The main difference is we have had success with smaller sample input, on the scale of 2×10^8 cells (~10 mg of total protein). Samples are not fractionated and the entire population of N-terminal labeled peptides is analyzed directly in a single SRM run. Skyline software is then used to determine peak area for each identified peptide in each sample. Overall sample intensity must be normalized to account for differences in labeling efficiency and MS conditions across runs. Prior to subtiligase labeling, we typically spike in purified proteins not endogenously present in the sample as internal normalization standards. Using SRM to monitor recombinant caspase cleavage of substrates in cell lysates, we were also able to derive plots of substrate appearance versus time, analogous to more traditional enzymology experiments but capable of tracking hundreds of substrates simultaneously (Agard et al., 2012; Fig. 13.5).

4.4. Applications to human plasma and serum

N-terminal labeling by subtiligase can be modified for human plasma and serum samples to reveal insights into the complex proteolysis in the circulation (Wildes & Wells, 2010). Blood collection tubes should be centrifuged as soon as possible after sampling to separate plasma/serum from cellular components. Plasma/serum can be stored at −80 °C or used immediately for labeling. The labeling process is somewhat simplified compared to cell culture samples. To plasma/serum volumes of 0.5–2 mL, add 1 *M* bicine pH 8.5 to a final concentration of 100 m*M*. Then, add AEBSF to a final

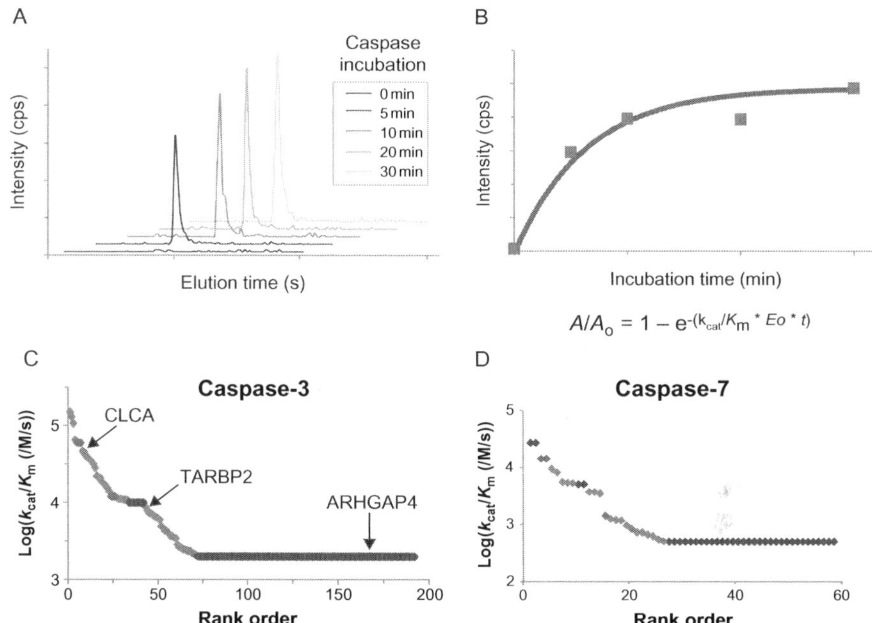

Figure 13.5 Monitoring kinetics of recombinant caspase cleavage. (A) SRM transitions show increase in intensity across time course after caspase addition. (B) Peptide intensities are fit to pseudo-first-order kinetic equations to determine kinetic efficiency (k_{cat}/K_m) for each substrate. (C and D) Rank order of catalytic efficiencies for substrates of caspases-3 and -7 span at least two orders of magnitude. Caspase-3 plot indicates substrates with rapid, medium, and slow cleavage. *Figure adapted from Agard et al. (2012) with permission of the authors.*

concentration of 1 m*M* to inhibit plasma serine proteases. Incubate at RT for 10 min. Add DTT to a final concentration of 2 m*M* and ester peptide to a final concentration of 1 m*M*. Vortex briefly. Add subtiligase enzyme to a final concentration of 1 μ*M* and incubate at RT for 1 h. For plasma samples, we have had success removing excess ester with NAP-25 chromatography columns (GE Healthcare) per manufacturer protocol with equilibration buffer of 50 m*M* bicine pH 8.0. After elution in 2.5 mL of 50 m*M* bicine pH 8.0, add 7.5 mL of 8 *M* Guanidine HCl to desalted sample. Reduce and alkylate thiols as in the protocols above. Then, add NeutrAvidin beads, typically at a slurry volume similar to the initial plasma volume. Sample preparation now proceeds identically as from Step 5 in Section 4.1.4. Enriched N-terminal peptides can be used for either Discovery or Targeted MS. Of note, general plasma proteomic studies are often confounded by the extremely high abundance of serum albumin, as signal intensity from albumin precludes detection of biologically interesting changes in low-abundance proteins (Anderson & Anderson, 2002). Fortunately, subtiligase labeling leads to extremely limited pull down of albumin, enabling detection of even very low-abundance plasma proteins (Fig. 13.6).

4.5. Application of N-terminal labeling by subtiligase to any biological sample

While we have focused on intact cells, cellular lysates, and human plasma, subtiligase-based enrichment can easily be applied to study proteolysis in virtually any biological sample. The general protocol would be highly similar to those shown above. The key step is obtaining the total protein sample in a buffer compatible with subtiligase labeling: nondenaturing conditions (i.e., no free detergent or denaturant) and pH ~8.5. Similar to plasma labeling, labeling of biological fluids (CSF, urine, etc.) can likely be achieved without any significant sample manipulation. Tissue samples can be processed similarly to cell culture samples. Alternatively, we have had success using trichloroacetic acid precipitation of total protein following by resuspension in guanidine-containing buffer. Buffer is then exchanged with desalting

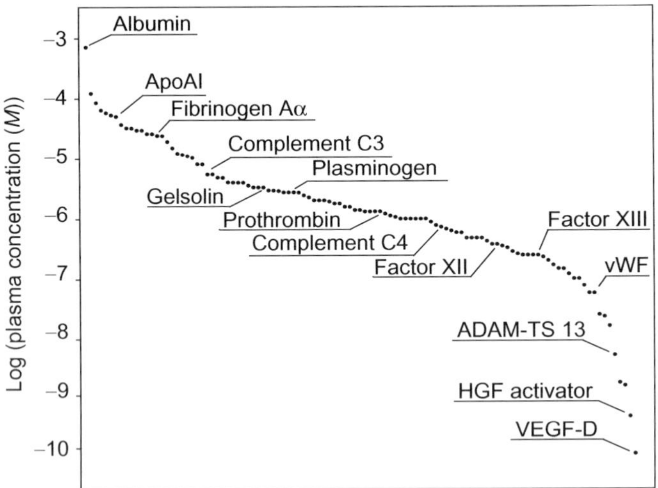

Figure 13.6 Plasma N-terminomics. Proteins identified with free N-termini in plasma demonstrate over six-order of magnitude range of abundance, demonstrating ability of subtiligase labeling to track low-abundance plasma proteins. *Figure adapted from Wildes and Wells (2010) with permission of the authors.*

columns into subtiligase-compatible conditions. From as little as 40 μg of starting protein, we can identify ∼50–200 proteolytically cleaved peptides released into the culture media after cellular apoptosis (Wiita, Hsu, Lu, Esensten, & Wells, 2014).

5. LIMITATIONS TO THE SUBTILIGASE LABELING METHOD

Subtiligase labeling has many advantages in proteolysis research: (i) it is an unbiased, enzymatically driven method without chemical protein modification, (ii) it can identify thousands of peptides over six orders of magnitude in abundance from complex biological samples, (iii) it can be combined with highly reproducible, label-free quantification, and (iv) single labeling with positive enrichment allows not only identification of the target, but reveals the precise site of proteolysis. However, there are limitations to this approach.

As described earlier, subtiligase labeling efficiency is the biggest current limitation of application to systems of interest. To counter this inefficiency, we use large amounts of starting material. This is relatively easy for cell culture-based experiments, but can be more complicated for animal or human samples. Additionally, there are a few N-terminal amino acids, notably proline, valine, and isoleucine, that subtiligase is relatively slow to ligate *in vitro* (Chang, Jackson, Burnier, & Wells, 1994). Nonetheless, we have found that subtiligase can indeed label these N-terminal amino acids in our cell-based experiments. Our N-terminomics data show that N-terminal labeling is only a small bias and does not significantly affect the utility of the method.

For Discovery experiments, these methods suffer the same limitation as all MS experiments: there can be incomplete overlap between peptides identified in one run to another, even under the same conditions. This is due to the stochastic nature of sampling of low-abundance peptides by the MS instrument. This limitation may be circumvented through the use of biological and technical replicates. A particular limitation for our strategy is that short peptides labeled by subtiligase will not be precipitated with other proteins (in Step 4 of protocol in Section 4.1.4) and will be discarded. However, semi-tryptic peptides can rescue many of these. Furthermore, the tryptic digestion of biotinylated proteins may lead to N-terminal peptides either too long or too short to be identified by MS.

Of note, we have provided the subtiligase plasmid and expression strains under a standard materials transfer agreement to more than 30 laboratories.

6. SUMMARY OF FINDINGS FROM SUBTILIGASE-BASED N-TERMINOMICS

As discussed in Section 1, there are multiple methods for enrichment and identification of proteolytically cleaved peptides in biological samples (Rogers & Overall, 2013). The subtiligase method has been extensively applied to the study of caspases. These studies showed that caspases can cleave more than 1000 cellular substrates during apoptosis, greatly expanding the scope of caspase biology. Initial studies revealed that caspases have a general preference for disordered structural elements in substrates (Mahrus et al., 2008). Further experiments across multiple cell lines allowed us to compile over 1700 caspase cleavage sites in nearly 1300 substrates as well as over 6000 noncaspase proteolytic sites (Crawford et al., 2013). With this large database, publicly available at http://wellslab.ucsf.edu/degrabase, we identified conserved motifs of caspase cleavage (Fig. 13.7) as well as sequence features of noncaspase endoproteases in nonapoptotic cells (Crawford et al., 2013). Furthermore, by combining recombinant enzyme purification with subtiligase labeling, we characterized specific substrates of the inflammatory caspases-1, -4, and -5 as well as the apoptotic caspases-3, -7, -8, and -9 (Agard et al., 2012; Agard, Maltby, & Wells, 2010). A combination of Forward and Reverse experiments allowed us to find evolutionarily conserved relationships between caspase cut site, protein substrates, and pathway-level relationships across organisms (Crawford et al., 2012). With a similar experimental combination, we probed proteolytic cleavage in human plasma which may relate to disease signatures (Wildes & Wells, 2010). The discovery of a broad range of caspase catalytic efficiencies across hundreds of substrates in parallel (Agard et al., 2012), was facilitated by the combination of label-free quantification with N-terminomics to determine catalytic efficiencies of natural substrates in complex mixtures. Furthermore, we showed that quantitative signatures of caspase cleavage can be used to monitor chemotherapeutic effects in cancer cells (Shimbo et al., 2012; Wiita et al., 2013, 2014). In summary, this method has revealed extensive information about proteolytic cleavage during apoptosis.

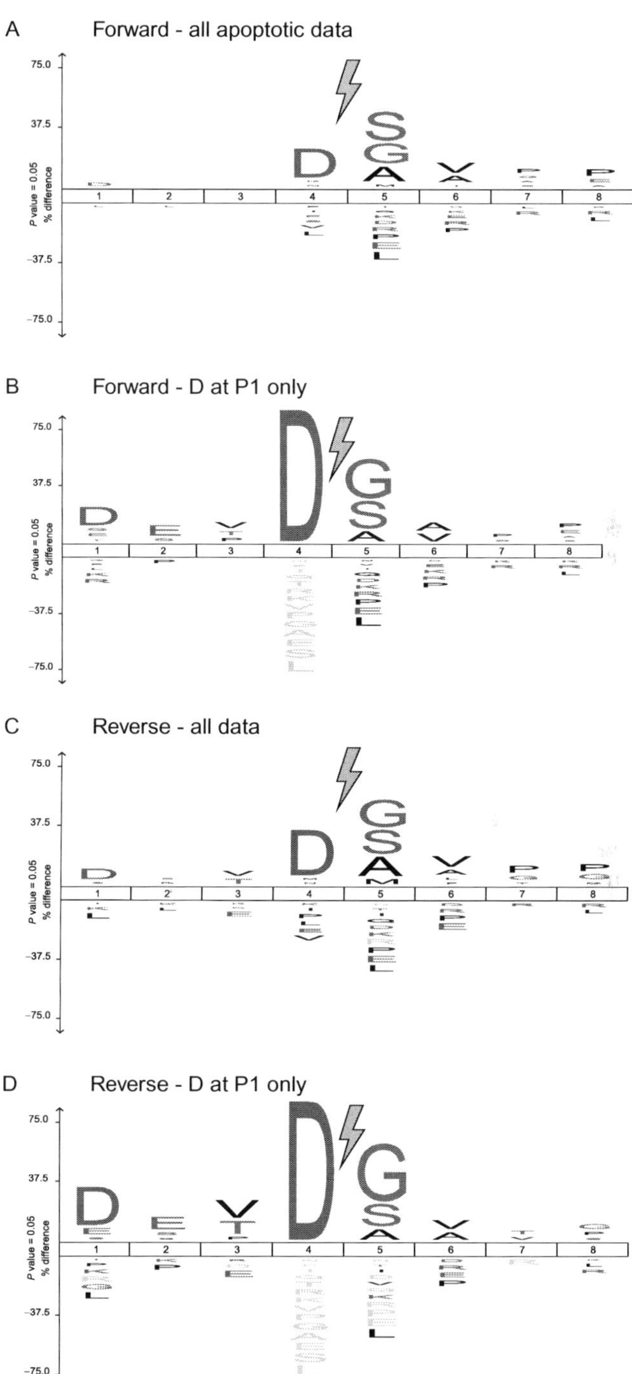

Figure 13.7 See legend on next page.

7. FUTURE DIRECTIONS

To broaden the range of applications of our methods, we are using protein engineering to improve subtiligase labeling efficiency and decrease the need for large amounts of starting material. In tandem, new highly sensitive mass spectrometers will allow for both more comprehensive substrate identification and potentially label-free quantification with less protein input (Gallien et al., 2012; Hebert et al., 2014; Peterson, Russell, Bailey, Westphall, & Coon, 2012). Both of these advances will allow for in-depth analysis of many more recombinant proteases and biological systems. In intact cells, caspases have only been fully profiled for substrate generation during apoptosis; their substrate profiles in processes such as differentiation or nonapoptotic cell stress (Kuranaga, 2012) have yet to be elucidated. Others have recently shown that there is significant cross-talk between protein phosphorylation and caspase cleavage (Dix et al., 2012). N-terminomics can further be combined with new enrichment methods to investigate the relation between caspase cleavage and other posttranslational modifications, such as ubiquitination and lysine acetylation (Mertins et al., 2013). Alternatively, isolating intracellular organelles or secreted domains from cell membrane proteins will allow one to monitor proteolysis specific to different cellular perturbations in different cellular compartments. This knowledge may lead to information relevant to therapeutic and diagnostic development. This subtiligase technique is poised for wide use in proteolysis research.

Figure 13.7 Sequence features of identified N-termini. Aggregate plots across all N-termini shown using IceLogo, where amino acids favored at a given site are above the baseline while those disfavored are below (Colaert, Helsens, Martens, Vandekerckhove, & Gevaert, 2009). Cleavage occurs between position P1 (left of lightning bolt) and P1′. (A) Across all peptides identified in Forward Discovery experiments in apoptotic cells, Asp at P1 is highly enriched. (B) Focusing on only peptides shown in (A) with Asp at P1, a signature of caspase cleavage, we identify the canonical D-E-V-D cleavage motif for caspases from P4 to P1 site. (C and D) This motif is also conserved in N-termini identified in cell lysate incubated with recombinant caspase-3 during Reverse Discovery experiments.

ACKNOWLEDGMENTS

This project was supported by Grants from National Institutes of Health R01 GM081051, R01 GM097316, and RO1 CA154802 (J. A. W.); The Rogers Family Foundation (J. A. W.), an award from the Sandler Program in Biological Sciences (J. A. W.); and the UCSF Stephen and Nancy Grand Multiple Myeloma Translational Initiative (J. A. W.). A. P. W. is a Damon Runyon Fellow supported by the Damon Runyon Cancer Research Foundation (DRG 111-12). J. E. S is supported by National Institutes of Health Training Grant T32 GM007175 and the Global Healthy Living Foundation. Mass spectrometry was performed at the Bio-Organic Biomedical Mass Spectrometry Resource at UCSF, which is supported by grants from the National Center for Research Resources (5P41RR001614) and the National Institute of General Medical Sciences (8P41GM103481 and 1S10RR015804) from the National Institutes of Health. We thank our long time collaborator Professor Alma Burlingame and his group for providing expert advice and open access to the UCSF Mass Spectrometry Center.

REFERENCES

Abrahmsen, L., Tom, J., Burnier, J., Butcher, K. A., Kossiakoff, A., & Wells, J. A. (1991). Engineering subtilisin and its substrates for efficient ligation of peptide bonds in aqueous solution. *Biochemistry, 30*, 4151–4159.

Aebersold, R., & Mann, M. (2003). Mass spectrometry-based proteomics. *Nature, 422*, 198–207.

Agard, N. J., Mahrus, S., Trinidad, J. C., Lynn, A., Burlingame, A. L., & Wells, J. A. (2012). Global kinetic analysis of proteolysis via quantitative targeted proteomics. *Proceedings of the National Academy of Sciences of the United States of America, 109*, 1913–1918.

Agard, N. J., Maltby, D., & Wells, J. A. (2010). Inflammatory stimuli regulate caspase substrate profiles. *Molecular and Cellular Proteomics, 9*, 880–893.

Agard, N. J., & Wells, J. A. (2009). Methods for the proteomic identification of protease substrates. *Current Opinion in Chemical Biology, 13*, 503–509.

Anderson, N. L., & Anderson, N. G. (2002). The human plasma proteome: History, character, and diagnostic prospects. *Molecular and Cellular Proteomics, 1*, 845–867.

Bantscheff, M., Lemeer, S., Savitski, M. M., & Kuster, B. (2012). Quantitative mass spectrometry in proteomics: Critical review update from 2007 to the present. *Analytical and Bioanalytical Chemistry, 404*, 939–965.

Bantscheff, M., Schirle, M., Sweetman, G., Rick, J., & Kuster, B. (2007). Quantitative mass spectrometry in proteomics: A critical review. *Analytical and Bioanalytical Chemistry, 389*, 1017–1031.

Black, R. A., Kronheim, S. R., & Sleath, P. R. (1989). Activation of interleukin-1 beta by a co-induced protease. *FEBS Letters, 247*, 386–390.

Braisted, A. C., Judice, J. K., & Wells, J. A. (1997). Synthesis of proteins by subtiligase. *Methods in Enzymology, 289*, 298–313.

Bushell, M., Stoneley, M., Kong, Y. W., Hamilton, T. L., Spriggs, K. A., Dobbyn, H. C., et al. (2006). Polypyrimidine tract binding protein regulates IRES-mediated gene expression during apoptosis. *Molecular Cell, 23*, 401–412.

Chang, T. K., Jackson, D. Y., Burnier, J. P., & Wells, J. A. (1994). Subtiligase: A tool for semisynthesis of proteins. *Proceedings of the National Academy of Sciences of the United States of America, 91*, 12544–12548.

Colaert, N., Helsens, K., Martens, L., Vandekerckhove, J., & Gevaert, K. (2009). Improved visualization of protein consensus sequences by iceLogo. *Nature Methods, 6*, 786–787.

Crawford, E. D., Seaman, J. E., Agard, N., Hsu, G. W., Julien, O., Mahrus, S., et al. (2013). The DegraBase: A database of proteolysis in healthy and apoptotic human cells. *Molecular and Cellular Proteomics*, *12*, 813–824.

Crawford, E. D., Seaman, J. E., Barber, A. E., 2nd., David, D. C., Babbitt, P. C., Burlingame, A. L., et al. (2012). Conservation of caspase substrates across metazoans suggests hierarchical importance of signaling pathways over specific targets and cleavage site motifs in apoptosis. *Cell Death and Differentiation*, *19*, 2040–2048.

Dix, M. M., Simon, G. M., & Cravatt, B. F. (2008). Global mapping of the topography and magnitude of proteolytic events in apoptosis. *Cell*, *134*, 679–691.

Dix, M. M., Simon, G. M., Wang, C., Okerberg, E., Patricelli, M. P., & Cravatt, B. F. (2012). Functional interplay between caspase cleavage and phosphorylation sculpts the apoptotic proteome. *Cell*, *150*, 426–440.

Gallien, S., Duriez, E., Crone, C., Kellmann, M., Moehring, T., & Domon, B. (2012). Targeted proteomic quantification on quadrupole-orbitrap mass spectrometer. *Molecular and Cellular Proteomics*, *11*, 1709–1723.

Galluzzi, L., Aaronson, S. A., Abrams, J., Alnemri, E. S., Andrews, D. W., Baehrecke, E. H., et al. (2009). Guidelines for the use and interpretation of assays for monitoring cell death in higher eukaryotes. *Cell Death and Differentiation*, *16*, 1093–1107.

Hebert, A. S., Richards, A. L., Bailey, D. J., Ulbrich, A., Coughlin, E. E., Westphall, M. S., et al. (2014). The one hour yeast proteome. *Molecular and Cellular Proteomics*, *13*, 339–347.

Huttenhain, R., Soste, M., Selevsek, N., Rost, H., Sethi, A., Carapito, C., et al. (2012). Reproducible quantification of cancer-associated proteins in body fluids using targeted proteomics. *Science Translational Medicine*, *4*, 142ra194.

Impens, F., Colaert, N., Helsens, K., Plasman, K., Van Damme, P., Vandekerckhove, J., et al. (2010). MS-driven protease substrate degradomics. *Proteomics*, *10*, 1284–1296.

Jackson, D. Y., Burnier, J., Quan, C., Stanley, M., Tom, J., & Wells, J. A. (1994). A designed peptide ligase for total synthesis of ribonuclease A with unnatural catalytic residues. *Science*, *266*, 243–247.

Kapp, E., & Schutz, F. (2007). Overview of tandem mass spectrometry (MS/MS) database search algorithms. *Current Protocols in Protein Science* Chapter 25, Unit 25.2.

Kleifeld, O., Doucet, A., auf dem Keller, U., Prudova, A., Schilling, O., Kainthan, R. K., et al. (2010). Isotopic labeling of terminal amines in complex samples identifies protein N-termini and protease cleavage products. *Nature Biotechnology*, *28*, 281–288.

Klingler, D., & Hardt, M. (2012). Profiling protease activities by dynamic proteomics workflows. *Proteomics*, *12*, 587–596.

Kuranaga, E. (2012). Beyond apoptosis: Caspase regulatory mechanisms and functions in vivo. *Genes to Cells*, *17*, 83–97.

Lange, P. F., Huesgen, P. F., & Overall, C. M. (2012). TopFIND 2.0—Linking protein termini with proteolytic processing and modifications altering protein function. *Nucleic Acids Research*, *40*, D351–D361.

Li, X. J., Hayward, C., Fong, P. Y., Dominguez, M., Hunsucker, S. W., Lee, L. W., et al. (2013). A blood-based proteomic classifier for the molecular characterization of pulmonary nodules. *Science Translational Medicine*, *5*, 207ra142.

Liebler, D. C., & Zimmerman, L. J. (2013). Targeted quantification of proteins by mass spectrometry. *Biochemistry*, *52*, 3797–3806.

Lopez-Otin, C., & Matrisian, L. M. (2007). Emerging roles of proteases in tumour suppression. *Nature Reviews Cancer*, *7*, 800–808.

Lucast, L. J., Batey, R. T., & Doudna, J. A. (2001). Large-scale purification of a stable form of recombinant tobacco etch virus protease. *Biotechniques*, *30*, 544.

MacLean, B., Tomazela, D. M., Shulman, N., Chambers, M., Finney, G. L., Frewen, B., et al. (2010). Skyline: An open source document editor for creating and analyzing targeted proteomics experiments. *Bioinformatics*, *26*, 966–968.

Mahrus, S., Trinidad, J. C., Barkan, D. T., Sali, A., Burlingame, A. L., & Wells, J. A. (2008). Global sequencing of proteolytic cleavage sites in apoptosis by specific labeling of protein N termini. *Cell*, *134*, 866–876.

Mertins, P., Qiao, J. W., Patel, J., Udeshi, N. D., Clauser, K. R., Mani, D. R., et al. (2013). Integrated proteomic analysis of post-translational modifications by serial enrichment. *Nature Methods*, *10*, 634–637.

Nahnsen, S., Bielow, C., Reinert, K., & Kohlbacher, O. (2013). Tools for label-free peptide quantification. *Molecular and Cellular Proteomics*, *12*, 549–556.

Nikolov, M., Schmidt, C., & Urlaub, H. (2012). Quantitative mass spectrometry-based proteomics: An overview. *Methods in Molecular Biology*, *893*, 85–100.

Peterson, A. C., Russell, J. D., Bailey, D. J., Westphall, M. S., & Coon, J. J. (2012). Parallel reaction monitoring for high resolution and high mass accuracy quantitative, targeted proteomics. *Molecular and Cellular Proteomics*, *11*, 1475–1488.

Pham, V. C., Pitti, R., Anania, V. G., Bakalarski, C. E., Bustos, D., Jhunjhunwala, S., et al. (2012). Complementary proteomic tools for the dissection of apoptotic proteolysis events. *Journal of Proteome Research*, *11*, 2947–2954.

Picotti, P., & Aebersold, R. (2012). Selected reaction monitoring-based proteomics: Workflows, potential, pitfalls and future directions. *Nature Methods*, *9*, 555–566.

Polevoda, B., & Sherman, F. (2003). N-terminal acetyltransferases and sequence requirements for N-terminal acetylation of eukaryotic proteins. *Journal of Molecular Biology*, *325*, 595–622.

Rawlings, N. D., Barrett, A. J., & Bateman, A. (2012). MEROPS: The database of proteolytic enzymes, their substrates and inhibitors. *Nucleic Acids Research*, *40*, D343–D350.

Rogers, L., & Overall, C. M. (2013). Proteolytic post translational modification of proteins: Proteomic tools and methodology. *Molecular and Cellular Proteomics*, *12*, 3532–3542.

Schilling, O., Huesgen, P. F., Barre, O., & Overall, C. M. (2011). Identification and relative quantification of native and proteolytically generated protein C-termini from complex proteomes: C-terminome analysis. *Methods in Molecular Biology*, *781*, 59–69.

Shao, W., Yeretssian, G., Doiron, K., Hussain, S. N., & Saleh, M. (2007). The caspase-1 digestome identifies the glycolysis pathway as a target during infection and septic shock. *Journal of Biological Chemistry*, *282*, 36321–36329.

Shimbo, K., Hsu, G. W., Nguyen, H., Mahrus, S., Trinidad, J. C., Burlingame, A. L., et al. (2012). Quantitative profiling of caspase-cleaved substrates reveals different drug-induced and cell-type patterns in apoptosis. *Proceedings of the National Academy of Sciences of the United States of America*, *109*, 12432–12437.

Smeekens, S. P., Montag, A. G., Thomas, G., Albigesrizo, C., Carroll, R., Benig, M., et al. (1992). Proinsulin processing by the subtilisin-related proprotein convertases furin, PC2, and PC3. *Proceedings of the National Academy of Sciences of the United States of America*, *89*, 8822–8826.

Staes, A., Van Damme, P., Helsens, K., Demol, H., Vandekerckhove, J., & Gevaert, K. (2008). Improved recovery of proteome-informative, protein N-terminal peptides by combined fractional diagonal chromatography (COFRADIC). *Proteomics*, *8*, 1362–1370.

Tanco, S., Lorenzo, J., Garcia-Pardo, J., Degroeve, S., Martens, L., Aviles, F. X., et al. (2013). Proteome-derived peptide libraries to study the substrate specificity profiles of carboxypeptidases. *Molecular and Cellular Proteomics*, *12*, 2096–2110.

Thiede, B., Treumann, A., Kretschmer, A., Sohlke, J., & Rudel, T. (2005). Shotgun proteome analysis of protein cleavage in apoptotic cells. *Proteomics*, *5*, 2123–2130.

Tonge, R., Shaw, J., Middleton, B., Rowlinson, R., Rayner, S., Young, J., et al. (2001). Validation and development of fluorescence two-dimensional differential gel electrophoresis proteomics technology. *Proteomics*, *1*, 377–396.

Wang, M., Weiss, M., Simonovic, M., Haertinger, G., Schrimpf, S. P., Hengartner, M. O., et al. (2012). PaxDb, a database of protein abundance averages across all three domains of life. *Molecular and Cellular Proteomics*, *11*, 492–500.

Wejda, M., Impens, F., Takahashi, N., Van Damme, P., Gevaert, K., & Vandenabeele, P. (2012). Degradomics reveals that cleavage specificity profiles of caspase-2 and effector caspases are alike. *Journal of Biological Chemistry*, *287*, 33983–33995.

Wiita, A. P., Ziv, E., Wiita, P. J., Urisman, A., Julien, O., Burlingame, A. L., et al. (2013). Global cellular response to chemotherapy-induced apoptosis. *eLife*, *2*, e01236.

Wiita, A. P., Hsu, G. W., Lu, C. M., Esensten, J. H., & Wells, J. A. (2014). Circulating proteolytic signatures of cell death in humans identified by N-terminal labeling. *Proceedings of the National Academy of Sciences of the United States of America*, In press.

Wildes, D., & Wells, J. A. (2010). Sampling the N-terminal proteome of human blood. *Proceedings of the National Academy of Sciences of the United States of America*, *107*, 4561–4566.

Xu, G., & Jaffrey, S. R. (2010). N-CLAP: Global profiling of N-termini by chemoselective labeling of the alpha-amine of proteins. *Cold Spring Harbor Protocols*, *2010* pdb.prot5528.

Xu, G., Shin, S. B., & Jaffrey, S. R. (2009). Global profiling of protease cleavage sites by chemoselective labeling of protein N-termini. *Proceedings of the National Academy of Sciences of the United States of America*, *106*, 19310–19315.

Xu, G., Shin, S. B., & Jaffrey, S. R. (2011). Chemoenzymatic labeling of protein C-termini for positive selection of C-terminal peptides. *ACS Chemical Biology*, *6*, 1015–1020.

Yang, F., Shen, Y., Camp, D. G., 2nd., & Smith, R. D. (2012). High-pH reversed-phase chromatography with fraction concatenation for 2D proteomic analysis. *Expert Review of Proteomics*, *9*, 129–134.

Yoshihara, H. A., Mahrus, S., & Wells, J. A. (2008). Tags for labeling protein N-termini with subtiligase for proteomics. *Bioorganic and Medicinal Chemistry Letters*, *18*, 6000–6003.

Zeeberg, B. R., Qin, H., Narasimhan, S., Sunshine, M., Cao, H., Kane, D. W., et al. (2005). High-throughput GoMiner, an 'industrial-strength' integrative gene ontology tool for interpretation of multiple-microarray experiments, with application to studies of common variable immune deficiency (CVID). *BMC Bioinformatics*, *6*, 168.

CHAPTER FOURTEEN

Complementary Methods for the Identification of Substrates of Proteolysis

Victoria C. Pham, Veronica G. Anania, Qui T. Phung, Jennie R. Lill[1]
Department of Protein Chemistry, Genentech Inc., South San Francisco, California, USA
[1]Corresponding author: e-mail address: jlill@gene.com

Contents

Abstract

Proteolysis describes the cleavage of proteins into smaller components, which *in vivo* occurs typically to either activate or impair the functionality of cellular proteins. Proteolysis can occur during cellular homeostasis or can be induced due to external stress stimuli such as heat, biological or chemical insult, and is mediated by the activity of cellular enzymes, namely, proteases. Proteolytic cleavage of proteins can influence protein activation by exposing an active site or disrupting inhibitor binding. Conversely, proteolytic cleavage of many proteins has also been shown to lead to protein degradation resulting in inactivation of the substrate. Thousands of proteolytic events are known to take place in regulated cellular processes such as apoptosis and pyroptosis, however, their individual contribution to these processes remains poorly understood. Additionally, many cellular homeostatic processes are regulated by proteolytic events, however, in some cases, few proteolytic substrates have been identified. To gain further insight into the mechanism of action of these cellular processes, and to characterize biomarkers of cell death and other pathological indications, it is imperative to utilize a complete arsenal of tools for studying proteolysis events *in vivo* and *in vitro*. In this chapter,

Methods in Enzymology, Volume 544
ISSN 0076-6879
http://dx.doi.org/10.1016/B978-0-12-417158-9.00014-5

we focus on alternative methodologies to N-terminomics for profiling substrates of proteolysis and describe an additional suite of tools including orthogonal biophysical separation techniques such as COFRADIC or GASSP, and affinity capture tools that can enrich for newly formed C-termini (C-terminomics) generated as a result of caspase-mediated proteolysis.

1. INTRODUCTION

Proteolysis is a key regulatory event that controls intracellular and extracellular signaling through irreversible changes in a protein's structure hence greatly altering its functionality (Doucet, Butler, Rodriguez, Prudova, & Overall, 2008; Wysocka & Lesner, 2013). This can occur in a much faster time frame than other types of posttranslational modifications and is therefore a very effective mechanism for responding rapidly to cellular insult or physiological cue. There are many classes of mammalian proteases that cover a variety of physiological roles from coagulation (Wysocka & Lesner, 2013), metabolism (Kojro & Postina, 2009), to cell death (McIlwain, Berger, & Mak, 2013) and due to disruption of these critical cellular processes multiple drug discovery initiatives are currently ongoing with the goal to either activate or inhibit these proteolysis events (Ciechanover, 2012). To cover this breadth of physiological control, proteolytic enzymes span across a wide range of enzymatic lineages including the serine hydrolase superfamily which is comprised of some 200 members, the more modest yet critical cysteine-dependent aspartate-directed proteases (Caspases) which are a family of cysteine proteases playing essential roles in various types of cell death and inflammation, in addition to various other subfamilies of proteolytic enzymes. Proteolysis substrates related to key disease indications or cellular fates, in addition to being drug target candidates can also serve as biomarkers that may be employed for *in vivo* or *in vitro* assays (e.g., PARP1 cleavage is indicative of caspase-3 activation and is a signature of apoptotic cell death). These substrates may allow further insight into the biological mechanisms triggering the cell fate of interest.

Due to interest in understanding the role of proteolysis in disease and also the importance of generating biomarkers that can act as key indicators of biological activity, it is important to be able to catalogue the substrates of various types of intracellular and extracellular proteolytic signaling events. Proteomic based tools to decipher substrates of proteolysis typically fall into one of three categories: N-terminomic approaches, C-terminomic approaches, or differential migration of proteins subjected to proteolysis using analytical methodologies such as SDS-PAGE or chromatography. In this chapter, we review the latter two approaches, describe the methodologies in detail, and highlight the benefits and limitations of these techniques.

2. COMBINED FRACTIONAL DIAGONAL CHROMATOGRAPHY

N-terminal combined fractional diagonal chromatography (COFRADIC) is a negative enrichment peptide separation technique that has successfully been employed to identify substrates of proteolysis. The basic principles of COFRADIC date back to 1966 when Brown and Hartley employed diagonal paper chromatography to identify disulfide bridges within bovine chymotrypsinogen A (Brown & Hartley, 1966). Following separation of a complex peptide mixture using electrophoresis, free cysteine residues were oxidized to cysteic acid using perfomic acid vapor. This induced a strong negative charge that could be exploited to separate free, undisulfide bonded cysteine-containing peptides from the rest of a complex proteome. Orthogonal electrophoresis to the first direction of separation was employed to reveal the cysteic acid "fingerprint." All noncysteic acid containing peptides migrated on a diagonal line, while any peptides containing free cysteines converted to cysteic acid migrated below the diagonal line owing to their increased negative charge thereby allowing them to successfully map disulfide bridges.

This approach has been adopted for many bioanalytical applications including characterization of protein proteolytic processing, phosphopeptide mapping, mapping N-glycosylation moieties, ATP-binding peptide mapping, and nitrotyrosinyl peptide mapping (Gevaert et al., 2007, 2005; Ghesquiere et al., 2006; Hanoulle et al., 2006). The term COFRADIC was first coined by Vandekerckhove and colleagues who developed a gel-free diagonal reverse phase chromatography technique to characterize N-terminal peptides from *E. coli* (Gevaert et al., 2003), an adaptation of their previous methionyl peptide capture method (Gevaert et al., 2002). This approach was then extended to identify proteolysis substrates during apoptosis where 93 caspase cleavage sites were identified in 71 proteins, demonstrating the utility of COFRADIC for studying substrates of numerous other proteases (Van Damme et al., 2005). To date, the COFRADIC method has been further applied to characterize proteolytic substrates of proteases such as caspase 1 and caspase 7 (Lamkanfi et al., 2008), granzyme B (Van Damme et al., 2009), and the mitochondrial serine protease high-temperature requirement protein A2 (HtrA2/Omi) (Vande Walle et al., 2007).

The COFRADIC peptide-sorting principle employed to separate N-terminal peptides from internal peptides is based on increasing the hydrophobicity of internal, non-N-terminal peptides. First, the undigested proteome is trideutero-acetylated to block all free amines (in principle this could be performed using other acylation modifications including iTRAQ, tandem mass tag (TMT), dimethylation, etc.). This distinguishes between *in vivo* blocked (acetylated) and *in vivo* free (e.g., trideutero-acetylated) protein N-termini. Tryptic cleavage is inhibited at blocked lysine residues, therefore generating Arg-C specific peptides. The peptide mixture is separated by RP-HPLC and collected in a small number of fractions. Following tryptic digestion, internal peptides have unblocked N-termini that can be reacted with 2,4,6-trinitrobenzenesulfonic acid (TNBS) to significantly increase their hydrophobicity. A second round of RP-HPLC on the same column will separate non-TNBS-modified peptides (true N-termini and neo-N-termini resulting from protease activity) from TNBS-modified peptides (internal peptides). Endogenous unblocked N-termini can be differentiated from protease-generated N-termini using bioinformatics tools (such as MS/MS search algorithms) or manual analysis of the data. Additionally, this method can be combined with *in vivo* or *in vitro* metabolic labeling (such as stable isotopic labeling of amino acids in cell culture (SILAC)) or differential quantitative chemical labeling (*in vitro*) to directly compare multiple proteomes in one experiment, thereby avoiding run-to-run variability (Gevaert et al., 2007).

COFRADIC offers several advantages over other proteolysis substrate capture methods in that it is inexpensive and can be applied to any protease. Enzymes other than trypsin (such as Glu-C) can easily be applied to allow diversity of protease profiling. It also provides information regarding the specific site of protease cleavage that can be useful for validation follow-up experiments and for determining cleavage consensus motifs. One caveat, however, is that high background levels resulting from incomplete derivatization of internal peptides can greatly impact sensitivity. Additionally, substrates are identified by a single N-terminal peptide, which may be inappropriately sized for LC-MS/MS detection. Perhaps the most inconvenient aspect of COFRADIC is the enormous amount of instrument time consumed (>100 fractions, 60 min each), however, a recent study published utilizes a sample pooling strategy that reduces the number of samples to be analyzed threefold (Staes et al., 2011). TNBS reacts nearly quantitatively with internal peptides; however, Proline-starting peptides are refractory toward TNBS modification, the latter category explaining the majority of unreacted internal peptides. Implementing a strong cation exchange prefractionation step in the enrichment procedure can greatly reduce this redundancy. Even with these limitations in mind, N-terminal COFRADIC remains a relatively inexpensive, versatile technology that can be implemented by most proteomics laboratories to investigate proteolytic activity. Figure 14.1 shows a flowchart for COFRADIC analyses.

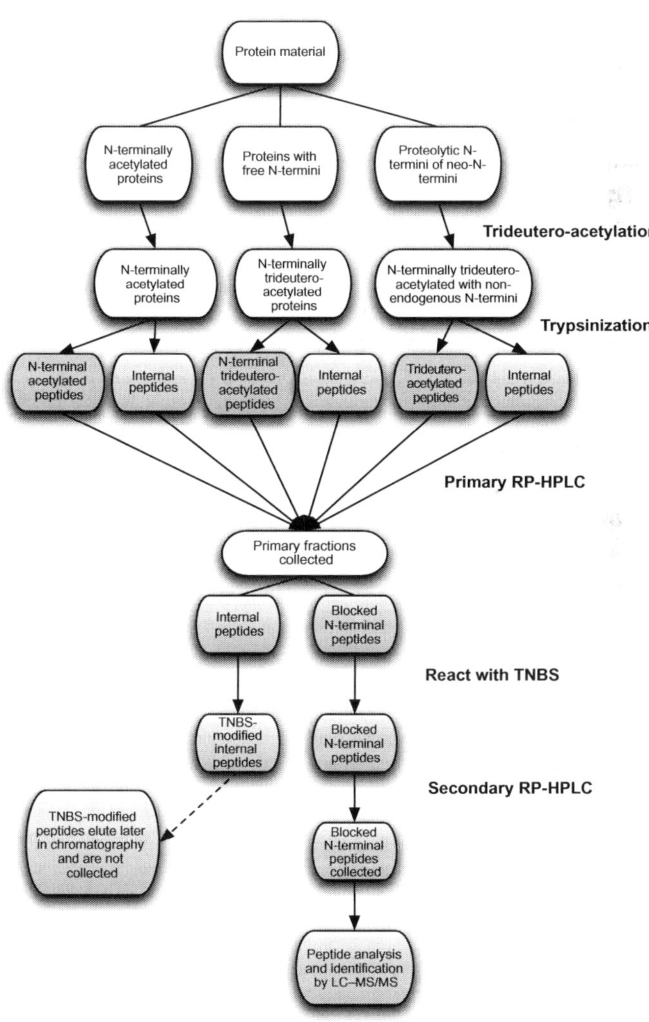

Figure 14.1 General schematic for the COFRADIC method.

2.1. COFRADIC method

For identification of caspase substrates during Fas-induced apoptosis (Van Damme et al., 2005), 25 million cells were lysed for 15 min in 2.5 mL of 0.7% CHAPS in a hypotonic buffer supplemented with protease inhibitors. Lysed cells were centrifuged for 30 min at 13,000 rpm at 4 °C and insoluble pellets were discarded. The resulting supernatants were desalted and proteins were reduced for 60 min at 30 °C in 1.5 m*M* tris (2-carboxyethyl)phosphine (TCEP) and alkylated for 60 min using 3 m*M* iodoacetamide. The protein material was again desalted and free amines were acetylated in 4.5 m*M* sulfo-*N*-hydroxysuccinimide acetate for 90 min at 30 °C. Partial acetylation of serine and threonine was reverted by adding 2 μL hydroxylamine and a final desalting step was performed in 3.5 mL of protein digestion buffer (50 m*M* ammonium bicarbonate at pH 7.6). Samples were digested with trypsin overnight at 37 °C, dried, and redissolved in 25 μL of 0.1 *M* KH_2PO_4 (pH 4.5), redried, and apoptotic peptides were reconstituted in 100 μL [^{18}O]-rich water (93.7% [$H_2^{18}O$] (w/w) pure) or 100 μL regular water for control peptides. Samples were left overnight at 37 °C to allow for incorporation of [^{18}O]-atoms and trypsin was inactivated by reductive alkylation in denaturing conditions.

One mg of total peptide material of apoptotic and control samples were mixed in a 1:1 ratio and separated on a RP-HPLC column over a 100 min gradient. Peptides eluting between 24 and 80 min were collected into 14 fractions of 4 min each. Samples were dried and redissolved in 50 μL of 50 m*M* sodium borate (pH 9.5). TNBS (150 nmol) was added to each sample and incubated at 37 °C for 60 min. This reaction was repeated four times to assure complete modification of internal peptides. Samples were acidified and subjected individually to a secondary RP-HPLC run using the same solvent gradient as the primary run. Eight secondary fractions were collected from the original 4 min interval (30 s increments), yielding a total of 112 subfractions for LC-MS/MS analysis. Since internal peptides are more hydrophobic after TNBS treatment, they no longer elute in the original 4-min interval and are therefore eliminated from the fractions collected. All 112 fractions are dried down, reconstituted in 0.05% formic acid (FA) in 98/2 (v/v) water/acetonitrile (ACN) and analyzed by LC-MS/MS.

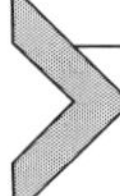

3. PROTEIN TOPOGRAPHY AND MIGRATION ANALYSIS PLATFORM

N-terminomic techniques for capturing N-terminal peptides allow for the identification of proteolytic substrates, however, they rely on the detection of a single N-terminal peptide. This may be problematic if that peptide possesses chemical-physical properties (including limitations on size or ionization compatibility) that hinder detection using a typical mass spectrometric analysis. Furthermore these techniques provide limited topographical data or information regarding the magnitude of proteolytic cleavage that is required to interpret the functional consequence of the proteolytic event. To address these limitations, Dix, Simon, and Cravatt (2008) developed the protein topography and migration analysis platform (PROTOMAP) to profile substrates of proteolysis.

The PROTOMAP technique is based on the differential molecular weight migration of intact proteins versus the molecular weight of cleaved proteins generated due to proteolysis after separation by SDS-PAGE. After separation, bands from the gel are excised and analyzed by LC-MS/MS. The integration between protein gel migration, sequence coverage, and spectral count information obtained from LC-MS/MS analyses reveals global changes in the protein's abundance in a complex biological sample. A summary of the PROTOMAP methodology is shown in Fig. 14.2. The protein migration rate and proteomic data are integrated to generate the "peptograph" where the vertical dimension shows the protein migration and the horizontal dimension correlates with protein sequence coverage from N- to C-terminus. Parental proteins (red) in the control migrate at higher MW with full sequence coverage typically from N- to C-terminus, whereas substrate proteins (blue) in the treated sample show a shift in gel migration from higher to lower MW and sequence coverage is truncated. The extent of proteolytic cleavage is semiquantitatively determined based on spectral counts.

The caspase-mediated proteolytic system was employed to evaluate the sensitivity and utility of PROTOMAP since caspase-induced proteolysis events are well documented (Brockstedt et al., 1998; Gerner et al., 2000;

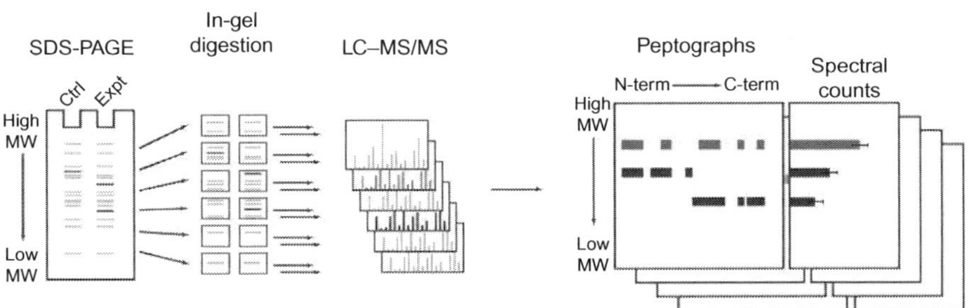

Figure 14.2 A schematic of the PROTOMAP protocol (Dix et al., 2008).

Schmidt et al., 2007; Thiede, Kretschmer, & Rudel, 2006; Van Damme et al., 2005). Using this approach, Dix et al. identified 261 cleavage substrates induced via intrinsic apoptosis among which 91 of them were previously identified caspase substrates and 170 were novel. These results suggested that the PROTOMAP approach could be more sensitive than the N-terminal labeling technique where Van Damme et al reported 50 proteins that underwent caspase-mediated cleavage in apoptotic cells (Gevaert et al., 2005). This maybe however due to the evolution of more sensitive mass spectrometers and also the more synchronous and uniform induction of apoptosis by staurosporine (STS) rather than anti-Fas. The peptograph in Fig. 14.3 shows an example of PARP1 (a known caspase-3 substrate) cleavage in STS-treated cells from the particulate fraction. Dix et al. also employed PROTOMAP to study cross talk between caspase activity and phosphorylation in apoptotic cells (Dix et al., 2012). Here, over 500 apoptotic-specific phosphorylation events were identified whose sites were enriched on caspase-cleaved proteins and clustered around sites of caspase proteolysis. This study revealed a potential new mechanism for caspase-directed proteolysis when it was demonstrated that caspase cleavages itself can potentially expose new sites for phosphorylation, or, conversely, phosphorylation when situated at the +3 position of cleavage sites can directly act to promote substrate proteolysis by caspase 8.

PROTOMAP is a robust technique that is not limited to the analysis of caspase-mediated proteolysis but can be applied to any proteolytic event. Although PROTOMAP was demonstrated to be more sensitive than some

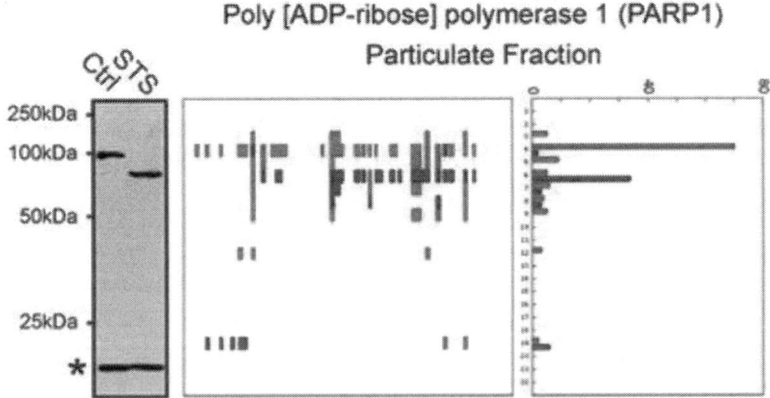

Figure 14.3 An example of a substrate of proteolysis identified using the PROTOMAP protocol. Here, PARP1 is identified in the particular fraction as a cleaved protein product (Dix et al., 2008).

of the existing substrate profiling methods, this technique has its own limitations. While PROTOMAP reveals the proteolytic substrates and the magnitude of cleavage this technique is not designed to decipher the precise site of cleavage (although in many cases this can be inferred).

3.1. PROTOMAP method

3.1.1 Cell culture and apoptosis induction

Jurkat cells were grown to a density of 1×10^6 cells/mL; the cells were treated with 1 μ*M* of STS at 37 °C for 4 h prior to lysis to trigger the intrinsic apoptosis signaling.

3.1.2 SDS-PAGE, tryptic digestion, and mass spectrometry analysis

Cytosolic proteins (100 μg) were separated on a 10% SDS-PAGE. Gel bands were excised from top to bottom. The gel pieces were washed with 100 m*M* ammonium bicarbonate followed by reduction in 10 m*M* TCEP in water at 65 °C for 0.5 h and alkylation in 55 m*M* iodoacetamide in water in the dark for 0.5 h. The gel pieces were dehydrated in 50:50 ACN: 100 m*M* ammonium bicarbonate and dried to completeness. Gel bands were then rehydrated in trypsin solution at a concentration of 10 ng/μL. Upon reswelling of the gel pieces 25 m*M* of ammonium bicarbonate was added to a final volume of 200 μL and digestion was performed at 37 °C overnight. The supernatant was transferred to another tube and the peptides were further extracted using 5% of FA in ACN. The combined extracts were dried and reconstituted in buffer A (0.1% FA/5% ACN/H_2O). The digest mixture was loaded onto a 100 μm (inner diameter) fused silica capillary column containing 10 cm of C_{18} resin. Peptides were separated using a 2-h gradient from 5% to 100% B (0.1% FA/80% ACN) with a flow-rate of 0.25 μL/min and the eluent introduced directly into an LTQ ion trap mass spectrometer (ThermoFisher, San Jose, CA). The LTQ was operated in data-dependent mode where one full MS scan (400–1800 m/z) was followed by seven MS/MS scans of the most intense ions. These experiments were performed in at least three biological replicates.

3.1.3 Data analysis

MS/MS data was searched via Sequest using a concatenated target/decoy human IPI database. Static and variable modifications included carbamidomethyl cysteine (+57 Da) and oxidized methionine (+16 Da), respectively. SEQUEST data from each band was filtered and sorted with DTASelect with the following parameters: peptides were required to be tryptic on at least one terminus and the C- terminal residue was allowed to be lysine, arginine, or aspartate. The minimum required deltaCN was 0.8 and peptides in the +1, +2, and +3 charge-states were required to have minimum XCorr values of 1.8, 2.5, and 3.5, respectively. Filtered proteomic data was organized and assembled into peptographs using three custom Perl scripts which are available at http://www.scripps.edu/chemphys/cravatt/protomap.

4. GLOBAL ANALYZER OF SILAC-DERIVED SUBSTRATES OF PROTEOLYSIS

Based upon the PROTOMAP approach, global analyzer of SILAC-derived substrates of proteolysis (GASSP) utilizes gel migration of intact proteins versus cleaved proteins to profile substrates of proteolysis. GASSP analysis incorporates quantitative metabolic labeling in combination with molecular weight migration information in order to profile the relative abundance of the intact protein compared to its proteolytic products. The metabolic labeling method employed utilizes SILAC whereby the light sample is grown in regular tissue culture medium and the heavy sample is grown in media where the amino acids lysine and arginine have been replaced with the C^{13} N^{15} counterparts. The GASSP technique in combination with bioinformatics and statistical analyses allows for the discovery of new proteolytic substrates (Pham et al., 2012).

A typical GASSP workflow is shown in Fig. 14.4. The untreated (SILAC heavy) and the proteolytic stimulated (SILAC light) samples are combined in a 1:1 ratio and are then separated by SDS-PAGE. Reverse labeling can also be performed to further validate results. The gel is excised from top to bottom into 20 gel regions and in-gel tryptic digestion is performed followed by mass spectrometric analysis. Each identified protein is quantified within each gel region in order to reveal the relative abundance between the untreated and treated samples. Proteolytic processing of a substrate will result in the cleaved product migrating at a lower molecular weight and therefore should be present in gel regions lower than that of the unprocessed, full-length protein. Figure 14.4 demonstrates how upregulated, downregulated, nonresponders, and substrates of proteolysis can be identified using the GASSP protocol. The latter part of Fig. 14.4 shows the bioinformatics analytical workflow required to identify caspase substrates in a semiautomated manner.

In order to evaluate the GASSP platform for global substrate profiling Pham et al. employed etoposide-treated Jurkat cells as the intrinsic apoptotic

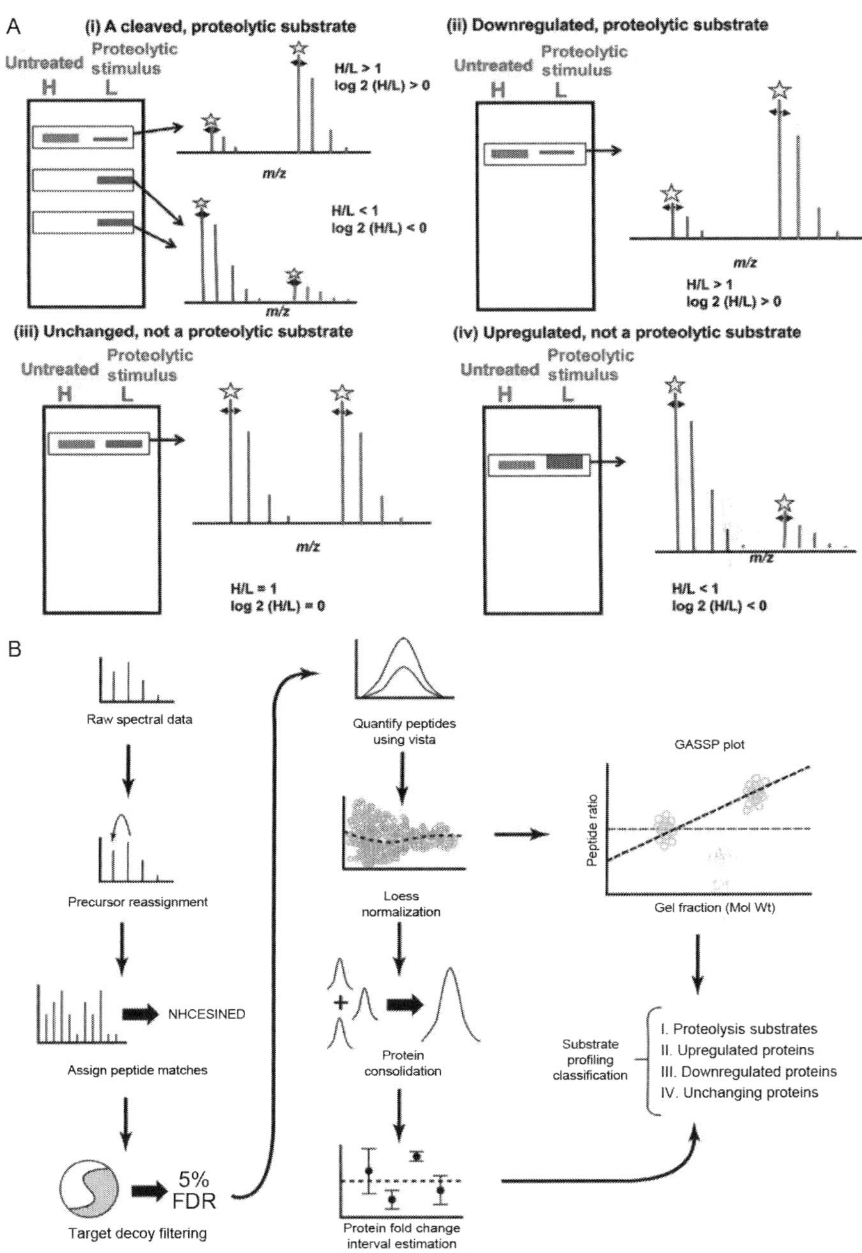

Figure 14.4 Workflow design for GASSP and associated bioinformatics/biostatistical processing. (A) Cells are grown in either heavy (Arg 6C134N15 and Lys 6C132N15) stable isotope labeled or light unlabeled media. Here, heavy cells remain untreated and light cells are treated with a proteolytic stimulus. Samples are mixed together in a 1:1 protein

(Continued)

model. Etoposide is a cytotoxic agent that induces DNA damage by forming complexes between DNA and topoisomerase eventually leading to apoptotic cell death (Tafani, Karpinich, Serroni, Russo, & Farber, 2006). In this study, GASSP was used to identify 360 etoposide-induced caspase substrates, of which, 200 were novel. This study also used GASSP to differentiate BAX-independent from BAX-dependent proapoptotic receptor agonist (PARA) induced apoptotic events (Anderson et al., 2004; Pham et al., 2012). Here, the isogenic HCT116 cell system was employed whereby the proapoptotic Bcl-2 associated X protein (BAX) −/+ heterozygous (which displays a WT phenotype) and BAX −/− variants were compared. In healthy mammalian cells, the majority of BAX is localized in the cytosol. However, upon initiation of apoptosis, BAX undergoes a conformational

Figure 14.4—Cont'd ratio and separated by SDS-PAGE. Bands were excised and digested with trypsin followed by tandem mass spectrometric analysis. (i) For proteins that have undergone proteolysis, a negative log2 ratio is observed at the expected molecular weight suggesting that degradation or proteolysis has occurred. In addition, a positive log2 ratio is observed at a lower molecular weight indicating the presence of a cleavage product. (ii) demonstrates a downregulated protein whereby the protein is identified at its expected molecular weight but shows a positive log2 ratio due to less sample being present in the treated sample. Note this can be due to proteolysis but without identification of the other proteolytic cleavage products this cannot be fully determined. (iii) An example of an unchanged protein whereby the protein was identified at its expected molecular weight with a log2 ratio of 0. (iv) An example of an upregulated protein whereby the protein is quantitated to have a negative log2 ratio at its expected molecular weight. (B) Raw spectral data was loaded into an external MySQL database and precursor ion masses were reassigned to their mono-isotopic mass where necessary. Peptide-spectral matches were then assigned using the Mascot algorithm and the resulting peptide data was filtered to an overall FDR of 5%. SILAC peaks were then quantified using the VistaQuant algorithm as described in the text and the normalized log ratios for each nonredundant peak were summed for both labeled and unlabeled species to generate the overall relative abundance ratio at the protein level. Peptide log-ratios for each protein were averaged by BAX genotype via a linear mixed effects model to estimate an overall log fold change and to capture the variation due to biological replicates and multiple peptides within a replicate. Standard errors were estimated as described in the methods and then used to construct confidence intervals. These confidence intervals were a basis to select proteins that appeared to be upregulated, downregulated, or unchanging. The GASSP plot for each protein was then generated evaluating each protein as a possible protease substrate by plotting on the vertical axis its observed peptide log-ratio values against the gel positions in which those peptides where detected. A positive trend in log-ratios from higher MW regions to lower MW regions was expected for substrates. Patterns were summarized statistically for each protein via a robust linear model as described in the methods.

shift and becomes incorporated into the outer mitochondrial membrane, releasing mitochondrial components leading to fully triggered apoptosis. GASSP analysis revealed 132 proteins as BAX-independent substrates of PARA-induced proteolysis, 61 of which were previously reported in the CASBAH database. (Pham et al., 2012).

GASSP has proven to be a robust technique for global proteolytic substrate profiling; it is also complementary to N-terminomics or C-terminomics (discussed below) which may not be suitable for specific proteolytic substrates. For example, as previously mentioned, PARP1 is a known substrate of caspase 3 that has a consensus sequence (in italics), *DEVD*GVDEVAKKKSK. Caspase cleavage of PARP1 results in the generation of the neo N-terminus GVDEVAKKKSK that would likely be missed by mass spectrometric analysis using the trypsin-based N-terminal capture methods because the resulting peptide, GVDEVAK, is quite small (using a typical m/z range of 375–1800 Da). PARP1 however, was readily observed as a substrate using GASSP analysis because it does not depend on the detection of one specific peptide. GASSP analysis is not limited solely for the identification of proteolytic substrates, it can also be applied to monitor proteins being up- or downregulated upon a stimulation of interest (see Fig. 14.4). Another advantage of GASSP is that it requires a minimal amount of starting material (50–100 μg) making this technique very attractive in a scenario where sample amount is limiting. Other groups have since adopted similar approaches for profiling proteolysis substrates, for example, Mann et al. employed SILAC in combination with 1D SDS-PAGE shifts to study apoptotic proteolysis(Stoehr, Schaab, Graumann, & Mann, 2013). This again shows the validity of this type of approach for profiling caspase and other proteolytic enzyme substrates.

4.1. GASSP method

4.1.1 Tissue culture

Jurkat cells were cultured and expanded in SILAC RPMI 1640 supplemented with L-glutamine, L-proline and 10% dialyzed fetal bovine serum. The heavy isotope medium was supplemented with heavy isotope L-lysine at 50 μg/mL and heavy isotope L-arginine at 40 μg/mL. The light isotopic media was supplemented with both of these amino acids containing natural isotopes. Once cells reached a density of $\sim 1 \times 10^6$/mL, they were treated with etoposide at a final concentration of 50 μM at 37 °C for 14 h to trigger intrinsic apoptosis.

4.1.2 SDS-PAGE, tryptic digestion, and mass spectrometry analysis

The light (untreated) and heavy (etoposide treated) lysates (or vice versa) were mixed at a 1:1 ratio and proteins were separated on a 4–12% Bis-Tris gel using 1 × MOPS running buffer at 125 V for 40 min. Protein bands were visualized using Simply Blue staining solution (Invitrogen, Carlsbad). Bands were excised from top to bottom of the gel into 20 gel regions. In-gel tryptic digestions were performed as described in Section 4.1.2. Extracts were combined and dried down completely in the SpeedVac.

Peptides were reconstituted in solvent A, 2% ACN/0.1% FA/water, and were injected via an auto-sampler for separation by reverse phase chromatography on a NanoAcquity UPLC system (Waters, Dublin, CA). Peptides were loaded onto a Symmetry® C_{18} column (1.7 μm BEH-130, 0.1 × 100 mm, Waters, Dublin, CA) with a flow rate of 1 μL/min and a gradient of 2–25% Solvent B (B is 0.1% FA/2% water/ACN) applied over 60 min with a total analysis time of 90 min. Peptides were eluted directly into an Advance Captive Spray ionization source (Michrom BioResources/Bruker, Auburn, CA) with a spray voltage of 1.4 kV and were analyzed using an LTQ Orbitrap mass spectrometer (ThermoFisher, San Jose, CA). Precursor ions were analyzed in the FTMS at 60,000 resolution. MS/MS was performed in the LTQ with the instrument operated in data-dependent mode whereby the top eight most abundant ions were subjected for fragmentation.

4.1.3 Bioinformatics

MS/MS data was searched using the Mascot search algorithm (Perkins, Pappin, Creasy, & Cottrell, 1999) (Matrix Sciences, London, UK) against a concatenated forward-reverse target-decoy database (Elias & Gygi, 2010) consisting of all known *Homo sapiens* proteins and common contaminant sequences from UniProt (TrEMBL/SwissProt) release 2010_12. Spectra were assigned using a precursor mass tolerance of 50 ppm and fragment ion tolerance of 0.8 Da. Variable modifications included oxidation of methionine residues (+15.994 Da), lysine (+8.014a Da), and arginine (+10.0082 Da). The search was performed using trypsin specificity with up to two miscleavages C-terminal to R, K, and D. Peptide assignments were first filtered to a 5% false discovery rate (FDR) at the peptide level using a linear discriminant analysis. Filtered data was quantified at the peptide level using the VistaQuant algorithm as previously described (Elias & Gygi, 2010). Relative abundance ratios at the protein level were calculated from summed abundance measurements of peptides with VistaQuant Confidence Scores ≥ 90 (Bakalarski et al., 2008).

4.1.4 Informatic processing

A full description of the downstream informatics analysis on data normalization or how substrate candidates were identified can be found in the supplemental method section in the link below:

http://pubs.acs.org/doi/suppl/10.1021/pr300035k/suppl_file/pr300035k_si_016.pdf

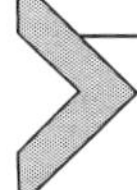

5. IMMUNO-AFFINITY ENRICHMENT OF C-TERMINAL ASPARTIC ACID CONTAINING PEPTIDES FOR THE IDENTIFICATION OF CASPASE SUBSTRATES

Immuno-affinity purification combined with mass spectrometry has become an important technique regularly employed in proteomics. Recent developments in proteomics have extended the use of antibodies beyond enrichment of specific peptides or proteins (Anderson et al., 2004; Moritz et al., 2010) to the enrichment of motif-specific posttranslationally modified substrates (Rush et al., 2005). By employing motif-based antibodies, intricate and stoichiometrically low-level posttranslational signaling events have been revealed providing insight into unique biological mechanisms (Pham et al., 2012; Moritz et al., 2010). In addition to using this technique to study phosphorylation, acetylation, or ubiquitination-mediated events, this protocol has also been applied for the isolation of peptides containing a C-terminal aspartic acid residue (Pham et al., 2012) allowing enrichment of potential caspase-specific proteolytic products. Two caspase motif antibodies are currently available, anticaspase-6/8 and anticaspase-2/3/7, motif respectively. Peptides captured with the C-terminal aspartic antibodies not only provide potential caspase substrate identification but also provide direct evidence about the site(s) of proteolytic cleavage (i.e., a peptide with a C-terminal aspartate in the P1 position).

C-terminal immuno-affinity capture approach is highly complementary to other techniques like N-terminomics or differential gel-based approaches (Pham et al., 2012). When studying proteolytic events in Jurkat cells, Pham et al. demonstrated that postinduction of intrinsic apoptosis via stimulation with etoposide, 360 substrates were identified via GASSP compared to 157 via C-terminal aspartic acid enrichment. Importantly, there was minimal overlap between substrates identified in both techniques. Some peptides derived after C-terminal immuno-affinity capture of tryptically cleaved caspase substrates are too short or too long for optimal mass spectrometric detection (as discussed previously for COFRADIC). However, in many cases these same substrates yielded an ideal GASSP identification due to distinct gel migration between the intact and caspase-processed form of the protein. Conversely, many substrates generating caspase cleavage sites close to the N- or C-terminus do not show distinct gel migration between the full-length and cleaved protein form and therefore may not be ideally suited for identification via GASSP (or PROTOMAP) methods. However, many of the substrates that were unidentifiable by GASSP could indeed be characterized using the C-terminal aspartic acid peptide immuno-affinity enrichment. For quantitative comparative analysis, samples can be isotopically, chemically labeled using approaches such as reductive methylation (Pham et al., 2012; Ross et al., 2004), isobaric tags for relative and absolute quantitation (iTRAQ), and TMT. With reductive methylation, quantitation can be performed at the precursor ion (MS1) level, comparing area under the curve of peaks labeled with different heavy, isotopically labeled dimethylation reagents. With iTRAQ and TMT, quantitative information can be obtained from the reporter ion signal acquired in the tandem mass spectral data (MS2) or second order MS2, MS3 spectra. The advantage with iTRAQ and TMT labeling is that they provide the ability to multiplex samples to compare up to 8 or 10 sample conditions in one analysis. Quantitation using the *in vivo* stable isotopic labeling of amino acids in cell culture (SILAC) approach is also compatible with the C-terminal aspartic acid immuno-affinity method with the caveat that tryptic peptides in combination with caspase-mediated proteolysis may result in the generation of non-arginine or lysine containing peptides and hence a potential loss of the quantitative signature. To overcome this technical problem, other enzymes can be employed such as endoproteinase Lys-N that would allow for retention of the isotopic moiety. However, this can also limit sensitivity by generating peptides that are too large, and therefore, chemical labeling is advised for most quantitative approaches.

Similar to many of the techniques developed for proteolytic substrate profiling, the C-terminal aspartic acid enrichment approach has some limitations. Since the enrichment is for peptides containing a C-terminal aspartic acid, the identification of potential caspase substrates is based on only one peptide, unless there are multiple caspase cleavage sites on the substrate. In addition, the antibodies maybe biased toward recognizing motifs from specific caspases. Substrates cleaved by caspases within motifs that do not fit the consensus of these antibodies may go undetected. As alluded to earlier, most proteomic analyses employing a bottom-up approach utilize enzymatic digestion of the sample to generate mass spectrometric compatible peptides. A further caveat to this technique is the generation of artifactual peptides resulting from biochemical or physiochemical cleavage of labile Asp-Pro signatures. Asp-Pro bonds can easily hydrolyze under acidic conditions, therefore when searching the data, any peptides containing an Asp residue at its C-terminus followed by a flanking Proline residue should not be completely discounted, as they could be genuine, however should be considered as possible nonbiological artifacts. Unlike other techniques such as GASSP that are unbiased toward proteolytic enzyme specificity, C-terminal aspartic acid immune-affinity enrichment is limited to the detection of specific class of proteases, those including but not limited to caspases, caspase-like enzymes, or granzyme B which also cleave at aspartic acid residues.

5.1. Immuno-affinity capture of caspase substrates method

For identification of caspase substrates, cells were grown to a density of 1×10^6 cells/mL, harvested and lysed in 20 m*M* HEPES (pH 8.0), 9 *M* urea, 1 m*M* sodium orthovanadate, 2.5 m*M* sodium pyrophosphate, and 1 m*M* β-glycerophosphate. Lysates were sonicated with 2×30 s bursts at 30 watts with a microtip sonicator followed by centrifugation at $16{,}000 \times g$ for 15 min at 15 °C. Bradford protein concentration reading was performed on the cleared supernatant and 12 mg of each sample was reduced with 4.5 m*M* dithioreitol for 20 min at 60 °C and alkylated with 10 m*M* iodoacetamide for 15 min at room temperature in the dark. Samples were diluted to a final concentration of 20 m*M* HEPES (pH 8.0) and 2 *M* urea followed digestion with 400 μL of 1 mg/mL Trypsin-TPCK overnight at room temperature. Resultant tryptic peptides were desalted with C_{18} SepPak cartridges and chemically labeled using reductive methylation during the solid phase extraction process. Samples were eluted with 40% ACN/ 0.1% TFA and if performing relative quantitation between two chemically labeled samples, combined into one tube for lyophilization for 36 h. After lyophilization, samples were reconstituted with 1.4 mL IAP buffer (cell signaling technology) and incubated with C-terminal aspartic acid motif-specific antibodies for 2 h at 4 °C. Peptides were eluted off antibody resin with 55 μL of 0.15% TFA after 10 min incubation at room temperature and subsequently eluted again with 45 μL of 0.15% TFA. The immunoprecipitation eluate was subjected to C_{18} STAGE tip desalting and downstream

LC-MS/MS analysis. For mass spectrometric analysis, desalted peptides were reconstituted in 2% ACN/0.1% FA/water and loaded onto a Symmetry® C_{18} column (1.7 μm BEH-130, 0.1 × 100 mm, Waters, Dublin, CA) using a NanoAcquity UPLC system (Waters, Dublin, CA). Peptides were loaded onto the column with a flow rate of 1 μL/min and a gradient of 2–25% Solvent B (B is 0.1% FA/2% water/ACN) applied over 85 min with a total analysis time of 120 min. Peptides were eluted directly into an Advance CaptiveSpray ionization source (Michrom BioResources/Bruker, Auburn, CA) with a spray voltage of 1.4 kV and were analyzed using an LTQ Orbitrap Velos mass spectrometer (ThermoFisher, San Jose, CA). Precursor ions were analyzed in the Orbitrap at 60,000 FWHM resolution. MS/MS was performed in the LTQ with the instrument operated in data-dependent mode whereby the top 15 most abundant ions were subjected for collision-induced dissociation fragmentation. From the LC-MS/MS analysis peptides containing a C-terminal aspartic acid can be identified and quantified (Fig. 14.5).

6. SUMMARY

The methodologies described in this chapter offer a set of highly complementary tools to the various N-terminomic approaches for cataloguing substrates of proteolysis. COFRADIC, PROTOMAP, and GASSP each offer a nonenzyme restricted methodology for the analysis of proteolysis substrates. The C-terminal immuno-affinity capture is specifically aimed at capturing peptides ending in a C-terminal aspartic acid residue (predominantly resulting from caspase cleavage). Each technique has its various advantages and limitations and therefore the researcher should distinguish which methodology, or set of methodologies is best suited to their experimental design and hypothesis under investigation. If sample is minimal, affinity capture techniques maybe suboptimal as these typically require >10 mg material for the purification step. In such cases, PROTOMAP or GASSP may offer a solution for the identification of proteolytic substrates as <50 μg material is required.

The majority of the protocols described in this chapter are designed for studying a binary sample set, however in many scenarios four or more conditions need to be examined in parallel in order to draw a biological conclusion. Here the C-terminal immuno-affinity capture technique is amenable to multiplexed quantitation tools such as TMT or iTRAQ labeling, and if sample quantity is a not limiting factor, this may be an effective strategy.

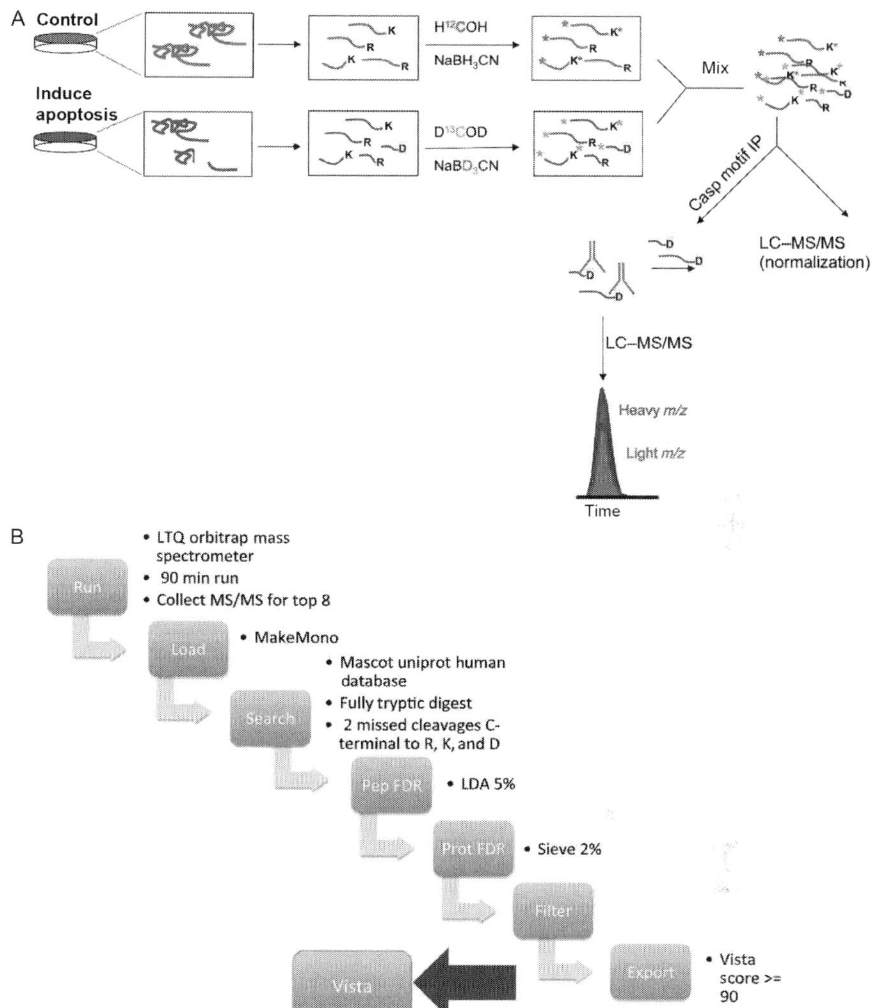

Figure 14.5 The workflow for performing caspase substrate profiling using immunoaffinity capture. (A) The sample preparation and mass spectrometric analysis and (B) the bioinformatics workflow (Pham et al., 2012).

Additionally, if SILAC labeling is not amenable for the cell type of interest (e.g., human tissue or primary cell lines which often prove difficult to gain full isotopic metabolic labeling) then GASSP will not be ideally suited as the tool for substrate identification, and instead, alternative N-terminomic or C-terminomic approaches should be adopted.

GASSP and PROTOMAP are also limited by the relative abundance of the proteolytic products, and without further fractionation or increased mass spectrometric analysis time, these methods are also limited to the identification of the top 6000 or so most abundant proteins in the proteome.

All in all, the plethora of complementary techniques now available to researchers interested in defining proteolytic events provides multiple strategies to help define the substrates of interest. This will allow progress in the identification of various disease states dictated by regulation or disregulation of proteolysis and will also allow researchers to identify diagnostic and prognostic biomarkers for research and therapeutic purposes.

REFERENCES

Anderson, N. L., Anderson, N. G., Haines, L. R., Hardie, D. B., Olafson, R. W., & Pearson, T. W. (2004). Mass spectrometric quantitation of peptides and proteins using stable isotope standards and capture by anti-peptide antibodies (SISCAPA). *Journal of Proteome Research, 3*(2), 235–244.

Bakalarski, C. E., Elias, J. E., Villén, J., Haas, W., Gerber, S. A., Everley, P. A., et al. (2008). The impact of peptide abundance and dynamic range on stable-isotope-based quantitative proteomic analyses. *Journal of Proteome Research*, 7(11), 4756–4765.

Brockstedt, E., Rickers, A., Kostka, S., Laubersheimer, A., Dörken, B., Wittmann-Liebold, B., et al. (1998). Identification of apoptosis-associated proteins in a human Burkitt lymphoma cell line. Cleavage of heterogeneous nuclear ribonucleoprotein A1 by caspase 3. *Journal of Biological Chemistry, 273*(43), 28057–28064. Erratum in: *Journal of Biological Chemistry,* 1998, *273*(50), 33884.

Brown, J. R., & Hartley, B. S. (1966). Location of disulphide bridges by diagonal paper electrophoresis. The disulphide bridges of bovine chymotrypsinogen A. *The Biochemical Journal, 101*(1), 214–228.

Ciechanover, A. (2012). Intracellular protein degradation: From a vague idea thru the lysosome and the ubiquitin-proteasome system and onto human diseases and drug targeting. *Biochimica et Biophysica Acta, 1824*(1), 3–13.

Dix, M. M., Simon, G. M., & Cravatt, B. F. (2008). Global mapping of the topography and magnitude of proteolytic events in apoptosis. *Cell, 134*(4), 679–691.

Dix, M. M., Simon, G. M., Wang, C., Okerberg, E., Patricelli, M. P., & Cravatt, B. F. (2012). Functional interplay between caspase cleavage and phosphorylation sculpts the apoptotic proteome. *Cell, 150*(2), 426–440.

Doucet, A., Butler, G. S., Rodriguez, D., Prudova, A., & Overall, C. M. (2008). Metadegradomics: Toward in vivo quantitative degradomics of proteolytic post-translational modifications of the cancer proteome. *Molecular & Cellular Proteomics*, 7(10), 1925–1951.

Elias, J. E., & Gygi, S. P. (2010). Target-decoy search strategy for mass spectrometry-based proteomics. *Methods in Molecular Biology, 604*, 55–71.

Gerner, C., Frohwein, U., Gotzmann, J., Bayer, E., Gelbmann, D., Bursch, W., et al. (2000). The Fas-induced apoptosis analyzed by high throughput proteome analysis. *Journal of Biological Chemistry, 275*(50), 39018–39026.

Gevaert, K., Goethals, M., Martens, L., Van Damme, J., Staes, A., Thomas, G. R., et al. (2003). Exploring proteomes and analyzing protein processing by mass spectrometric identification of sorted N-terminal peptides. *Nature Biotechnology, 21*(5), 566–569.

Gevaert, K., Impens, F., Van Damme, P., Ghesquière, B., Hanoulle, X., & Vandekerckhove, J. (2007). Applications of diagonal chromatography for proteome-wide characterization of protein modifications and activity-based analyses. *The FEBS Journal, 274*(24), 6277–6289.

Gevaert, K., Staes, A., Van Damme, J., De Groot, S., Hugelier, K., Demol, H., et al. (2005). Global phosphoproteome analysis on human HepG2 hepatocytes using reversed-phase diagonal LC. *Proteomics*, *5*(14), 3589–3599.

Gevaert, K., Van Damme, J., Goethals, M., Thomas, G. R., Hoorelbeke, B., Demol, H., et al. (2002). Chromatographic isolation of methionine-containing peptides for gel-free proteome analysis: Identification of more than 800 Escherichia coli proteins. *Molecular & Cellular Proteomics*, *1*(11), 896–903.

Ghesquiere, B., Goethals, M., Van Damme, J., Staes, A., Timmerman, E., Vandekerckhove, J., et al. (2006). Improved tandem mass spectrometric characterization of 3-nitrotyrosine sites in peptides. *Rapid Communications in Mass Spectrometry*, *20*(19), 2885–2893.

Hanoulle, X., Van Damme, J., Staes, A., Martens, L., Goethals, M., Vandekerckhove, J., et al. (2006). A new functional, chemical proteomics technology to identify purine nucleotide binding sites in complex proteomes. *Journal of Proteome Research*, *5*(12), 3438–3445.

Kojro, E., & Postina, R. (2009). Regulated proteolysis of RAGE and AbetaPP as possible link between type 2 diabetes mellitus and Alzheimer's disease. *Journal of Alzheimer's Disease*, *16*(4), 865–878.

Lamkanfi, M., Kanneganti, T. D., Van Damme, P., Vanden Berghe, T., Vanoverberghe, I., Vandekerckhove, J., et al. (2008). Targeted peptidecentric proteomics reveals caspase-7 as a substrate of the caspase-1 inflammasomes. *Molecular & Cellular Proteomics*, 7(12), 2350–2363.

McIlwain, D. R., Berger, T., & Mak, T. W. (2013). Caspase functions in cell death and disease. *Cold Spring Harbor Perspectives in Biology*, *5*(4), a008656.

Moritz, A., Li, Y., Guo, A., Villén, J., Wang, Y., MacNeill, J., et al. (2010). Akt-RSK-S6 kinase signaling networks activated by oncogenic receptor tyrosine kinases. *Science Signaling*, *3*(136), ra64.

Perkins, D. N., Pappin, D. J., Creasy, D. M., & Cottrell, J. S. (1999). Probability-based protein identification by searching sequence databases using mass spectrometry data. *Electrophoresis*, *20*(18), 3551–3567.

Pham, V. C., Pitti, R., Anania, V. G., Bakalarski, C. E., Bustos, D., Jhunjhunwala, S., et al. (2012). Complementary proteomic tools for the dissection of apoptotic proteolysis events. *Journal of Proteome Research*, *11*(5), 2947–2954.

Ross, P. L., Huang, Y. N., Marchese, J. N., Williamson, B., Parker, K., Hattan, S., et al. (2004). Multiplexed protein quantitation in Saccharomyces cerevisiae using amine-reactive isobaric tagging reagents. *Molecular & Cellular Proteomics*, *3*(12), 1154–1169.

Rush, J., Moritz, A., Lee, K. A., Guo, A., Goss, V. L., Spek, E. J., et al. (2005). Immunoaffinity profiling of tyrosine phosphorylation in cancer cells. *Nature Biotechnology*, *23*(1), 94–101.

Staes, A., Impens, F., Van Damme, P., Ruttens, B., Goethals, M., Demol, H., et al. (2011). Selecting protein N-terminal peptides by combined fractional diagonal chromatography. *Nature Protocols*, *6*(8), 1130–1141.

Schmidt, F., Hustoft, H. K., Strozynski, M., Dimmler, C., Rudel, T., & Thiede, B. (2007). Quantitative proteome analysis of cisplatin-induced apoptotic Jurkat T cells by stable isotope labeling with amino acids in cell culture, SDS-PAGE, and LC-MALDI-TOF/TOF MS. *Electrophoresis*, *28*(23), 4359–4368.

Stoehr, G., Schaab, C., Graumann, J., & Mann, M. (2013). A SILAC-based approach identifies substrates of caspase-dependent cleavage upon TRAIL-induced apoptosis. *Molecular & Cellular Proteomics*, *12*(5), 1436–1450.

Tafani, M., Karpinich, N. O., Serroni, A., Russo, M. A., & Farber, J. L. (2006). Re-evaluation of the distinction between type I and type II cells: The necessary role of the mitochondria in both the extrinsic and intrinsic signaling pathways upon Fas receptor activation. *Journal of Cellular Physiology, 208*(3), 556–565.

Thiede, B., Kretschmer, A., & Rudel, T. (2006). Quantitative proteome analysis of CD95 (Fas/Apo-1)-induced apoptosis by stable isotope labeling with amino acids in cell culture, 2-DE and MALDI-MS. *Proteomics, 6*(2), 614–622.

Van Damme, P., Martens, L., Van Damme, J., Hugelier, K., Staes, A., Vandekerckhove, J., et al. (2005). Caspase-specific and nonspecific in vivo protein processing during Fas-induced apoptosis. *Nature Methods, 2*(10), 771–777.

Van Damme, P., Maurer-Stroh, S., Plasman, K., Van Durme, J., Colaert, N., Timmerman, E., et al. (2009). Analysis of protein processing by N-terminal proteomics reveals novel species-specific substrate determinants of granzyme B orthologs. *Molecular & Cellular Proteomics, 8*(2), 258–272.

Vande Walle, L., Van Damme, P., Lamkanfi, M., Saelens, X., Vandekerckhove, J., Gevaert, K., et al. (2007). Proteome-wide identification of HtrA2/Omi substrates. *Journal of Proteome Research, 6*(3), 1006–1015.

Wysocka, M., & Lesner, A. (2013). Future of protease activity assays. *Current Pharmaceutical Design, 19*(6), 1062–1067.

CHAPTER FIFTEEN

Phosphoplipid Scrambling on the Plasma Membrane

Jun Suzuki, Shigekazu Nagata[1]
Department of Medical Chemistry, Graduate School of Medicine, Kyoto University, Kyoto, Japan
[1]Corresponding author: e-mail address: snagata@mfour.med.kyoto-u.ac.jp

Contents

Abstract

Phospholipids are asymmetrically distributed at plasma membrane in normal cells, and scrambled in various biological situations such as blood clotting and apoptotic cell death. Recent studies revealed that phospholipid scrambling is mediated by at least two independent mechanisms. An eight transmembrane-containing protein TMEM16F Ca^{2+}-dependently promotes the phospholipid scrambling, which is required for the PS exposure in activated platelets during blood clotting. On the other hand, a six transmembrane-containing protein Xk-related family protein 8 is activated by caspases during apoptosis and promotes the phospholipid scrambling, thus exposing PS as an "eat-me-signal." In this chapter, we describe procedures for assaying phospholipid scrambling.

1. INTRODUCTION

Programmed cell death or apoptosis is characterized by DNA fragmentation, nuclear condensation, cell shrinkage, blebbing, and phosphatidylserine (PS) exposure (Nagata & Golstein, 1995; Vaux & Korsmeyer, 1999).

Methods in Enzymology, Volume 544
ISSN 0076-6879
http://dx.doi.org/10.1016/B978-0-12-417158-9.00015-7

The anionic phospholipid PS is normally restricted to the inner side of the plasma membrane by an ATP-dependent flippase; P4-type ATPases are candidate flippases that have been demonstrated to have lipid transporting activities in cells (Balasubramanian & Schroit, 2003). During the process of apoptosis, flippase activity has been shown to be inhibited, although flippase inhibition alone is not sufficient for PS externalization (Bevers & Williamson, 2010). A putative protein known as a calcium-dependent scramblase, which transports lipids bi-directionally and nonspecifically, must be activated (Bevers & Williamson, 2010). Following scramblase activation, PS and phosphatidylethanolamine (PE), which are initially located on the cytoplasmic side, are exposed to the cell surface, while phosphatidylcholine (PC) and sphingomyelin (SM), which are initially located on the extracellular side, are internalized. Once PS is exposed to the cell surface, it functions as an "eat-me-signal," allowing apoptotic cells to be recognized and engulfed by phagocytes (Nagata, Hanayama, & Kawane, 2010; Ravichandran & Lorenz, 2007). The phospholipid scramblase also regulates PS exposure in activated platelets, to provide a scaffold for coagulation factors, whose activation is required to control bleeding (Zwaal, Comfurius, & Bevers, 1992). Although phospholipid scrambling resulting in PS exposure is important in many physiological situations, the molecular mechanism mediating this process was completely unknown until recently. Over the past few years, studies have begun to clarify the details involved in membrane lipid scrambling. In particular, we recently showed that apoptotic cells and activated platelets utilize different protein(s) in promoting PS exposure.

An eight transmembrane domain-containing protein, TMEM16F, was identified as an essential component involved in regulating calcium-dependent phospholipid scrambling (Suzuki, Umeda, Sims, & Nagata, 2010). TMEM16F belongs to the 10-member TMEM 16 family. Mutations in *TMEM16F* have been identified in patients with Scott syndrome and are associated with defective calcium activation-induced PS exposure, resulting in impaired thrombin generation and mild bleeding (Castoldi, Collins, Williamson, & Bevers, 2011; Suzuki et al., 2010). Similarly, lymphocytes derived from *TMEM16F* knock-out mice are defective in calcium-induced phospholipid scrambling such as PS externalization and PC internalization (Suzuki, Fujii, et al., 2013; Yang et al., 2012; Fig. 15.1); however, these cells exhibit normal PS exposure following apoptotic stimulation (Suzuki, Fujii, et al., 2013), as also observed in lymphocytes derived from a Scott syndrome patient (Williamson et al., 2001). These results suggest that calcium- and

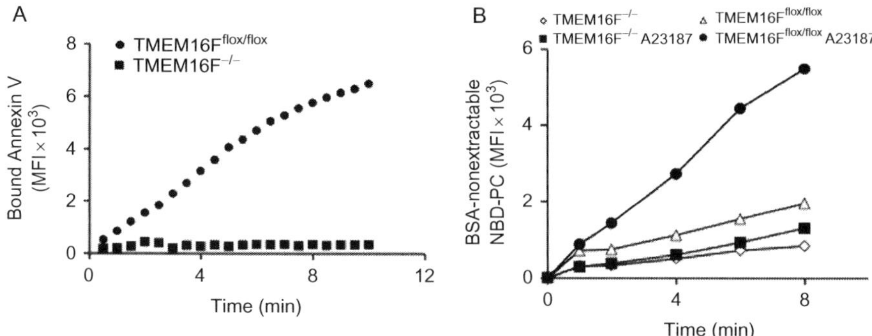

Figure 15.1 Ca^{2+}-induced scrambling of phospholipids promoted by TMEM16F. (A) Ca^{2+} ionophore-induced PS exposure. The wild-type and TMEM16F$^{-/-}$ IFETs were treated at 20 °C with 3.0 μ*M* A23187, and Annexin V binding to the cells was monitored by flow cytometry for 10 min. (B) Ca^{2+} ionophore-induced lipid internalization. The wild-type and TMEM16F$^{-/-}$ IFETs were treated at 15 °C with 250 n*M* A23187 in the presence of NBD-PC. Using aliquots of the reaction mixture, the BSA-nonextractable level of NBD-PC in the Sytox Blue-negative population was determined at the indicated time by flow cytometry and shown in mean fluorescence intensity. *Reprinted from Suzuki, Fujii, et al. (2013).*

apoptosis-induced phospholipid scrambling may be differentially regulated. Indeed, a six transmembrane domain-containing protein, Xk-related protein 8 (Xkr8) has been shown to regulate apoptosis-dependent phospholipid scrambling when it is cleaved by caspases (Suzuki, Denning, Imanishi, Horvitz, & Nagata, 2013). The leukemia cell lines Raji and PLB985 are defective in apoptotic PS exposure (Fadeel et al., 1999; Fadok, de Cathelineau, Daleke, Henson, & Bratton, 2001) due to a limited expression of *Xkr8* resulting from hypermethylation in the promoter region (Suzuki, Denning, et al., 2013). The exogenous expression of Xkr8 rescues phospholipid scrambling such as PS and PE externalization, and PC and SM internalization in these cells, confirming the importance of Xkr8 in this process (Fig. 15.2). CED-8, the only Xkr8 homolog in *C. elegans*, also regulates PS exposure and promotes apoptotic cell engulfment (Suzuki, Denning, et al., 2013), indicating that the CED-8/Xkr8 homologs share an evolutionarily conserved role in regulating phospholipid scrambling. In contrast, lymphocytes from *Xkr8* knock-out mice exhibit normal PS exposure upon calcium ionophore stimulation, indicating that apoptosis- and calcium-induced PS exposure are differentially regulated by Xkr8 and TMEM16F, respectively.

The following sections describe cell-based assays that we have developed for calcium- and apoptosis-induced phospholipid scrambling.

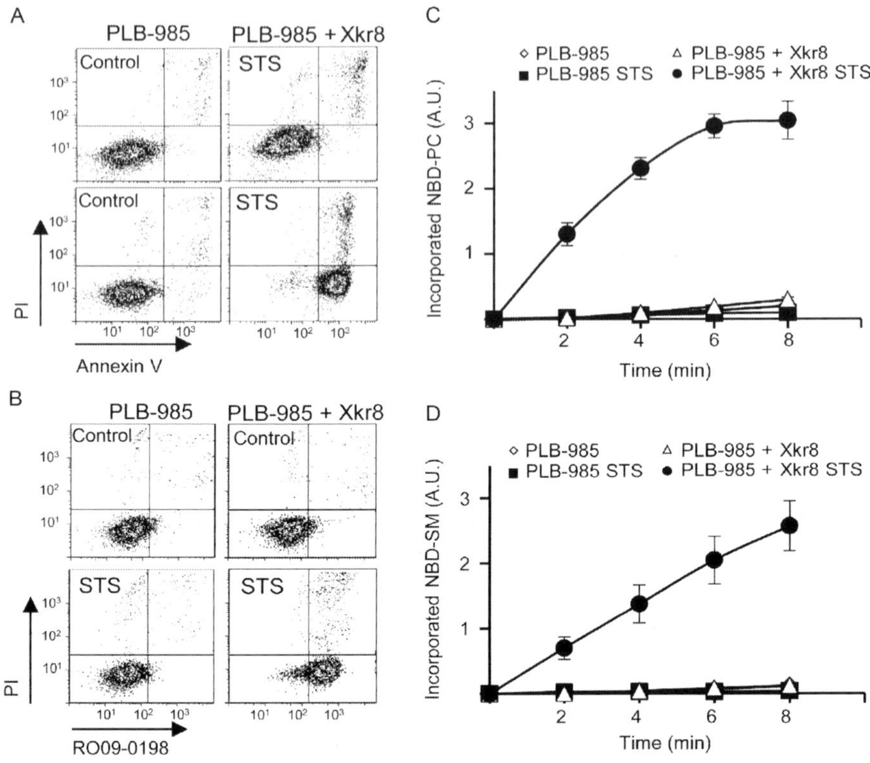

Figure 15.2 Apoptosis-induced scrambling of phospholipids promoted by Xkr8. (A and B) Apoptotic exposure of PS and PE. PLB-985 and PLB-985 transformants expressing Xkr8 were treated with STS, stained with Annexin V-Cy5 (A) or biotin-RO peptide/Streptavidin-APC (B), and PI, and analyzed by flow cytometry. (C and D) Scrambling of PC and SM in apoptotic cells. PLB-985 and PLB-985 transformants expressing Xkr8 were treated with STS and incubated with NBD-PC (C) or NBD-SM (D), and the incorporated lipids were analyzed by flow cytometry. The fluorescence intensity in the Sytox Blue-negative fraction is shown in arbitrary units. *Reprinted from Suzuki, Denning, et al. (2013).*

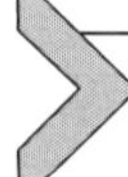

2. ASSAYS FOR CALCIUM-INDUCED PHOSPHOLIPID SCRAMBLING

Early studies showed that the activation of platelets with collagen plus thrombin facilitates PS exposure, leading to the enhanced conversion of prothrombin to thrombin by coagulation factors (Bevers, Comfurius, & Zwaal, 1983). At the same time PS exposure is increased, cell surface SM is decreased, suggesting that lipids are transported or scrambled from one side of the membrane to the other. This observation was confirmed by monitoring nitrobenzoxadiazole (NBD)-conjugated lipids in red blood cells and platelets stimulated with calcium (Smeets, Comfurius, Bevers, & Zwaal, 1994; Williamson et al., 1995). Here, we describe PS and PE exposure assays, and NBD-PC and NBD-SM incorporation assays using activated lymphocytes. Experimental conditions such as cell numbers and reaction volumes can be changed, but the ratio of components is important. Incubation time and temperature should be optimized depending on the cell types.

2.1. Cells, reagents, and buffers

2.1.1 Cells

Any types of cells can be used for the phospholipid-scrambling assays. The cells listed here are routinely used in our laboratory.

1. Ba/F3 cells (Palacios & Steinmetz, 1985) are maintained in RPMI1640 containing 10% fetal calf serum (FCS), 45 units/ml mouse Interleukin (IL)-3, and 50 μM β-mercaptoethanol.
2. PLB985 cells (Tucker, Lilly, Heck, & Rado, 1987) are maintained in RPMI1640 containing 10% FCS and 50 μM β-mercaptoethanol.
3. Immortalized fetal thymocytes (IFETs) (Imao & Nagata, 2013; Suzuki, Fujii, et al., 2013) are cultured in Dulbecco's Modified Eagle Medium (DMEM) containing 10% FCS, 1 × nonessential amino acids (Gibco), 10 mM Hepes–NaOH buffer (pH 7.4), 50 μM β-mercaptoethanol, and GlutaMax (Gibco).

2.1.2 Reagents

1. *PS exposure*: Annexin V-Cy5 (Biovision), milk fat globule EGF factor 8 (MFG-E8)-FITC (Hematologic Technologies), propidium iodide (PI) (Sigma), and Ca^{2+} ionophore A23187 (Sigma).
2. *PE exposure*: Biotin-labeled RO09-0198 (Dr. Umeda's Laboratory, Kyoto University) and Streptavidin-APC (BD Biosciences).
3. *Lipid incorporation*: 1-Oleoyl-2-{6-[(7-nitro-2-1,3-benzoxadiazol-4-yl) amino]hexanoyl}-*sn*-glycero-3-phosphocholine (NBD-PC) (Avanti) and *N*-[6-[(7-nitro-2-1,3-benzoxadiazol-4-yl)amino]hexanoyl]-sphingosine-1-phosphocholine (NBD-SM) (Avanti). NBD-lipids in chloroform were dried with nitrogen gas to avoid oxidization and resuspended in DMSO to a final concentration of 1 mM. NBD-lipids in DMSO can be stored at −20 °C for at least 1 month.

2.1.3 Buffers

1. *Annexin buffer:* 2.5 m*M* $CaCl_2$, 140 m*M* NaCl, and 10 m*M* Hepes–NaOH (pH 7.4).
2. *Lipid incorporation buffer:* Hank's balanced salt solution (HBSS) (Gibco) and 1 m*M* $CaCl_2$.
3. *Lipid extraction buffer:* HBSS, 1 m*M* $CaCl_2$, 5 mg/ml fatty acid-free Bovine Serum Albumin (BSA) (Sigma), and 500 n*M* Sytox Blue (Molecular Probes).

2.2. Protocol for the phospholipid exposure assay

PS exposure can be evaluated using the PS-binding proteins Annexin V (Koopman et al., 1994) or MFG-E8 (Hanayama et al., 2002). PE exposure can be detected using the PE-binding peptide RO09-0198 (Emoto, Toyama-Sorimachi, Karasuyama, Inoue, & Umeda, 1997). Both PS and PE exposure can be initiated by stimulating with the calcium-ionophore A23187. In both cases, necrotic cells or membrane-broken cells can be identified by staining with PI or Sytox Blue.

2.2.1 PS exposure

1. Transfer 2×10^6 cells to a 15-ml tube.
2. Centrifuge ($300 \times g$, 2 min, room temperature) and wash cells with 1 ml Phosphate-Buffered Saline (PBS).
3. Centrifuge and wash cells with 1 ml Annexin buffer.
4. Centrifuge and resuspend cells in 2 ml Annexin buffer.
5. Transfer 1 ml of the cell suspension to a 1.5-ml tube.
6. Add 1 μl Annexin V-Cy5 (1:1000) and 10 μl of 500 μg/ml PI (final concentration: 5 μg/ml).
   *Annexin V can be replaced with MFG-E8-FITC (final concentration: 83 ng/ml).
7. Incubate in a heat block at 20 °C for 3 min.
8. Transfer 500 μl of the stained cells to a Fluorescence Activated Cell Sorting (FACS) tube with a cap filter.
9. Stimulate cells with 5 μl of 300 μ*M* A23187 (final concentration: 3 μ*M*).
   *The A23187 concentration should be optimized for different cell types by titrating from 0.5 to 10 μ*M*. The optimal A23187 concentration was found to be 0.5 μ*M* for Ba/F3 pro-B cells, 10 μ*M* for PLB985 myeloid progenitor cells, and 3 μ*M* for IFETs.
10. Analyze stained and activated cells by FACSAria at 20 °C.

2.2.2 PE exposure

At step 6, in the PS exposure assay, cells are resuspended in Annexin buffer with RO09-0198-biotin, Streptavidin-APC (1 μg/ml), and 5 μg/ml PI (the RO09-0198-biotin and Streptavidin-APC are mixed in Annexin buffer prior to use). The RO09-0198-biotin concentration should be optimized.

2.3. Protocol for the phospholipid incorporation assay

Lipid incorporation can be observed using fluorescence-conjugated lipids (NBD-lipids) (McIntyre & Sleight, 1991), and internalized lipids can be measured following the specific removal of outer layer lipids using fatty acid-free BSA (Haest, Plasa, & Deuticke, 1981). The ATP-dependent aminophospholipid translocases, or flippases, have the ability to incorporate PS and PE, but not PC (Colleau, Hervé, Fellmann, & Devaux, 1991), whereas scramblases nonspecifically incorporate phospholipids (Williamson, Kulick, Zachowski, Schlegel, & Devaux, 1992). Thus, NBD-PC is more suitable than NBD-PS to assay the scramblase. Necrotic cells or membrane-broken cells can be detected by staining with Sytox Blue. It should be noted that A23187 possesses a weak fluorescence spectrum similar to Sytox Blue.

2.3.1 NBD-labeled phospholipid incorporation

1. Transfer 4×10^6 cells to a 15-ml tube.
2. Centrifuge ($300 \times g$, 2 min, room temperature) and wash cells with 1 ml HBSS.
3. Centrifuge and wash cells with 1 ml HBSS-containing 1 m*M* $CaCl_2$ (HBSS/$CaCl_2$).
4. Centrifuge and resuspend cells in 2 ml cold HBSS/$CaCl_2$.
5. Transfer 600 μl of the washed cells to two 1.5-ml tubes (one for DMSO and one for A23187).
6. Incubate in an aluminum block on ice for 7 min.
7. Add 600 μl of 200 n*M* NBD-PC dissolved in HBSS/$CaCl_2$ to the cell suspension (NBD-PC final concentration: 100 n*M*).
   *The NBD-PC concentration should be optimized for different cell types by titrating from 0.1 to 1 μ*M*.
8. Place the cell suspension on ice for 3 min.
9. For the time = 0 min sample, mix 200 μl of the cell suspension with 200 μl of the chilled lipid extraction buffer, and keep on ice.
10. To the remaining 1 ml cell suspension kept at step 8, add 5 μl DMSO or 5 μl 50 μ*M* A23187 (to a final concentration of 250 n*M*) to 1 ml cell suspension.

*The A23187 concentration should be optimized for different cell types by titrating from 0.25 to 1 μ*M*.

11. Incubate in a heat block at 15 °C.

*The incubation temperature can be increased up to 25 °C, but the cells may lyse at higher temperatures, depending on the cell type and the labeled lipid used. At lower temperature (15–25 °C), scramblase-independent incorporation (such as endocytosis) may be prevented.

12. Collect 150-μl samples at 1, 2, 4, 6, and 8 min and add to 150 μl of the chilled lipid extraction buffer or the same buffer without BSA, and keep on ice.

13. Analyze the Sytox Blue-negative population of the applied cells for the incorporated NBD-PC by flow cytometry. The BSA-nonextractable NBD-PC signal represents incorporated lipids that have been internalized. The BSA-nontreated NBD-PC signal represents total lipids, including those attached to the cell surface and those that have been internalized.

3. ASSAYS FOR APOPTOSIS-INDUCED PHOSPHOLIPID SCRAMBLING

PS exposed on the cell surface was first suggested to be an "eat-me-signal" by studies in which red blood cells were labeled with the fluorescence-conjugated PS (NBD-PS); unincorporated NBD-PS remaining on the surface of the red blood cells was found to facilitate their uptake by macrophages (Tanaka & Schroit, 1983). Approximately 10 years later, apoptotic lymphocytes exposing PS were shown to be engulfed by phagocytes in a PS-dependent manner (Fadok et al., 1992). When PS is exposed to the cell surface, PC or SM located on the outer surface of the membrane becomes internalized, suggesting the existence of proteins that transport lipids bidirectionally and nonspecifically (Williamson et al., 2001). Here, we describe PS and PE exposure assays and NBD-PC and NBD-SM incorporation assays using apoptotic leukocytes. Experimental conditions such as cell numbers and reaction volumes can be changed, but the ratio of components is important. Incubation time and temperature should be optimized depending on the cell types.

3.1. Cells, reagents, and buffer

3.1.1 Cells

Any types of cells that can undergo apoptosis can be used for the apoptotic phospholipid-scrambling assays. The cells listed here are routinely used in our laboratory.

1. WR19L cells (ATCC, #TIB-52) expressing mouse Fas (Ogasawara et al., 1993) are maintained in RPMI1640 containing 10% FCS and 50 μ*M* β-mercaptoethanol.
2. Raji cells (ATCC, #CCL-86) are maintained in RPMI1640 containing 10% FCS and 50 μ*M* β-mercaptoethanol.
3. Culture condition of PLB985 cells is described in Section 2.1.1.

3.1.2 Reagents and buffers

Most of the reagents and buffers used here were described in Sections 2.1.2 and 2.1.3. The reagents used specifically in this section include: Fas ligand and staurosporine (STS) (Kyowa Hakko Kirin, Co.) to induce apoptosis.

3.1.2.1 Fas ligand

The leucine zipper-tagged human Fas ligand (Shiraishi et al., 2004) was prepared by introducing its expression vector into COS-7 cells. In brief, 150 μg of the expression vector was mixed with 3×10^7 COS-7 cells in 3 ml of KPBS (30.8 m*M* NaCl, 121 m*M* KCl, 8.1 m*M* Na_2HPO_4, 1.46 m*M* KH_2PO_4) and subjected to electroporation using a Gene-pulser (960 μF, 0.23 kV) (Bio-Rad). After electroporation, the cells in ten 10-cm plates were cultured for 3 days in 10 ml of DMEM containing 1% FCS, and the culture supernatant was collected. The cells were further cultured for 2 days with fresh medium, and the conditioned medium was combined with the medium collected at day 3 (a total of 150 ml). A 225 ml aliquot of saturated ammonium sulfate was added to the conditioned medium at 4 °C to obtain 60% saturation of ammonium sulfate, and the proteins were precipitated by centrifugation at $15{,}000 \times g$ for 30 min. The precipitated proteins were dissolved in 4 ml of PBS, dialyzed against PBS, filtered through a 0.22-μm filter, and stored at −80 °C until use. The biological activity of Fas ligand was determined with WR19L cells expressing mouse Fas as described (Tanaka, Suda, Yatomi, Nakamura, & Nagata, 1997), and one unit is defined as the dilution that gives a half-maximum response.

3.2. Protocol for the phospholipid exposure assay

Like the Ca^{2+}-induced PS exposure, PS exposed during apoptosis can be detected using the PS-binding proteins Annexin V (Koopman et al., 1994) or MFG-E8 (Hanayama et al., 2002), while PE exposure is detected with RO09-0198 (Emoto et al., 1997). Apoptotic cells later undergo secondary necrosis, and these cells are stained by Annexin V and PI (or Sytox Blue).

3.2.1 PS exposure

1. Transfer 5×10^5 cells in 500 μl medium to a 24-well plate.
2. Stimulate cells with adequate apoptotic stimuli.
   WR19L cells expressing mouse Fas are stimulated with 10 units/ml Fas ligand for 60 min.
   Raji cells are stimulated with 400 units/ml Fas ligand for 2 h.
   PLB985 cells suspended in PBS are exposed to 2000 J/m^2 UV irradiation and cultured in medium for 2–3 h, or treated with 10 μ*M* STS for 2–4 h.
3. Transfer 230 μl of the cell suspension to a 96-well round-bottom plate.
4. Centrifuge (300 × *g*, 2 min, 4 °C) and wash cells with 200 μl PBS.
5. Centrifuge and wash cells with 200 μl Annexin buffer.
6. Centrifuge and resuspend cells in 100 μl Annexin buffer containing Annexin V-Cy5 (1:1000).
7. Incubate cells on ice for 15 min.
8. Add 100 μl Annexin buffer containing 10 μg/ml PI.
9. Incubate cells on ice for 2 min.
10. Analyze stained cells with flow cytometry.

3.2.2 PE exposure

The cells are treated with apoptotic stimuli as described above (steps 1–5).

1. Centrifuge and resuspend cells in 100 μl Annexin buffer with biotin-RO09-0198.
2. Incubate cells on ice for 15 min.
3. Add 100 μl Annexin buffer, centrifuge, and wash cells with 200 μl Annexin buffer.
4. Centrifuge and resuspend cells in 100 μl Annexin buffer with Streptavidin-APC (1 μg/ml).
5. Incubate cells on ice for 15 min.

6. Add 100 μl Annexin buffer, centrifuge, and wash cells with 200 μl Annexin buffer.
7. Centrifuge and resuspend cells in 200 μl Annexin buffer containing 5 μg/ml PI.
8. Analyze stained cells with flow cytometry.

3.3. Protocol for the phospholipid incorporation assay

Like the Ca^{2+}-induced scrambling of phospholipids, the scrambling of phospholipids during apoptosis can be monitored using NBD-PC, and internalized lipids can be distinguished from those bound to the cells by the specific removal of outer layer lipids using fatty acid-free BSA (Haest et al., 1981). The apoptotic PS exposure is caspase dependent (Martin, Finucane, Amarante-Mendes, O'Brien, & Green, 1996), and when cells undergo the secondary necrosis, phospholipid scrambling may occur at the membrane-broken sites in a scramblase-independent fashion. Thus, the timing of assays of scramblase activity in apoptotic cells is critical. It is recommended to perform Annexin V/PI staining of apoptotic cells as described in Section 3.2.1 to confirm that cells are in an early apoptotic phase (Annexin V positive and PI negative) prior to the phospholipid incorporation assay.

3.3.1 NBD-labeled phospholipid incorporation

Cells were stimulated with apoptotic stimuli as described above (see Section 3.2.1).

1. Transfer 2×10^6 cells to a 15-ml tube.
2. Centrifuge ($300 \times g$, 2 min, room temperature) and wash cells with 1 ml HBSS.
3. Centrifuge and wash cells with 1 ml HBSS/$CaCl_2$.
4. Centrifuge and resuspend cells in 1 ml cold HBSS/$CaCl_2$.
5. Transfer 600 μl of the cell suspension to a 1.5-ml tube.
6. Incubate in an aluminum block on ice for 7 min.
7. Add 600 μl of 200 n*M* NBD-lipids in HBSS/$CaCl_2$ to the cell suspension (NBD-lipids final concentration: 100 n*M*).
   *NBD-lipid concentration should be optimized for different cell types by titrating from 0.1 to 1 μ*M*.
8. Place the cell suspension on ice for 3 min.
9. For the time = 0 min sample, mix 200 μl of the cell suspension with 200 μl of the chilled lipid extraction buffer, and keep on ice.
10. Place the remaining 1 ml cell suspension at step 8 onto a heat block and incubate at 20 °C.

11. Take 150-μl aliquots at 1, 2, 4, 6, and 8 min and mix with 150 μl of lipid extraction buffer or the same buffer without BSA, and keep on ice.
12. Analyze the Sytox Blue-negative population for the incorporation of fluorescent lipids by flow cytometry.

REFERENCES

Balasubramanian, K., & Schroit, A. J. (2003). Aminophospholipid asymmetry: A matter of life and death. *Annual Review of Physiology, 65*, 701–734.

Bevers, E. M., Comfurius, P., & Zwaal, R. F. (1983). Changes in membrane phospholipid distribution during platelet activation. *Biochimica et Biophysica Acta, 736*, 57–66.

Bevers, E. M., & Williamson, P. L. (2010). Phospholipid scramblase: An update. *FEBS Letters, 584*, 2724–2730.

Castoldi, E., Collins, P. W., Williamson, P. L., & Bevers, E. M. (2011). Compound heterozygosity for 2 novel TMEM16F mutations in a patient with Scott syndrome. *Blood, 117*, 4399–4400.

Colleau, M., Hervé, P., Fellmann, P., & Devaux, P. F. (1991). Transmembrane diffusion of fluorescent phospholipids in human erythrocytes. *Chemistry and Physics of Lipids, 57*, 29–37.

Emoto, K., Toyama-Sorimachi, N., Karasuyama, H., Inoue, K., & Umeda, M. (1997). Exposure of phosphatidylethanolamine on the surface of apoptotic cells. *Experimental Cell Research, 232*, 430–434.

Fadeel, B., Gleiss, B., Högstrand, K., Chandra, J., Wiedmer, T., Sims, P. J., et al. (1999). Phosphatidylserine exposure during apoptosis is a cell-type-specific event and does not correlate with plasma membrane phospholipid scramblase expression. *Biochemical and Biophysical Research Communications, 266*, 504–511.

Fadok, V. A., de Cathelineau, A., Daleke, D. L., Henson, P. M., & Bratton, D. L. (2001). Loss of phospholipid asymmetry and surface exposure of phosphatidylserine is required for phagocytosis of apoptotic cells by macrophages and fibroblasts. *Journal of Biological Chemistry, 276*, 1071–1077.

Fadok, V. A., Voelker, D. R., Campbell, P. A., Cohen, J. J., Bratton, D. L., & Henson, P. M. (1992). Exposure of phosphatidylserine on the surface of apoptotic lymphocytes triggers specific recognition and removal by macrophages. *Journal of Immunology, 148*, 2207–2216.

Haest, C. W., Plasa, G., & Deuticke, B. (1981). Selective removal of lipids from the outer membrane layer of human erythrocytes without hemolysis. Consequences for bilayer stability and cell shape. *Biochimica et Biophysica Acta, 649*, 701–708.

Hanayama, R., Tanaka, M., Miwa, K., Shinohara, A., Iwamatsu, A., & Nagata, S. (2002). Identification of a factor that links apoptotic cells to phagocytes. *Nature, 417*, 182–187.

Imao, T., & Nagata, S. (2013). Apaf-1- and caspase-8-independent apoptosis. *Cell Death and Differentiation, 20*, 343–352.

Koopman, G., Reutelingsperger, C. P., Kuijten, G. A., Keehnen, R. M., Pals, S. T., & van Oers, M. H. (1994). Annexin V for flow cytometric detection of phosphatidylserine expression on B cells undergoing apoptosis. *Blood, 84*, 1415–1420.

Martin, S. J., Finucane, D. M., Amarante-Mendes, G. P., O'Brien, G. A., & Green, D. R. (1996). Phosphatidylserine externalization during CD95-induced apoptosis of cells and cytoplasts requires ICE/CED-3 protease activity. *Journal of Biological Chemistry, 271*, 28753–28756.

McIntyre, J. C., & Sleight, R. G. (1991). Fluorescence assay for phospholipid membrane asymmetry. *Biochemistry, 30*, 11819–11827.

Nagata, S., & Golstein, P. (1995). The Fas death factor. *Science, 267*, 1449–1456.

Nagata, S., Hanayama, R., & Kawane, K. (2010). Autoimmunity and the clearance of dead cells. *Cell, 140*, 619–630.

Ogasawara, J., Watanabe-Fukunaga, R., Adachi, M., Matsuzawa, A., Kasugai, T., Kitamura, Y., et al. (1993). Lethal effect of the anti-Fas antibody in mice. *Nature, 364*, 806–809.

Palacios, R., & Steinmetz, M. (1985). Il-3-dependent mouse clones that express B-220 surface antigen, contain Ig genes in germ-line configuration, and generate B lymphocytes in vivo. *Cell, 41*, 727–734.

Ravichandran, K. S., & Lorenz, U. (2007). Engulfment of apoptotic cells: Signals for a good meal. *Nature Reviews Immunology*, 7, 964–974.

Shiraishi, T., Suzuyama, K., Okamoto, H., Mineta, T., Tabuchi, K., Nakayama, K., et al. (2004). Increased cytotoxicity of soluble Fas ligand by fusing isoleucine zipper motif. *Biochemical and Biophysical Research Communications, 322*, 197–202.

Smeets, E. F., Comfurius, P., Bevers, E. M., & Zwaal, R. F. (1994). Calcium-induced transbilayer scrambling of fluorescent phospholipid analogs in platelets and erythrocytes. *Biochimica et Biophysica Acta, 1195*, 281–286.

Suzuki, J., Denning, D. P., Imanishi, E., Horvitz, H. R., & Nagata, S. (2013). Xk-related protein 8 and CED-8 promote phosphatidylserine exposure in apoptotic cells. *Science, 341*, 403–406.

Suzuki, J., Fujii, T., Imao, T., Ishihara, K., Kuba, H., & Nagata, S. (2013). Calcium-dependent phospholipid scramblase activity of TMEM16 protein family members. *Journal of Biological Chemistry, 288*, 13305–13316.

Suzuki, J., Umeda, M., Sims, P. J., & Nagata, S. (2010). Calcium-dependent phospholipid scrambling by TMEM16F. *Nature, 468*, 834–838.

Tanaka, Y., & Schroit, A. J. (1983). Insertion of fluorescent phosphatidylserine into the plasma membrane of red blood cells. Recognition by autologous macrophages. *Journal of Biological Chemistry, 258*, 11335–11343.

Tanaka, M., Suda, T., Yatomi, T., Nakamura, N., & Nagata, S. (1997). Lethal effect of recombinant human Fas ligand in mice pretreated with Propionibacterium acnes. *Journal of Immunology, 58*, 2303–2309.

Tucker, K. A., Lilly, M. B., Heck, L., Jr., & Rado, T. A. (1987). Characterization of a new human diploid myeloid leukemia cell line (PLB-985) with granulocytic and monocytic differentiating capacity. *Blood, 70*, 372–378.

Vaux, D. L., & Korsmeyer, S. J. (1999). Cell death in development. *Cell, 96*, 245–254.

Williamson, P., Bevers, E. M., Smeets, E. F., Comfurius, P., Schlegel, R. A., & Zwaal, R. F. (1995). Continuous analysis of the mechanism of activated transbilayer lipid movement in platelets. *Biochemistry, 34*, 10448–10455.

Williamson, P., Christie, A., Kohlin, T., Schlegel, R. A., Comfurius, P., Harmsma, M., et al. (2001). Phospholipid scramblase activation pathways in lymphocytes. *Biochemistry, 40*, 8065–8072.

Williamson, P., Kulick, A., Zachowski, A., Schlegel, R. A., & Devaux, P. F. (1992). Ca^{2+} induces transbilayer redistribution of all major phospholipids in human erythrocytes. *Biochemistry, 31*, 6355–6360.

Yang, H., Kim, A., David, T., Palmer, D., Jin, T., Tien, J., et al. (2012). TMEM16F forms a Ca^{2+}-activated cation channel required for lipid scrambling in platelets during blood coagulation. *Cell, 151*, 111–122.

Zwaal, R. F., Comfurius, P., & Bevers, E. M. (1992). Platelet procoagulant activity and microvesicle formation. Its putative role in hemostasis and thrombosis. *Biochimica et Biophysica Acta, 1180*, 1–8.

CHAPTER SIXTEEN

Studying Apoptosis in the Zebrafish

Peter M. Eimon[1]

Department of Electrical Engineering and Computer Science, Massachusetts Institute of Technology (MIT), Cambridge, Massachusetts, USA

[1]Corresponding author: e-mail address: eimon@mit.edu

Contents

Abstract

Zebrafish (*Danio rerio*) have been extensively used to study apoptotic cell death during normal development and under a wide range of experimental manipulations. A number of features make zebrafish a particularly powerful model organism: (1) embryos are small in size, develop rapidly outside the mother, and are optically transparent; (2) tools are readily available for rapid knockdown and overexpression of genes; and (3) embryos can be arrayed into multiwell plates and are permeable to a wide range of drugs and small molecules. The molecular machinery underlying the intrinsic and extrinsic apoptosis pathways appears to be highly conserved between zebrafish and mammals. In this chapter, techniques are described for detecting apoptotic cells *in situ* in both fixed and live zebrafish embryos. Methods for inducing and inhibiting apoptosis and for functionally manipulating genes involved in apoptotic signaling are also discussed.

Methods in Enzymology, Volume 544
ISSN 0076-6879
http://dx.doi.org/10.1016/B978-0-12-417158-9.00016-9

1. INTRODUCTION

Zebrafish (*Danio rerio*) have been used as a model organism for decades, with adoption increasing dramatically following the publication of several large forward genetic screens in the 1990s (Driever et al., 1996; Haffter & Nusslein-Volhard, 1996). Zebrafish embryos possess features that make them well suited for vertebrate genetics, developmental biology, and high-throughput *in vivo* drug screening. A single pair of adult zebrafish can produce 100+ embryos per week. The embryos themselves are small in size (0.7 mm in diameter at fertilization), optically transparent, and develop rapidly outside the mother, allowing early developmental processes to be observed. Embryos complete gastrulation by ~10 hours postfertilization (hpf), and most organs become functional between 3 and 5 days postfertilization (dpf) (Westerfield, 1995). Numerous studies have shown that zebrafish possess a high degree of genetic and physiological homology to mammals and often respond similarly to drugs and other bioactive chemicals (Eimon & Rubinstein, 2009; Kokel & Peterson, 2011; Milan, Peterson, Ruskin, Peterson, & MacRae, 2003; Rihel et al., 2010).

Many useful tools have been developed to expand the range of experiments possible in zebrafish. Among the most widely used are microinjection techniques that allow genes of interest to be transiently overexpressed (using either synthetic mRNA or DNA expression constructs) and inhibited (primarily using morpholino antisense oligonucleotides) (Nasevicius & Ekker, 2000) in the embryo. Efficient methods are available for generating stable transgenic lines that express genes of interest (often fluorescent reporter proteins) under the control of tissue-specific promoters. More recently, the advent of zinc-finger nucleases (Amacher, 2008; McCammon & Amacher, 2010), transcription activator-like effector nucleases (TALENs) (Huang et al., 2011; Sander et al., 2011), and clustered regularly interspaced short palindromic repeats (CRISPRs) (Hwang et al., 2013) has enabled precise targeting of genes for mutation and, in some cases, genome editing (Zu et al., 2013). These advances, in combination with traditional *N*-ethyl-*N*-nitrosourea mutagenesis approaches, have resulted in a large and well-characterized collection of zebrafish mutant lines, many of which are available to the research community through the Zebrafish International Resource Center (ZIRC), the European Zebrafish Resource Center (EZRC), and the National BioResource Project Zebrafish. Large-scale initiatives are currently underway through the Sanger Institute Zebrafish

Mutation Project (http://www.sanger.ac.uk/Projects/D_rerio/zmp/) to generate additional mutant lines in a more systematic manner.

Apoptotic signaling mechanisms in both mammals and zebrafish can be divided into two key pathways: the cell-intrinsic (or mitochondrial) pathway and the cell-extrinsic (or death receptor) pathway (Eimon & Ashkenazi, 2010). The intrinsic pathway is regulated through the Bcl-2 gene family and can be engaged by a variety of stimuli, including growth factor withdrawal and cellular damage. The extrinsic pathway is activated by ligation of death receptors on the cell surface and triggers apoptosis through a death-inducing signaling complex. Both pathways converge on a common set of effector caspases (caspase-3, -6, and -7), which carry out programmed cell death. The molecular machinery of both pathways appears to be broadly conserved between zebrafish and mammals (Table 16.1; see Eimon & Ashkenazi (2010) for a more detailed review).

2. DETECTING AND QUANTIFYING APOPTOSIS IN ZEBRAFISH

Many assays that were originally developed to detect apoptotic cells in cell culture and other model organisms have been successfully adapted for use on zebrafish embryos. In live embryos, apoptosis can be detected using a variety of vital dyes (Section 2.1) or through the use of transgenic lines expressing a genetically encoded reporter of apoptosis (Section 2.2). In fixed embryos, apoptotic cells can be visualized in whole-mount preparations using either the terminal dUTP nick end-labeling (TUNEL) method (Section 2.3) or an antibody against the cleaved, active form of caspase-3 (Section 2.4). Finally, caspase activity can be quantified in zebrafish embryo lysates based on hydrolysis of fluorometric or luminogenic substrates (Section 2.5).

2.1. Vital dyes for *in vivo* labeling of apoptotic cells

Apoptotic cells can be visualized in living zebrafish embryos using a variety of vital dyes. By far the most common is acridine orange, a metachromatic fluorescent cationic dye that preferentially stains apoptotic cells and fluoresces green upon binding to DNA (Darzynkiewicz et al., 1992; Furutani-Seiki et al., 1996; Negron & Lockshin, 2004). Because acridine orange staining can be performed rapidly on live embryos in multiwell plates, it is particularly useful in large-scale mutagenesis and chemical screens. Results obtained with acridine orange or other vital dyes should

Table 16.1 Mutant lines and morpholinos

Gene family	Mammalian gene (common aliases)	Zebrafish homolog	Aliases	Reference sequence	Mutant lines[a]	Published MOs
TNF	FASLG (CD95L/ TNFSF6)	faslg	FasL	NM_001042701		
	TNF (TNFSF2)	tnfa	TNF1, tnf-alpha, TNF-a_v1, TNF-a_v3	NM_212859	sa16250	Qi et al. (2010) and Matthews et al. (2009)
	TNF (TNFSF2)	tnfb	TNF2, tnf, Tnf-alpha, TNF-a_v2	NM_001024447		Qi et al. (2010)
	TNFSF10 (Apo2L/ TRAIL)	tnfsf10l	DL1a, TRAIL-like	NM_131843		
	TNFSF10 (Apo2L/ TRAIL)	tnfsf10	DL2	NM_001002593		
	TNFSF10 (Apo2L/ TRAIL)	tnfsf10l3	DL1b	NM_001042713		
	TNFSF10 (Apo2L/ TRAIL)	tnfsf10l4	DL3	NM_001013283		

	TNFSF15 (TL1A)	si:ch211-158d24.4	tnfsf15	NM_001123259		
	Unknown	lta	TNF-N	NM_001024821		
TNFR (death domain)	FAS (CD95/ TNFRSF6)	fas		XM_685355	sa11234	
	TNFRSF1A (TNFR1)	tnfrsf1a	TNFR1	NM_213190	sa15813	Espin et al. (2013) and Bates, Akerlund, Mittge, and Guillemin (2007)
	TNFRSF1B	tnfrsf1b				Espin et al. (2013)
	TNFRSF10A and B (DR4 and DR5)	hdr	ZH-DR	NM_194391		Eimon et al. (2006) and Kwan et al. (2006)
	TNFRSF10A and B (DR4 and DR5)	tnfrsfa	OTR, tnf	NM_131840		Eimon et al. (2006)
	TNFRSF21 (DR6)	tnfrsf21	DR6	NM_001042688		Tam et al. (2012)
Death domain adaptors	CRADD (RAIDD)	cradd		NM_001006066		
	DEDD	zgc:92202		NM_001002639		
	EDARADD	edaradd		XM_690046		
	FADD	fadd		XM_001923858		Eimon et al. (2006)
	MADD	madd		XM_005168954	sa12918	

Continued

Table 16.1 Mutant lines and morpholinos—cont'd

Gene family	Mammalian gene (common aliases)	Zebrafish homolog	Aliases	Reference sequence	Mutant lines	Published MOs
	PIDD (LRDD)	pidd		XM_694585	sa1707	
	RIPK1 (RIP)	ripk1l	rip1	NM_001043350		Roca and Ramakrishnan (2013)
	TRADD	tradd		NM_131607		Eimon et al. (2006)
Caspase (apoptotic)	CASP2	casp2		NM_001042695	la019585Tg, sa12064	Sidi et al. (2008)
	CASP3	casp3a	casp3	NM_131877		Yamashita, Mizusawa, Hojo, and Yabu (2008)
	CASP3	casp3b	casp3l	NM_001048066	sa15886	
	CASP4	caspa	caspy	NM_131505		Masumoto et al. (2003)
	CASP4	caspb	caspy2	NM_152884		
	CASP4	caspbl		NM_001145592		
	CASP4	zgc:171731		NM_001109712		
	CASP4	si:ch211-233g6.8		XM_003197776		
	CASP6	casp6	casp6a	NM_001020497	sa10094, sa1603	
	CASP6	casp6l1	casp6b	NM_001005973		
	CASP6	casp6l2	casp6c	NM_001039980		
	CASP7	casp7	casp7a	NM_001020607		

	CASP8	casp8	casp8a	NM_131510		Sidi et al. (2008)
	CASP8	casp8l1	casp8a-like	NM_001098619		
	CASP8	casp8l2	casp8b, casp10-like	XM_680338		
	CASP9	casp9		NM_001007404	sa11164	Sidi et al. (2008)
	CFLAR (c-FLIP)	cflara	FLIP, Clarp1	NM_194399	sa11641, sa2599	
	Unknown	zgc:194469	card-casp8	NM_001136252		
Bcl2 (prosurvival)	BCL2	bcl2	zBlp2	NM_001030253		Kratz et al. (2006)
	BCL2L1 (Bcl-xL)	bcl2l1	zBlp1, zfBLP1, zfBcl-xL, bcl2l	NM_131807	la010762Tg, sa15253	Kratz et al. (2006)
	BCL2L10	bcl2l10	zNR13, Nrz, NR-13, mcl1l	NM_194398		Arnaud et al. (2006)
	MCL1	mcl1a	zfMLP1, zfMcl-1a	NM_131599	sa10132, sa12089	Kratz et al. (2006)
	MCL1	mcl1b		NM_194394		Kratz et al. (2006)
Bcl2 (proapoptotic)	BAD	bada		XM_005161364		
	BAD	badb		NM_001270595		Jensen, Gitlin, and Carayannopoulos (2006)
	BAX	baxa	zBax1, bax	NM_131562	sa14515	Kratz et al. (2006), 16282981

Continued

Table 16.1 Mutant lines and morpholinos—cont'd

Gene family	Mammalian gene (common aliases)	Zebrafish homolog	Aliases	Reference sequence	Mutant lines	Published MOs
	BAX	baxb	zBax2	NM_001013296		Kratz et al. (2006)
	BBC3 (PUMA)	bbc3	zPuma	NM_001045472	la027618Tg	Kratz et al. (2006) and Sidi et al. (2008)
	BCL2L11 (BIM)	bcl2l11	BIM	NM_001135791		
	BCL2L13 (BCL-RAMBO)	bcl2l13		NM_001044891	sa10792	
	BID	bida	zBid	NM_001079826	la020320Tg, sa10083	Kratz et al. (2006)
	BIK	bik		NM_001045038		Kratz et al. (2006)
	BMF	bmf1		NM_001045224		Kratz et al. (2006)
	BMF	bmf2		NM_001045473	la028280Tg	
	BNIP1	bnip1a		XM_684156		Nishiwaki et al. (2013)
	BNIP1	bnip1b		XM_001333689		Nishiwaki et al. (2013)
	BNIP2	bnip2		NM_201218		
	BNIP2	LOC100332015		XM_002666916		
	BNIP3L	bnip3la	nix	NM_001012242		
	BNIP3L	bnip3lb	nip3a	NM_205571		

	BNIP subfamily	bnipl		NM_001128394		
	BNIP subfamily	bnip4	nip3b	NM_212693		
	BOK	boka	zBok1, bok	NM_001003612		
	BOK	bokb	zBok2	NM_201185	sa13282	
	PMAIP1 (NOXA)	pmaip1	zNoxa	NM_001045474		Kratz et al. (2006)
IAP	BIRC2 (cIAP1)	birc2	IAP1, birc3	NM_194395	hi2026aTg, la021887Tg, s805	Santoro, Samuel, Mitchell, Reed, and Stainier (2007)
	BIRC5	birc5a	bir5a, survivin, survivin 1	NM_194397	hi1326Tg, hi2111Tg, hi3018cTg	Ma et al. (2007) and Delvaeye et al. (2009)
	BIRC5	birc5b	survivin 2	NM_145195	p1aiue	Ma et al. (2007) and Delvaeye et al. (2009)
	BIRC6 (APOLLON)	birc6		XM_005158707	la019994Tg, sa10215, sa12006	
	BIRC7 (MLIAP)	birc7		NM_001098768	sa10528	
	XIAP (BIRC4)	xiap	birc4	NM_194396	la012549Tg, la018928Tg, sa2739	

Continued

Table 16.1 Mutant lines and morpholinos—cont'd

Gene family	Mammalian gene (common aliases)	Zebrafish homolog	Aliases	Reference sequence	Mutant lines	Published MOs
Other	APAF1	apaf1		NM_001045243	la024463Tg, la024464Tg, sa10293, sa16025	
	CYCS (cytochrome *c*)	cycsb		NM_001002068		
	TP53	tp53	p53, brp53, drp53	NM_131327	zdf1, etc.	Langheinrich, Hennen, Stott, and Vacun (2002)
	DIABLO (SMAC)	diabloa		NM_200346		
	DIABLO (SMAC)	diablob		NM_001243034		
	TNFAIP3 (A20)	tnfaip3		XM_687830		

[a]Names of all mutant alleles are from ZFIN.

be confirmed using a more rigorous assay such as TUNEL. The following protocol describes how to remove zebrafish embryos from their protective chorions and stain apoptotic cells using acridine orange.

2.1.1 Required materials for acridine orange staining

- Embryo medium:
  - 5.03 m*M* NaCl
  - 0.17 m*M* KCl
  - 0.33 m*M* $CaCl_2 \cdot 2H_2O$
  - 0.33 m*M* $MgSO_4 \cdot 7H_2O$
  - 0.1% (w/v) Methylene blue (prevents fungal growth)

  Note: Embryo medium is typically made up as a 60×-stock solution. If methylene blue is used, it should be added at the time that the 1× working solution is prepared.
- Acridine orange (Sigma A6014)
- 16 mg mL^{-1} buffered tricaine (ethyl 3-aminobenzoate methanesulfonate/MS-222; Sigma A5040)
- Large Petri dish
- Glass beaker
- (Optional) Two pair of dissecting forceps (e.g., Dumont #5 forceps; Fine Science Tools 11251-30)
- (Optional) 1 mg mL^{-1} Pronase (Protease, Type XIV: Bacterial, from *Streptomyces griseus*; Sigma P5147) resuspended in embryo medium.

2.1.2 Method

- Standard protocols for zebrafish husbandry and embryonic culture, along with a detailed developmental staging series, have been published elsewhere and will not be covered in detail here (Nüsslein-Volhard & Dahm, 2002; Westerfield, 1995).
- Dechorionation of embryos (option 1, manual)
  - Carefully grab the chorion with one pair of forceps.
  - Using another pair of forceps, grab the chorion at a second location approximately 90° distant from the first.
  - Pierce the chorion with either set of forceps to create a small tear.
  - Carefully enlarge the tear by pulling apart the chorion with both forceps until the embryo is released. Be careful not to pinch or compress the embryo.
- Dechorionation of embryos (option 2, enzymatic)
  - Pronase can be prepared as a 10 mg mL^{-1} stock in deionized water and frozen as single-use aliquots at −20 °C until needed.

- Thaw the pronase and dilute to 1 mg mL^{-1} in embryo medium.
- Transfer the embryos into a Petri dish, remove the embryo medium, and add the diluted pronase.
- Gently swirl the embryos while monitoring them under a dissecting microscope.
- Once a few embryos have dropped out of their chorions, transfer all embryos to a 500-mL beaker filled with embryo medium. Swirl and allow the embryos to settle to the bottom.
- Rinse at least three times with 300 mL of fresh embryo medium, swirling frequently until all the remaining embryos have been released from their chorions.

- Zebrafish embryos will develop normally outside their chorions when raised in embryo medium at 28 °C. Dechorionated embryos younger than 24 hpf should only be handled in glass dishes or plastic dishes covered in 1% agarose to avoid damage.
- Transfer the embryos to a Petri dish-containing 5 μg mL^{-1} acridine orange in embryo medium. *Note:* Acridine orange can be prepared as a 5-mg mL^{-1} stock (1000 ×) in deionized water and stored in the dark at 4 °C.
- Incubate the embryos for 15–30 min in the dark at room temperature.
- Transfer embryos to a large Petri dish and rinse twice for several minutes in 20 mL of embryo medium.
- Image the staining immediately under fluorescent light using a GFP or FITC filter. Embryos may be anesthetized using 16 mg mL^{-1} tricaine prior to imaging if desired (Westerfield, 1995). *Note:* Analysis of acridine orange staining in zebrafish embryos is complicated by autofluorescence and/or nonspecific uptake by the yolk, iridophores, hatching gland, and neuromasts (Rodriguez & Driever, 1997).
- A protocol has been developed for quantitative analysis of apoptosis based on acridine orange staining (Tucker & Lardelli, 2007).

In addition to acridine orange, other vital dyes have been used to characterize apoptotic cells in live zebrafish embryos. These include: (1) Hoechst 33342 (Molecular Probes), a cell-permeant nuclear counterstain often used to distinguish condensed pyknotic nuclei in apoptotic cells (Negron & Lockshin, 2004); (2) propidium iodide (Sigma P4170), which is taken up by both apoptotic and necrotic cells but excluded by live cells (Negron & Lockshin, 2004); (3) DePsipher (Trevigen 6300-100-K), a cationic dye used to visualize the collapse of the electrochemical gradient across the mitochondrial membrane during mitochondrial permeability transition (Negron &

Lockshin, 2004); and (4) Alexa Fluor 488 Annexin V conjugate (Molecular Probes A13201), used to detect the externalization of phosphatidylserine (PS), one of the earliest indicators of apoptosis (Negron & Lockshin, 2004).

2.2. Transgenic apoptosis reporter lines

Imaging of apoptosis in live zebrafish has been facilitated by a transgenic line that expresses a genetically encoded fluorescent reporter of cell death (van Ham, Mapes, Kokel, & Peterson, 2010). This reporter line exploits the fact that PS becomes exposed on the cell surface during apoptosis, where it serves as a signal for phagocytic engulfment (Fadok, Bratton, Frasch, Warner, & Henson, 1998; Fadok, Bratton, & Henson, 2001). Annexin V (A5) is a phospholipid-binding protein with high affinity for PS (Vermes, Haanen, Steffens-Nakken, & Reutelingsperger, 1995) that is widely used for detecting apoptotic cells in flow cytometry and fluorescence microscopy. van Ham et al. (2010) demonstrated that secreted A5 fused to yellow fluorescent protein (YFP; secA5-YFP) binds to the surface of apoptotic cells in live embryos and allows individual cells to be tracked from the onset of apoptosis through the disappearance of cell corpses. The secA5-YFP transgene is under the control of the UAS enhancer, so expression can be driven in any cell type of interest through crosses with appropriate transgenic lines expressing the GAL4 activator under tissue-specific promoters. The secA5-YFP reporter line has been used in combination with morpholino antisense oligonucleotides (discussed in Section 3.2) to dissect the molecular pathways underlying clearance of apoptotic cells *in vivo* (van Ham, Kokel, & Peterson, 2012).

2.3. Whole-mount TUNEL assay

Apoptotic cells can be visualized in fixed zebrafish embryos using the TUNEL method, which detects fragmented DNA characteristic of apoptotic cells (Furutani-Seiki et al., 1996; Gavrieli, Sherman, & Ben-Sasson, 1992). According to some reports, use of TUNEL on zebrafish embryos older than 48 hpf can be complicated by background staining and nonspecific labeling of periderm cells (Rodriguez & Driever, 1997). The following whole-mount TUNEL protocol is adapted from Parsons et al. (2002). Visualization of both TUNEL staining and antiactive caspase-3 staining (Section 2.4) can be facilitated by using nonpigmented zebrafish lines such as *casper* (White et al., 2008) and *nacre* (Lister, Robertson, Lepage, Johnson, & Raible, 1999).

2.3.1 Required materials

- PBST: 1 × PBS + 0.1% Tween 20
- 4% (w/v) paraformaldehyde in PBST, pH 7.4 (prepare fresh before use)
- Methanol
- Proteinase K: Prepare a 10-mg mL^{-1} stock solution in PBST and store at −20 °C. Dilute to 10 μg mL^{-1} for embryo permeabilization.
- Ethanol:acetic acid, 2:1 (v/v)
- Blocking solution: 1 × PBT, 2% (v/v) sheep serum, 2 mg mL^{-1} BSA (Sigma A3059)
- ApopTag Plus Peroxidase *In Situ* Apoptosis Detection Kit (Millipore S7101) or equivalent (see Section 2.3.3)
- Alkaline phosphatase buffer (NTMT): 100 m*M* NaCl; 100 m*M* Tris HCl, pH 9.5; 50 m*M* $MgCl_2$; 0.1% Tween 20
- NBT/BCIP stock solution (Roche 11681451001)
- 5 mL screw-cap glass vials
- Glass Pasteur pipettes and a 10-mL handheld pipette pump (e.g., VWR 53502-233)

2.3.2 Method

2.3.2.1 Day 1

- Dechorionate live embryos of the desired stage(s) either manually or using pronase (see Section 2.1) and fix them in fresh 4% paraformaldehyde overnight at 4 °C or 1–2 h at room temperature. Very young embryos can be fixed in their chorions and removed manually using dissecting forceps following several brief washes in PBST.
- Wash the embryos twice for 2–5 min in PBST at room temperature.
- Transfer the embryos in PBST to 5 mL screw-cap glass vials using either a transfer pipette or a large diameter glass Pasteur pipette. Once the embryos have settled to the bottom, remove most of the PBST and refill with 100% methanol. Wash once with fresh methanol and dehydrate the embryos for at least 2 h at −20 °C. *Note:* Fixed embryos can be stored in methanol at −20 °C for several months.
- Rehydrate the embryos by successive 2–5 min washes in the following:
    - 75% methanol + 25% PBST
    - 50% methanol + 50% PBST
    - 25% methanol + 75% PBST.
- Wash the embryos twice for 2–5 min in PBST at room temperature.
- Permeabilize the embryos by incubating for 5–15 min in 10 μg mL^{-1} proteinase K in PBST at room temperature. *Note:* The developmental stage of the embryos, the length of fixation, and variability between

batches of proteinase K can all influence the time required for permeabilization. In general, younger embryos should be permeabilized for shorter times, and the activity of each new batch of proteinase K should be verified.

- Stop the proteinase K digestion and refix the embryos by incubating for 20 min in 4% paraformaldehyde at room temperature.
- Wash the embryos twice for 2–5 min in PBST at room temperature.
- Post-fix the embryos in precooled ethanol:acetic acid (2:1) for 10 min at −20 °C. Drain, but do not allow to dry.
- Wash the embryos twice for 2–5 min in PBST at room temperature.
- Remove PBST and incubate the embryos for 1 h at room temperature in ApopTag Equilibration Buffer (reagents for this and subsequent end-labeling steps are from the Millipore ApopTag *In Situ* Apoptosis Detection Kit, S7101. See Section 2.3.3). Embryos may be transferred from glass vials to microcentrifuge tubes to reduce the working volume and conserve reagents during end-labeling steps.
- Remove the ApopTag Equilibration Buffer and incubate the embryos overnight at 37 °C in working strength ApopTag TdT Enzyme (70% ApopTag Reaction Buffer + 30% ApopTag TdT Enzyme; mix by vortexing). Working strength ApopTag TdT Enzyme should be made fresh and stored on ice for no more than 6 h.

2.3.2.2 Day 2

- Stop the end-labeling reaction by washing the embryos for 10 min in working strength ApopTag Stop/Wash Buffer (1 mL Stop/Wash Buffer + 34 mL H_2O).
- Transfer the embryos back to 5 mL glass vials and block in blocking solution at room temperature for 3–4 h.
- Remove blocking solution and add 1:5000 anti-digoxigenin-AP Fab fragment (Roche 11093274910) diluted in blocking solution. Incubate for 2–4 h at room temperature or overnight at 4 °C with gentle agitation (40 rpm on a horizontal orbital shaker).
- If not incubating overnight in antibody, wash the embryos twice for 10–15 min in PBST at room temperature with gentle agitation and then store overnight at 4 °C.

2.3.2.3 Day 3

- Finish the PBST washes. Total wash time should be at least 2 h and a minimum of six changes of PBST.

- Wash the embryos three times for 5 min with gentle agitation in fresh NTMT buffer.
- Prepare fresh staining solution: 10 μL of NBT/BCIP (5-bromo-4-chloro-3-indolyl-phosphate/nitro blue tetrazolium) stock solution in 1 mL of NTMT buffer. The staining solution should be yellow in color and must be kept in the dark. Use ~300 μL per batch of embryos. After adding the staining solution to the embryos, incubate at room temperature in a dark location. Monitor the color reaction periodically under a dissecting microscope and stop when apoptotic cells begin to turn dark blue. Staining time will vary from 10 min to several hours.
- Stop the staining reaction by washing the embryos several times in PBST.
- Refix embryos overnight in 4% paraformaldehyde.

2.3.3 Alternate methods

As an alternative to chromogenic staining with anti-digoxigenin-AP and NBT/BCIP, detection can be done using fluorescent secondary antibodies (Yager, Ikegami, Rivera-Bennetts, Zhao, & Brooker, 1997) or fluorescent TUNEL cell death detection reagents such as TMR Red (Roche 12156792910) (van Ham et al., 2010). This allows TUNEL to be combined with other fluorescent stains such as DAPI, Hoechst 33258, or ethidium bromide. Embryos may also be embedded and serially sectioned prior to TUNEL staining using standard protocols (Sullivan-Brown, Bisher, & Burdine, 2011).

2.4. Whole-mount antiactive caspase-3 assay

Executioner caspases (e.g., caspase-3 and -7) are present in the cytosol as inactive proenzymes that become activated following proteolytic processing at conserved aspartic residues to produce a large and small subunit (Boatright & Salvesen, 2003). Activated caspase-3 can be directly detected in fixed zebrafish embryos using a monoclonal antibody that does not recognize the inactive full-length form of caspase-3, only the large activated fragment resulting from cleavage adjacent to the Asp175 residue (Eimon et al., 2006).

2.4.1 Required materials

- Methanol
- PBST: 1 × PBS + 0.1% Tween 20

- 4% paraformaldehyde: 4% (w/v) paraformaldehyde in PBST, pH 7.4 (prepare fresh before use)
- Acetone (prechilled to −20 °C)
- IHC blocking solution: 1 × PBST, 5% (v/v) fetal bovine serum, 2 mg mL^{-1} BSA (Sigma A3059)
- 5 mL screw-cap glass vials (e.g., VWR 66011-041)
- Glass Pasteur pipettes and a 10-mL handheld pipette pump (e.g., VWR 53502-233)

2.4.2 Method

- Dechorionate live embryos of the desired stage(s) either manually or using pronase (see Section 2.1), fix them in 4% paraformaldehyde, and then dehydrate them in 100% methanol following the protocol in Section 2.3.
- Rehydrate the embryos by successive 2–5 min washes in the following:
  - 75% methanol + 25% PBST
  - 50% methanol + 50% PBST
  - 25% methanol + 75% PBST
- Wash the embryos twice for 2–5 min in PBST at room temperature.
- Wash the embryos once for 5 min in deionized water.
- Remove as much of the water as possible and permeabilize the embryos by incubating in prechilled acetone for 7–10 min at −20 °C.
- Remove the acetone and wash the embryos in deionized water for 5 min at room temperature.
- Wash the embryos twice for 2–5 min in PBST at room temperature with gentle agitation.
- Block embryos for 2 h at room temperature in IHC blocking solution.
- Prepare a 1:500 dilution of the antiactive caspase-3 antibody (BD Pharmingen 559565) in IHC blocking solution and incubate the embryos overnight at 4 °C; gentle agitation may be used during the incubation step but is not essential. To minimize antibody usage, the embryos can be transferred from glass vials to 1.5 mL microcentrifuge tubes. This will enable incubation volumes of 50–100 μL. *Note:* Several embryos should be incubated in IHC blocking solution without the anti-active caspase-3 antibody to serve as a no-primary control.
- Transfer the embryos back to 5 mL glass vials and wash three times for 20 min in PBST at room temperature with gentle agitation (40 rpm on a horizontal orbital shaker).

- Block embryos for 2 h at room temperature in IHC blocking solution.
- Prepare a dilution of an appropriate anti-rabbit IgG secondary antibody in IHC blocking solution and incubate the embryos for 2 h at room temperature or overnight at 4 °C. Secondary antibodies conjugated to fluorescent dyes can be used to enable fluorescence microscopy. As before, the embryos can be transferred to the 1.5-mL microcentrifuge tubes to minimize the reagent volume if desired. Secondary antibodies that have been used successfully for this assay include:
  - F(ab′)2 Fragment Goat anti-rabbit IgG (H + L) (Jackson ImmunoResearch; 1:500 dilution) (Eimon et al., 2006; Kratz et al., 2006)
  - Goat anti-rabbit IgG (H + L) (Molecular Probes; 1:200 dilution) (Jette et al., 2008)
  - Biotinylated anti-rabbit IgG (H + L) (Vector Laboratories; 1:1000 dilution) (Porreca, De Felice, Fagman, Di Lauro, & Sordino, 2012)
  - Goat anti-rabbit IgG (H + L) (GE Healthcare) (Yamashita et al., 2008)
- Transfer the embryos back to the 5-mL glass vials and wash three times for 20 min in PBST at room temperature with gentle agitation to remove excess secondary antibody.
- Following the wash steps, antiactive caspase-3 staining can be visualized by fluorescence microscopy. Staining is stable for several days when the embryos are stored at 4 °C in PBST.
- Caspase-3 activity can be quantified using image analysis software, keeping in mind that fluorescence intensity is likely to be influenced both by the number of apoptotic cells and the level of caspase-3 activity within cells (Sorrells et al., 2012).

2.4.3 Alternate method

The above protocol is intended for fluorescent imaging. An alternative method involves detection of the antiactive caspase-3 antibody using a secondary antibody conjugated to peroxidase instead of a fluorescent dye, followed by visualization with a chromogenic substrate such as DAB (Sigma) (Porreca et al., 2012). One advantage of this technique is that it can be easily combined with standard *in situ* hybridization protocols (Thisse & Thisse, 2008) to detect gene expression and caspase-3 activity in the same sample. When using peroxidase-conjugated antibodies, zebrafish embryos should be incubated with 10% H_2O_2 for 30 min immediately prior to the blocking step in order to inactivate endogenous peroxidases.

2.5. Fluorometric and luminogenic caspase substrate assays

Another useful technique for assessing apoptosis in zebrafish embryos is to directly measure the protease activity of caspases. This is done using caspase-specific substrates that release fluorogenic or luminogenic moieties upon proteolytic cleavage. Several commercially available caspase activity assays have been applied to zebrafish, including Caspase-Glo 8 and Caspase-Glo 9 Assays (Promega) (Geiger et al., 2006), Z-DEVD-AMC (a caspase-3/7 substrate; Molecular Probes) (Lee et al., 2008), Ac-LEHD-AFC and Ac-DEVD-AFC (substrates for caspase-9 and caspase-3/7, respectively) (Mei, Zhang, Li, Lin, & Gui, 2008), and a number of acetyl-substrates linked to 4-methyl-coumaryl-7-amide (substrates tested include DEVD, YVAD, LEHD, and IETD; Peptide Institute; Yabu, Kishi, Okazaki, & Yamashita, 2001). Protocols vary depending on the assay used, but typically involve the following steps:

- Wash dechorionated embryos in PBS, pooling anywhere from 5 to 20 embryos per experimental condition.
- Homogenize the embryos in an appropriate lysis buffer (e.g., 20 m*M* Hepes/KOH pH 7.5, 250 m*M* sucrose, 50 m*M* KCl, 2.5 m*M* $MgCl_2$, and 1 m*M* dithiothreitol) (Yabu et al., 2001) supplemented with a protease inhibitor cocktail. Embryos can be efficiently homogenized by passage through a 27-gauge needle.
- Centrifuge the lysate at 10,000 × *g* for 15 min at 4 °C and collect the supernatant.
- At this step, protein concentration can be determined using a Bradford assay or similar protocol in order to normalize results based on total protein levels.
- Follow the manufacturer's recommendation when performing the caspase activity assay, as protocols vary based on the kit used. The following protocol is taken from the Z-DEVD-AMC Substrate Caspase-3 Assay Kit (Molecular Probes; Lee et al., 2008).
- Transfer ~200 μg of embryo lysate into a microwell plate for reaction with 100 μ*M* caspase-3 substrate (Z-DEVD-AMC) in Caspase Reaction Buffer (10 m*M* PIPES, pH 7.4, 2 m*M* EDTA, 0.1% CHAPS). Use an equivalent volume of the lysis buffer alone as a no-enzyme control to determine the background fluorescence of the substrate. If desired, add 1 μL of 1 m*M* Ac-DEVD-CHO inhibitor to selected samples as an additional control and incubate at room temperature for 10 min. The remaining samples (without inhibitor) should be stored on ice during this time.

- Cover the microplate and incubate the samples for 30 min–2 h at room temperature.
- The release of the fluorescent moiety is measured through change in fluorescence using a standard fluorescence microplate reader and appropriate excitation and emission filters or settings (e.g., excitation/emission ~342/441 nm for Z-DEVD-AMC). Because the assay is continuous, observations can be made at multiple time points in order to verify that linear initial rates are being measured.

Although most caspase activity assays are carried out on zebrafish embryo lysates as described earlier, Caspase-Glo assays (Promega) have also been performed on intact embryos in microwell plates (Geiger et al., 2006). Prior to imaging, embryos are plated into 96-well plates (10 embryos per well) along with the Caspase-Glo substrate, which is prepared following the manufacturer's instructions. After incubation for 2 h at room temperature, the microplate along with the embryos is transferred to a plate-reading luminometer, images are acquired, and activity is analyzed using imaging software.

3. APPROACHES TO MANIPULATING APOPTOSIS PATHWAYS IN ZEBRAFISH

Zebrafish offer a convenient model system to study the *in vivo* effects of activation or inhibition of specific components of the apoptotic machinery, particularly in the context of early development. This can be especially valuable when studying genes that result in lethal embryonic phenotypes when knocked out in mice (Eimon et al., 2006; Kratz et al., 2006; Rinkenberger, Horning, Klocke, Roth, & Korsmeyer, 2000; Varfolomeev et al., 1998; Yeh et al., 1998, 2000) and genes that are involved in regulation of normal developmental apoptosis programs (Porreca et al., 2012). Additionally, the ability to overexpress and knock down combinations of genes in a relatively high-throughput low-cost manner makes the zebrafish an attractive model for conducting epitasis analysis of apoptotic pathways *in vivo* (Espin et al., 2013; Jette et al., 2008; Nishiwaki et al., 2013; Rodriguez-Aznar & Nieto, 2011; Sidi et al., 2008). Genes of interest are commonly overexpressed or inhibited by microinjection of synthetic mRNA (Section 3.1) and morpholino antisense oligonucleotides (Section 3.2), respectively.

3.1. mRNA microinjection

Transient overexpression can be achieved by microinjection of either synthetic mRNA or plasmid DNA. In theory, injection of plasmid DNA constructs allows overexpression to be targeted to a specific organ of interest, provided a tissue-specific promoter has been identified. In practice, DNA constructs are typically expressed in a limited number of cells and in a highly mosaic pattern (Nüsslein-Volhard & Dahm, 2002). Several strategies have been developed to increase the efficiency of plasmid DNA expression, including incorporating the transcriptional activator Gal4-VP16 to amplify transgene expression (Koster & Fraser, 2001) and adding Tol2 sites to the backbone vector and coinjecting synthetic mRNA encoding the Tol2 transposase enzyme (Fisher et al., 2006; Kwan et al., 2007). In spite of these improvements, most transient expression experiments in zebrafish are still done using *in vitro* transcribed synthetic mRNAs. Unlike DNA constructs, injected mRNA is distributed uniformly throughout the developing embryo and is, therefore, expressed ubiquitously during the first 3–5 dpf. Although this approach sacrifices the spatiotemporal control of expression that can be achieved using DNA constructs, it compensates by producing overexpression phenotypes that are robust, uniform, and reproducible.

3.1.1 RNA synthesis: Preparation of template

By far the most common vector used for *in vitro* transcription of synthetic mRNA for zebrafish overexpression experiments is pCS2+. Numerous variants of this vector are available through Addgene (e.g., Addgene 17095). The pCS2+ vector contains a number of useful features, including: (1) a multiple cloning site (MCS) located immediately downstream of the SP6 promoter; (2) an SV40 late polyadenylation (polyA) signal sequence (Okayama & Berg, 1983), which has been shown to be highly effective in stabilizing mRNA transcripts in zebrafish (Kwan et al., 2007); and (3) a second MCS immediately downstream of the SV40 poly(A) sequence for linearization. Versions of pCS2+ have been generated that are compatible with the widely used Gateway cloning system (Villefranc, Amigo, & Lawson, 2007). Typically, when a gene is subcloned into the pCS2+ vector, a strong Kozak consensus sequence (ACCATG; start codon underlined) is included.

Once the full-length open-reading frames from the genes of interest have been subcloned into pCS2+, a linearized template for *in vitro* transcription is generated using the following protocol:

- Linearize 10 μg of the pCS2+ expression vector using an appropriate restriction enzyme from the MCS downstream of the SV40 poly(A) signal sequence.
- Purify the linearized template using a column-based PCR purification kit (e.g., Qiagen 28104 or equivalent). Alternatively, for added stringency in removing nucleases prior to mRNA synthesis, the following protocol may be used in place of the column-based approach:
  - Bring the total sample volume to 100 μL, add 5 μL of 10% SDS and 1 μL of 20 mg mL^{-1} Proteinase K Solution (e.g., Ambion AM2546), and incubate for 30 min at 55 °C.
  - Bring the total sample volume to 200 μL with nuclease-free water, add an equal volume of phenol/chloroform/isoamyl alcohol (25:24:1, v/v/v, pH 6.7; Life Technologies 15593-031), vortex briefly to mix, and centrifuge 5 min at maximum speed.
  - Transfer the upper aqueous layer to a new microcentrifuge tube (being careful not to include any phenol) and re-extract as above with an equal volume of chloroform.
  - Transfer the supernatant to a new microcentrifuge tube, add 20 μL of 5 *M* nuclease-free ammonium acetate (Ambion AM9070G), and 200 μL of 100% ethanol. Mix well by inverting the tube several times and chill at −20 °C for 15 min.
  - Pellet the DNA by centrifugation for 15 min at maximum speed. Remove the supernatant, add 70% ethanol, and centrifuge for 5 min at maximum speed.
  - Remove as much of the supernatant as possible and leave the tube open to air-dry for 10 min or until no ethanol is visible.
  - Resuspend the purified template in 10–30 μL of nuclease-free water by vortexing and a brief 10 min incubation at 37 °C.
- Verify the concentration of the purified template using a spectrophotometer.

3.1.2 RNA synthesis: In vitro *transcription*

- Generate synthetic mRNA using the purified template and the mMessage mMachine SP6 kit (Ambion AM1340).
  +10 μL 2× NTP/Cap mix
  +2 μL 10× reaction buffer
  +2 μg linearized template DNA
  +2 μL enzyme mix
  - Bring to 20 μL with nuclease-free water
  - Incubate 2 h at 37 °C.

  - Add 1 μL DNaseI and incubate for an additional 15 min at 37 °C.
- Purify mRNA using either commercial purification columns (e.g., Qiagen RNeasy Kit 74104 or Ambion NucAway spin columns AM10070) or the following lithium-chloride precipitation protocol:

  +115 μL of nuclease-free water

  +15 μL ammonium acetate stop solution (5 *M* ammonium acetate, 100 m*M* EDTA, included with mMessage mMachine Kit)
  - Mix thoroughly and extract as above with an equal volume of phenol/chloroform/isoamyl alcohol followed by an equal volume of chloroform.
  - Transfer the supernatant to a new tube and precipitate the mRNA by adding one volume of isopropanol and mixing well.
  - Chill the mixture for at least 15 min at –20 °C.
  - Centrifuge at 4 °C for 15 min at maximum speed to pellet the mRNA.
  - Carefully remove the supernatant and resuspend the mRNA in nuclease-free water.
- The size and integrity of synthetic mRNAs should be assessed by electrophoresis on either a standard agarose gel (0.8% agarose and 0.5 × TAE) or a denaturing agarose gel (1.5% agarose, 1 × MOPS, 5 m*M* sodium acetate, 0.1 m*M* EDTA, and 6% formaldehyde; run in 1 × MOPS running buffer).
- Store mRNA frozen in single-use aliquots at −70 °C.

Working stocks of mRNAs for injection into zebrafish embryos are often prepared in near-isotonic solutions such as Danieau's (see below), which was originally developed for cell transplantation work. For visualization of the injection, 0.1% phenol red can be added to the working stock. The concentration of the working stock should be determined such that the desired amount of mRNA can be delivered in a volume of no more than 5 nL. The total amount of mRNA to inject depends on the potency of the gene(s) being assayed. Typical amounts range from less than 10 pg up to 500 + pg.

Danieau's solution (30 ×)

- 1740 m*M* NaCl
- 21 m*M* KCl
- 12 m*M* $MgSO_4 \cdot 7H_2O$
- 18 m*M* $Ca(NO_3)_2$
- 150 m*M* Hepes buffer

  Store at 4 °C. The pH is 7.6.

3.1.3 Microinjection of zebrafish embryos

A number of detailed microinjection protocols are available and should be referred to as a supplement to the basic steps outlined later (Bill, Petzold, Clark, Schimmenti, & Ekker, 2009; McKinney & Weinstein, 2008; Nüsslein-Volhard & Dahm, 2002; Rosen, Sweeney, & Mably, 2009).

3.1.3.1 Prior to injection

- Prepare a microinjection plate to hold the embryos during the injection procedure. Diagrams for injection plate molds have been published (Westerfield, 1995); molds are also commercially available (Adaptive Science Tools TU-1).
- Pull a 1.0-mm OD glass capillary using a Sutter Instruments micropipette puller to produce two microinjection needles. Immediately prior to loading, the tip of the needle is broken using dissecting forceps under a high-power dissecting microscope to give an opening between 0.05 and 0.15 mm.
- Set up the microinjection apparatus. Microinjectors can be divided into two categories: Pressure-driven injectors (e.g., the PV820 Pneumatic PicoPump from WPI), which use pneumatic pumps and must be recalibrated to deliver a defined volume for each new microinjection needle, and piston-driven injectors (e.g., the Nanoliter 2000 from WPI), which use displacement of mineral oil to deliver reproducible volumes independent of the size of the needle opening.

3.1.3.2 On the day of injection

- Load the mRNA into the needle and attach it to the microinjector (needles are typically backfilled for pressure-driven injectors and front filled for piston-driven ones). *Note:* Do not leave the needle tip exposed to the air for too long or it may clog due to evaporation. Once loaded, the tip should be immersed in a Petri dish-containing embryo medium or mineral oil when not in use.
- Collect the embryos as soon as possible after spawning (remove any that are dead or unfertilized) and place them into the microinjection plate.
- Inject the embryos by inserting the needle through the chorion and into the yolk cytoplasm, placing the tip just below the developing blastomeres. Because the membrane that separates the blastomeres from the yolk does not form until the 16-cell stage, mRNA injected up to the 8-cell stage will be ubiquitously expressed.

- Transfer the embryos back into a Petri dish-containing fresh embryo medium, removing any that were damaged during the injection process, and incubate at 28 °C.

Monitor embryos periodically to access the overexpression phenotype, making sure to change the embryo medium daily and promptly remove any dead or dying embryos. Overexpression phenotypes can be assessed using the protocols in Section 2 or other standard methods that have been optimized for use on zebrafish, including quantitative real-time PCR (Lan, Tang, Un San Leong, & Love, 2009), whole-mount *in situ* hybridization (Thisse & Thisse, 2008), cell transplantation (Li, White, & Zon, 2011), and the wide range of mutant and tissue-specific transgenic lines available through resources like ZIRC.

3.2. Morpholinos

Morpholino oligonucleotides (manufactured by GeneTools) are the most widely used antisense knockdown tool in zebrafish embryos (Heasman, 2002; Nasevicius & Ekker, 2000). Morpholinos are short (typically 25 subunit) oligonucleotides comprised of standard nucleic acid bases (A, C, G, and T) positioned on a backbone in which the ribose/deoxyribose rings have been replaced by morpholine rings, and the standard phosphodiester linkages have been replaced by phosphorodiamidate linkages (Summerton & Weller, 1997). Because of their unnatural backbones, morpholinos are highly resistant to degradation by nucleases (Hudziak et al., 1996). Morpholinos, like synthetic mRNAs, are delivered immediately following fertilization using a microinjection apparatus. Once inside the cell, a morpholino specifically binds to and functionally inhibits its selected target by one of two mechanisms: (1) blocking mRNA translation or (2) preventing proper pre-mRNA splicing.

3.2.1 Translation blocking

Translation-blocking morpholinos bind to the start codon or 5′ UTR of a target transcript (typically anywhere between position −50 and +25) and sterically block progression of the ribosomal initiation complex. Unlike siRNA antisense technologies, morpholinos generally do not cause degradation of the targeted mRNA transcript. This means that RT-PCR is not a suitable means for ascertaining the efficacy of translation-blocking morpholinos. Ideally, the level of knockdown should be determined using an antibody to the protein of interest, although this is not always feasible. As an alternative, the knockdown efficiency can be assessed by coinjecting the morpholino and *in vitro* transcribed mRNA encoding a version of the target gene containing a tag (e.g., HA, FLAG, GFP).

3.2.2 Splice blocking

Splice-blocking morpholinos target splice-donor or splice-acceptor sites and block proper pre-mRNA processing (Fig. 16.1). This typically results in lost exons or intron inclusions that encode a truncated, nonfunctional protein product. The activity of splice-blocking morpholinos can be detected by RT-PCR, since successful splice modification will result in a mobility shift or complete loss of the wild-type transcript (Draper, Morcos, & Kimmel,

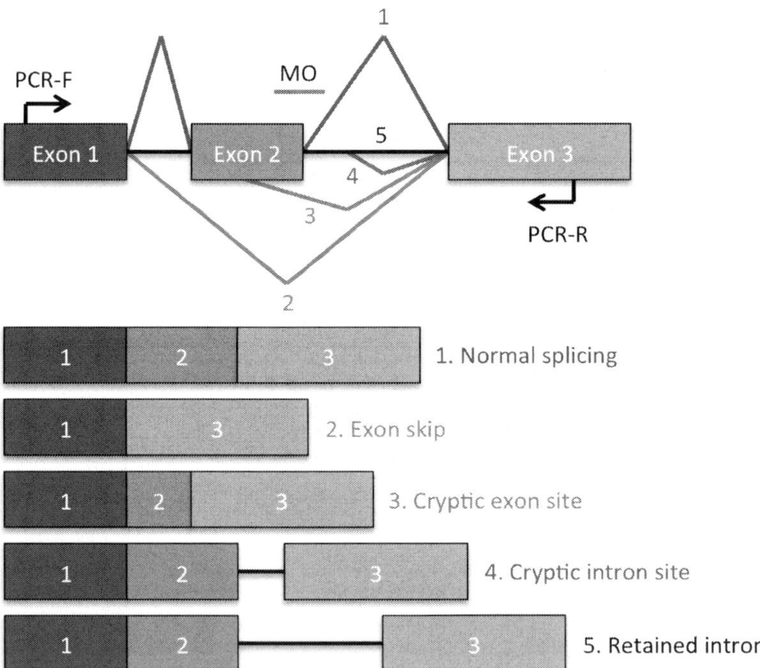

Figure 16.1 Potential outcomes when using splice-blocking morpholinos to inhibit gene function. Splice-blocking morpholinos can be targeted to either exon/intron or intron/exon boundaries. The most common outcome is a skipped exon, but other possibilities include activation of cryptic splice sites and retention of an entire intron. The resultant product is typically nonfunctional due to deletion of functional domains and frame shifts resulting in premature stop codons. The diagram illustrates a three exon transcript in which the splice-donor site at the 3′ end of exon 2 has been targeted and shows five possible outcomes (below) that may result from specific binding of the morpholino. RT-PCR primers (PCR-F and PCR-R) flanking the affected exons can be used to detect and analyze the activity of the splice-blocking morpholino.

2001). Because of their mode of action, splice-blocking morpholinos only target zygotic transcripts. In contrast, translation-blocking morpholinos can inhibit both zygotic and maternal mRNA transcripts.

The following section describes the basic steps of morpholino design, handling, injection, and analysis in zebrafish embryos. More detailed protocols can be found in several recent reviews (Bill et al., 2009; Eisen & Smith, 2008). Information on published morpholino sequences that target apoptosis-related genes can be found in Table 16.1. ZFIN gene pages (http://zfin.org) also serve as valuable resources for published morpholinos (Sprague et al., 2008).

3.2.3 Designing morpholinos

For a detailed discussion of the considerations involved in successful morpholino design, see the excellent review by Bill et al. (2009). Optimal morpholino oligonucleotides are 25 bases in length, have 40–60% GC-content, contain no more than three contiguous guanine residues and no more than nine total, and lack significant self-complementarity (i.e. no more than 16 contiguous intrastrand hydrogen bonds). A BLAST search should always be performed to confirm that the proposed target sequence is not found elsewhere in the genome.

For translation-blocking morpholinos, the target site should be located either entirely within the 5′ UTR or extend at most about 30 bases upstream of the start codon. The efficiency of morpholinos drops off rapidly when the target site is moved further upstream of the translation initiation site. One effective strategy for designing translation-blocking morpholinos is to start by analyzing the first 25 bases of coding sequence and then progressively moving the target window upstream until a site is found that satisfies the basic requirements for an optimized morpholino. This approach has the advantage of minimizing the amount of target sequence that falls within the UTR, which is more likely to contain single nucleotide polymorphisms than the coding region.

Design considerations for splice-blocking morpholinos are more complex (Morcos, 2007). Obviously, the target gene must be composed of at least two exons in order to utilize this approach and the Ensembl Genome Browser (http://uswest.ensembl.org/Danio_rerio/Info/Index) or other appropriate databases should be queried to determine if alternatively spliced mRNAs are known. Splice-blocking morpholinos are designed to target either splice-donor or splice-acceptor sites and may lead to a variety of outcomes, including exon skipping, intron insertion, and activation of cryptic splice sites (Fig. 16.1). Targeting the splice-donor site of the first exon or the splice acceptor site of the last exon will often lead to inclusion of the entire associated intron. In contrast, targeting either splice-donor or -acceptor sites of internal exons typically results in exon skipping. The activation of cryptic splice sites is another possible outcome and leads to partial inclusion of introns or partial deletion of exons. All of these outcomes typically cause a frameshift resulting in premature stop codons and the resultant mRNA may be eliminated through nonsense-mediated decay. RT-PCR should be carried out using primers that flank the targeted splice site to quantify efficacy of the knockdown. Sequencing of the PCR product may also be carried out to confirm the precise nature of the splice alteration(s) induced by the morpholino. It has been reported that knockdown efficiency can be increased by coinjection of two morpholinos, one targeting the splice junction and the other targeting the splice branch site (Morcos, 2007).

3.2.4 Storage and handling of morpholinos

Morpholino oligonucleotides should be resuspended at a concentration of 1–3 m*M* in cell culture grade distilled water (DEPC-free). Heating the stock for 10 min at 65 °C and vortexing will encourage full resuspension. It is a good practice to verify the concentration of all stocks using a spectrophotometer and the following protocol:

- Blank the spectrophotometer using 0.1 N HCl.
- Dilute the morpholino stock at least 1:20 in 0.1 N HCl.
- Measure the absorbance at 265 nm.
- Calculate the concentration of the morpholino using the Beer-Lambert law: $C = A/\varepsilon$. C, concentration; A, absorbance (multiplied by the dilution factor); ε, molar absorptivity at 265 nm (which can be found on the morpholino product information sheet). *Note:* This calculation assumes a spectrophotometer with a 1-cm path length (or a spectrophotometer such as the NanoDrop, with software that automatically normalizes data to the standard 1 cm path).

Although many labs store morpholinos at 4 °C or −20 °C, GeneTools recommends storing stocks at room temperature in tightly sealed tubes wrapped with parafilm. For secure longer-term storage (months to years), lyophilization in small glass vials is strongly recommended (Bill et al., 2009). Chilling the morpholinos or subjecting them to multiple freeze-thaw cycles may, in some instances, cause them to drop out of solution and become associated with the walls of the container. This will lead to a loss of activity over time. For this reason, morpholinos should never be kept on ice on the bench top.

3.2.5 *Microinjection of morpholinos*

Microinjection of morpholinos is carried out using the same basic protocol as microinjection of mRNAs (Section 3.2) (Bill et al., 2009). Morpholinos are typically injected at the 1–8 cell stage at doses ranging from 1 to 10 ng per embryo. The first time a new morpholino is used a dose–response experiment should be carried out to optimize the phenotype-to-toxicity ratio. For genes with mutant lines available, optimization can be done by selecting the lowest dose that reliably phenocopies the mutant phenotype. For other genes, optimization may involve quantifying knockdown using RT-PCR (for splice-blocking morpholinos) or antibodies (80% or greater knockdown for mRNA or protein is commonly used as the benchmark for an "effective" morpholino), or by assessing the specificity of the knockdown phenotype in combination with an mRNA rescue experiment, as discussed later.

3.3. Special considerations when using morpholinos to target apoptotic pathways

As with all antisense techniques, morpholinos have the potential to trigger nonspecific "off-target" effects (Ekker & Larson, 2001; Wright, Leslie, Ariza-McNaughton, & Lewis, 2004). Therefore, it is critical to employ appropriate controls in all morpholino experiments (Bill et al., 2009; Eisen & Smith, 2008). One of the most commonly reported off-target phenotypes—and one that is of particular concern when studying apoptotic pathways—is extensive neural death peaking at ~1 dpf. This phenotype has been reported for both translation- and splice-blocking morpholinos and may be associated with as many as 15–20% of all morpholinos (Ekker & Larson, 2001). Although the underlying mechanism remains unclear, the phenotype has been characterized in some detail (Robu et al., 2007). It appears to be mediated through tp53-induced apoptosis and can be attenuated by coinjection of a morpholino targeting the endogenous tp53 transcript. Because neuronal apoptosis is such a common off-target effect, some labs choose to routinely coinject a tp53-targeting morpholino. This can be a good solution, since tp53 is not required for normal development in zebrafish (Berghmans et al., 2005), and the tp53 morpholino does not appear to interfere with most gene-specific phenotypes (Bill et al., 2009). It does, however, have obvious implications for experiments targeting apoptotic pathways and must be used with due caution. At least some specific cell death phenotypes, such as the elevated levels of apoptosis seen in the tails of embryos injected with a chordin-targeting morpholino, are still observed in embryos coinjected with the tp53 morpholino (Robu et al., 2007).

When using morpholinos to evaluate a gene with an unknown phenotype, at least two separate morpholinos must be designed and shown to produce the same outcome. Often this is done using one translation-blocking morpholino and one splice-blocking morpholino, although any two non-overlapping targets are acceptable. The specificity of the observed phenotype can be further confirmed by preparing a dose–response curve for each morpholino and then coinjecting both at concentrations just below their lowest-observed-effect levels and showing that they act synergistically to reproduce the expected phenotype.

A second common control is to demonstrate that the observed phenotype is not seen when injecting a negative control morpholino that does not target the gene of interest. A number of possible control morpholino strategies have been proposed. One common approach is to use a standard control morpholino sold by GeneTools. More stringent options include using a fully scrambled sequence that preserves the overall base composition of the targeting morpholino, an invert of the target sequence, or a five-base mismatch morpholino. Alternatively, morpholinos that specifically target other genes may sometimes be used as a negative control, provided they do not impact the normal development of the organ(s) being studied (Eisen & Smith, 2008).

The final, and most reliable, control is to show reversal of the phenotype when the morpholino is coinjected with an *in vitro* transcribed mRNA that has been rendered "immune" to inhibition. This type of control is particularly critical when using morpholinos to assess apoptotic processes that rely on tp53 signaling and, therefore, prohibit the use of the tp53-targeted morpholino discussed above. There are several options for removing the morpholino target sequence from the synthetic mRNA for "rescue" experiments. The simplest case is splice-blocking morpholinos, which will have no effect on synthetic mRNAs that lack introns. For translation-blocking morpholinos with target sites in the 5′ UTR, the open-reading frame minus the UTR is often cloned into a standard *in vitro* transcription vector such as pCR2+ (see Section 3.2). Alternatively, if the morpholino target sequence is contained entirely within the coding sequence, a number of silent mutations can be introduced into the rescue construct. Once an appropriate mRNA has been engineered, it should be stored, handled, and microinjected as discussed in Section 3.2. At a minimum, rescue experiments should contain the following experimental groups: (1) morpholino alone, (2) mRNA alone, and (3) morpholino + mRNA. For added stringency, a negative control mRNA may be included in the "morpholino alone" condition (e.g., an mRNAs encoding a fluorescent protein) and one of the negative control morpholinos discussed above may be included in the "mRNA alone" condition to keep total mRNA and morpholino concentrations constant. One final note of caution: since injected mRNA is distributed more-or-less uniformly throughout the developing embryo, rescue experiments are only feasible when the gene being investigated is also either uniformly distributed or has a mild and reproducible overexpression phenotype. These criteria are not always possible to achieve, particularly for genes that activate core elements of the apoptotic machinery and induce widespread cell death.

4. MUTANT LINES AND MORPHOLINOS

See Table 16.1.

5. SUMMARY

Zebrafish offer unique advantages for the study of apoptosis. Among the most compelling is the ability to observe apoptotic processes *in vivo* at cellular (and in some cases subcellular) resolution in an intact vertebrate organism. This capability is even more powerful when combined with techniques for rapidly inhibiting and activating specific components of the apoptotic machinery. Future apoptosis research in zebrafish will continue to exploit these strengths while also drawing on powerful emerging tools and resources. In particular, although loss-of-function mutant lines have not yet been identified for many of the apoptosis-related genes listed in Table 16.1, this situation is expected to change dramatically in the coming years as tools such as TALENs and CRISPRs become more widely applied to genome editing in zebrafish. In addition, it is likely that researchers will continue to develop sophisticated transgenic lines expressing genetically encoded reporters of apoptosis and tools for precise activation and/or inhibition of components of the apoptotic machinery.

ACKNOWLEDGMENTS

The author wishes to thank the NIH Transformative Research Award (R01 NS073127), the Massachusetts Institute of Technology, and Professor Mehmet Fatih Yanik for funding and support.

REFERENCES

Amacher, S. L. (2008). Emerging gene knockout technology in zebrafish: Zinc-finger nucleases. *Briefings in Functional Genomics & Proteomics*, 7(6), 460–464. http://dx.doi.org/10.1093/bfgp/eln043, eln043 [pii].

Arnaud, E., Ferri, K. F., Thibaut, J., Haftek-Terreau, Z., Aouacheria, A., Le Guellec, D., et al. (2006). The zebrafish bcl-2 homologue Nrz controls development during somitogenesis and gastrulation via apoptosis-dependent and -independent mechanisms. *Cell Death and Differentiation*, *13*(7), 1128–1137. http://dx.doi.org/10.1038/sj.cdd.4401797, 4401797 [pii].

Bates, J. M., Akerlund, J., Mittge, E., & Guillemin, K. (2007). Intestinal alkaline phosphatase detoxifies lipopolysaccharide and prevents inflammation in zebrafish in response to the gut microbiota. *Cell Host & Microbe*, *2*(6), 371–382. http://dx.doi.org/10.1016/j.chom.2007.10.010.

Berghmans, S., Murphey, R. D., Wienholds, E., Neuberg, D., Kutok, J. L., Fletcher, C. D., et al. (2005). tp53 mutant zebrafish develop malignant peripheral nerve sheath tumors. *Proceedings of the National Academy of Sciences of the United States of America*, *102*(2), 407–412. http://dx.doi.org/10.1073/pnas.0406252102, 0406252102 [pii].

Bill, B. R., Petzold, A. M., Clark, K. J., Schimmenti, L. A., & Ekker, S. C. (2009). A primer for morpholino use in zebrafish. *Zebrafish*, *6*(1), 69–77. http://dx.doi.org/10.1089/zeb.2008.0555.

Boatright, K. M., & Salvesen, G. S. (2003). Mechanisms of caspase activation. *Current Opinion in Cell Biology*, *15*(6), 725–731.

Darzynkiewicz, Z., Bruno, S., Del Bino, G., Gorczyca, W., Hotz, M. A., Lassota, P., et al. (1992). Features of apoptotic cells measured by flow cytometry. *Cytometry*, *13*(8), 795–808. http://dx.doi.org/10.1002/cyto.990130802.

Delvaeye, M., De Vriese, A., Zwerts, F., Betz, I., Moons, M., Autiero, M., et al. (2009). Role of the 2 zebrafish survivin genes in vasculo-angiogenesis, neurogenesis, cardiogenesis and hematopoiesis. *BMC Developmental Biology*, *9*, 25. http://dx.doi.org/10.1186/1471-213X-9-25, 1471-213X-9-25 [pii].

Draper, B. W., Morcos, P. A., & Kimmel, C. B. (2001). Inhibition of zebrafish fgf8 pre-mRNA splicing with morpholino oligos: A quantifiable method for gene knockdown. *Genesis*, *30*(3), 154–156.

Driever, W., Solnica-Krezel, L., Schier, A. F., Neuhauss, S. C., Malicki, J., Stemple, D. L., et al. (1996). A genetic screen for mutations affecting embryogenesis in zebrafish. *Development*, *123*, 37–46.

Eimon, P. M., & Ashkenazi, A. (2010). The zebrafish as a model organism for the study of apoptosis. *Apoptosis*, *15*(3), 331–349. http://dx.doi.org/10.1007/s10495-009-0432-9.

Eimon, P. M., Kratz, E., Varfolomeev, E., Hymowitz, S. G., Stern, H., Zha, J., et al. (2006). Delineation of the cell-extrinsic apoptosis pathway in the zebrafish. *Cell Death and Differentiation*, *13*(10), 1619–1630. http://dx.doi.org/10.1038/sj.cdd.4402015, 4402015 [pii].

Eimon, P. M., & Rubinstein, A. L. (2009). The use of in vivo zebrafish assays in drug toxicity screening. *Expert Opinion on Drug Metabolism & Toxicology*, *5*(4), 393–401. http://dx.doi.org/10.1517/17425250902882128.

Eisen, J. S., & Smith, J. C. (2008). Controlling morpholino experiments: Don't stop making antisense. *Development*, *135*(10), 1735–1743,. http://dx.doi.org/10.1242/dev.001115.

Ekker, S. C., & Larson, J. D. (2001). Morphant technology in model developmental systems. *Genesis*, *30*(3), 89–93.

Espin, R., Roca, F. J., Candel, S., Sepulcre, M. P., Gonzalez-Rosa, J. M., Alcaraz-Perez, F., et al. (2013). TNF receptors regulate vascular homeostasis in zebrafish through a caspase-8, caspase-2 and P53 apoptotic program that bypasses caspase-3. *Disease Models & Mechanisms*, *6*(2), 383–396. http://dx.doi.org/10.1242/dmm.010249.

Fadok, V. A., Bratton, D. L., Frasch, S. C., Warner, M. L., & Henson, P. M. (1998). The role of phosphatidylserine in recognition of apoptotic cells by phagocytes. *Cell Death and Differentiation*, *5*(7), 551–562. http://dx.doi.org/10.1038/sj.cdd.4400404.

Fadok, V. A., Bratton, D. L., & Henson, P. M. (2001). Phagocyte receptors for apoptotic cells: Recognition, uptake, and consequences. *Journal of Clinical Investigation*, *108*(7), 957–962. http://dx.doi.org/10.1172/JCI14122.

Fisher, S., Grice, E. A., Vinton, R. M., Bessling, S. L., Urasaki, A., Kawakami, K., et al. (2006). Evaluating the biological relevance of putative enhancers using Tol2 transposon-mediated transgenesis in zebrafish. *Nature Protocols*, *1*(3), 1297–1305. http://dx.doi.org/10.1038/nprot.2006.230.

Furutani-Seiki, M., Jiang, Y. J., Brand, M., Heisenberg, C. P., Houart, C., Beuchle, D., et al. (1996). Neural degeneration mutants in the zebrafish, Danio rerio. *Development*, *123*, 229–239.

Gavrieli, Y., Sherman, Y., & Ben-Sasson, S. A. (1992). Identification of programmed cell death in situ via specific labeling of nuclear DNA fragmentation. *Journal of Cell Biology*, *119*(3), 493–501.

Geiger, G. A., Parker, S. E., Beothy, A. P., Tucker, J. A., Mullins, M. C., & Kao, G. D. (2006). Zebrafish as a "biosensor"? Effects of ionizing radiation and amifostine on embryonic viability and development. *Cancer Research*, *66*(16), 8172–8181. http://dx.doi.org/10.1158/0008-5472.CAN-06-0466, 66/16/8172 [pii].

Haffter, P., & Nusslein-Volhard, C. (1996). Large scale genetics in a small vertebrate, the zebrafish. *International Journal of Developmental Biology*, *40*(1), 221–227.

Heasman, J. (2002). Morpholino oligos: Making sense of antisense? *Developmental Biology*, *243*(2), 209–214. http://dx.doi.org/10.1006/dbio.2001.0565.

Huang, P., Xiao, A., Zhou, M., Zhu, Z., Lin, S., & Zhang, B. (2011). Heritable gene targeting in zebrafish using customized TALENs. *Nature Biotechnology*, *29*(8), 699–700. http://dx.doi.org/10.1038/nbt.1939.

Hudziak, R. M., Barofsky, E., Barofsky, D. F., Weller, D. L., Huang, S. B., & Weller, D. D. (1996). Resistance of morpholino phosphorodiamidate oligomers to enzymatic degradation. *Antisense & Nucleic Acid Drug Development*, *6*(4), 267–272.

Hwang, W. Y., Fu, Y., Reyon, D., Maeder, M. L., Tsai, S. Q., Sander, J. D., et al. (2013). Efficient genome editing in zebrafish using a CRISPR-Cas system. *Nature Biotechnology*, *31*(3), 227–229. http://dx.doi.org/10.1038/nbt.2501.

Jensen, P. J., Gitlin, J. D., & Carayannopoulos, M. O. (2006). GLUT1 deficiency links nutrient availability and apoptosis during embryonic development. *Journal of Biological Chemistry*, *281*(19), 13382–13387. http://dx.doi.org/10.1074/jbc.M601881200, M601881200 [pii].

Jette, C. A., Flanagan, A. M., Ryan, J., Pyati, U. J., Carbonneau, S., Stewart, R. A., et al. (2008). BIM and other BCL-2 family proteins exhibit cross-species conservation of function between zebrafish and mammals. *Cell Death and Differentiation*, *15*(6), 1063–1072. http://dx.doi.org/10.1038/cdd.2008.42, cdd200842 [pii].

Kokel, D., & Peterson, R. T. (2011). Using the zebrafish photomotor response for psychotropic drug screening. *Methods in Cell Biology*, *105*, 517–524. http://dx.doi.org/10.1016/B978-0-12-381320-6.00022-9.

Koster, R. W., & Fraser, S. E. (2001). Tracing transgene expression in living zebrafish embryos. *Developmental Biology*, *233*(2), 329–346. http://dx.doi.org/10.1006/dbio.2001.0242.

Kratz, E., Eimon, P. M., Mukhyala, K., Stern, H., Zha, J., Strasser, A., et al. (2006). Functional characterization of the Bcl-2 gene family in the zebrafish. *Cell Death and Differentiation*, *13*(10), 1631–1640. http://dx.doi.org/10.1038/sj.cdd.4402016, 4402016 [pii].

Kwan, K. M., Fujimoto, E., Grabher, C., Mangum, B. D., Hardy, M. E., Campbell, D. S., et al. (2007). The Tol2kit: A multisite gateway-based construction kit for Tol2 transposon transgenesis constructs. *Developmental Dynamics*, *236*(11), 3088–3099. http://dx.doi.org/10.1002/dvdy.21343.

Kwan, T. T., Liang, R., Verfaillie, C. M., Ekker, S. C., Chan, L. C., Lin, S., et al. (2006). Regulation of primitive hematopoiesis in zebrafish embryos by the death receptor gene. *Experimental Hematology*, *34*(1), 27–34.

Lan, C. C., Tang, R., Un San Leong, I., & Love, D. R. (2009). Quantitative real-time RT-PCR (qRT-PCR) of zebrafish transcripts: Optimization of RNA extraction, quality control considerations, and data analysis. *Cold Spring Harbor Protocols*. *2009*(10). http://dx.doi.org/10.1101/pdb.prot5314, pdb.prot5314.

Langheinrich, U., Hennen, E., Stott, G., & Vacun, G. (2002). Zebrafish as a model organism for the identification and characterization of drugs and genes affecting p53 signaling. *Current Biology*, *12*(23), 2023–2028, S0960-9822(02)01319-2 [pii].

Lee, K. C., Goh, W. L., Xu, M., Kua, N., Lunny, D., Wong, J. S., et al. (2008). Detection of the p53 response in zebrafish embryos using new monoclonal antibodies. *Oncogene*, *27*(5), 629–640. http://dx.doi.org/10.1038/sj.onc.1210695, 1210695 [pii].

Li, P., White, R. M., & Zon, L. I. (2011). Transplantation in zebrafish. *Methods in Cell Biology*, *105*, 403–417. http://dx.doi.org/10.1016/B978-0-12-381320-6.00017-5.

Lister, J. A., Robertson, C. P., Lepage, T., Johnson, S. L., & Raible, D. W. (1999). Nacre encodes a zebrafish microphthalmia-related protein that regulates neural-crest-derived pigment cell fate. *Development*, *126*(17), 3757–3767.

Ma, A., Lin, R., Chan, P. K., Leung, J. C., Chan, L. Y., Meng, A., et al. (2007). The role of survivin in angiogenesis during zebrafish embryonic development. *BMC Developmental Biology*, 7, 50. http://dx.doi.org/10.1186/1471-213X-7-50, 1471-213X-7-50 [pii].

Masumoto, J., Zhou, W., Chen, F. F., Su, F., Kuwada, J. Y., Hidaka, E., et al. (2003). Caspy, a zebrafish caspase, activated by ASC oligomerization is required for pharyngeal arch development. *Journal of Biological Chemistry*, *278*(6), 4268–4276.

Matthews, R. P., Lorent, K., Manoral-Mobias, R., Huang, Y., Gong, W., Murray, I. V., et al. (2009). TNFalpha-dependent hepatic steatosis and liver degeneration caused by mutation of zebrafish S-adenosylhomocysteine hydrolase. *Development*, *136*(5), 865–875. http://dx.doi.org/10.1242/dev.027565.

McCammon, J. M., & Amacher, S. L. (2010). Using zinc finger nucleases for efficient and heritable gene disruption in zebrafish. *Methods in Molecular Biology*, *649*, 281–298. http://dx.doi.org/10.1007/978-1-60761-753-2_18.

McKinney, M. C., & Weinstein, B. M. (2008). Chapter 4. Using the zebrafish to study vessel formation. *Methods in Enzymology*, *444*, 65–97. http://dx.doi.org/10.1016/S0076-6879(08)02804-8.

Mei, J., Zhang, Q. Y., Li, Z., Lin, S., & Gui, J. F. (2008). C1q-like inhibits p53-mediated apoptosis and controls normal hematopoiesis during zebrafish embryogenesis. *Developmental Biology*, *319*(2), 273–284. http://dx.doi.org/10.1016/j.ydbio.2008.04.022.

Milan, D. J., Peterson, T. A., Ruskin, J. N., Peterson, R. T., & MacRae, C. A. (2003). Drugs that induce repolarization abnormalities cause bradycardia in zebrafish. *Circulation*, *107*(10), 1355–1358.

Morcos, P. A. (2007). Achieving targeted and quantifiable alteration of mRNA splicing with Morpholino oligos. *Biochemical and Biophysical Research Communications*, *358*(2), 521–527. http://dx.doi.org/10.1016/j.bbrc.2007.04.172.

Nasevicius, A., & Ekker, S. C. (2000). Effective targeted gene 'knockdown' in zebrafish. *Nature Genetics*, *26*(2), 216–220.

Negron, J. F., & Lockshin, R. A. (2004). Activation of apoptosis and caspase-3 in zebrafish early gastrulae. *Developmental Dynamics*, *231*(1), 161–170.

Nishiwaki, Y., Yoshizawa, A., Kojima, Y., Oguri, E., Nakamura, S., Suzuki, S., et al. (2013). The BH3-only SNARE BNip1 mediates photoreceptor apoptosis in response to vesicular fusion defects. *Developmental Cell*, *25*(4), 374–387. http://dx.doi.org/10.1016/j.devcel.2013.04.015.

Nüsslein-Volhard, C., & Dahm, R. (2002). *Zebrafish: A practical approach* (1st ed.). Oxford, NY: Oxford University Press.

Okayama, H., & Berg, P. (1983). A cDNA cloning vector that permits expression of cDNA inserts in mammalian cells. *Molecular and Cellular Biology*, *3*(2), 280–289.

Parsons, M. J., Pollard, S. M., Saude, L., Feldman, B., Coutinho, P., Hirst, E. M., et al. (2002). Zebrafish mutants identify an essential role for laminins in notochord formation. *Development*, *129*(13), 3137–3146.

Porreca, I., De Felice, E., Fagman, H., Di Lauro, R., & Sordino, P. (2012). Zebrafish bcl2l is a survival factor in thyroid development. *Developmental Biology*, *366*(2), 142–152. http://dx.doi.org/10.1016/j.ydbio.2012.04.013.

Qi, F., Song, J., Yang, H., Gao, W., Liu, N. A., Zhang, B., et al. (2010). Mmp23b promotes liver development and hepatocyte proliferation through the tumor necrosis factor pathway in zebrafish. *Hepatology*, *52*(6), 2158–2166. http://dx.doi.org/10.1002/hep.23945.

Rihel, J., Prober, D. A., Arvanites, A., Lam, K., Zimmerman, S., Jang, S., et al. (2010). Zebrafish behavioral profiling links drugs to biological targets and rest/wake regulation. *Science*, *327*(5963), 348–351. http://dx.doi.org/10.1126/science.1183090, 327/5963/348 [pii].

Rinkenberger, J. L., Horning, S., Klocke, B., Roth, K., & Korsmeyer, S. J. (2000). Mcl-1 deficiency results in peri-implantation embryonic lethality. *Genes and Development*, *14*(1), 23–27.

Robu, M. E., Larson, J. D., Nasevicius, A., Beiraghi, S., Brenner, C., Farber, S. A., et al. (2007). p53 activation by knockdown technologies. *PLoS Genetics*, *3*(5), e78. http://dx.doi.org/10.1371/journal.pgen.0030078, 06-PLGE-RA-0378R3 [pii].

Roca, F. J., & Ramakrishnan, L. (2013). TNF dually mediates resistance and susceptibility to mycobacteria via mitochondrial reactive oxygen species. *Cell*, *153*(3), 521–534. http://dx.doi.org/10.1016/j.cell.2013.03.022.

Rodriguez, M., & Driever, W. (1997). Mutations resulting in transient and localized degeneration in the developing zebrafish brain. *Biochemistry and Cell Biology*, *75*(5), 579–600.

Rodriguez-Aznar, E., & Nieto, M. A. (2011). Repression of Puma by scratch2 is required for neuronal survival during embryonic development. *Cell Death and Differentiation*, *18*(7), 1196–1207. http://dx.doi.org/10.1038/cdd.2010.190.

Rosen, J. N., Sweeney, M. F., & Mably, J. D. (2009). Microinjection of zebrafish embryos to analyze gene function. *Journal of Visualized Experiments*. *2009*(25). http://dx.doi.org/10.3791/1115.

Sander, J. D., Cade, L., Khayter, C., Reyon, D., Peterson, R. T., Joung, J. K., et al. (2011). Targeted gene disruption in somatic zebrafish cells using engineered TALENs. *Nature Biotechnology*, *29*(8), 697–698. http://dx.doi.org/10.1038/nbt.1934.

Santoro, M. M., Samuel, T., Mitchell, T., Reed, J. C., & Stainier, D. Y. (2007). Birc2 (cIap1) regulates endothelial cell integrity and blood vessel homeostasis. *Nature Genetics*, *39*(11), 1397–1402. http://dx.doi.org/10.1038/ng.2007.8, ng.2007.8 [pii].

Sidi, S., Sanda, T., Kennedy, R. D., Hagen, A. T., Jette, C. A., Hoffmans, R., et al. (2008). Chk1 suppresses a caspase-2 apoptotic response to DNA damage that bypasses p53, Bcl-2, and caspase-3. *Cell*, *133*(5), 864–877. http://dx.doi.org/10.1016/j.cell.2008.03.037, S0092-8674(08)00502-3 [pii].

Sorrells, S., Carbonneau, S., Harrington, E., Chen, A. T., Hast, B., Milash, B., et al. (2012). Ccdc94 protects cells from ionizing radiation by inhibiting the expression of p53. *PLoS Genetics*, *8*(8), e1002922. http://dx.doi.org/10.1371/journal.pgen.1002922.

Sprague, J., Bayraktaroglu, L., Bradford, Y., Conlin, T., Dunn, N., Fashena, D., et al. (2008). The Zebrafish Information Network: The zebrafish model organism database provides expanded support for genotypes and phenotypes. *Nucleic Acids Research*, *36*(Database issue), D768–D772. http://dx.doi.org/10.1093/nar/gkm956.

Sullivan-Brown, J., Bisher, M. E., & Burdine, R. D. (2011). Embedding, serial sectioning and staining of zebrafish embryos using JB-4 resin. *Nature Protocols*, *6*(1), 46–55. http://dx.doi.org/10.1038/nprot.2010.165.

Summerton, J., & Weller, D. (1997). Morpholino antisense oligomers: Design, preparation, and properties. *Antisense & Nucleic Acid Drug Development*, 7(3), 187–195.

Tam, S. J., Richmond, D. L., Kaminker, J. S., Modrusan, Z., Martin-McNulty, B., Cao, T. C., et al. (2012). Death receptors DR6 and TROY regulate brain vascular development. *Developmental Cell*, *22*(2), 403–417. http://dx.doi.org/10.1016/j.devcel.2011.11.018.

Thisse, C., & Thisse, B. (2008). High-resolution in situ hybridization to whole-mount zebrafish embryos. *Nature Protocols*, *3*(1), 59–69. http://dx.doi.org/10.1038/nprot.2007.514.

Tucker, B., & Lardelli, M. (2007). A rapid apoptosis assay measuring relative acridine orange fluorescence in zebrafish embryos. *Zebrafish*, *4*(2), 113–116. http://dx.doi.org/10.1089/zeb.2007.0508.

van Ham, T. J., Kokel, D., & Peterson, R. T. (2012). Apoptotic cells are cleared by directional migration and elmo1-dependent macrophage engulfment. *Current Biology*, *22*(9), 830–836. http://dx.doi.org/10.1016/j.cub.2012.03.027.

van Ham, T. J., Mapes, J., Kokel, D., & Peterson, R. T. (2010). Live imaging of apoptotic cells in zebrafish. *FASEB Journal*, *24*(11), 4336–4342. http://dx.doi.org/10.1096/fj.10-161018.

Varfolomeev, E. E., Schuchmann, M., Luria, V., Chiannlkulchai, N., Beckmann, J. S., Mett, I. L., et al. (1998). Targeted disruption of the mouse caspase-8 gene ablates cell death induction by the TNF receptors, Fas/Apo1, and DR3 and is lethal prenatally. *Immunity*, *9*, 267–276.

Vermes, I., Haanen, C., Steffens-Nakken, H., & Reutelingsperger, C. (1995). A novel assay for apoptosis. Flow cytometric detection of phosphatidylserine expression on early apoptotic cells using fluorescein labelled Annexin V. *Journal of Immunological Methods*, *184*(1), 39–51.

Villefranc, J. A., Amigo, J., & Lawson, N. D. (2007). Gateway compatible vectors for analysis of gene function in the zebrafish. *Developmental Dynamics*, *236*(11), 3077–3087. http://dx.doi.org/10.1002/dvdy.21354.

Westerfield, M. (1995). *The zebrafish book: A guide for the laboratory use of zebrafish (*Danio rerio*)*. Eugene, OR: University of Oregon Press.

White, R. M., Sessa, A., Burke, C., Bowman, T., LeBlanc, J., Ceol, C., et al. (2008). Transparent adult zebrafish as a tool for in vivo transplantation analysis. *Cell Stem Cell*, *2*(2), 183–189. http://dx.doi.org/10.1016/j.stem.2007.11.002.

Wright, G. J., Leslie, J. D., Ariza-McNaughton, L., & Lewis, J. (2004). Delta proteins and MAGI proteins: An interaction of Notch ligands with intracellular scaffolding molecules and its significance for zebrafish development. *Development*, *131*(22), 5659–5669. http://dx.doi.org/10.1242/dev.01417.

Yabu, T., Kishi, S., Okazaki, T., & Yamashita, M. (2001). Characterization of zebrafish caspase-3 and induction of apoptosis through ceramide generation in fish fathead minnow tailbud cells and zebrafish embryo. *Biochemistry Journal*, *360*(Pt 1), 39–47.

Yager, T. D., Ikegami, R., Rivera-Bennetts, A. K., Zhao, C., & Brooker, D. (1997). High-resolution imaging at the cellular and subcellular levels in flattened whole mounts of early zebrafish embryos. *Biochemistry and Cell Biology*, *75*(5), 535–550.

Yamashita, M., Mizusawa, N., Hojo, M., & Yabu, T. (2008). Extensive apoptosis and abnormal morphogenesis in pro-caspase-3 transgenic zebrafish during development. *Journal of Experimental Biology*, *211*(Pt 12), 1874–1881. http://dx.doi.org/10.1242/jeb.012690.

Yeh, W. C., de la Pompa, J. L., McCurrach, M. E., Shu, H. B., Elia, A. J., Shahinian, A., et al. (1998). FADD: Essential for embryo development and signaling from some, but not all, inducers of apoptosis. *Science*, *279*, 1954–1958.

Yeh, W.-C., Itie, A., Elia, A., Ng, M., Shu, H.-B., Wakeham, A., et al. (2000). Requirement for casper (c-FLIP) in regulation of death receptor-induced apoptosis and embryonic development. *Immunity*, *12*, 633–642.
Zu, Y., Tong, X., Wang, Z., Liu, D., Pan, R., Li, Z., et al. (2013). TALEN-mediated precise genome modification by homologous recombination in zebrafish. *Nature Methods*, *10*(4), 329–331. http://dx.doi.org/10.1038/nmeth.2374.

AUTHOR INDEX

Note: Page numbers followed by "*f*" indicate figures, "*t*" indicate tables, and "*np*" indicate footnotes.

C

E

F

G

H

I

J

K

L

M

N

Q

R

S

U

V

W

X

Y

Z

SUBJECT INDEX

Note: Page numbers followed by "*f*" indicate figures and "*t*" indicate tables.

C

G

H

N

O

P

T

V

X

Z

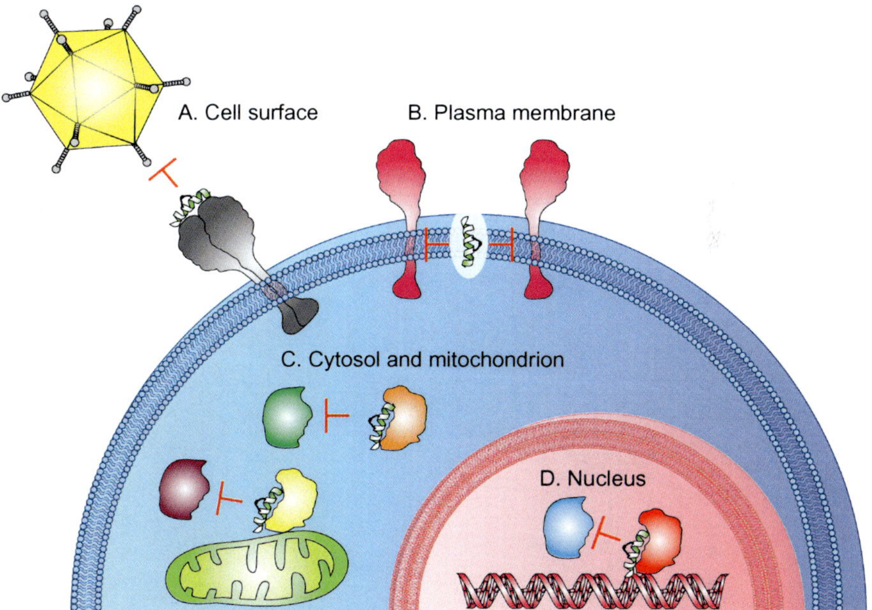

Susan Lee *et al.*, Figure 2.1 The peptide α-helix is a ubiquitous secondary structural motif that mediates a host of biomedically relevant protein interactions. Hydrocarbon-stapled peptides modeled after bioactive helices can be generated to target and modulate protein interactions from the surface to the inner nuclear core of the cell.

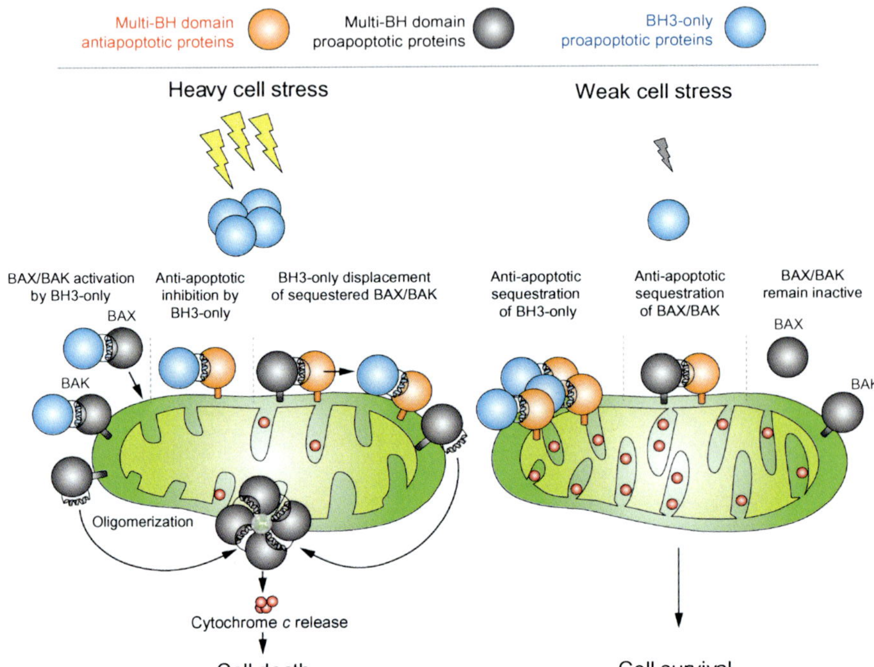

Susan Lee *et al.*, Figure 2.2 BCL-2 family proteins regulate the life and death decision of stressed cells. If mitochondrial antiapoptotic proteins (orange) can effectively harness their C-terminal binding pockets to trap and sequester the BH3-signaling helices of proapoptotic proteins, cell survival prevails. However, with increased cellular stress, proapoptotic signals overwhelm the antiapoptotic reserve. The BH3 domain helices of BH3-only proteins (blue) can directly activate the essential executioner proteins BAX and BAK (gray) and also release trapped forms from antiapoptotic inhibition by targeting the antiapoptotic groove.

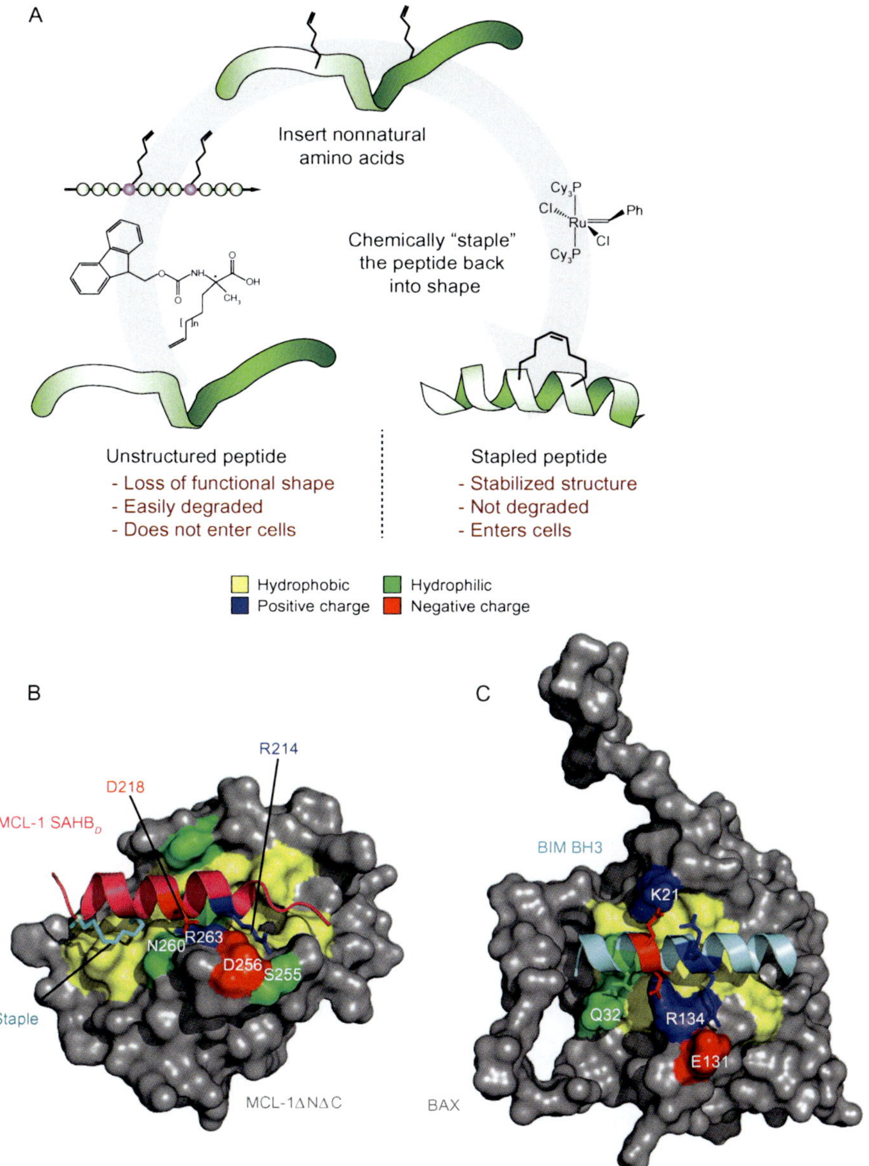

Susan Lee *et al.*, Figure 2.3 Synthetic overview of all-hydrocarbon stapling and its applications in developing α-helical peptides for structural studies of protein interactions. (A) Pairs of α,α-disubstituted nonnatural amino acids bearing olefin tethers are substituted into the peptide sequence at discrete locations (e.g., *i*, *i*+4), followed by ruthenium-catalyzed ring-closing metathesis (RCM) to generate "stapled peptides." (B) Crystal structure of a stapled MCL-1 BH3 helix in complex with antiapoptotic MCL-1 (Stewart, Fire, Keating, & Walensky, 2010). (C) Calculated model structure of a BIM BH3 helix engaging the N-terminal trigger site of full-length BAX, as derived from paramagnetic relaxation enhancement NMR analyses using full-length ^{15}N-BAX- and MTSL-labeled BIM SAHBs (aa 145–164) (Gavathiotis et al., 2008).

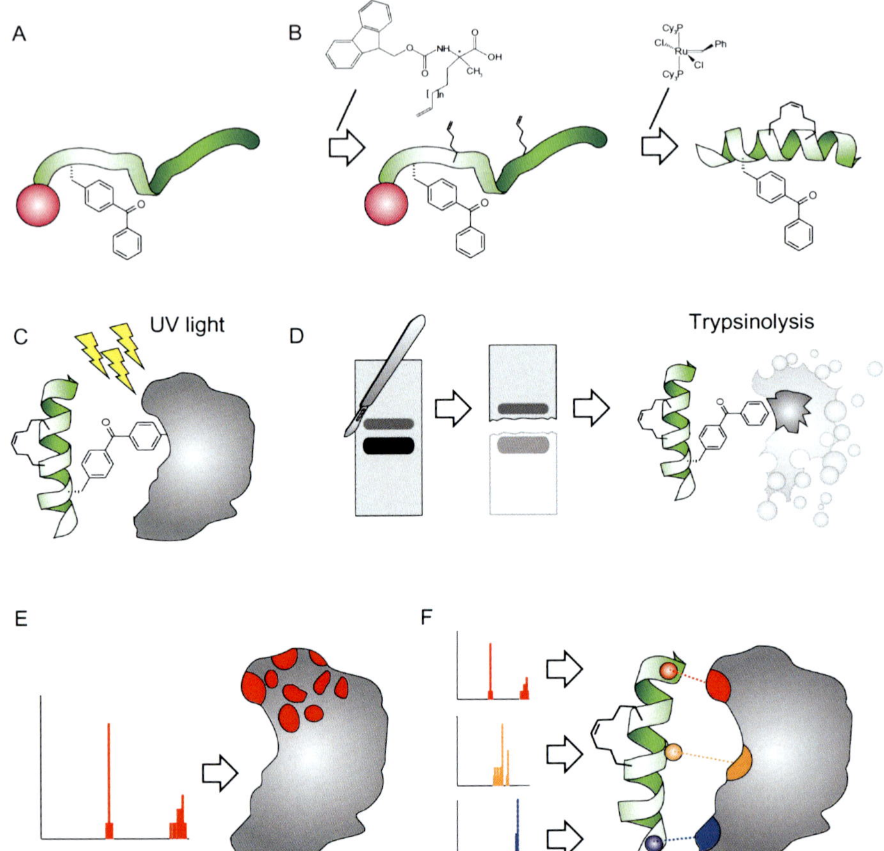

Susan Lee *et al.*, Figure 2.4 Design of photoreactive stabilized alpha-helices (pSAHs) for binding site identification by mass spectrometry. A photoreactive Bpa residue (A) and a pair of stapling amino acids (B) are inserted into the peptide template followed by RCM to generate a pSAH. Upon exposure to UV light, the bound pSAH covalently crosslinks to the target protein (C) and, following electrophoresis of the mixture, crosslinked protein is excised from the gel and subjected to in-gel digestion with trypsin (D). LC–MS/MS analysis identifies the explicit sites of covalent modification, which when mapped on to the protein structure reveal the region of pSAH interaction (E). Top scoring crosslinks from a series of experiments employing sequentially placed Bpa residues can provide interaction restraints for calculating model structures (F).

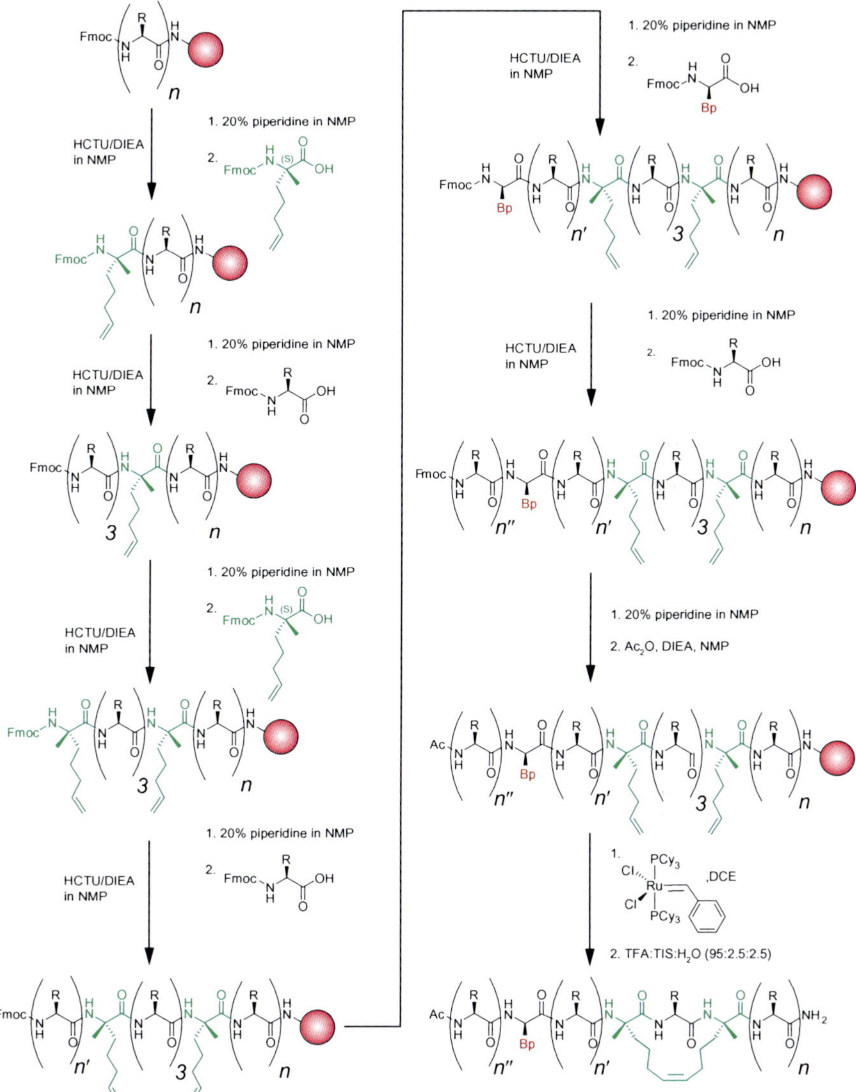

Susan Lee *et al.*, Figure 2.5 Synthetic steps for the automated production of pSAHs by Fmoc-based solid-phase peptide synthesis and ruthenium-catalyzed RCM. Alternatives to N-terminal acetylation include capping with FITC or biotin, depending on the desired application. Bp, 4-benzoylphenyl.

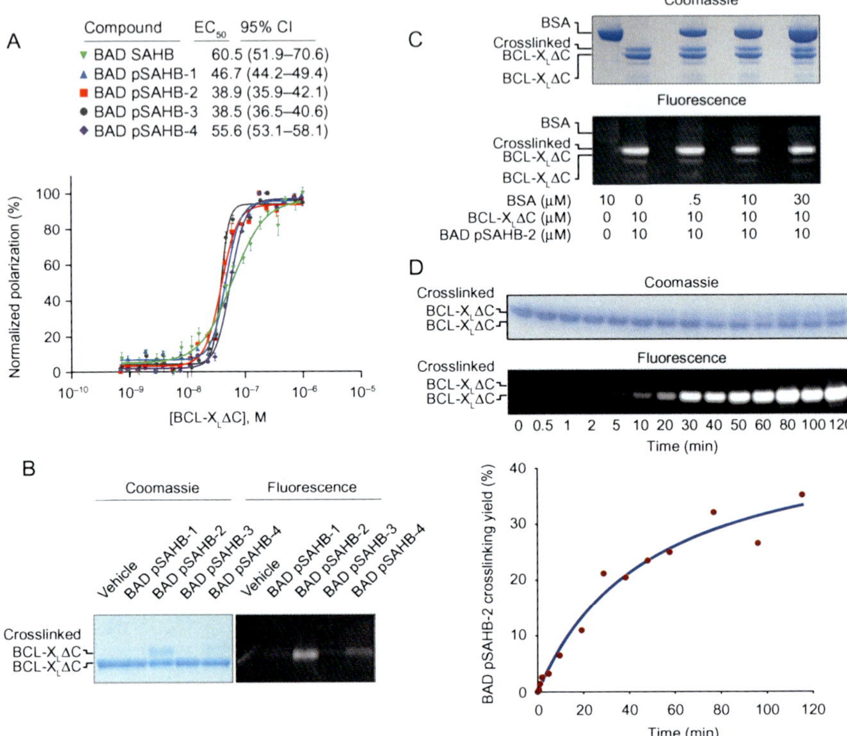

Susan Lee *et al.*, Figure 2.6 pSAHs bind and selectively crosslink to their protein targets, as exemplified by the photocrosslinking of BAD pSAHBs to antiapoptotic BCL-$X_L\Delta C$ (Braun et al., 2010). (A) A series of FITC-BAD pSAHBs retain nanomolar binding affinity to BCL-$X_L\Delta C$ as measured by FP analysis. (B) FITC-BAD pSAHBs exhibit a spectrum of BCL-$X_L\Delta C$-crosslinking efficiency upon UV exposure, with BAD pSAHB-2 demonstrating the greatest reactivity toward BCL-$X_L\Delta C$ (left, coomassie stain; right, fluorescence scan). (C) The selectivity of BAD pSAHB-2 is reflected by the absence of crosslinking to BSA and the lack of effect of added BSA on the BCL-$X_L\Delta C$-crosslinking efficiency of FITC-BAD pSAHB-2. (D) Time course for photocrosslinking of FITC-BAD pSAHB-2 to BCL-$X_L\Delta C$ as monitored by coomassie stain (top) and fluorescence scan (middle). A plot of crosslinking yield, calculated based on densitometry, demonstrates the time-dependent production of crosslinked BCL-$X_L\Delta C$ (bottom).

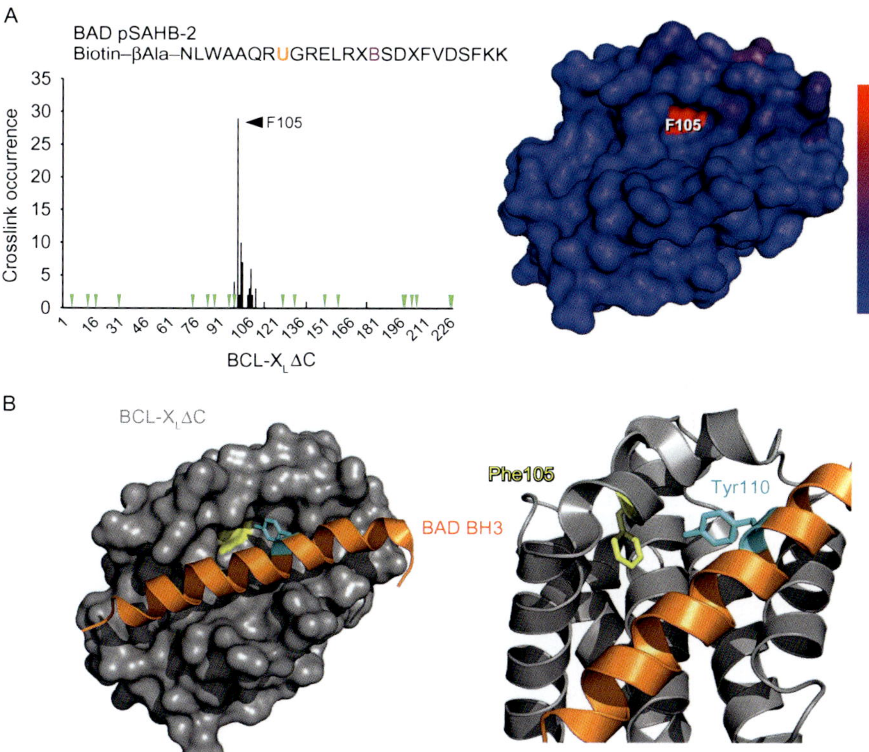

Susan Lee *et al.*, Figure 2.7 pSAHs localize helix/target interaction sites with high fidelity, as exemplified by the capacity of BAD pSAHB-2 to accurately map the BAD BH3 interaction site on BCL-X$_L$ΔC (Braun et al., 2010). (A) The plot (left) depicts the frequency of crosslinked sites identified across the BCL-X$_L$ΔC polypeptide sequence. Mapping of the crosslinked residues onto the BCL-X$_L$ΔC structure (right) revealed their striking colocalization within a circumscribed region of the canonical BH3-binding pocket, with the frequency of occurrence reflected by the color scale. Green arrowheads, trypsin digestion sites. (B) The most abundant crosslink, located between BAD pSAHB-2 Bpa110 and BCL-X$_L$ΔC F105, precisely matches the structurally defined interaction between BAD BH3 Y110 (cyan) and BCL-X$_L$ΔC F105 (yellow) at the BH3-binding pocket of BCL-X$_L$ΔC (PDB ID 2BZW). U, Bpa; X, stapling amino acid; B, norleucine.

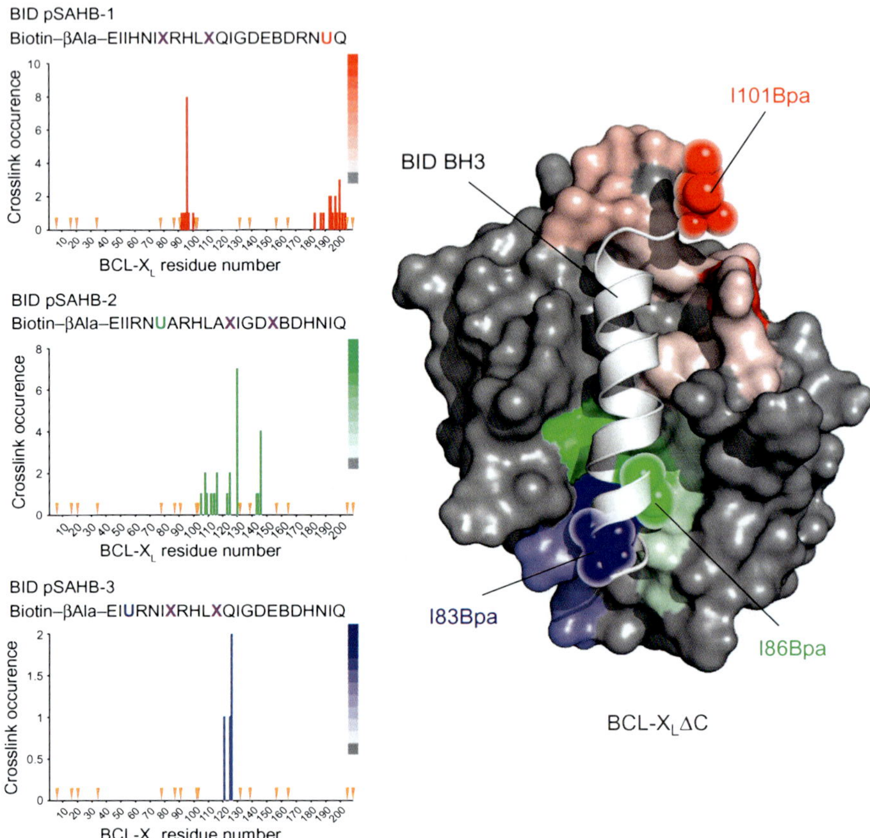

Susan Lee *et al.*, Figure 2.8 The high regiospecificity of pSAH crosslinking enables the calculation of model structures of helix/target complexes, as exemplified by a panel of BID pSAHBs crosslinked to BCL-X$_L$ΔC (Leshchiner et al., 2013). The plots (left) and the corresponding mapping (right) reflect the frequency of crosslinked sites across the BCL-X$_L$ΔC polypeptide sequence for three distinct BID pSAHBs. The covalently modified residues for each pSAHB construct maps to a highly circumscribed subregion of the canonical BH3-binding pocket of BCL-X$_L$ΔC, facilitating the calculation of a model structure of the complex (right) using CNS within HADDOCK 2.0. Orange arrowheads, tryptic digestion sites; U, Bpa; X, stapling amino acid; B, norleucine.

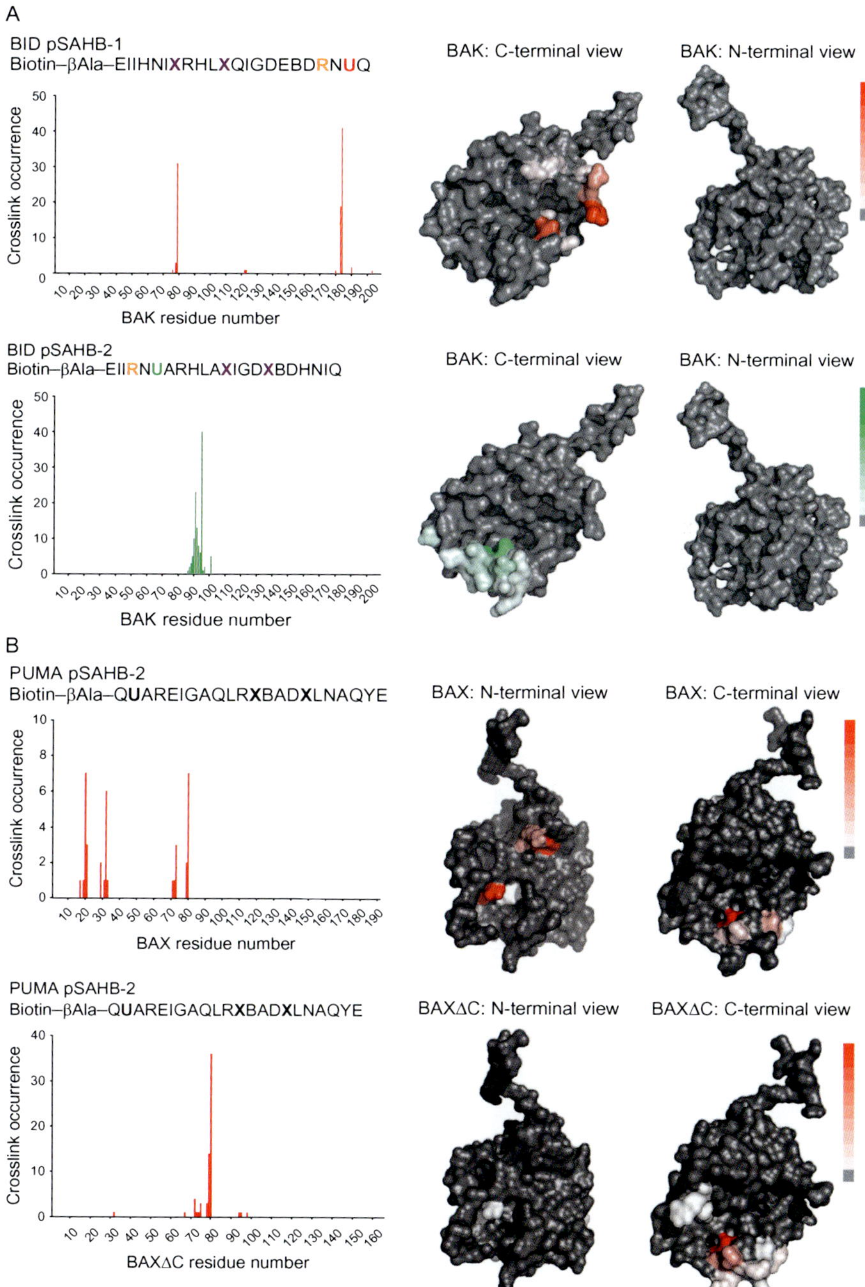

Susan Lee *et al.*, Figure 2.9 pSAHs identify novel sites of BCL-2 family protein interaction. BID, PUMA, and phospho-BAD pSAHBs revealed BH3 interaction sites on (A) full-length BAK (Leshchiner et al., 2013), (B) full-length versus C-terminally truncated BAX

(Continued)

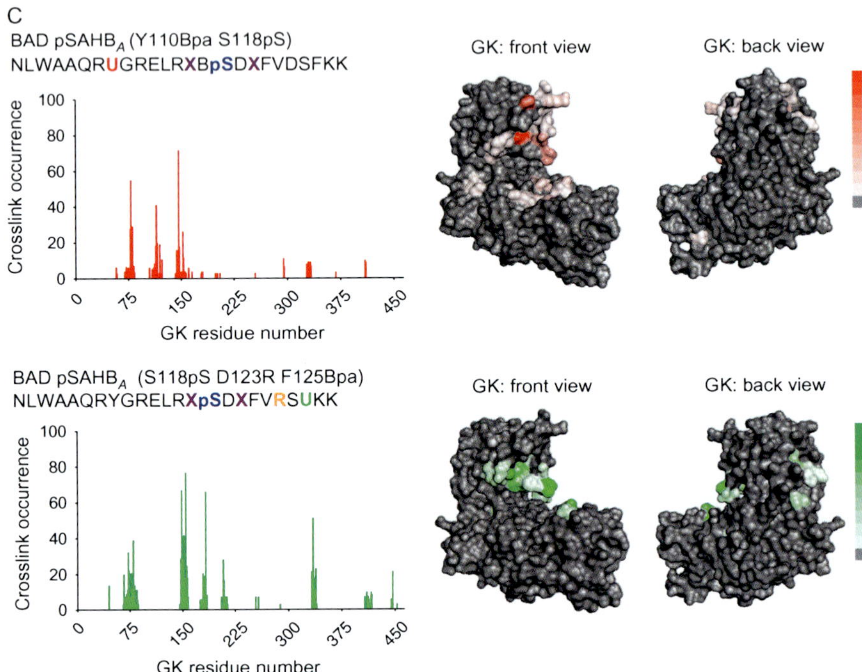

Figure 2.9—Cont'd (i.e., BAX vs. BAXΔC) (Edwards et al., 2013), and (C) glucokinase (Szlyk et al., 2014), respectively. U, Bpa; X, stapling amino acid; B, norleucine; pS, phosphoserine; orange R, single Arg substitution to facilitate tryptic digestion of pSAHBs into shorter and more identifiable fragments by MS.

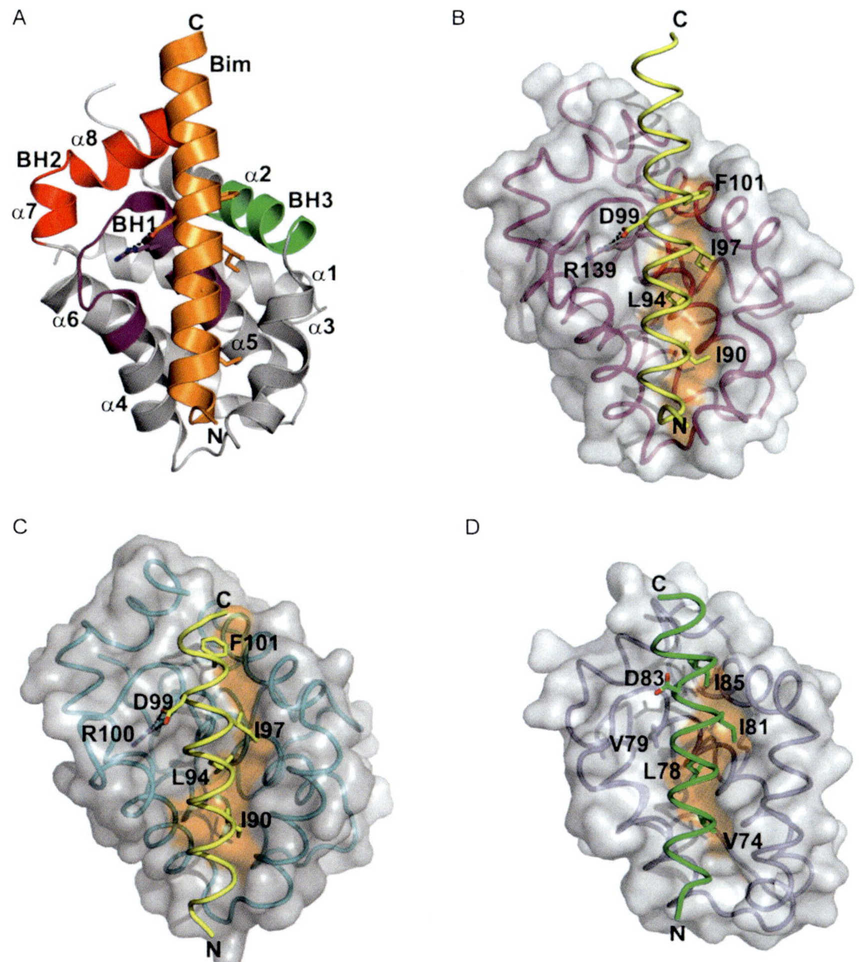

Marc Kvansakul and Mark G. Hinds, Figure 3.3 Structures of BH3-motifs bound to mammalian and viral prosurvival Bcl-2 proteins. (A) Ribbon representation of mouse Bcl-x_L:Bim BH3 complex (pdb: 1PQ1). Bim-BH3 (orange) binds as a helix in a groove provided by the BH1 (violet), BH2 (red); and BH3 (green) motifs of Bcl-x_L that corresponds to helices α2–α5 and α8. (B) Surface representation of Bcl-x_L:Bim BH3 in (A) with Bim depicted as a tube (green) and the highly conserved residues of Bim indicated. The conserved Asp–Arg interaction between Bim and Arg R139 in the BH1 motif of Bcl-x_L is also shown. (C) Interactions of BHRF1:Bim BH3 (pdb:2WH6) complex. (D) Interactions of M11L:Bak BH3 complex (pdb:1JBY). Molecular surfaces are shaded gray and the surface from residues within 4 Å from the key BH3-only residues in orange.

Marc Kvansakul and Mark G. Hinds, Figure 3.4 Mimics of BH3-only proteins bind in the BH3-binding groove. (A) Structure of an ABT-199 analog bound to Bcl-2 (pdb: 4LXD) (Souers et al., 2013). The drug binds in the same groove as BH3-only proteins but the conserved Arg in the BH1 motif of Bcl-2 does not make an ionic interaction with the acidic acylsulfonomide of the ABT-199 analog. The structure is aligned as those in Fig. 3.3 and Bcl-2 is depicted as a surface representation. (B) ABT-199 analog.

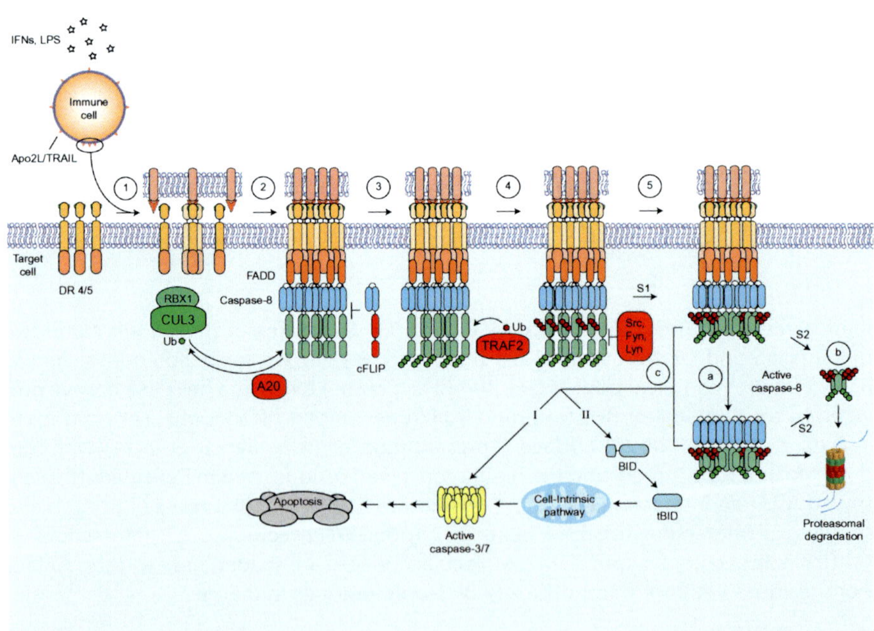

Pradeep Nair *et al.*, Figure 5.1

1. Death receptors 4 and 5, expressed on the surface of target cells, are ligated by Apo2L/TRAIL, presented on the surface of activated immune cells.

(Continued)

Figure 5.1—Cont'd

2. Multiple ligand–receptor complexes cluster together at the cell membrane, triggering recruitment of the adaptor protein FADD and the apoptosis-initiating protease caspase-8 to the DISC. An enzymatically inactive protein related to caspase-8, cFLIP, can also be recruited to the DISC, inhibiting caspase-8 activation.
3. In epithelial cells, the E3 ligase complex CUL3/RBX1 catalyzes polyubiquitination (likely via K63 chains) of lysine 461 on the p10 domain of caspase-8, leading to further clustering and activation of the caspase. The DUB enzyme A20 can reverse this ubiquitination.
4. The RING domain-containing adaptor protein TRAF2 mediates K48 ubiquitination on the p18 region of caspase-8. This targets caspase-8 for proteasomal degradation upon proteolytic processing and release from the plasma membrane into the cytosol, thereby setting a "shutoff timer" for caspase-8 activation.
5. S1 cleavage removes the linker between the p10 and the p18 domains of caspase-8, leading to the formation of a catalytically active $(p43/p10)_2$ heterodimer. S1 cleavage can be inhibited through phosphorylation on the linker region between the p10 and the p18 domains, catalyzed by members of the Src-family tyrosine kinases. Once S1 cleavage occurs, the pathway can proceed in three distinct directions:
   - **a.** The $(p43/p10)_2$ heterodimer can dissociate from the membrane-anchored DISC, leading to a cytosolic pool of active caspase-8. The pre-added K48-ubiquitin on p18 (within p43) targets the p43 fragment for proteasomal degradation, thereby limiting the duration of caspase-8 activity.
   - **b.** In either the membrane-associated or the cytosolic active caspase-8 pools, S2 cleavage can occur between the p18 and the N-terminal DEDs of caspase-8, releasing a $(p18/p10)_2$ heterodimer that no longer oligomerizes due to its separation from the DEDs. The pre-added K48-ubiquitin on p18 targets this fragment for proteasomal degradation, thereby limiting duration of activity.
   - **c.** Active caspase-8, either at the cell membrane or in the cytosol, can proceed to proteolytically cleave downstream substrates. In type I cells, caspase-8 activation is sufficient to directly cleave and activate executioner caspases, such as caspases-3 and -7 and commit the cell to apoptosis. In type II cells, less caspase-8 activity is generated and amplification of the signal is needed to drive sufficient effector caspase activity. This can be achieved through caspase-8-mediated cleavage of the BH3-only protein BID to form truncated BID (tBID). tBID triggers the cell-intrinsic (mitochondrial) pathway, augmenting the activation of executioner caspases and committing the cell to apoptotic demise.

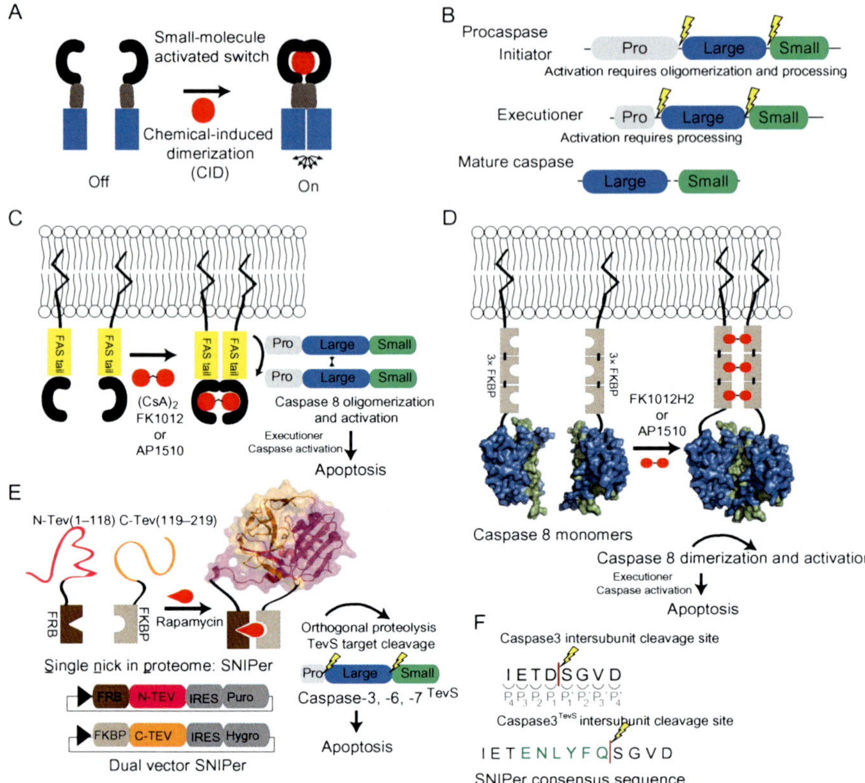

Charles W. Morgan *et al.*, Figure 8.1 Schematic overview of chemical genetic strategies for selective caspase activation. (A) Chemical-induced dimerization (CID) utilizes dimerization domains, dependent on cell-permeable small molecules, to control the proximity of signal transduction domains (blue). The chemical genetic strategy of CID creates an orthogonal circuit in the cell, thereby enabling the control of the ON/OFF state of signal transduction. (B) Procaspases are composed of three domains; a prodomain, a large subunit, and a small subunit. Initiators have a long prodomain and require both oligomerization and processing for activation. The executioners are predimerized and they have a short prodomain and proteolysis at the large–small domain junction is sufficient for activation. (C) Utilizing a CID-based approach, FAS tails are dimerized by the addition of the symmetric small molecule, AP1510 or FK1012, thereby reconstituting the caspase-8 (PDB ID 1QTN) activation scaffold and ultimately resulting in apoptosis. (D) Direct caspase-8 activation via CID. (E) The engineered split-tobacco etch viral (TEV) variant, single nick in proteome (SNIPer), used for inducible and selective cleavage of the executioner caspase isoforms (TEV structure PDB ID 1LVM). Each SNIPer half is expressed from a single plasmid that coexpresses an IRES-driven drug-resistance marker for stable cell line engineering. The addition of rapamycin causes the heterodimerization of FKBP and FRB, thereby rescuing TEV protease activity. (F) The TEV protease consensus sequence (green) is inserted into the caspase-processing site to generate a TevS allele, susceptible to SNIPer-mediated proteolysis.

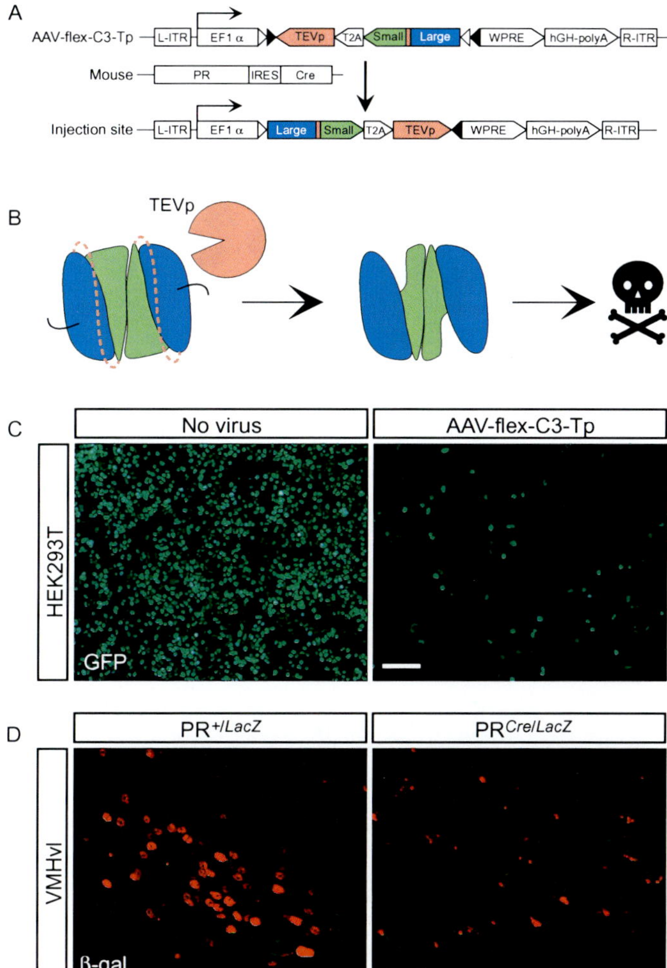

Charles W. Morgan *et al.*, Figure 8.3 Utilizing AAV-flex-C3-Tp. (A) Strategy to ablate PR-expressing cells *in vivo*. (B) Within Cre+ cells, the TEV protease will cleave its consensus recognition sequence (red) between the large (blue) and small (green) subunits of the caspase-3 dimer, activating it and leading to apoptosis. (C) The AAV-flex-C3-Tp encoding plasmid was transfected into HEK293T cells expressing the fusion protein Cre-GFP. Cell death was evaluated after 1 week. (D) AAV-flex-C3-Tp was stereotaxically targeted to the ventrolateral region of the VMH (VMHvl) of adult $PR^{+/LacZ}$ and $PR^{Cre/LacZ}$ mice which harbor the transgenes nuclear LacZ (PR^{LacZ} allele) or Cre recombinase (PR^{Cre} allele) inserted into the PR locus by homologous recombination. Cell death was evaluated after 4 weeks. Scale bars represent 100 μm (C) and 25 μm (D). *(C) and (D) are reproduced from Yang et al. (2013).*

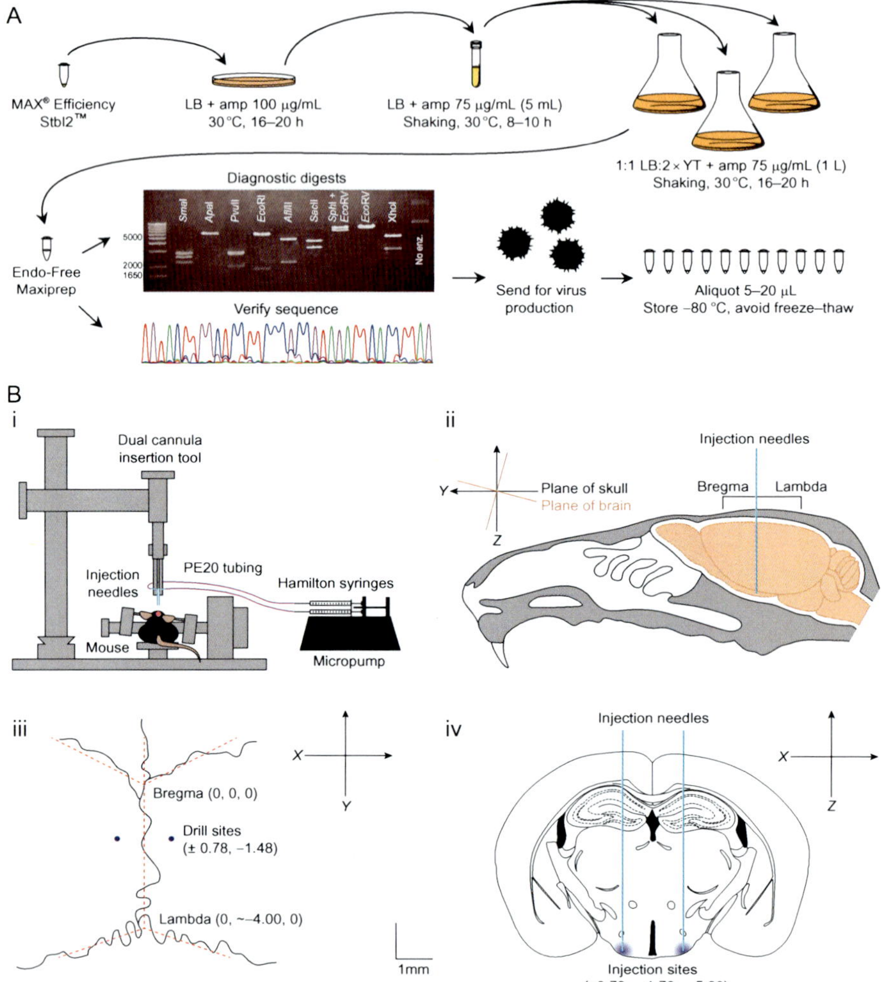

Charles W. Morgan *et al.*, Figure 8.4 Overview of AAV virus production and injection. (A). Summary of the protocol used for AAV plasmid growth and verification for virus production. Maxipreps originating from the same colony were pooled for diagnostics. (B) Schematic of virus injection. (i) Stereotax and micropump setup: injection needles are connected to Hamilton syringes by PE20 tubing. The syringes are depressed steadily and simultaneously by a micropump. (ii) Orientation of the brain within the cranial cavity. Note the 15° slant of the skull relative to the brain which is common in inbred strains. (iii) Cranial sutures with bregma and lambda marked. (iv) Coronal section through the VMHvl where we injected the virus. Note that the *y* coordinate is different from the drilling site in (iii) because of the 15° slant. Scale bar represents 1 mm (iii and iv). *Figure (iv) was modified from Paxinos Brain Atlas (Paxinos & Franklin, 2004).*

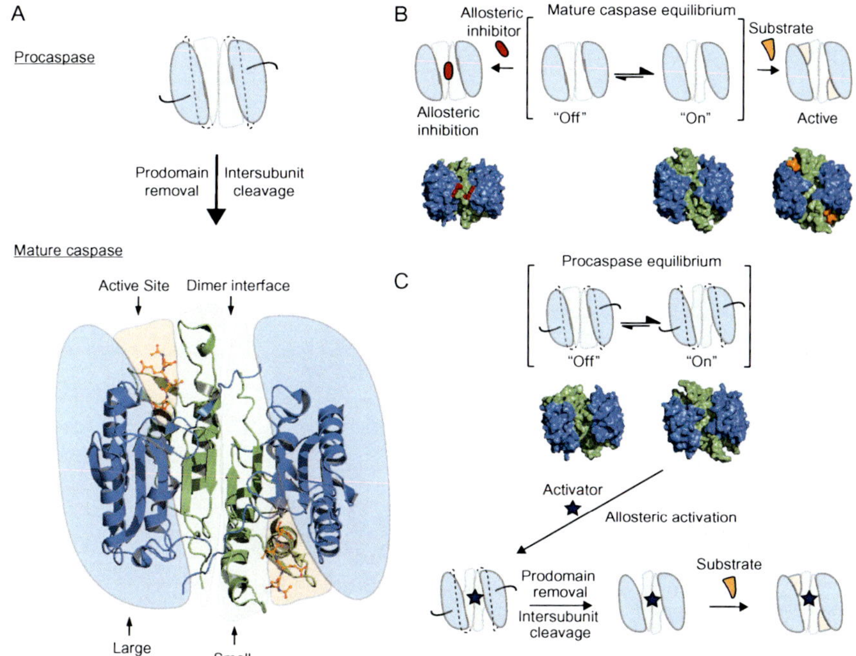

Charles W. Morgan *et al.*, Figure 8.5 Mechanism of caspase regulation. (A) Executioner procaspase maturation requires proteolysis of the N-terminal prodomain (black lines) and cleavage of the intersubunit linker (dashed lines) between the large and small subunits, colored in blue and green, respectively. The crystal structure of mature caspase-7 is shown (PDB ID 1F1J) highlighting the dimer interface, the large and small subunits and the Ac-DEVD-CHO peptide occupying the active site (orange). (B) Mechanism of allosteric inhibition of caspases, showing the hotspot for allosteric binding located at the dimer interface. The surface representation of experimentally determined X-ray structures of caspase-7 is shown below the cartoons. From left to right, the allosteric-site ligand-bound structure of caspase-7 complexed with DICA, the ligand-free apo caspase-7 structure in an "on" conformation, and the active-site ligand-bound structure of caspase-7 in complex with the Ac-DEVD-CHO peptide (PDB ID 1SHJ, 1K86, 1F1J, respectively). (C) Mechanism of allosteric activation of procaspases. The brackets denote the open and closed active-site equilibrium, with the surface of the X-ray structures of procaspase-3 in the "on" and "off" states (PDB ID 4JR0 and 4JQY, respectively). Upon binding of a hypothetical allosteric activator (dark blue star), the proenzyme is locked into an "on" conformation, allowing possible proteolytic cleavage of the prodomain and intersubunit linkers.

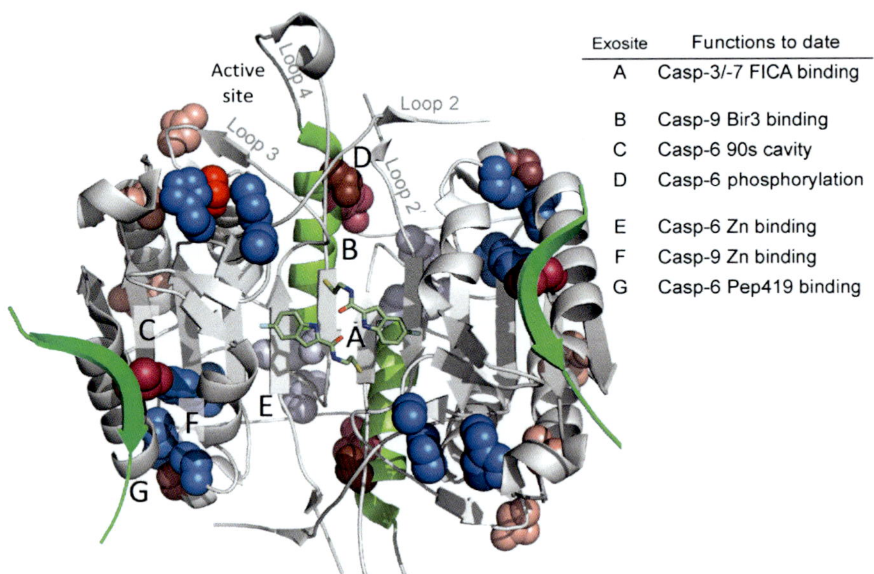

Kevin Dagbay *et al.*, Figure 9.1 Current map of global apoptotic caspase allostery. Exosites that have been identified by the methods included in this chapter are annotated on a canonical caspase structure (gray cartoon). Red, phosphorylation; blue, zinc binding; green, chemical ligands. Various shades of red are used to denote sites of phosphorylation in different caspases. The green caspase helix that forms exosite B is highlighted in the chemical ligands to indicate that this is the region that interacts with the BIR3 region of XIAP- and BIR3-derived peptide inhibitors. *Adapted from Velazquez-Delgado (2012) with permission.*

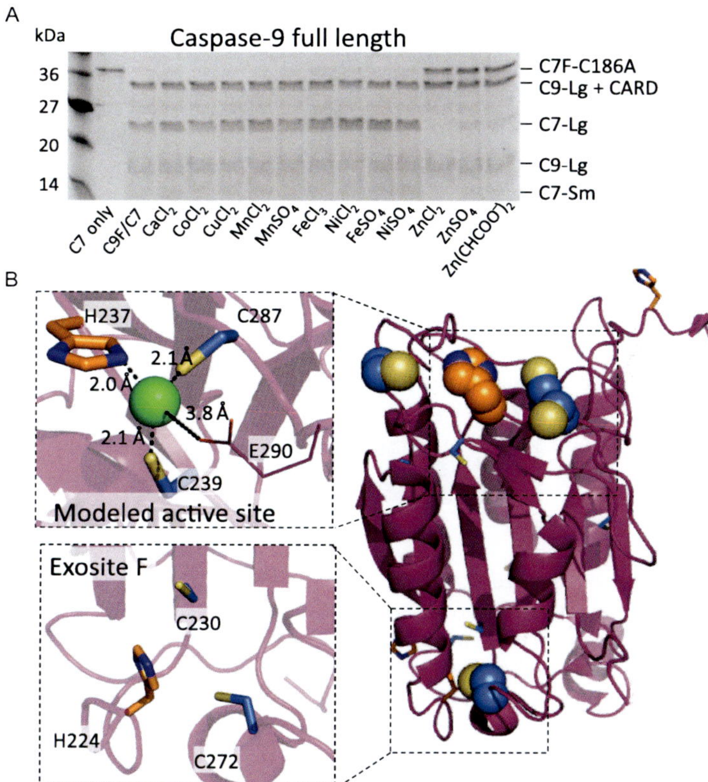

Kevin Dagbay *et al.*, Figure 9.3 Zinc binds and inhibits both active and allosteric sites in caspase-9. (A) Zinc is the predominant metal cation to inhibit full-length caspase-9 (C9 FL) as monitored by cleavage of a natural caspase-9 substrate, the caspase-7 zymogen (C7 C186A) to the caspase-7 large (C7 Lg) and small (C7 Sm) subunits. (B) Location of the conserved active-site and 210's helix exosite-ligand clusters on caspase-9 (PDB ID 1JXQ). A model of caspase-9 active-site ligand interactions with a modeled zinc ion was obtained by altering the H237, C239, and C287 rotamers in PyMol. *Adapted from Huber and Hardy (2012) with permission.*

A

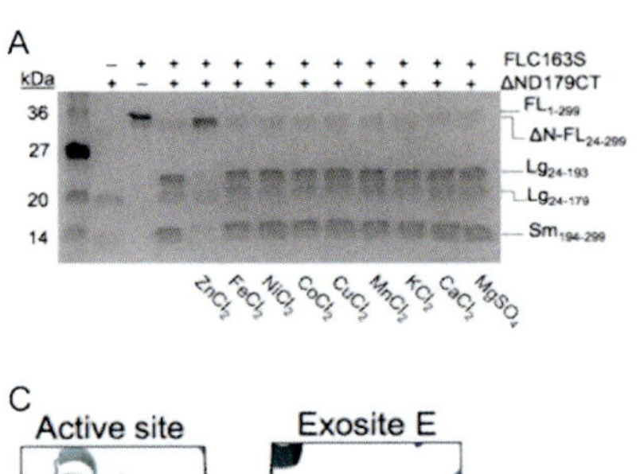

B

	Zinc : Monomer
Caspase-3 (Wild type)	3.2 ± 0.4
Caspase-6 (Wild type)	1.0 ± 0.2
Active site C163S	0.8 ± 0.1
Exosite E K36A	0.4 ± 0.1
E244A	0.4 ± 0.3
H287A	0.2 ± 0.2
E244A/H287A	0.3 ± 0.2
Exosite F H108A/C148A	1.2 ± 0.3
Caspase-7 (Wild type)	0.7 ± 0.1
Caspase-9 (Wild type)	1.8 ± 0.3
Active site C287A	0.8 ± 0.1
H237A	0.8 ± 0.1
C287A /C239S	0.7 ± 0.3
Exosite F C272A	0.7 ± 0.3
Active + F C272A /C287A	0.2 ± 0.2

C

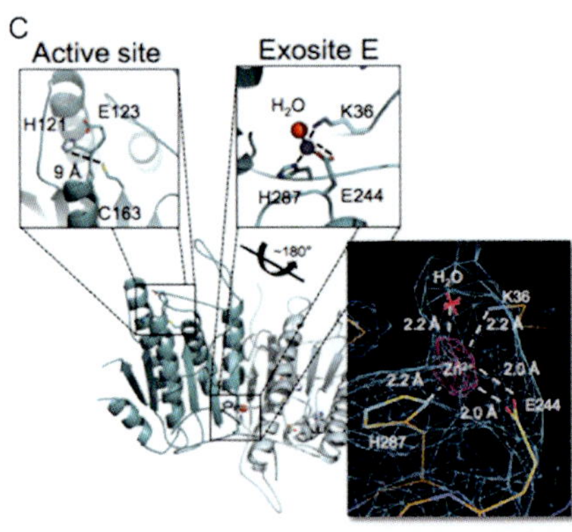

Kevin Dagbay *et al.*, Figure 9.4 Zinc binds an allosteric site to inhibit caspase-6. (A) Metal inhibition was tested in an electrophoretic mobility gel-based assay as fragments produced from active caspase-6-(ΔND179CT) mediated cleavage of the full-length C163S zymogen (substrate). Fragments from cleavage include the following: full length (FL), FL lacking the N-terminal prodomain (ΔN-FL), large (Lg), and small subunits (Sm) with the amino acids present in those bands (subscripts) labeled. Zinc is the only metal cation that inhibits caspase-6 activity. (B) The zinc-binding stoichiometry for various caspases, as well as mutants designed to ablate zinc binding are shown. Zinc binding was measured by inductively coupled plasma–optical emission spectroscopy. (C) The structure of zinc-bound caspase-6 (ribbons, PDB ID 4FXO). An anomalous difference map calculated from data collected above the zinc absorbance edge is contoured at 5σ (inset, purple mesh), clearly indicating the location of zinc at the allosteric site (black inset). The side chains and water serving as ligands are drawn as sticks. The 2Fo-Fc electron density map (blue) into which the structure was build is contoured at 1σ. The boxed regions show active site, which contains residues appropriate for metal binding including H121, E126, and C163. In this structure, these residues are not properly positioned to coordinate zinc. The caspase-6 zinc-binding exosite is shown with the side chain ligands for zinc in sticks (K36, E244, H287), zinc (blue), and the water molecule (red). *Adapted from Velazquez-Delgado and Hardy (2012b) with permission.*

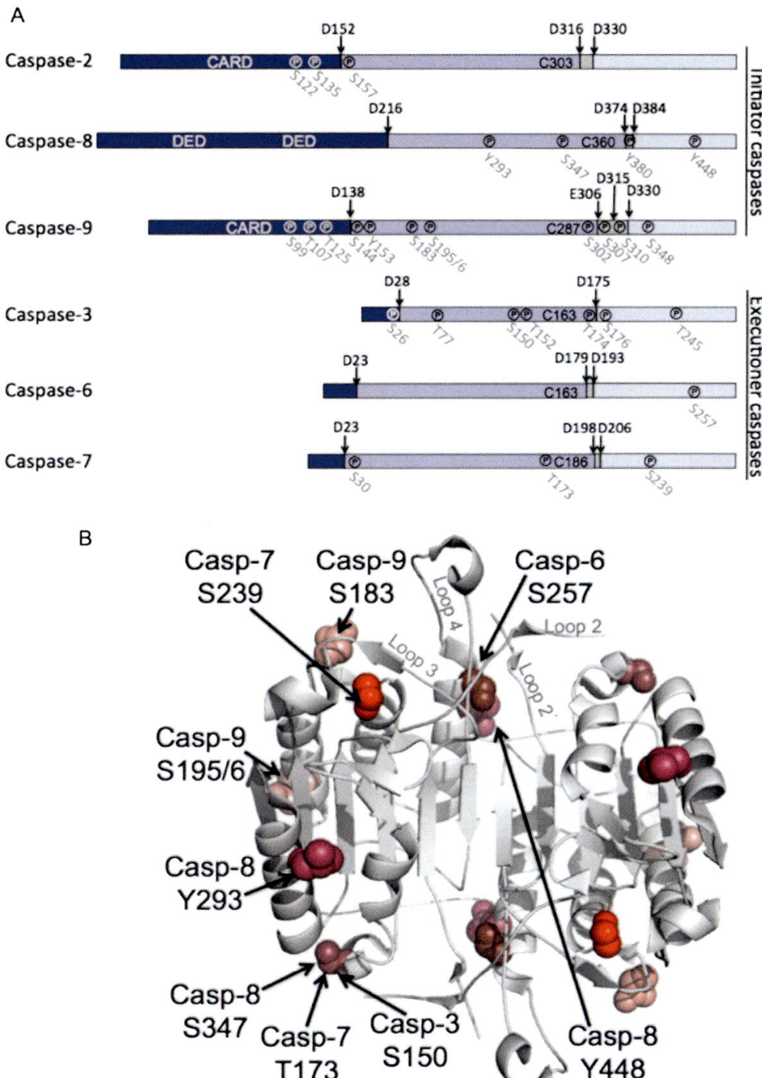

Kevin Dagbay *et al.*, Figure 9.5 Confirmed sites of phosphorylation in apoptotic caspases. (A) This map represents the major, confirmed sites of phosphorylation in the apoptotic caspases. We recognize that some other sites of phosphorylation have been reported in various databases. (B) Structural distribution of functionally critical sites of phosphorylation. *(A) Adapted from Velazquez-Delgado and Hardy (2012a) with permission. (B) Adapted from Velazquez-Delgado (2012) with permission.*

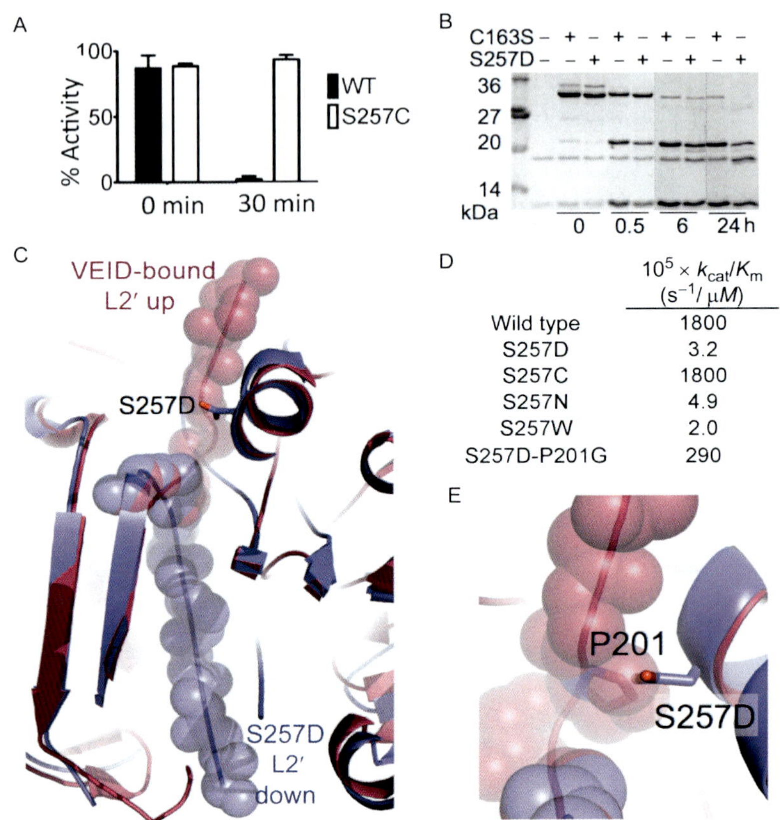

	$10^5 \times k_{cat}/K_m$ ($s^{-1}/\mu M$)
Wild type	1800
S257D	3.2
S257C	1800
S257N	4.9
S257W	2.0
S257D-P201G	290

Kevin Dagbay *et al.*, Figure 9.6 S257D is a robust phosphomimetic that elucidated the mechanism of allosteric inhibition by phosphorylation. (A) WT caspase-6 is inhibited by 30 min incubation with ARK5 kinase, whereas the S257C unphosphorylatable mutant is unaffected by ARK5. The S257D phosphomimetic is inactive similar to phosphorylated WT caspase-6. (B) The S257D active-site loops are ordered differently from the C163S zymogen, leading to different rates of cleavage of S257D from C163S. (C) In the structure of S257D (PDB ID 3S8E), the L2′ loop is in a down position (inactive, blue) in contrast to the VEID (substrate)-bound structure in which the L2′ loop is in the up position (active, red). (D) Mutations at S257 show that size is more important than charge in S257 inhibition and that any residue larger than cysteine or serine results in inactivation. Removal of P201 by mutation to glycine restores activity to S257D indicating that a steric clash between P201 and S257D leads to loss of activity. (E) The steric clash between P201 and S257D mirrors that of phosphoserine 257. *Adapted from Velazquez-Delgado and Hardy (2012a) with permission.*

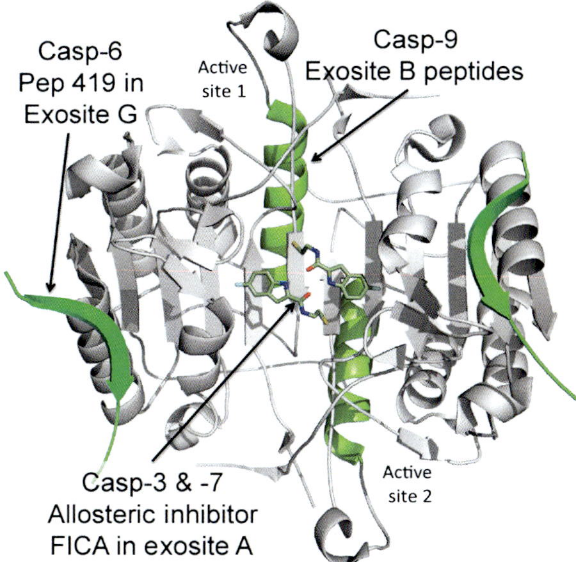

Kevin Dagbay *et al.*, Figure 9.7 Sites for binding of chemical ligands that affect caspase function. The green caspase helix that forms exosite B is highlighted since it is the region that interacts with the BIR3 region of XIAP- and BIR3-derived peptide inhibitors. *Adapted from Velazquez-Delgado (2012) with permission.*

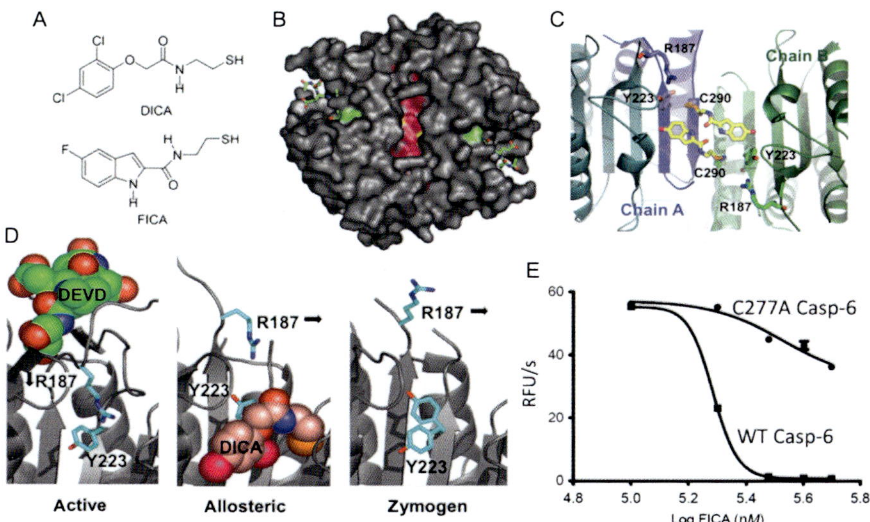

Kevin Dagbay *et al.*, Figure 9.8 Allosteric inhibition at exosite A. (A) The structures of DICA and FICA. (B) The allosteric pocket at the caspase-7 (PDB ID 1F1J) dimer interface exosite A (pink). The FICA/DICA-binding residues C290 on each half of the dimer are shown as yellow patches at the bottom of the pocket. Active sites are denoted by DEVD active-site inhibitor (green sticks) bound to the catalytic cysteine (green patch). (C) The crystal structure of the FICA (yellow) bound caspase-7 exosite A at C290. (D) Conformations of Y233 and R187 shown for active, allosteric inhibitor (DICA, salmon spheres) bound and zymogen caspase-7. DICA bound at the allosteric site prevents R187 and Y233 from pointing down to generate a properly ordered substrate-binding pocket. (E) FICA inhibits wild-type (WT) caspase-6 much more efficiently than C277A caspase-6, indicating that FICA binds and inhibits caspase-6 at the same site and likely by the same mechanism as observed for caspase-7. *Panels A–D are adapted from Hardy et al. (2004a) with permission. Panel E adapted from Velazquez-Delgado (2012) with permission.*

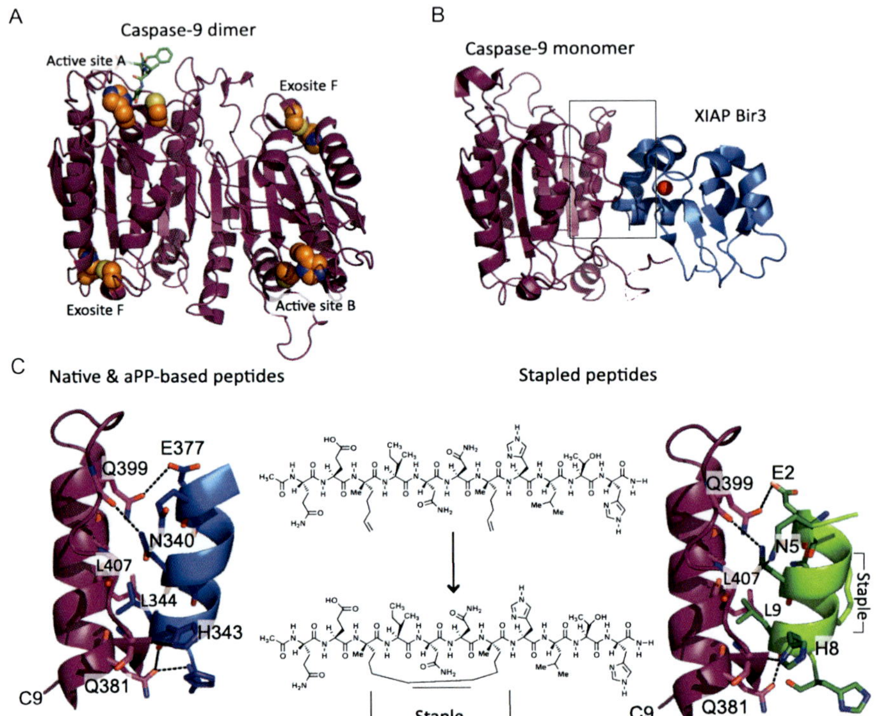

Kevin Dagbay *et al.*, Figure 9.9 Allosteric inhibition of caspase-9 at the dimer interface. (A) Caspase-9 is a dimer composed of A and B chains, each of which has the same amino acid sequence and each possessing the residues required to form one active site. In the structure of mature caspase-9 (PDB ID 1JXQ), only one active site (active site A) is observed in a catalytically competent conformation. Green sticks mark peptide inhibitor in the A-chain active site. For reference, the active site and a zinc-binding exosite F are drawn as orange spheres. (B) XIAP BIR3 domain prevents dimerization and activation of caspase-9 (PDB ID 1NW9). (C) Three classes of BIR3-mimicking peptides produced by novel synthetic schemes developed are shown. *Adapted from Huber et al. (2012) with permission.*

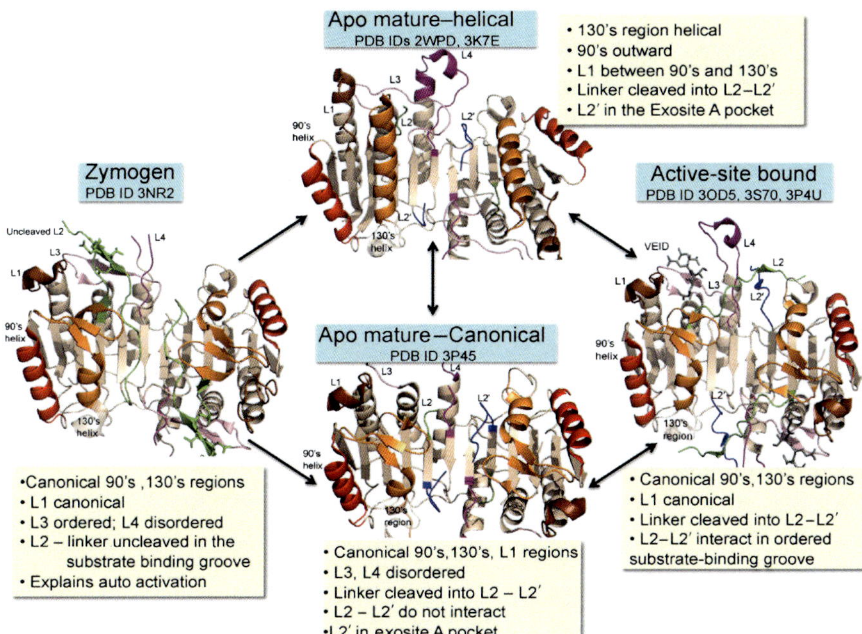

Kevin Dagbay *et al.*, Figure 9.11 Conformational changes during caspase-6 lifecycle. Crystal structures of the caspase-6 are available for zymogen, mature and active-site bound forms. The zymogen of caspase-6 exists in a conformation that explains the structural basis of self-processing. The mature form exists in a conformational equilibrium between a helical and a canonical conformation, though in the canonical conformation far more regions of the protein are disordered. The equilibrium seems to favor the helical form at all pHs. The active-site bound form attains a structure that is extremely similar to all other active-site bound caspases. The colored regions highlight conformational changes that occur during the transitions between these states in the 90's helix (red), 130's region (orange), L1 loop (brown), L2 (green), L2′ (blue), L3 pink, and L4 (purple). Significant structural characteristics are denoted for each conformation (yellow boxes).

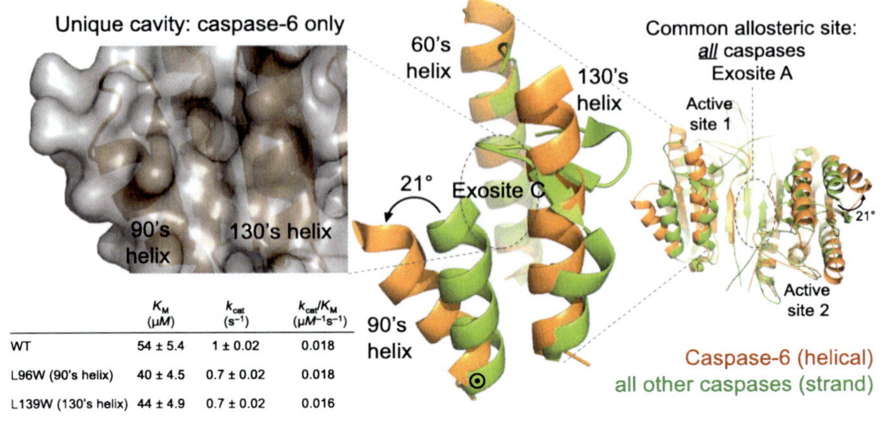

	K_M (μM)	k_{cat} (s^{-1})	k_{cat}/K_M ($\mu M^{-1}s^{-1}$)
WT	54 ± 5.4	1 ± 0.02	0.018
L96W (90's helix)	40 ± 4.5	0.7 ± 0.02	0.018
L139W (130's helix)	44 ± 4.9	0.7 ± 0.02	0.016
L96W/L139W	47 ± 2.0	$4.6 \pm 4.2 \times 10^{-3}$	9.8×10^{-5}

Kevin Dagbay *et al.*, Figure 9.12 Caspase-6 (orange, PDB ID 3NKF) exists in an extended helical conformation that cannot be attained by any other caspase, which all have the same canonical strand structure as caspase-7 (green, PDB ID 1F1J) in the 90's helix. A cavity formed when the 90's helix rotates away from the 130's helix can function as a unique allosteric site at exosite C, as shown by insertion of tryptophan residues designed to hold the 90's helix in an open conformation. *Adapted from Vaidya et al. (2011) with permission.*

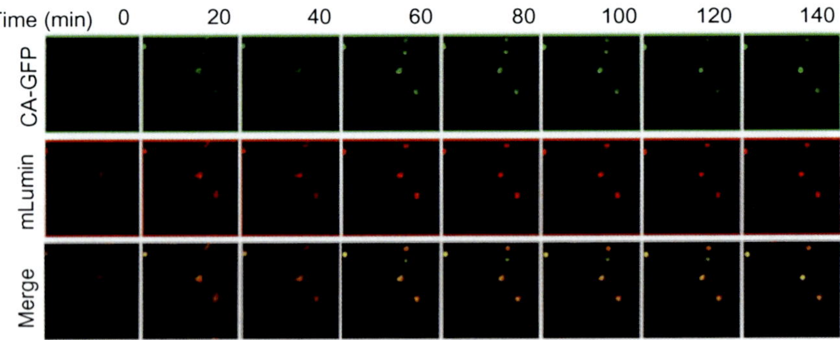

Samantha B. Nicholls and Bradley T. Hyman, Figure 10.2 Dark-to-bright response of CA-GFP in H4 human neuronal gliomal cell line. CA-GFP was transiently transfected in a vector containing an IRES followed by a constitutively bright red fluorescent protein, mLumin. There is a green response which increases in intensity and colocalizes with the red in the H4 cells treated with 1 mM staurosprine, a known apoptosis inducer, as the cell begins to bleb and undergo apoptosis.

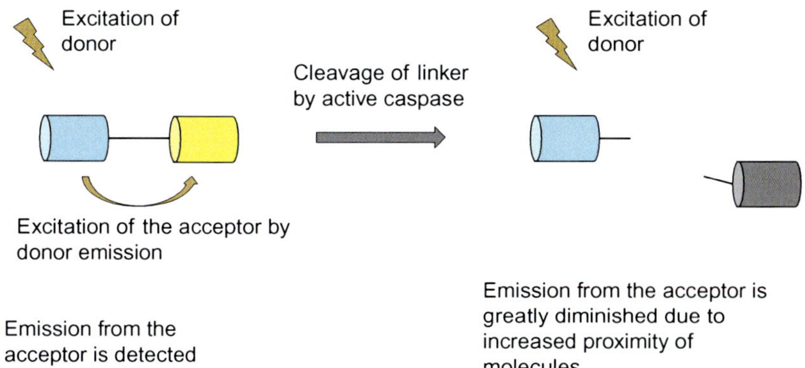

Samantha B. Nicholls and Bradley T. Hyman, Figure 10.3 FRET response in caspase activatable fluorescent proteins. Regardless of which fluorescent proteins used, most of the FRET-based reporters described have a similar mechanism by which the donor and acceptor fluorescent proteins are held in close physical proximity by a caspase cleavable linker allowing for excitation of the acceptor. After cleavage of the linker by active caspase, the two fluorescent proteins diffuse further from each other and the signal from the acceptor is greatly diminished.

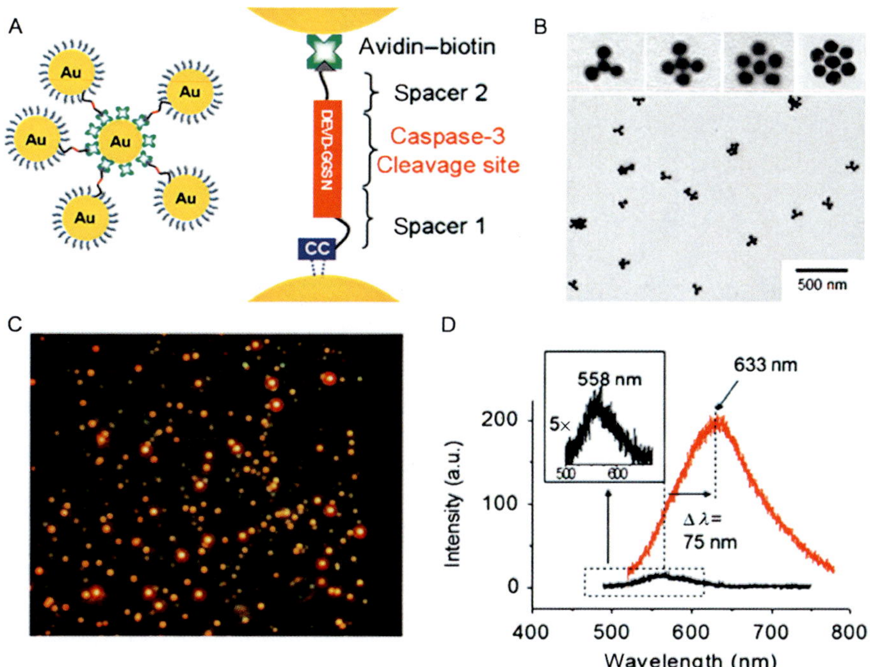

Cheryl Tajon *et al.*, Figure 11.2 Crown nanoparticle sensors. (A) Illustration of a crown nanoparticle containing an avidin-coated gold core nanoparticle with multiple biotinylated gold satellite nanoparticles. Linkers bearing the caspase-3 cleavage sequence DEVD tether the core and satellite nanoparticles together. (B) Transmission electron microscopy (TEM) shows different configurations of the crown nanoparticles. They contain either three to six satellite nanoparticles linked to the core nanoparticle. (C) A representative scattering image of crown nanoparticles by darkfield microscopy. Each red (dark gray in print) spot corresponds to a single crown nanoparticle. (D) Representative scattering spectra of crown (red) and monomeric gold nanoparticles (black). The crown nanoparticles exhibited an increased scattering intensity (~44×) and a red shift ($\Delta\lambda \sim 75$ nm), compared to monomeric gold nanoparticles. *Reprinted from Jun et al. (2009).*

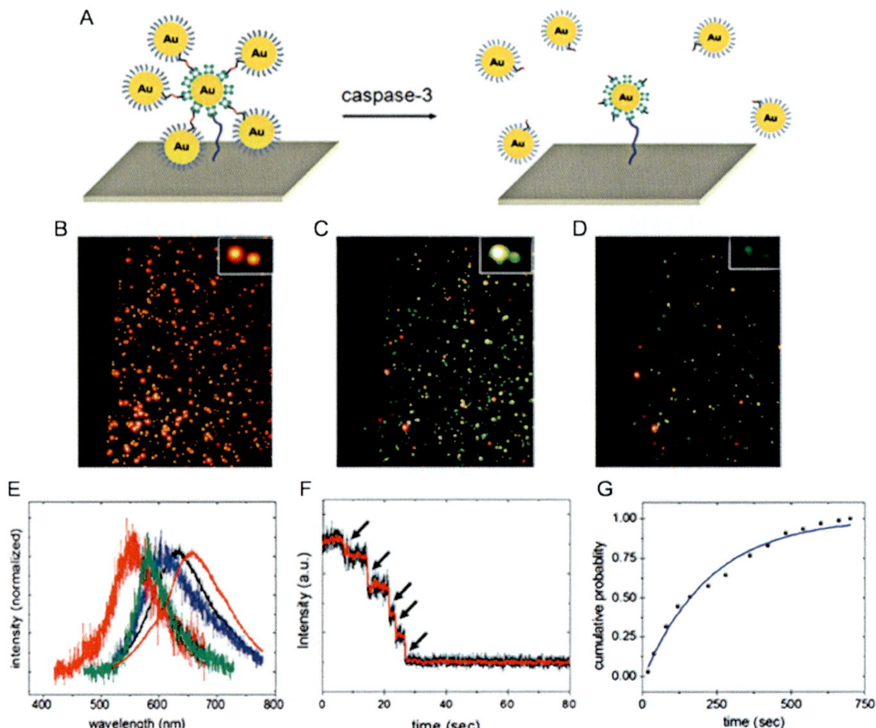

Cheryl Tajon *et al.*, Figure 11.3 *In vitro* caspase-3 activity reported by crown nanoparticles. (A) Illustration of immobilized crown nanoparticles on a glass flow chamber by an avidin–biotin linkage. After equilibration, recombinant caspase-3 was presented to the crown nanoparticles. Color and intensity changes were visualized by darkfield microscopy. (B–D) Scattering color changes were observed upon treatment with caspase-3, from red to yellow to green spots over time. (E) A representative spectral shift of a single crown nanoparticle upon exposure to caspase-3. At 654 nm, crown nanoparticles exhibited their scattering peak maximum. As the crown nanoparticles responded to caspase-3 proteolysis, blue shifts were observed over time. (F) A single crown nanoparticle intensity trace versus time was recorded at 100 Hz. Each arrow corresponds to the time at which caspase-3 mediated a cutting event of the crown nanoparticle substrate. (G) Progress curve showing cumulative probability of cutting events as a function of time that enabled extraction of a k_{cat} value (6.2 s^{-1}). *Reprinted from Jun et al. (2009).*

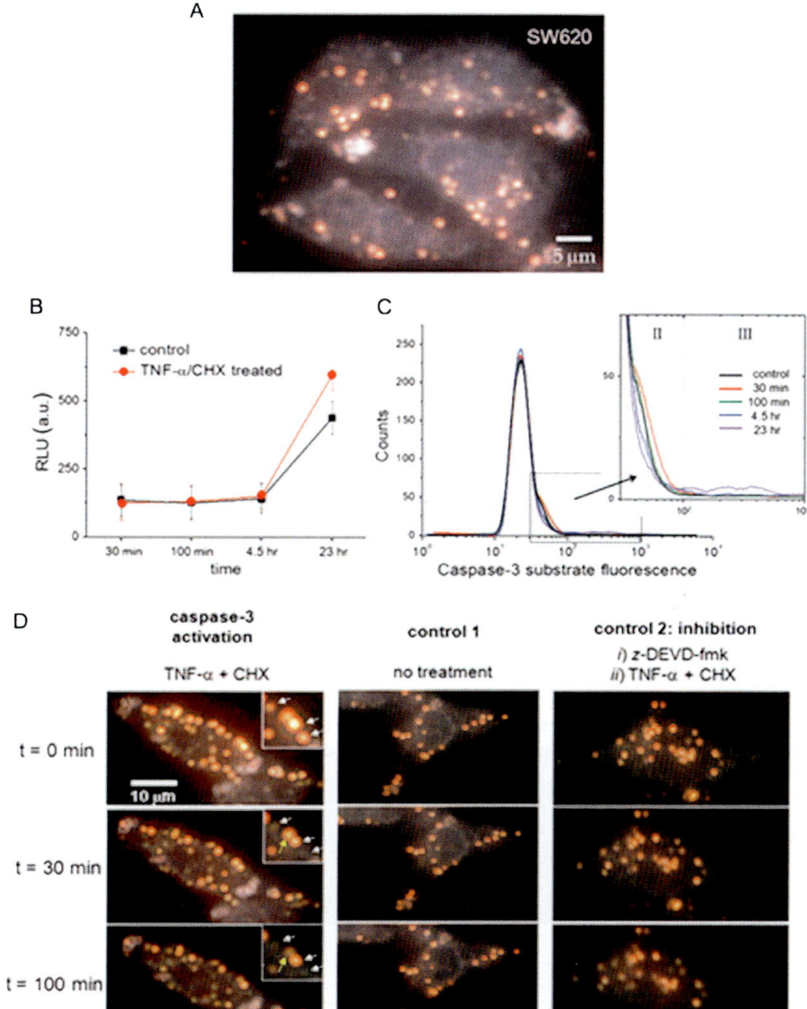

Cheryl Tajon *et al.*, Figure 11.4 Caspase-3 activation in live cells reported by Caspase Glo 3/7® assay, flow cytometry, and crown nanoparticles. (A) Crown nanoparticles were modified with the cell-penetration peptide, TAT, and delivered into SW620 colon cancer cells. Red-colored spots (dark gray in print) show the location of each individual crown nanoparticle. (B and C) Ensemble caspase-3 activity was monitored by either a luminescence assay (Caspase Glo 3/7®) or flow cytometry after treatment of cells with TNF-α/CHX. (B) With the luminescence assay, minimal caspase-3 activation was observed at early time points and was apparent only at 23 h after induction of cells. (C) Flow cytometry spectra show similar results to luminescence assay. Shoulder peaks may indicate caspase-3 activity at earlier time points. (D) (Left) After treatment of cells with TNF-α/CHX, caspase-3 activation was evident as early as 30 min. This is indicated by the change in scattering color and intensity across the time points. (Middle) Crown nanoparticles show no response in vehicle-treated cells. (Right) Pretreatment with inhibitor, z-DEVD-fmk, followed by induction with TNF-α/CHX, showed no response from crown nanoparticles. *Reprinted from Jun et al. (2009).*

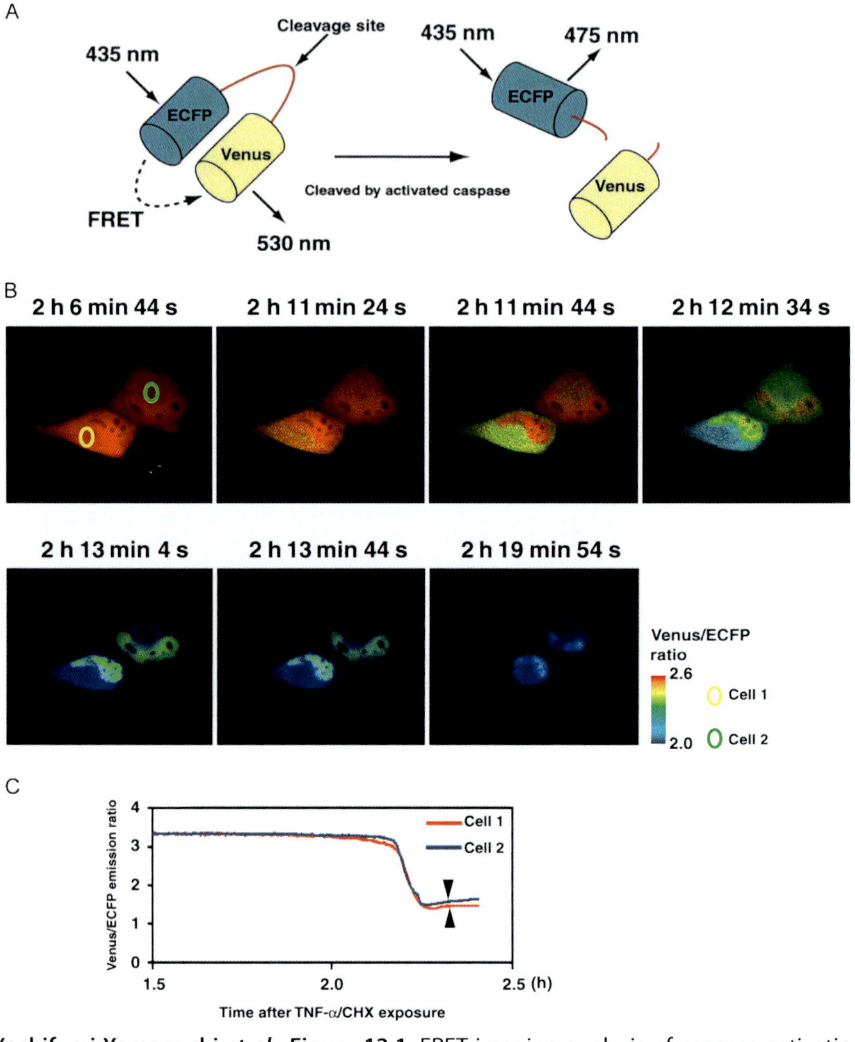

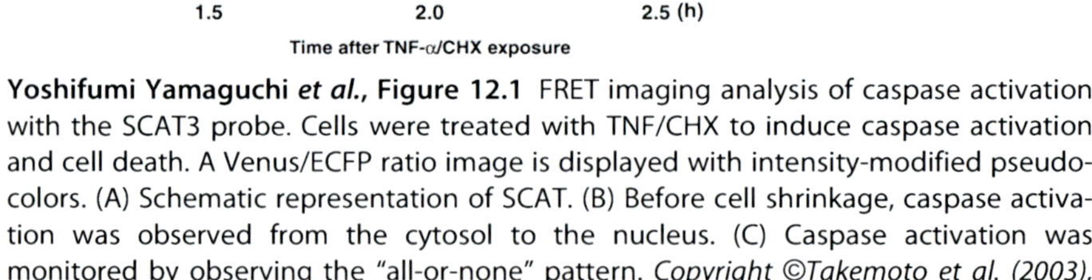

Yoshifumi Yamaguchi *et al.*, Figure 12.1 FRET imaging analysis of caspase activation with the SCAT3 probe. Cells were treated with TNF/CHX to induce caspase activation and cell death. A Venus/ECFP ratio image is displayed with intensity-modified pseudocolors. (A) Schematic representation of SCAT. (B) Before cell shrinkage, caspase activation was observed from the cytosol to the nucleus. (C) Caspase activation was monitored by observing the "all-or-none" pattern. *Copyright ©Takemoto et al. (2003), originally published in* Journal of Cell Biology. *http://dx.doi.org/10.1083/jcb.200207111.*

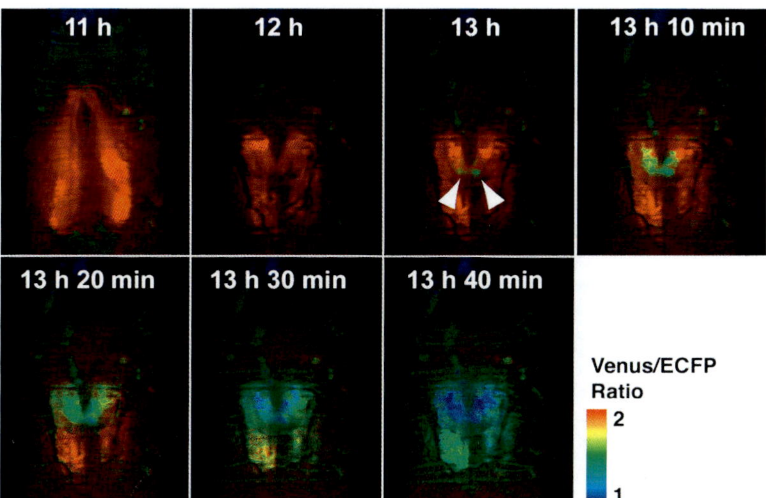

Yoshifumi Yamaguchi *et al.*, Figure 12.2 Local initiation and propagation of the caspase activities during programmed cell death in the salivary gland *in vivo*. Venus/ECFP ratio images of SCAT3-expressing salivary glands. *In vivo* live imaging analysis of SCAT3 was started from 10 to 11 h APF. Arrowheads indicate the symmetrical initiation of caspase activation. Time indicates APF. *Copyright* © National Academy of Sciences, USA *from Takemoto et al. (2007). http://dx.doi.org/10.1073/pnas.0702733104.*

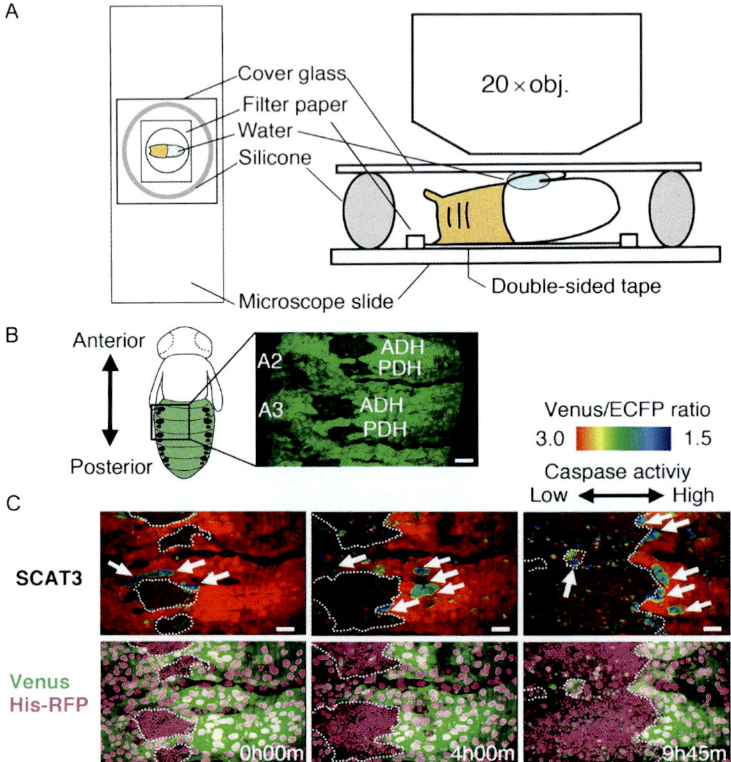

Yoshifumi Yamaguchi *et al.*, Figure 12.3 (A) Illustration of mounting a pupa for live imaging of the epidermal replacement. The abdominal part of the pupal case is removed, and the pupa is placed on the slide as dorsal histoblasts face the cover glass. (B) Schematic illustration (left) and confocal image (right) of the dorsal abdominal epidermis. Anterior dorsal histoblasts (ADH) and posterior dorsal histoblasts (PDH) exist in each segment (A1–A6). Green (gray in print) color represents LECs. (C) Caspase activation during epidermal replacement, revealed by SCAT3. Arrows indicate caspase-activated LECs. Dashed lines indicate the LEC/histoblast boundary. Pseudocolor ratio images are shown in the upper panels, and nucleus positions are labeled by His-RFP in the lower panels. *UAS-SCAT3* is expressed by *tsh-Gal4*. Scale bars, 50 μm. *Panels (B) and (C) are reprinted and modified from Nakajima et al. (2011). Copyright © American Society for Microbiology,* Molecular and Cellular Biology, *http://dx.doi.org/10.1128/MCB.01046-10.*

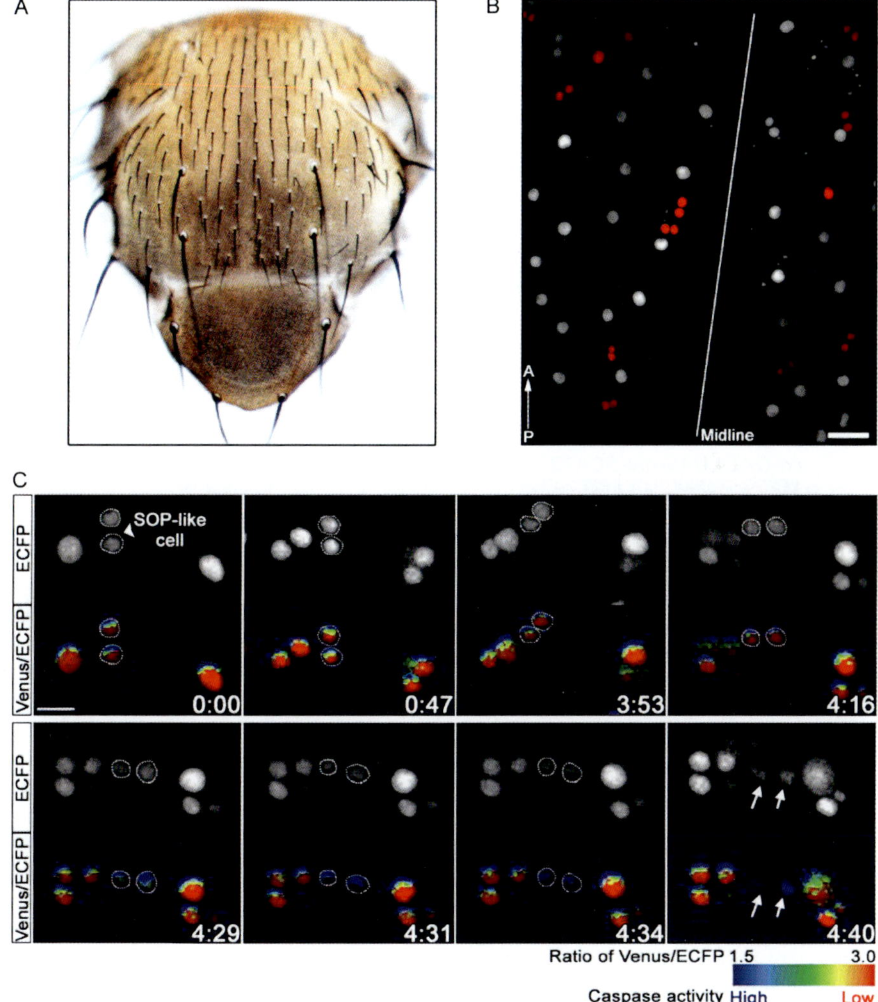

Yoshifumi Yamaguchi *et al.*, Figure 12.4 (A) Light-microscopy image of sensory organs on the notum of a wild-type fly. (B) Confocal imaging of nls-SCAT3 using a *neu-GAL4* driver. *Neuralized*-positive cells that were removed by apoptosis are marked in red. Scale bar represents 20 μm. (C) Caspase activation was detected prior to nuclear fragmentation (arrows) in *neuralized*-positive cells that appeared in aberrant positions or timing. The lower row of panels shows caspase activity images taken with nls-SCAT3 and shown in pseudocolor. Scale bar represents 10 μm. *Copyright ©Koto et al. (2011), originally published in* Current Biology, *http://dx.doi.org/10.1016/j.cub.2011.01.015.*

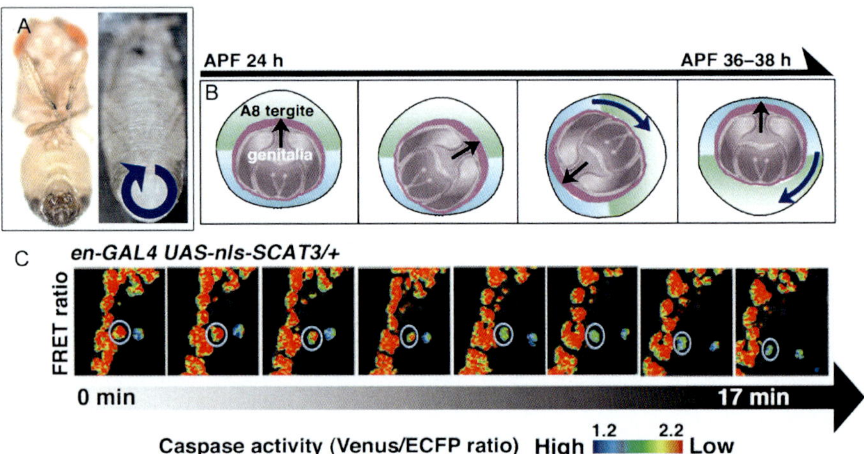

Yoshifumi Yamaguchi *et al.*, Figure 12.5 (A) Ventral view of an adult male (left) and pupae (right). The 360° clockwise rotation of the genitalia during the pupal stage is indicated by blue arrows (gray in print) (right). (B) Schematic illustration of male genitalia rotation in *Drosophila*. Cells in the posterior part of the A8 tergite (magenta) drive the rotation of the genitalia by 180° (pink–gray); then, the other cells in the A8 tergite (light-green: ventral and blue: dorsal) drive the rotation through the remaining 180° (dark-blue arrow), completing the rotation. The black arrow indicates the anus-to-penis direction on the genital plate. (C) Caspase activity was examined by imaging using a FRET-based probe, nls-SCAT3, and shown in pseudocolor. White circles indicate cells that underwent apoptosis; the pseudocolor gradually changed from red to blue. The genotype was *en-GAL4 UAS-nls-SCAT3/+*. *(C) is reprinted from Kuranaga et al. (2011),* Development, *http://dx.doi.org/10.1242/dev.058958.*

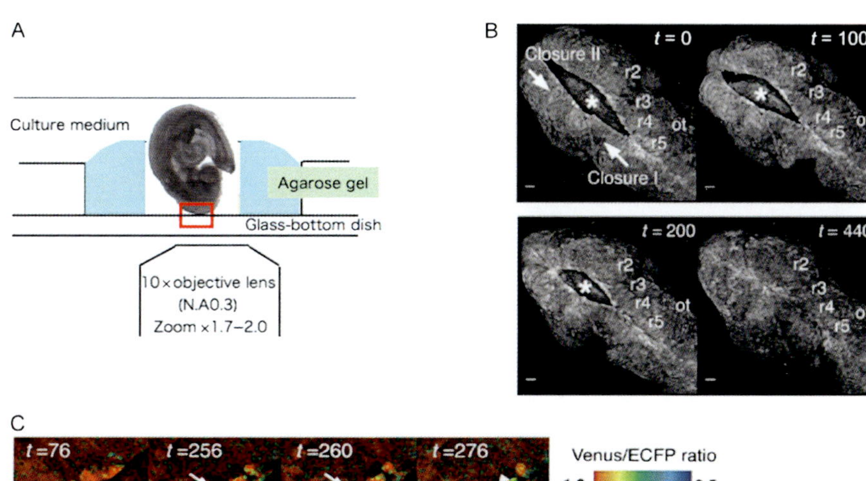

Yoshifumi Yamaguchi *et al.*, Figure 12.6 (A) System for time-lapse imaging of apoptosis in the cranial NTC. Embryos undergoing NTC (E8.5–9.0) were dissected from the uterus and yolk sac and placed into holes in 2% low-temperature melting agarose in glass-bottom dishes. The dishes were filled with culture medium and placed in a humidified incubator at 37 °C/5% CO_2 in order to visualize the hindbrain region. (B) Time-lapse observation of closure in hindbrain of a SCAT3-transgenic embryo (ECFP images). (C) Live imaging of apoptosis during the hindbrain closure. A cell showing activated caspase (arrows in *V/C* images; $t = 260$) shrank ($t = 260$) and became fragmented ($t = 276$; arrowheads) in the boundary domain before the completion of NTC. The same cell is circled by magenta (gray in print) in the ECFP images in the magnified images of the boxed areas. All scale bar represent 50 μm. *Copyright © Yamaguchi et al. (2011), originally published in* Journal of Cell Biology, *http://dx.doi.org/10.1083/jcb.201104057.*

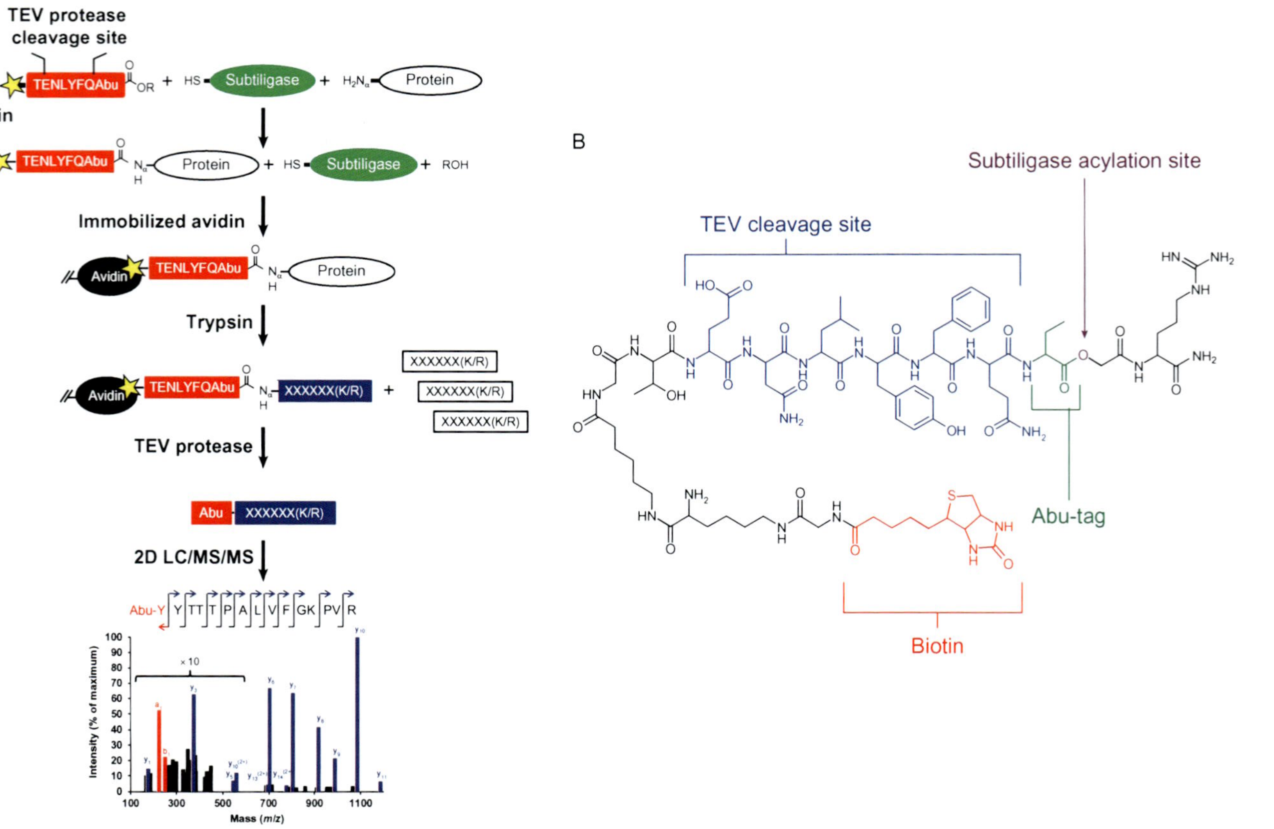

Arun P. Wiita *et al.*, Figure 13.1 An overview of the subtiligase N-terminal labeling method. (A) Proteins with free N-termini in a mixture are selectively tagged using the engineered enzyme, subtiligase. Whole protein samples are incubated with subtiligase and the peptide ester containing a biotin tag. After enzymatic labeling, free N-termini are captured on avidin beads. Proteins are digested by trypsin. The final N-terminal peptide is released from beads via TEV protease cleavage and identified by mass spectrometry. (B) The current peptide ester contains an ester subtiligase acylation site, Abu-tag for positive mass spectrometry identification, a TEV protease site and a biotin label. The peptide ester can be further modified for specific experimental needs.

CPI Antony Rowe
Eastbourne, UK
September 24, 2014